时代教育·国外高校优秀教材精选

Craig 土力学

（注释改编版·原书第8版）

[英] J. A. 纳珀特（J. A. Knappett）
R. F. 克雷格（R. F. Craig） 著
顾晓强 杨朔成 注释

机械工业出版社

Craig's Soil Mechanics 是欧美畅销的土力学教材之一，由英国邓迪大学（University of Dundee）R. F. 克雷格（R. F. Craig）教授所著，首版于1974年发行。第8版由邓迪大学的J. A. 纳珀特（J. A. Knappett）教授和R. F. 克雷格（R. F. Craig）教授共同完成，主要修订工作包括采用欧洲规范7改写全书，从土力学的基本概念和理论及其在岩土工程设计中的应用两方面重新组织本书框架，并强化了极限分析方法、原位测试、渗流分析、土体刚度、临界状态等主题内容。

《Craig 土力学》是根据我国土力学课程的教学内容和特点，在 *Craig's Soil Mechanics* 第8版基础上删减而成的。同时，为便于我国学生学习，由同济大学的顾晓强教授团队对重要术语、难句进行了大量注释。本书主要内容包括土的基本特性、渗流、有效应力、固结、土的剪切特性、地基勘察、原位测试、浅基础、挡土结构、自承式土体稳定性等。

本书可作为土木工程专业本科生和研究生的教材，也可作为大学教师和工程技术人员的参考书。

Craig's Soil Mechanics/by J. A. Knappett and R. F. Craig/ISBN 9780415561266

图书在版编目（CIP）数据

Craig 土力学：注释改编版：原书第8版/（英）J. A. 纳珀特（J. A. Knappett），（英）R. F. 克雷格（R. F. Craig）著；顾晓强，杨朔成注释. —北京：机械工业出版社，2019. 7
时代教育·国外高校优秀教材精选
ISBN 978-7-111-62807-1

Ⅰ. ①C… Ⅱ. ①J…②R…③顾…④杨… Ⅲ. ①土力学－高等学校－教材 Ⅳ. ①TU43

中国版本图书馆 CIP 数据核字（2019）第097671号

机械工业出版社（北京市百万庄大街22号 邮政编码100037）
策划编辑：李 帅 责任编辑：李 帅 马军平
责任校对：樊钟英 封面设计：张 静
责任印制：孙 炜
保定市中画美凯印刷有限公司印刷
2020年1月第1版第1次印刷
184mm×260mm·28印张·774千字
标准书号：ISBN 978-7-111-62807-1
定价：99.00元

电话服务	网络服务
客服电话：010-88361066	机 工 官 网：www. cmpbook. com
010-88379833	机 工 官 博：weibo. com/cmp1952
010-68326294	金 书 网：www. golden-book. com
封底无防伪标均为盗版	机工教育服务网：www. cmpedu. com

前　言

很荣幸被 Taylor & Francis 公司邀请来编写这本备受欢迎教材的第 8 版。在接受这项任务前，我从未意识到，要达到前七版的高水准需要耗费我如此多的精力和时间。本书初版于 1974 年发行，我个人认为在本书诞生 40 年这个时间节点，对其做一次重大更新是恰当的。当然，在进行大幅度修订的同时，我也尽力去维持本书最核心的特征——清晰且富有深度的说明。

我们对所有的原章节都进行了更新，并扩充了部分章节的内容，同时增加了一些新的章节，以满足当下工程和课程的需要。但本书的主要作用没有发生变化，依旧是满足土木工程专业本科生的需求，作为学生从课堂向工程实际转变过程中的重要参照。本书中的部分高阶主题内容扩展了本书的适用范围，使其能够为一些研究生课程提供参考。

主要变化如下：

- **将素材分为两个主要部分：**第一部分主要涉及土力学中的基本概念与理论，第二部分着力于岩土工程设计中所必需力学性质的确定。
- **关于原位测试的新章节：**着力于由各个试验可靠地确定参数，以及通过实际测点的数据得出对力学性能的解读。
- **关于基础特性及设计的新章节：**有关基础的内容现在分为 3 章（浅基础、深基础和高阶课题），这样有利于更灵活的课程设计。
- **极限状态设计（对应于欧洲规范 7）：**有关于岩土工程设计部分的章节均基于通用的极限状态设计理论体系（取代了原有的容许应力法）。更进一步的背景材料可参见欧洲规范 7。为了更好地使学生完成由大学到设计单位的转变，这一规范应用于本书的数值示例及章末习题。
- **补充案例学习：**采用了以往版本的案例，但新版中包含了极限状态设计理论应用于实际工程问题这一过程，以培养学生的工程判断力。
- **包含了极限状态分析方法：**随着基于这些方法的高端计算软件的应用越来越广泛，为了学生日后的职业生涯，我认为他们有必要在离开大学时对于这些方法有基本的了解。同时我们也为承载力系数和极限压力提供了更详细的背景材料，这在以往的版本中是没有的。

我要对邓迪大学的同事表示无比的感激，是他们给予了我完成这一版的时间，同时他们也在成书的过程中给予了我许多建设性的建议。我同样也要感谢 Taylor & Francis 公司的所有成员，有了他们的帮助我才能够完成这一令人生畏的任务。最后，感谢那些允许我使用他们的图标、数据和图像的人。

之前的土木工程师认为本书以往的版本是实用的，信息丰富和给人以启发。我衷心希望我主编的这一版本也能得到现在的及未来的土木工程同行同样的评价。

J. A. 纳珀特
邓迪大学
2011 年 7 月

Preface

When I was approached by Taylor & Francis to write the new edition of Craig's popular textbook, while I was honoured to be asked, I never realised how much time and effort would be required to meet the high standards set by the previous seven editions. Initially published in 1974, I felt that the time was right for a major update as the book approaches its fortieth year, though I have tried to maintain the clarity and depth of explanation which has been a core feature of previous editions.

All chapters have been updated, several extended, and new chapters added to reflect the demands of today's engineering students and courses. It is still intended primarily to serve the needs of the undergraduate civil engineering student and act as a useful reference through the transition into engineering practice. However, inclusion of some more advanced topics extends the scope of the book, making it suitable to also accompany many post-graduate level courses.

The key changes are as follows:

- **Separation of the material into two major sections**: the first deals with basic concepts and theories in soil mechanics, and the determination of the mechanical properties necessary for geotechnical design, which forms the second part of the book.
- **New chapter on in-situ testing**: focusing on the parameters that can be reliably determined using each test and interpretation of mechanical properties from digital data based on real sites.
- **New chapters on foundation behaviour and design**: coverage of foundations is now split into three separate sections (shallow foundations, deep foundations and advanced topics), for increased flexibility in course design.
- **Limit state design (to Eurocode 7)**: The chapters on geotechnical design are discussed wholly within a modern generic limit state design framework, rather than the out-dated permissible stress approach. More extensive background is provided on Eurocode 7, which is used in the numerical examples and end-of-chapter problems, to aid the transition from university to the design office.
- **Extended case studies**: building on those in previous editions, but now including application of the limit state design techniques in the book to these real-world problems, to start to build engineering judgement.
- **Inclusion of limit analysis techniques**: With the increasing prevalence and popularity of advanced computer software based on these techniques, I believe it is essential for students to leave university with a basic understanding of the underlying theory to aid their future professional development. This also provides a more rigorous background to the origin of bearing capacity factors and limit pressures, missing from previous editions.

I am immensely grateful to my colleagues at the University of Dundee for allowing me the time to complete this new edition, and for their constructive comments as it took shape. I would also like to express my gratitude to all those at Taylor & Francis who have helped to make such a daunting task achievable, and thank all those who have allowed reproduction of figures, data and images.

I hope that current and future generations of civil engineers will find this new edition as useful, informative and inspiring as previous generations have found theirs.

Jonathan Knappett
University of Dundee
July 2011

目 录

Contents

Part 1

Development of a mechanical model for soil

Chapter 1

Basic characteristics of soils

Learning outcomes

After working through the material in this chapter, you should be able to:

1 Understand how soil deposits are formed and the basic composition and structure of soils at the level of the micro-fabric[①] (Sections 1.1 and 1.2);

2 Describe (Sections 1.3 and 1.4) and classify (Section 1.5) soils based on their basic physical characteristics;

3 Determine the basic physical characteristics of a soil continuum[②] (i.e. at the level of the macro-fabric, Section 1.6);

4 Specify compaction[③] required to produce engineered fill materials with desired continuum properties for use in geotechnical constructions (Section 1.7).

① 微观结构。

② 视土为连续介质体。

③ 击实。

1.1 The origin of soils

To the civil engineer, soil is any uncemented or weakly cemented accumulation of mineral particles formed by the weathering of rocks as part of the rock cycle[④] (Figure 1.1), the void space between the particles containing water and/or air. Weak cementation can be due to carbonates or oxides precipitated between the particles, or due to organic matter[⑤]. Subsequent deposition and compression of soils, combined with cementation between particles, transforms soils into sedimentary rocks[⑥] (a process known as **lithification**[⑦]). If the products of weathering remain at their original location they constitute a **residual soil**. If the products are transported and deposited in a different location they constitute a **transported soil**, the agents of transportation being gravity, wind, water and glaciers. During transportation, the size and shape of particles can undergo change and the particles can be sorted into specific size ranges. Particle sizes in soils can vary from over 100mm to less than 0.001mm. In the UK, the size ranges are described as shown in Figure 1.2. In Figure 1.2, the terms 'clay', 'silt'[⑧] etc. are used to describe only the sizes of particles

④ 对于土木工程师而言，土是一种未胶结或弱胶结的矿物颗粒堆积体，其为岩石循环中岩石风化形成的产物。

⑤ 有机物。

⑥ 沉积岩。

⑦ 岩化。

⑧ 这里的“clay”表示“黏粒”，“silt”表示“粉粒”。

between specified limits. However, the same terms are also used to describe particular types of soil, classified according to their mechanical behaviour (see Section 1.5).

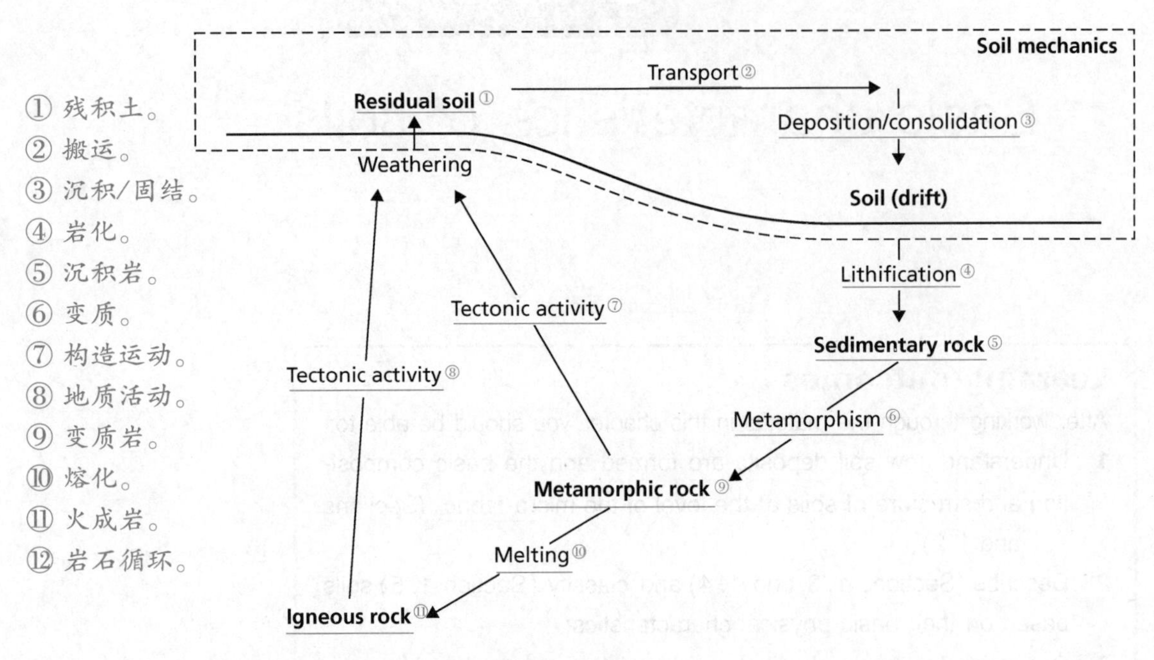

Figure 1.1 The rock cycle[12].

① 残积土。
② 搬运。
③ 沉积/固结。
④ 岩化。
⑤ 沉积岩。
⑥ 变质。
⑦ 构造运动。
⑧ 地质活动。
⑨ 变质岩。
⑩ 熔化。
⑪ 火成岩。
⑫ 岩石循环。

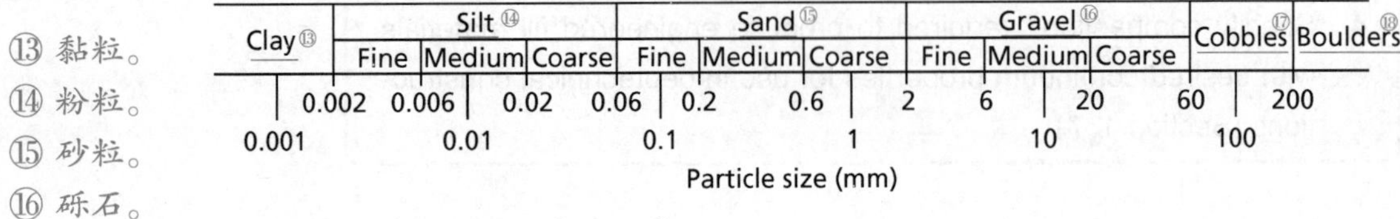

Figure 1.2 Particle size ranges[19].

⑬ 黏粒。
⑭ 粉粒。
⑮ 砂粒。
⑯ 砾石。
⑰ 卵石。
⑱ 漂石(块石)。
⑲ 颗粒尺寸范围。

The type of transportation and subsequent deposition of soil particles has a strong influence on the distribution of particle sizes at a particular location[20]. Some common depositional regimes are shown in Figure 1.3. In glacial regimes, soil material is eroded from underlying rock by the frictional and freeze – thaw action of glaciers. The material, which is typically very varied in particle size from clay to boulder-sized particles, is carried along at the base of the glacier and deposited as the ice melts; the resulting material is known as **(glacial) till**[21]. Similar material is also deposited as a terminal moraine[22] at the edge of the glacier. As the glacier melts, moraine is transported in the outwash[23]; it is easier for smaller, lighter particles to be carried in suspension[24], leading to a gradation in particle size with distance from the glacier as shown in Figure 1.3 (a). In warmer temperate climates the chief transporting action is water (i. e. rivers and seas), as shown in Figure 1.3 (b). The deposited material is known as **alluvium**[25], the composition of which depends on the speed of water flow. Faster-flowing rivers can carry larger particles in suspension, resulting in alluvium, which is a mixture of sand and gravel-sized

⑳ 土颗粒的搬运方式与随后沉积的种类对于在某一位置的土颗粒尺寸级配有很大的影响。
㉑ 冰碛土。
㉒ 终碛。
㉓ 冰水冲积物。
㉔ 土颗粒与水组成的悬浮液。
㉕ 冲积土。

particles, while slower-flowing water will tend to carry only smaller particles. At estuarine locations where rivers meet the sea, material may be deposited as a shelf or **delta**[1]. In arid (desert) environments[2] (Figure 1.3 (c)) wind is the key agent of transportation, eroding rock outcrops and forming a **pediment** (the desert floor) of fine wind-blown sediment (**loess**[3]). Towards the coast, a **playa**[4] of temporary evaporating lakes, leaving salt deposits, may also be formed. The large temperature differences between night and day additionally cause thermal weathering of rock outcrops, producing **scree**. These surface processes are geologically very recent, and are referred to as **drift deposits** on geological maps. Soil which has undergone significant compression/consolidation following deposition is typically much older and is referred to as **solid**, alongside rocks, on geological maps.[5]

① 三角洲。
② 干旱(沙漠)环境。
③ 黄土。
④ 河滩。
⑤ 在沉积后经历压缩/固结的土一般年代较久，在地质图中与岩石一样被当作固体。

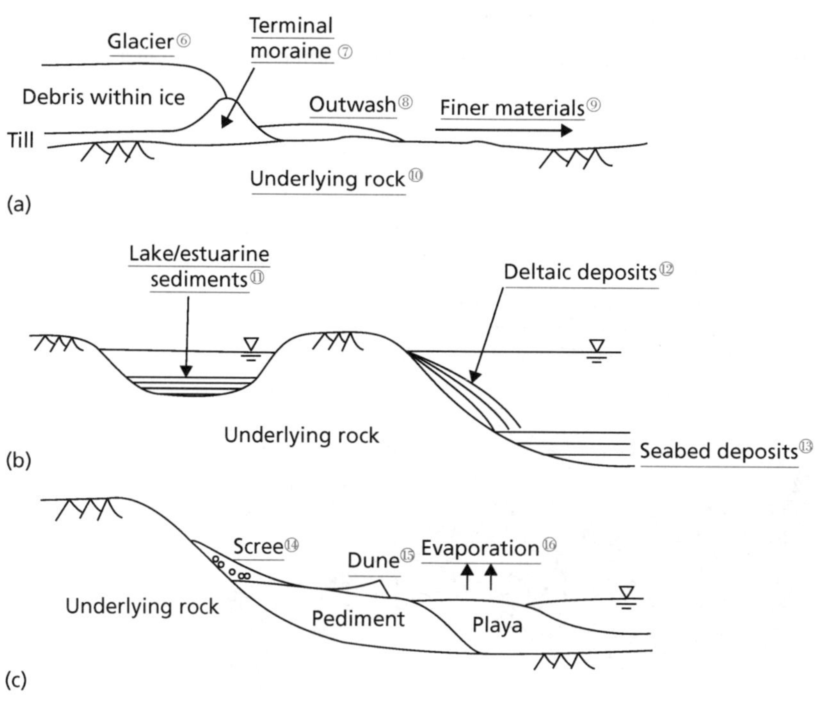

Figure 1.3 Common depositional environments: (a) glacial, (b) fluvial, (c) desert.[17]

⑥ 冰碛。
⑦ 终碛。
⑧ 冲蚀。
⑨ 更细物质。
⑩ 母岩。
⑪ 湖/河口沉积物。
⑫ 三角洲沉积。
⑬ 海床沉积。
⑭ 碎石。
⑮ 沙丘。
⑯ 蒸发。
⑰ 常见的沉积环境：
(a) 冰川
(b) 河流
(c) 沙漠

The relative proportions of different-sized particles within a soil are described as its **particle size distribution** (PSD)[18], and typical curves for materials in different depositional environments are shown in Figure 1.4. The method of determining the PSD of a deposit and its subsequent use in soil classification[19] is described in Sections 1.4 and 1.5.

At a given location, the subsurface materials will be a mixture of rocks and soils, stretching back many hundreds of millions of years in geological time. As a result, it is important to understand the full geological history of an area to understand the likely characteristics of the deposits that will be present at the surface, as the depositional regime may have changed significantly over geological time.[20] As

⑱ 颗粒级配。
⑲ 土的分类。
⑳ 综上所述，因为在不同地质年代沉积区会发生显著改变，为了了解将会成为表层的沉积土大概的特性，对这一地区的地质历史的研究非常重要。

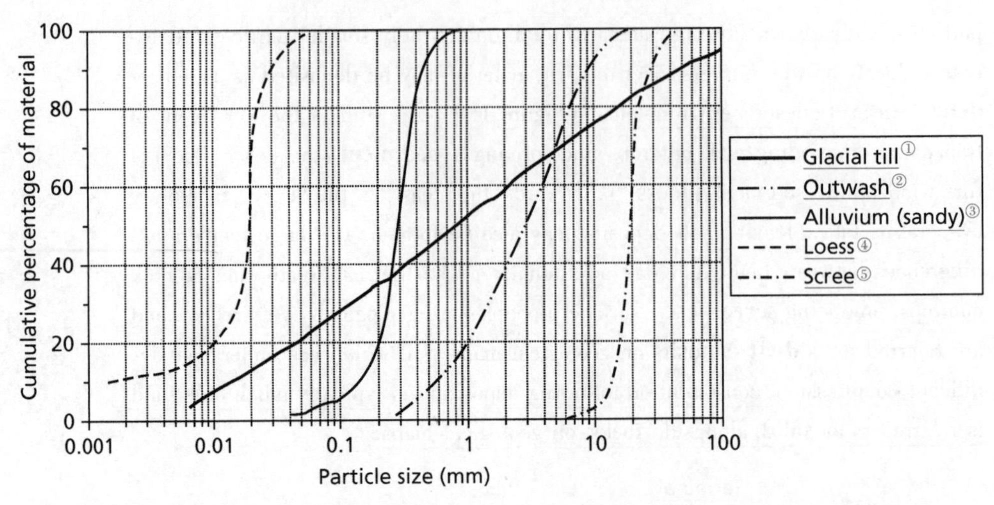

Figure 1.4 Particle size distributions of sediments from different depositional environments.⑥

① 冰碛土。
② 冲蚀土。
③ 冲积土。
④ 黄土。
⑤ 碎石。
⑥ 不同沉积环境中的沉积物颗粒尺寸分布。

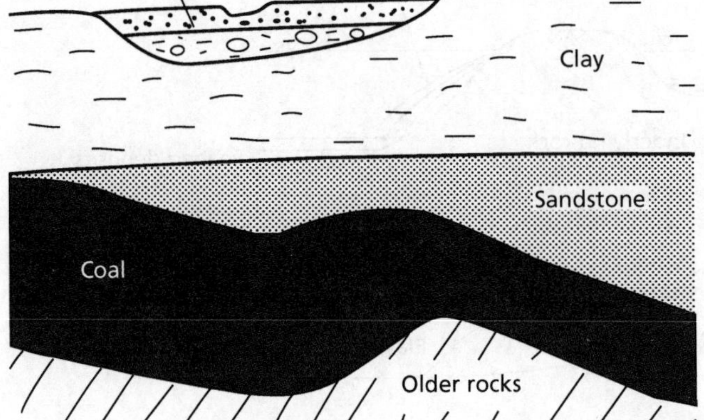

Figure 1.5 Typical ground profile in the West Midlands, UK.

⑦ 石炭纪。
⑧ 三叠纪。
⑨ 白垩纪。
⑩ 冰川时期。
⑪ 更新世。

an example, the West Midlands in the UK was deltaic in the Carboniferous period⑦ (~395 – 345 million years ago), depositing organic material which subsequently became coal measures. In the subsequent Triassic period⑧ (280 – 225 million years ago), due to a change in sea level sandy materials were deposited which were subsequently lithified to become Bunter sandstone. Mountain building during this period on what is now the European continent caused the existing rock layers to become folded. It was subsequently flooded by the North Sea during the Cretaceous/Jurassic periods⑨ (225 – 136 million years ago), depositing fine particles and carbonate material (Lias clay and Oolitic limestone). The Ice Ages⑩ in the Pleistocene period⑪ (1.5 – 2 million years ago) subsequently led to glaciation over all but the southernmost part of the UK, eroding some of the recently deposited softer rocks and depositing glacial till. The subsequent melting of the glaciers created river valleys, which deposited alluvium above the till. The geological history would therefore suggest that the surficial soil conditions are likely to consist of alluvium overlying till/clay overlying stronger rocks, as shown schematically in Figure 1.5. This example demonstrates the importance of engineering geology[①] in understanding ground conditions. A thorough introduction to this topic can be found in Waltham (2002).

① 工程地质。

1.2 The nature of soils

The destructive process in the formation of soil from rock may be either physical or chemical.[②] The physical process may be erosion by the action of wind, water or glaciers, or disintegration caused by cycles of freezing and thawing in cracks in the rock.[③] The resultant soil particles retain the same mineralogical composition as that of the parent rock[④] (a full description of this is beyond the scope of this text). Particles of this type are described as being of 'bulky' form, and their shape can be indicated by terms such as angular, rounded, flat and elongated. The particles occur in a wide range of sizes, from boulders, through gravels and sands, to the fine rock flour formed by the grinding action of glaciers. The structural arrangement of bulky particles (Figure 1.6) is described as **single grain**[⑤], each particle being in direct contact with adjoining particles without there being any bond between them. The state of the particles can be described as dense, medium dense or loose, depending on how they are packed together (see Section 1.5).

② 由岩石形成土的破坏过程可能是物理破坏或化学破坏。

③ 物理破坏可能是因为风、水或冰川造成的侵蚀，或冻融循环造成的瓦解。

④ 母岩。

⑤ 单粒。

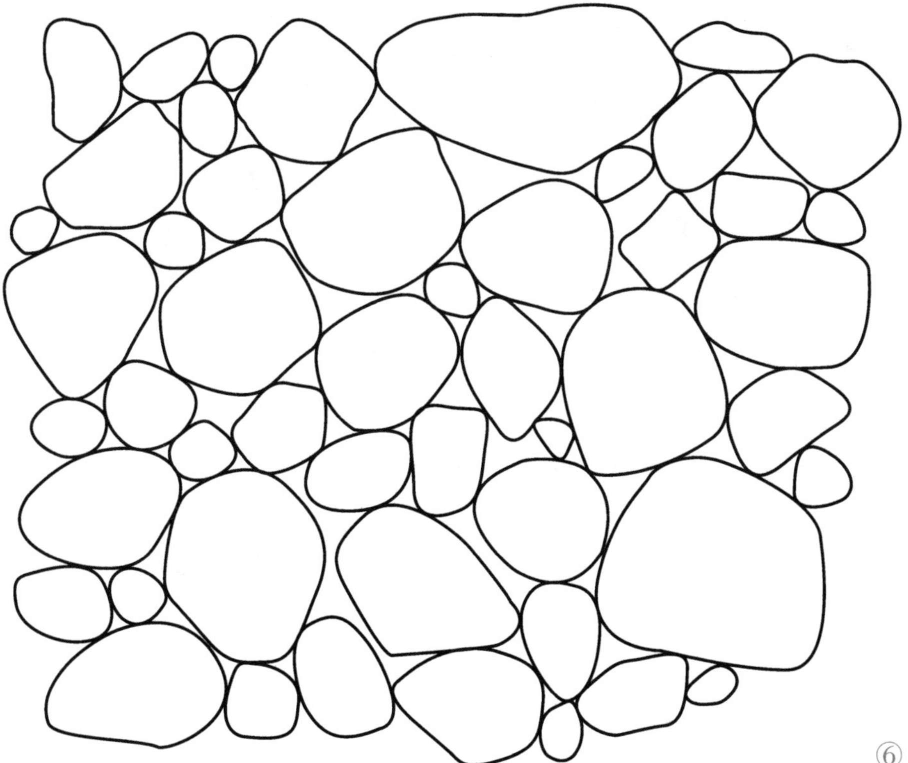

Figure 1.6 Single grain structure.[⑥]

⑥ 单粒结构。

Chemical processes result in changes in the mineral form of the parent rock due to the action of water (especially if it contains traces of acid or alkali), oxygen and carbon dioxide.① Chemical weathering results in the formation of groups of crystalline particles of **colloidal**② size (<0.002mm) known as clay minerals. The clay mineral kaolinite③, for example, is formed by the breakdown of feldspar④ by the action of water and carbon dioxide. Most clay mineral particles are of 'plate-like' form⑤, having a high specific surface⑥ (i. e. a high surface area to mass ratio), with the result that their structure is influenced significantly by surface forces. Long 'needle-shaped⑦' particles can also occur, but are comparatively rare.

The basic structural units of most clay minerals are a silicon – oxygen tetrahedron⑧ and an aluminium hydroxyl octahedron⑨, as illustrated in Figure 1.7 (a). There are valency imbalances in both units, resulting in net negative charges. The basic units therefore do not exist in isolation, but combine to form sheet structures.⑩ The tetrahedral units combine by the sharing of oxygen ions to form a silica sheet. The octahedral units combine through shared hydroxyl ions to form a gibbsite⑪ sheet. The silica sheet retains a net negative charge, but the gibbsite sheet is electrically neutral⑫. Silicon and aluminium may be partially replaced by other elements, this being known as **isomorphous substitution**⑬, resulting in further charge imbalance. The sheet structures are represented symbolically in Figure 1.7 (b). Layer structures then form by the bonding of a silica sheet with either one or two gibbsite sheets. Clay mineral particles consist of stacks of these layers, with different forms of bonding between the layers.

① 化学破坏是由于水（特别当其中有酸或碱时）、氧气与二氧化碳的共同作用导致母岩矿物的改变。
② 胶体的。
③ 高岭石。
④ 长石。
⑤ 片状。
⑥ 比表面积。
⑦ 针状。
⑧ 硅-氧四面体。
⑨ 铝-氢氧基八面体。
⑩ 在两种晶体单元中存在原子价的不平衡，导致净负电荷。因此基体单元并不孤立存在，它们结合在一起形成片状结构。
⑪ 三水铝矿，水铝氧。
⑫ 中性。
⑬ 同晶置换。

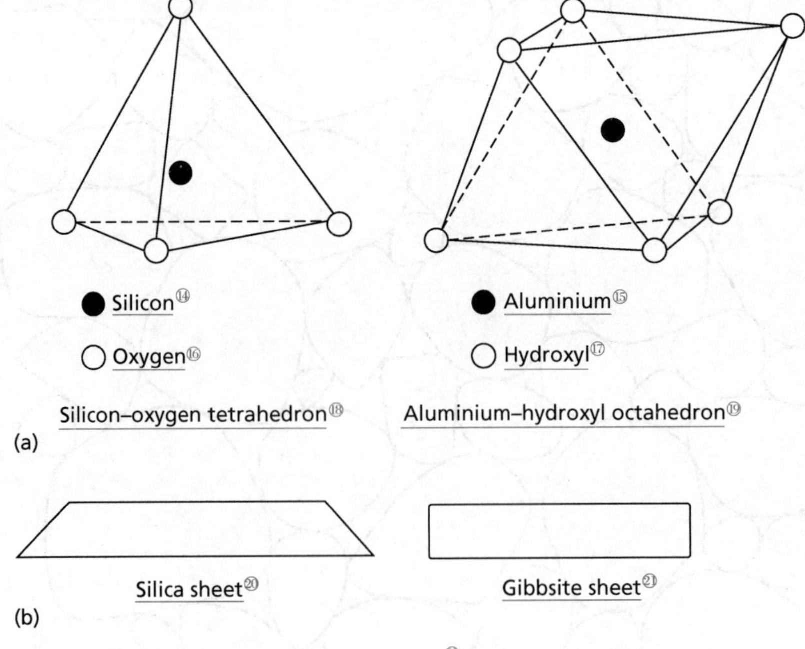

Figure 1.7 Clay minerals: basic units.㉒

⑭ 硅。
⑮ 铝。
⑯ 氧。
⑰ 氢氧基。
⑱ 硅-氧四面体。
⑲ 铝氧-氢氧基八面体。
⑳ 硅氧片层。
㉑ 三水铝片层。
㉒ 黏土矿物：基本单元。

The surfaces of clay mineral particles carry residual negative charges, mainly as a result of the isomorphous substitution of silicon or aluminium by ions of lower valency but also due to disassociation of hydroxyl ions. Unsatisfied charges due to 'broken bonds' at the edges of particles also occur. The negative charges result in cations present in the water in the void space being attracted to the particles. The cations are not held strongly and, if the nature of the water changes, can be replaced by other cations, a phenomenon referred to as **base exchange**[①].

① 阳离子交换。

Cations are attracted to a clay mineral particle because of the negatively charged surface, but at the same time they tend to move away from each other because of their thermal energy.[②] The net effect is that the cations form a dispersed layer adjacent to the particle, the cation concentration decreasing with increasing distance from the surface until the concentration becomes equal to that in the general mass of water in the void space of the soil as a whole. The term 'double layer' describes the negatively charged particle surface and the dispersed layer of cations. For a given particle, the thickness of the cation layer depends mainly on the valency and concentration of the cations: an increase in valency (due to cation exchange) or an increase in concentration will result in a decrease in layer thickness.

② 由于表面的负电荷，正电离子被吸引至黏土矿物颗粒上，但同时由于它们的热动能，它们又有互相远离的趋势。

Layers of water molecules are held around a clay mineral particle by hydrogen bonding and (because water molecules are dipolar) by attraction to the negatively charged surfaces. In addition, the exchangeable cations attract water (i. e. they become hydrated). The particle is thus surrounded by a layer of adsorbed water. The water nearest to the particle is strongly held and appears to have a high viscosity, but the viscosity decreases with increasing distance from the particle surface to that of 'free' water at the boundary of the adsorbed layer.[③] Adsorbed water molecules can move relatively freely parallel to the particle surface, but movement perpendicular to the surface is restricted.

③ 距颗粒最近的水被牢固地吸住，表现出较强的黏性，但随着从颗粒表面与吸附层边界自由水距离的增加，黏性随之减弱。

The structures of the principal clay minerals are represented in Figure 1. 8. Kaolinite consists of a structure based on a single sheet of silica combined with a single sheet of gibbsite. There is very limited isomorphous substitution. The combined silica – gibbsite sheets are held together relatively strongly by hydrogen bonding. A kaolinite particle may consist of over 100 stacks. Illite[④] has a basic structure consisting of a sheet of gibbsite between and combined with two sheets of silica. In the silica sheet, there is partial substitution of silicon by aluminium. The combined sheets are linked together by relatively weak bonding due to non-exchangeable potassium ions[⑤] held between them. Montmorillonite[⑥] has the same basic structure as illite. In the gibbsite sheet there is partial substitution of aluminium by magnesium and iron, and in the silica sheet there is again partial substitution of silicon by aluminium. The space between the combined sheets is occupied by water molecules and exchangeable cations other than potassium, resulting in a very weak bond. Considerable swelling of montmorillonite (and therefore of any soil of which it is a part) can occur due to additional water being adsorbed between the combined sheets.[⑦] This demonstrates that understanding the basic composition of a soil in terms of its

④ 伊利石。

⑤ 钾离子。

⑥ 蒙脱石。

⑦ 蒙脱石（存在于任何土中）会由于附加水被吸引至结合层间而产生可观的膨胀。

mineralogy can provide clues as to the geotechnical problems which may subsequently be encountered.

Forces of repulsion and attraction act between adjacent clay mineral particles. Repulsion occurs between the like charges of the double layers, the force of repulsion depending on the characteristics of the layers. An increase in cation valency or concentration will result in a decrease in repulsive force and vice versa. Attraction between particles is due to short-range van der Waals forces[①] (electrical forces of attraction between neutral molecules), which are independent of the double-layer characteristics, that decrease rapidly with increasing distance between particles. The net inter-particle forces influence the structural form of clay mineral particles on deposition. If there is net repulsion the particles tend to assume a face-to-face orientation, this being referred to as a **dispersed** structure[②]. If, on the other hand, there is net attraction the orientation of the particles tends to be edge-to-face or edge-to-edge, this being referred to as a **flocculated** structure[③]. These structures, involving interaction between single clay mineral particles, are illustrated in Figures 1.9 (a) and (b).

① 范德华力。

② 分散结构。

③ 絮状结构。

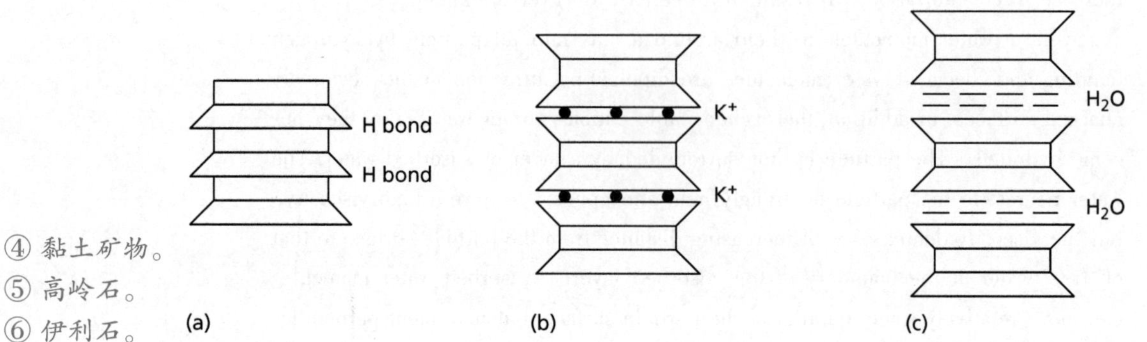

Figure 1.8 Clay minerals[④]: (a) kaolinite[⑤], (b) illite[⑥], and (c) montmorillonite[⑦].

④ 黏土矿物。
⑤ 高岭石。
⑥ 伊利石。
⑦ 蒙脱石。

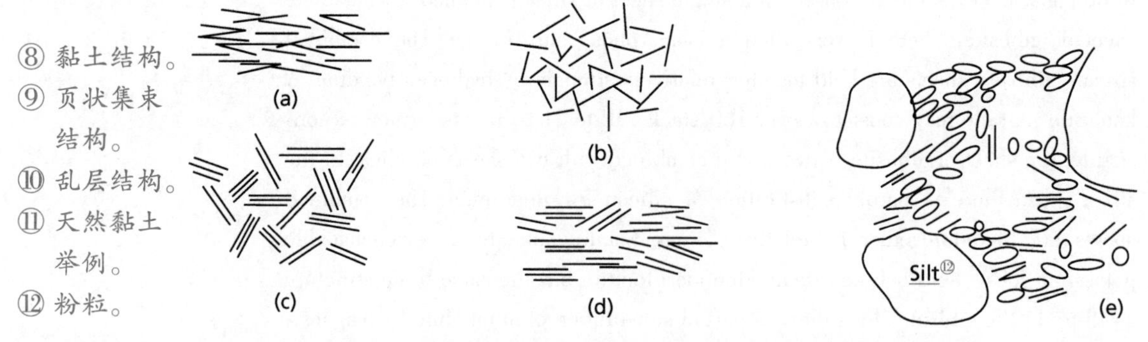

Figure 1.9 Clay structures[⑧]: (a) dispersed, (b) flocculated, (c) bookhouse[⑨], (d) turbostratic[⑩], (e) example of a natural clay[⑪].

⑧ 黏土结构。
⑨ 页状集束结构。
⑩ 乱层结构。
⑪ 天然黏土举例。
⑫ 粉粒。

In natural clays, which normally contain a significant proportion of larger, bulky particles, the structural arrangement can be extremely complex.[⑬] Interaction between single clay mineral particles is rare, the tendency being for the formation of

⑬ 在天然黏土中，一般包含了很大比例的大碎块，结构排列极复杂。

elementary **aggregations** of particles with a face-to-face orientation.[①] In turn, these elementary aggregations combine to form larger assemblages, the structure of which is influenced by the depositional environment[②]. Two possible forms of particle assemblage, known as the bookhouse and turbostratic structures, are illustrated in Figures 1.9 (c) and (d). Assemblages can also occur in the form of connectors or a matrix between larger particles. An example of the structure of a natural clay, in diagrammatical form, is shown in Figure 1.9 (e).

If clay mineral particles are present they usually exert a considerable influence on the properties of a soil, an influence out of all proportion to their percentage by weight in the soil.[③] Soils whose properties are influenced mainly by clay and silt size particles are commonly referred to as **fine-grained** (or fine) soils[④]. Those whose properties are influenced mainly by sand and gravel size particles are referred to as **coarse-grained** (or coarse) soils[⑤].

① 单个黏土颗粒间接触的情况极少，通常形式是由面-面接触形成的黏土颗粒聚合体组成。

② 沉积环境。

③ 黏土矿物颗粒的存在，会对土的特性有很大的影响，而影响程度取决于其在土中所占的比例（按质量计）。

④ 细粒土。

⑤ 粗粒土。

1.3 Plasticity of fine-grained soils[⑥]

⑥ 细粒土的塑性。

Plasticity is an important characteristic in the case of fine-grained soils, the term 'plasticity' describing the ability of a soil to undergo irrecoverable deformation without cracking or crumbling. In general, depending on its **water content**[⑦] (defined as the ratio of the mass of water in the soil to the mass of solid particles), a soil may exist in one of the liquid, plastic, semi-solid and solid states. If the water content of a soil initially in the liquid state is gradually reduced, the state will change from liquid through plastic and semi-solid, accompanied by gradually reducing volume, until the solid state is reached. The water contents at which the transitions between states occur differ from soil to soil. In the ground, most fine-grained soils exist in the plastic state. Plasticity is due to the presence of a significant content of clay mineral particles (or organic material) in the soil. The void space between such particles is generally very small in size with the result that water is held at negative pressure by capillary tension[⑧], allowing the soil to be deformed or moulded. Adsorption of water due to the surface forces on clay mineral particles may contribute to plastic behaviour. Any decrease in water content results in a decrease in cation layer thickness and an increase in the net attractive forces between particles.

⑦ 含水量。

⑧ 毛细张力。

The upper and lower limits of the range of water content over which the soil exhibits plastic behaviour are defined as the **liquid limit** (w_L)[⑨] and the **plastic limit** (w_P)[⑩], respectively. Above the liquid limit, the soil flows like a liquid (slurry); below the plastic limit, the soil is brittle and crumbly. The water content range itself is defined as the **plasticity index** (I_P)[⑪], i. e.:

$$I_P = w_L - w_P \tag{1.1}$$

However, the transitions between the different states are gradual, and the liquid and plastic limits must be defined arbitrarily. The natural water content (w) of a soil (adjusted to an equivalent water content of the fraction passing the 425-μm sieve)

⑨ 液限。

⑩ 塑限。

⑪ 塑性指数。

① 液性指数。

relative to the liquid and plastic limits can be represented by means of the **liquidity index** (I_L)[①], where

$$I_L = \frac{w - w_P}{I_P} \tag{1.2}$$

The relationship between the different consistency limits is shown in Figure 1.10.

② 黏粒比例。

③ 活性。

④ 高活性土（含有大量的蒙脱石）会对岩土工程产生特殊破坏。

The degree of plasticity of the clay-size fraction of a soil is expressed by the ratio of the plasticity index to the percentage of clay-size particles in the soil (the **clay fraction**[②]): this ratio is called the **activity**[③]. 'Normal' soils have an activity between 0.75 and 1.25, i.e. I_P is approximately equal to the clay fraction. Activity below 0.75 is considered inactive, while soils with activity above 1.25 are considered active. Soils of high activity have a greater change in volume when the water content is changed (i.e. greater swelling when wetted and greater shrinkage when drying). Soils of high activity (e.g. containing a significant amount of montmorillonite) can therefore be particularly damaging to geotechnical works.[④]

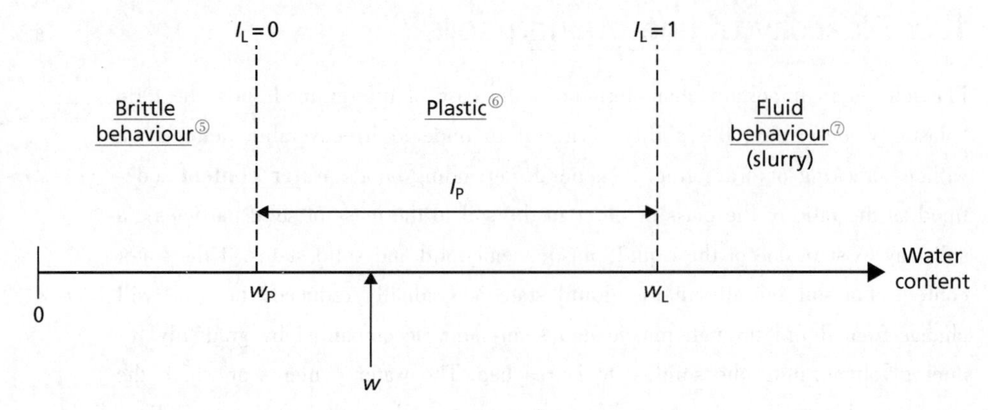

⑤ 脆性。

⑥ 塑性。

⑦ 液性。

Figure 1.10 Consistency limits for fine soils.[⑧]

⑧ 细粒的稠度界限。

Table 1.1 gives the activity of some common clay minerals, from which it can be seen that activity broadly correlates with the specific surface of the particles (i.e. surface area per unit mass), as this governs the amount of adsorbed water.

⑨ 缩限。

The transition between the semi-solid and solid states occurs at the **shrinkage limit**[⑨], defined as the water content at which the volume of the soil reaches its lowest value as it dries out.

⑩ 将土样完全烘干，使其产生裂缝破碎，这一操作用到土碾与橡胶板，期间不击碎单一颗粒，试验只采用通过 425μm 土工筛的土。

The liquid and plastic limits are determined by means of arbitrary test procedures. In the UK, these are fully detailed in BS 1377, Part 2 (1990). In Europe CEN ISO/TS 17892 – 12 (2004) is the current standard, while in the United States ASTM D4318 (2010) is used. These standards all relate to the same basic tests which are described below.

The soil sample is dried sufficiently to enable it to be crumbled and broken up, using a mortar and a rubber pestle, without crushing individual particles; only material passing a 425-μm sieve is typically used in the tests.[⑩] The apparatus for the liquid limit test consists of a penetrometer[⑪] (or 'fall-cone') fitted with a 30° cone of stainless steel, 35mm long: the cone and the sliding shaft to which it is attached

⑪ 贯入仪。

have a mass of 80g. This is shown in Figure 1.11 (a). The test soil is mixed with distilled water① to form a thick homogeneous paste, and stored for 24h. Some of the paste is then placed in a cylindrical metal cup, 55mm internal diameter by 40mm deep, and levelled off at the rim of the cup to give a smooth surface. The cone is lowered so that it just touches the surface of the soil in the cup, the cone being locked in its support at this stage. The cone is then released for a period of 5s, and its depth of penetration into the soil is measured. A little more of the soil paste is added to the cup and the test is repeated until a consistent value of penetration has been obtained. (The average of two values within 0.5mm or of three values within 1.0mm is taken.) The entire test procedure is repeated at least four times, using the same soil sample but increasing the water content each time by adding distilled water.② The penetration values should cover the range of approximately 15 – 25mm, the tests proceeding from the drier to the wetter state of the soil. Cone penetration is plotted against water content, and the best straight line fitting the plotted points is drawn. This is demonstrated in Example 1.1. The liquid limit is defined as the percentage water content (to the nearest integer) corresponding to a cone penetration of 20mm.③ Determination of liquid limit may also be based on a single test (the one-point method), provided the cone penetration is between 15 and 25mm.

① 蒸馏水。

② 整个试验步骤至少重复四次，试验过程中使用同一土样，但每次通过加蒸馏水来增加其含水量。

③ 土的液限定义为对应于圆锥沉入深度为 20mm 时的含水量。

Table 1.1 Activity of some common clay minerals④

Mineral group	*Specific surface (m^2/g)*[1]	*Activity*[2]
Kaolinite	10 – 20	0.3 – 0.5
Illite	65 – 100	0.5 – 1.3
Montmorillonite	Up to 840	4 – 7

Notes: 1 After Mitchell and Soga (2005). 2 After Day (2001).

④ 一些常见黏土矿物的活性。

An alternative method for determining the liquid limit uses the Casagrande apparatus (Figure 1.11 (b)), which is popular in the United States and other parts of the world (ASTM D4318). A soil paste is placed in a pivoting flat metal cup and divided by cutting a groove. A mechanism enables the cup to be lifted to a height of 10mm and dropped onto a hard rubber base. The two halves of the soil gradually flow together as the cup is repeatedly dropped. The water content of the soil in the cup is then determined; this is plotted against the logarithm of the number of blows, and the best straight line fitting the plotted points is drawn.⑤ For this test, the liquid limit is defined as the water content at which 25 blows are required to close the bottom of the groove. It should be noted that the Casagrande method is generally less reliable than the preferred penetrometer method, being more operator dependent and subjective.

⑤ 随着圆碟重复多次落下，两半土条逐渐合拢，即可测定土样的含水量，绘制含水量与下落次数的对数值关系曲线，并以直线拟合曲线。

For the determination of the plastic limit, the test soil is mixed with distilled water until it becomes sufficiently plastic to be moulded into a ball. Part of the soil

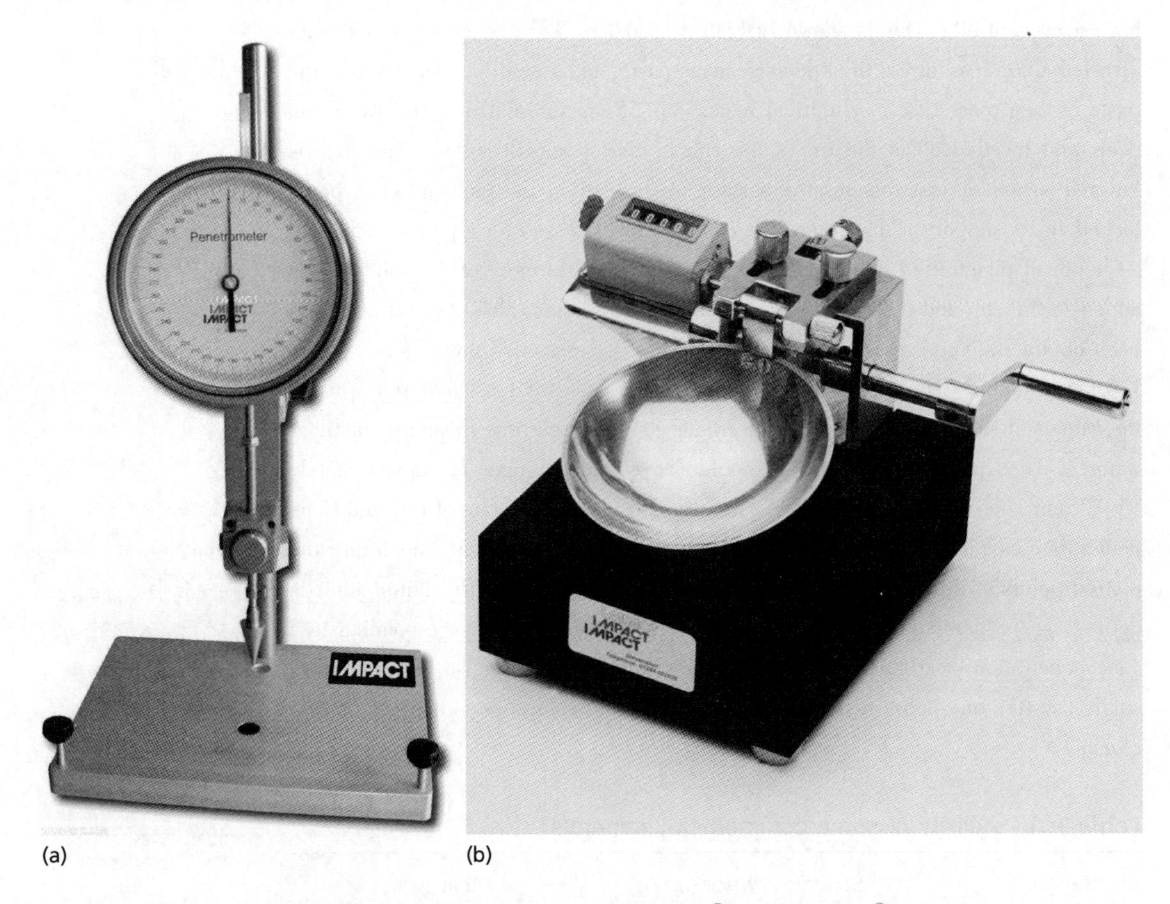

(a) (b)

Figure 1.11 Laboratory apparatus for determining liquid limit[①]: (a) fall-cone[②], (b) Casagrande apparatus[③] (images courtesy of Impact Test Equipment Ltd).

① 测定液限的实验仪器。
② 落锥。
③ 卡萨格兰德液限仪。

sample (approximately 2.5g) is formed into a thread, approximately 6mm in diameter, between the first finger and thumb of each hand. The thread is then placed on a glass plate and rolled with the tips of the fingers of one hand until its diameter is reduced to approximately 3mm: the rolling pressure must be uniform throughout the test. The thread is then remoulded between the fingers (the water content being reduced by the heat of the fingers) and the procedure is repeated until the thread of soil shears both longitudinally and transversely when it has been rolled to a diameter of 3mm. The procedure is repeated using three more parts of the sample, and the percentage water content of all the crumbled soil is determined as a whole. This water content (to the nearest integer) is defined as the plastic limit of the soil. The entire test is repeated using four other subsamples, and the average taken of the two values of plastic limit: the tests must be repeated if the two values differ by more than 0.5%. Due to the strongly subjective nature of this test, alternative methodologies have recently been proposed for determining w_P, though these are not incorporated within current standards. Further information can be found in Barnes (2009) and Sivakumar *et al.* (2009).

1.4 Particle size analysis[①]

① 粒径分析。

Most soils consist of a graded mixture of particles from two or more size ranges.[②] For example, clay is a type of soil possessing cohesion and plasticity which normally consists of particles in both the clay size and silt size ranges. **Cohesion**[③] is the term used to describe the strength of a clay sample when it is unconfined, being due to negative pressure in the water filling the void space, of very small size, between particles. This strength would be lost if the clay were immersed in a body of water. Cohesion may also be derived from cementation between soil particles. It should be appreciated that all clay-size particles are not necessarily clay mineral particles: the finest rock flour particles may be of clay size.

② 大多数土是由两种或两种以上粒径范围的颗粒混合而成的。

③ 黏聚力。

The particle size analysis of a soil sample involves determining the percentage by mass of particles within the different size ranges.[④] The particle size distribution of a coarse soil can be determined by the method of **sieving**[⑤]. The soil sample is passed through a series of standard test sieves having successively smaller mesh sizes. The mass of soil retained in each sieve is determined, and the cumulative percentage by mass passing each sieve is calculated. If fine particles are present in the soil, the sample should be treated with a deflocculating agent (e. g. a 4% solution of sodium hexametaphosphate) and washed through the sieves.

④ 土样的粒径分析包括确定不同尺寸范围的颗粒所占比例。

⑤ 筛分法。

The particle size distribution (PSD) of a fine soil or the fine fraction of a coarse soil can be determined by the method of **sedimentation**[⑥]. This method is based on Stokes' law, which governs the velocity at which spherical particles settle in a suspension: the larger the particles are the greater is the settling velocity, and vice versa[⑦]. The law does not apply to particles smaller than 0.0002mm, the settlement of which is influenced by Brownian motion[⑧]. The size of a particle is given as the diameter of a sphere which would settle at the same velocity as the particle. Initially, the soil sample is pretreated with hydrogen peroxide to remove any organic material. The sample is then made up as a suspension in distilled water to which a deflocculating agent has been added to ensure that all particles settle individually, and placed in a sedimentation tube. From Stokes' law it is possible to calculate the time, t, for particles of a certain 'size', D (the equivalent settling diameter), to settle to a specified depth in the suspension. If, after the calculated time t, a sample of the suspension is drawn off with a pipette at the specified depth below the surface, the sample will contain only particles smaller than the size D at a concentration unchanged from that at the start of sedimentation. If pipette samples are taken at the specified depth at times corresponding to other chosen particle sizes, the particle size distribution can be determined from the masses of the residues. An alternative procedure to pipette sampling is the measurement of the specific gravity of the suspension by means of a special hydrometer, the specific gravity depending on the mass of soil particles in the suspension at the time of measurement. Full details of the determination of particle size distribution by these methods are given in BS

⑥ 细粒土或粗粒土的细颗粒部分的颗粒级配可由沉降法测定。

⑦ 颗粒越大，沉降速度越快，反之亦然。

⑧ 布朗运动。

1377 – 2 (UK), CEN ISO/TS 17892 – 4 (Europe) and ASTM D6913 (US). Modern optical techniques can also be used to determine the PSD of a coarse soil. Single Particle Optical Sizing (SPOS) works by drawing a stream of dry particles through the beam of a laser diode. As each individual particle passes through the beam it casts a shadow on a light sensor which is proportional to its size (and therefore volume). The optical sizer automatically analyses the sensor output to determine the PSD by volume. Optical methods have been found to overestimate particle sizes compared to sieving (White, 2003), though advantages are that the results are repeatable and less operator dependent compared to sieving, and testing requires a much smaller volume of soil.

The particle size distribution of a soil is presented as a curve on a semilogarithmic plot, the ordinates being the percentage by mass of particles smaller than the size given by the abscissa. The flatter the distribution curve, the larger the range of particle sizes in the soil; the steeper the curve, the smaller the size range.① A coarse soil is described as **well graded**② if there is no excess of particles in any size range and if no intermediate sizes are lacking. In general, a well-graded soil is represented by a smooth, concave distribution curve. A coarse soil is described as **poorly graded**③ (a) if a high proportion of the particles have sizes within narrow limits (a uniform soil), or (b) if particles of both large and small sizes are present but with a relatively low proportion of particles of intermediate size (a **gap-graded**④ or step-graded soil). Particle size is represented on a logarithmic scale so that two soils having the same degree of uniformity are represented by curves of the same shape regardless of their positions on the particle size distribution plot. Examples of particle size distribution curves appear in Figure 1.4. The particle size corresponding to any specified percentage value can be read from the particle size distribution curve. The size such that 10% of the particles are smaller than that size is denoted by D_{10}. Other sizes, such as D_{30} and D_{60}, can be defined in a similar way. The size D_{10} is defined as the **effective size**⑤, and can be used to estimate the permeability of the soil (see Chapter 2). The general slope and shape of the distribution curve can be described by means of the **coefficient of uniformity**⑥ (C_u) and the **coefficient of curvature**⑦ (C_z), defined as follows:

$$C_u = \frac{D_{60}}{D_{10}} \tag{1.3}$$

$$C_z = \frac{D_{30}^2}{D_{60}D_{10}} \tag{1.4}$$

The higher the value of the coefficient of uniformity, the larger the range of particle sizes in the soil. A well-graded soil has a coefficient of curvature between 1 and 3.⑧ The sizes D_{15} and D_{85} are commonly used to select appropriate material for granular drains used to drain geotechnical works (see Chapter 2).

1.5 Soil description and classification⑨

It is essential that a standard language should exist for the description of soils. A

① 级配曲线越平缓，土中的粒径分布范围越大；级配曲线越陡，土中的粒径分布范围越小。
② 级配良好。
③ 级配不良。
④ 间断级配。
⑤ 有效粒径。
⑥ 不均匀系数。
⑦ 曲率系数。
⑧ 不均匀系数值越高，土中颗粒尺寸分布范围越大，级配良好土的曲率系数范围为1~3。
⑨ 土的描述与分类。

comprehensive description should include the characteristics of both the soil material and the in-situ soil mass. Material characteristics can be determined from disturbed samples of the soil – i. e. samples having the same particle size distribution as the in-situ soil but in which the in-situ structure has not been preserved. The principal material characteristics are particle size distribution (or grading) and plasticity, from which the soil name can be deduced. ① Particle size distribution and plasticity properties can be determined either by standard laboratory tests (as described in Sections 1.3 and 1.4) or by simple visual and manual procedures. Secondary material characteristics are the colour of the soil and the shape, texture and composition of the particles. Mass characteristics should ideally be determined in the field, but in many cases they can be detected in undisturbed samples – i. e. samples in which the in-situ soil structure has been essentially preserved. A description of mass characteristics should include an assessment of insitu compactive state (coarse-grained soils) or stiffness (fine-grained soils), and details of any bedding, discontinuities and weathering. The arrangement of minor geological details, referred to as the soil macro-fabric, should be carefully described, as this can influence the engineering behaviour of the insitu soil to a considerable extent. Examples of macro-fabric features are thin layers of fine sand and silt in clay, silt-filled fissures in clay, small lenses of clay in sand, organic inclusions, and root holes. The name of the geological formation, if definitely known, should be included in the description; in addition, the type of deposit may be stated (e. g. till, alluvium), as this can indicate, in a general way, the likely behaviour of the soil.

① 最主要的材料特性是粒径级配与塑性，由它们可推出该土的名称。

It is important to distinguish between soil description and soil classification. Soil description includes details of both material and mass characteristics, and therefore it is unlikely that any two soils will have identical descriptions. In soil classification, on the other hand, a soil is allocated to one of a limited number of behavioural groups on the basis of material characteristics only. ② Soil classification is thus independent of the in-situ condition of the soil mass. If the soil is to be employed in its undisturbed condition, for example to support a foundation, a full soil description will be adequate and the addition of the soil classification is discretionary. However, classification is particularly useful if the soil in question is to be used as a construction material when it will be remoulded – for example in an embankment. Engineers can also draw on past experience of the behaviour of soils of similar classification.

② 土的描述包括土的材料和质量的详细特性，因此任何两种土不大可能有相同的描述。另一方面，在土的工程分类中，通常只基于土的材料特性将土划分到数量有限的性状类似的组别。

Rapid assessment procedures③

③ 快速评估步骤。

Both soil description and classification require knowledge of grading and plasticity. This can be determined by the full laboratory procedure using standard tests, as described in Sections 1.3 and 1.4, in which values defining the particle size distribution and the liquid and plastic limits are obtained for the soil in question. Alternatively, grading and plasticity can be assessed using a rapid procedure which involves personal judgements based on the appearance and feel of the soil. The rapid procedure can be used in the field and in other situations where the use of the laboratory procedure is not possible or not justified. In the rapid procedure, the following indicators should be used.

Particles of 0.06mm, the lower size limit for coarse soils, are just visible to the naked eye, and feel harsh but not gritty when rubbed between the fingers; finer material feels smooth to the touch.[①] The size boundary between sand and gravel is 2mm, and this represents the largest size of particles which will hold together by capillary attraction when moist. A purely visual judgement must be made as to whether the sample is well graded or poorly graded, this being more difficult for sands than for gravels.

① 粒径为 0.06mm 时，裸眼可分辨，也是粗粒土的粒径下限，用手指摩擦时有粗糙感但无砂砾感，更细粒的材料摸起来会更顺滑。

If a predominantly coarse soil contains a significant proportion of fine material, it is important to know whether the fines are essentially plastic or non-plastic (i. e. whether they are predominantly clay or silt, respectively). This can be judged by the extent to which the soil exhibits cohesion and plasticity. A small quantity of the soil, with the largest particles removed, should be moulded together in the hands, adding water if necessary. Cohesion is indicated if the soil, at an appropriate water content, can be moulded into a relatively firm mass. Plasticity is indicated if the soil can be deformed without cracking or crumbling, i. e. without losing cohesion.[②] If cohesion and plasticity are pronounced, then the fines are plastic. If cohesion and plasticity are absent or only weakly indicated, then the fines are essentially non-plastic.

② 土的黏聚性是指土中含水量达到某一值土体可以被塑成一个相对坚固的整体。土的塑性是指土体不产生裂隙或破碎即黏聚性不消失。

The plasticity of fine soils can be assessed by means of the toughness and dilatancy tests, described below. An assessment of dry strength may also be useful. Any coarse particles, if present, are first removed, and then a small sample of the soil is moulded in the hand to a consistency judged to be just above the plastic limit (i. e. just enough water to mould); water is added or the soil is allowed to dry as necessary. The procedures are then as follows.

Toughness test[③]

③ 粗糙度测试。

A small piece of soil is rolled out into a thread on a flat surface or on the palm of the hand, moulded together, and rolled out again until it has dried sufficiently to break into lumps at a diameter of around 3mm. In this condition, inorganic clays of high liquid limit are fairly stiff and tough; those of low liquid limit are softer and crumble more easily. Inorganic silts produce a weak and often soft thread, which may be difficult to form and which readily breaks and crumbles.

Dilatancy test[④]

④ 膨胀性测试。

A pat of soil, with sufficient water added to make it soft but not sticky, is placed in the open (horizontal) palm of the hand. The side of the hand is then struck against the other hand several times. Dilatancy is indicated by the appearance of a shiny film of water on the surface of the pat; if the pat is then squeezed or pressed with the fingers, the surface becomes dull as the pat stiffens and eventually crumbles. These reactions are pronounced only for predominantly silt-size material and for very fine sands. Plastic clays give no reaction.

Dry strength test[1]

① 干燥强度测试。

A pat of soil about 6mm thick is allowed to dry completely, either naturally or in an oven. The strength of the dry soil is then assessed by breaking and crumbling between the fingers. Inorganic clays have relatively high dry strength; the greater the strength, the higher the liquid limit. Inorganic silts of low liquid limit have little or no dry strength, crumbling easily between the fingers.

Soil description details[2]

② 土体描述详情。

A detailed guide to soil description as used in the UK is given in BS 5930 (1999), and the subsequent discussion is based on this standard. In Europe the standard is EN ISO 14688 – 1 (2002), while in the United States ASTM D2487 (2011) is used. The basic soil types are boulders, cobbles, gravel, sand, silt and clay, defined in terms of the particle size ranges shown in Figure 1. 2; added to these are organic clay, silt or sand, and peat. These names are always written in capital letters in a soil description. Mixtures of the basic soil types are referred to as composite types.

A soil is of basic type sand or gravel (these being termed coarse soils) if, after the removal of any cobbles or boulders, over 65% of the material is of sand and gravel sizes.[3] A soil is of basic type silt or clay (termed fine soils) if, after the removal of any cobbles or boulders, over 35% of the material is of silt and clay sizes. However, these percentages should be considered as approximate guidelines, not forming a rigid boundary. Sand and gravel may each be subdivided into coarse, medium and fine fractions as defined in Figure 1. 2. The state of sand and gravel can be described as well graded, poorly graded, uniform or gap graded, as defined in Section 1. 4. In the case of gravels, particle shape (angular, sub-angular, sub-rounded, rounded, flat, elongated) and surface texture (rough, smooth, polished) can be described if necessary. Particle composition can also be stated. Gravel particles are usually rock fragments (e. g. sandstone, schist). Sand particles usually consist of individual mineral grains (e. g. quartz, feldspar). Fine soils should be described as either silt or clay: terms such as silty clay should not be used.

③ 基本土型砂或砾（俗称粗粒土）指去除卵石或漂石后的砂粒或碎石粒质量大于总质量65%的土。

Organic soils[4] contain a significant proportion of dispersed vegetable matter, which usually produces a distinctive odour and often a dark brown, dark grey or bluish grey colour. Peats consist predominantly of plant remains, usually dark brown or black in colour and with a distinctive odour. If the plant remains are recognisable and retain some strength, the peat is described as fibrous. If the plant remains are recognisable but their strength has been lost, they are pseudo-fibrous. If recognisable plant remains are absent, the peat is described as amorphous. Organic content is measured by burning a sample of soil at a controlled temperature to determine the reduction in mass which corresponds to the organic content. Alterna-

④ 有机质土。

tively, the soil may be treated with hydrogen peroxide (H_2O_2), which also removes the organic content, resulting in a loss of mass.

Composite types of coarse soil are named in Table 1.2, the predominant component being written in capital letters. Fine soils containing 35 – 65% coarse material are described as sandy and/or gravelly SILT (or CLAY). Deposits containing over 50% of boulders and cobbles are referred to as very coarse, and normally can be described only in excavations and exposures. Mixes of very coarse material with finer soils can be described by combining the descriptions of the two components – e.g. COBBLES with some FINER MATERIAL (sand); gravelly SAND with occasional BOULDERS.

① 粗粒土的组合类型。

Table 1.2 Composite types of coarse soil[①]

Slightly sandy GRAVEL	Up to 5% sand
Sandy GRAVEL	5 – 20% sand
Very sandy GRAVEL	Over 20% sand
SAND and GRAVEL	About equal proportions
Very gravelly SAND	Over 20% gravel
Gravelly SAND	5 – 20% gravel
Slightly gravelly SAND	Up to 5% gravel
Slightly silty SAND (and/or GRAVEL)	Up to 5% silt
Silty SAND (and/or GRAVEL)	5 – 20% silt
Very silty SAND (and/or GRAVEL)	Over 20% silt
Slightly clayey SAND (and/or GRAVEL)	Up to 5% clay
Clayey SAND (and/or GRAVEL)	5 – 20% clay
Very clayey SAND (and/or GRAVEL)	Over 20% clay

Note: Terms such as 'Slightly clayey gravelly SAND' (having less than 5% clay and gravel) and 'Silty sandy GRAVEL' (having 5 – 20% silt and sand) can be used, based on the above proportions of secondary constituents.

The state of compaction or stiffness of the in-situ soil can be assessed by means of the tests or indications detailed in Table 1.3.

Discontinuities such as fissures and shear planes, including their spacings, should be indicated. Bedding features, including their thickness, should be detailed. Alternating layers of varying soil types or with bands or lenses of other materials are described as **interstratified**[②]. Layers of different soil types are described as **interbedded**[③] or **inter-laminated**[④], their thickness being stated. Bedding surfaces that separate easily are referred to as **partings**. If partings incorporate other material, this should be described.

② 间层。
③ 互层。
④ 夹层。

Table 1.3 Compactive state and stiffness of soils[1]

Soil group	*Term(relative density – Section 1.6)*	*Field test or indication*
Coarse soils[2]	Very loose (0 – 20%)	Assessed on basis of N – value determined by means of Standard Penetration Test (SPT) – see Chapter 7
	Loose (20 – 40%)	
	Medium dense (40 – 60%)	
	Dense (60 – 80%)	For definition of relative density, see Equation (1.23)
	Very dense (80 – 100%)	
	Slightly cemented	Visual examination: pick removes soil in lumps which can be abraded
Fine soils[3]	Uncompact	Easily moulded or crushed by the fingers
	Compact	Can be moulded or crushed by strong finger pressure
	Very soft	Finger can easily be pushed in up to 25mm
	Soft	Finger can be pushed in up to 10mm
	Firm	Thumb can make impression easily
	Stiff	Thumb can make slight indentation
	Very stiff	Thumb nail can make indentation
	Hard	Thumb nail can make surface scratch
Organic soils[4]	Firm	Fibres already pressed together
	Spongy	Very compressible and open structure
	Plastic	Can be moulded in the hand and smears fingers

Some examples of soil description are as follows:

Dense, reddish-brown, sub-angular, well-graded SAND
Firm, grey, laminated CLAY with occasional silt partings 0.5 – 2.0mm (Alluvium)
Dense, brown, well graded, very silty SAND and GRAVEL with some COBBLES (Till)
Stiff, brown, closely fissured CLAY (London Clay)
Spongy, dark brown, fibrous PEAT (Recent Deposits)

Soil classification systems[5]

General classification systems, in which soils are placed into behavioural groups on the basis of grading and plasticity, have been used for many years. [6]The feature of these systems is that each soil group is denoted by a letter symbol representing main and qualifying terms. [7]The terms and letters used in the UK are detailed in Table 1.4. The boundary between coarse and fine soils is generally taken to be 35% fines (i. e. particles smaller than 0.06mm). The liquid and plastic limits are used to classify fine soils, employing the plasticity chart shown in Figure 1.12. The axes of the plasticity chart are the plasticity index and liquid limit; therefore, the

① 土的压缩状态与刚度。
② 粗粒土。
③ 细粒土。
④ 有机质土。
⑤ 土的分类体系。
⑥ 基于土的级配和塑性将其归入相应行为组的经典分类体系已运用了很多年。
⑦ 此分类体系的特点为每类土都由代表其的主要要素与定性要素标识。

plasticity characteristics of a particular soil can be represented by a point on the chart. Classification letters are allotted to the soil according to the zone within which the point lies. The chart is divided into five ranges of liquid limit. The four ranges I, H, V and E can be combined as an upper range (U) if closer designation is not required, or if the rapid assessment procedure has been used to assess plasticity. The diagonal line on the chart, known as the **A-line**, should not be regarded as a rigid boundary between clay and silt for purposes of soil description, as opposed to classification[①]. The A-line may be mathematically represented by

$$I_P = 0.73\ (w_L - 20) \tag{1.5}$$

① A 线区分了粉土和黏土的区域。A 线上方为黏土，下方为粉土。

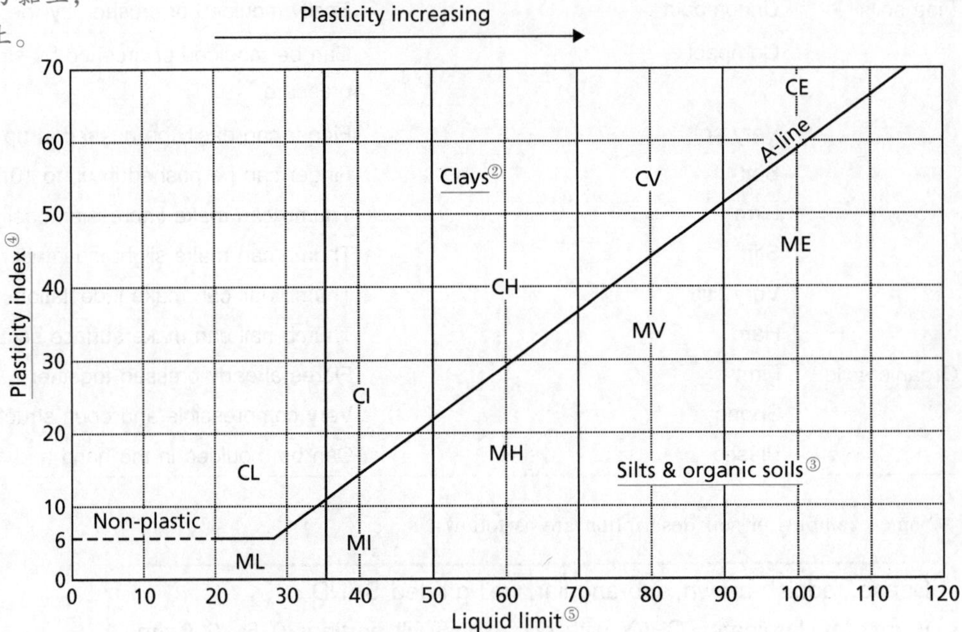

Figure 1.12 Plasticity chart[⑥]: British system (BS 1377 – 2: 1990).

② 黏土。
③ 粉土或有机质土。
④ 塑性指数。
⑤ 液限。
⑥ 塑性图。

Table 1.4 Descriptive terms for soil classification (BS 5930)

Main terms		*Qualifying terms*	
GRAVEL	G	Well graded	W
SAND	S	Poorly graded	P
		Uniform	Pu
		Gap graded	Pg
FINE SOIL, FINES	F	Of low plasticity (w_L < 35)	L
SILT (M-SOIL)	M	Of intermediate plasticity (w_L 35 – 50)	I
CLAY	C	Of high plasticity (w_L 50 – 70)	H
		Of very high plasticity (w_L 70 – 90)	V
		Of extremely high plasticity (w_L > 90)	E
		Of upper plasticity range (w_L > 35)	U
PEAT	Pt	Organic (may be a suffix to any group)	O

The letter denoting the dominant size fraction is placed first in the group symbol. If a soil has a significant content of organic matter, the suffix O is added as the last letter of the group symbol. A group symbol may consist of two or more letters, for example:

SW – well-graded SAND[①]
SCL – very clayey SAND (clay of low plasticity)[②]
CIS – sandy CLAY of intermediate plasticity[③]
MHSO – organic sandy SILT of high plasticity[④].

① 级配良好砂。
② 黏质砂（低塑性黏土）。
③ 中等塑性砂质黏土。
④ 含有机质的高塑性砂质粉土。

The name of the soil group should always be given, as above, in addition to the symbol, the extent of subdivision depending on the particular situation. If the rapid procedure has been used to assess grading and plasticity, the group symbol should be enclosed in brackets to indicate the lower degree of accuracy associated with this procedure.

The term FINE SOIL or FINES (F) is used when it is not required, or not possible, to differentiate between SILT (M) and CLAY (C). SILT (M) plots below the A-line and CLAY (C) above the A-line on the plasticity chart, i. e. silts exhibit plastic properties over a lower range of water content than clays having the same liquid limit.[⑤] SILT or CLAY is qualified as gravelly if more than 50% of the coarse fraction is of gravel size, and as sandy if more than 50% of the coarse fraction is of sand size. The alternative term M-SOIL is introduced to describe material which, regardless of its particle size distribution, plots below the A-line on the plasticity chart: the use of this term avoids confusion with soils of predominantly silt size (but with a significant proportion of clay-size particles), which plot above the A-line. Fine soils containing significant amounts of organic matter usually have high to extremely high liquid limits, and plot below the A-line as organic silt. Peats usually have very high or extremely high liquid limits.

⑤ 粉土（M）的点落在 A 线下方，黏土（C）的点则落在 A 线上方，例如粉土与黏土液限相同，前者可在较小范围的含水量之间表现出塑性特性。

Any cobbles or boulders (particles retained on a 63-mm sieve) are removed from the soil before the classification tests are carried out, but their percentages in the total sample should be determined or estimated.[⑥] Mixtures of soil and cobbles or boulders can be indicated by using the letters Cb (COBBLES) or B (BOULDERS) joined by a + sign to the group symbol for the soil, the dominant component being given first – for example:

⑥ 卵石或巨砾（不能通过 63mm 筛子的颗粒）在分类测试之前被从土中除去，但在总土样中它们所占百分比应该被确定或是预估。

GW + Cb – well-graded GRAVEL with COBBLES
B + CL – BOULDERS with CLAY of low plasticity.

A similar classification system, known as the Unified Soil Classification System (USCS), was developed in the United States (described in ASTM D2487), but with less detailed subdivisions. As the USCS method is popular in other parts of the world, alternative versions of Figure 1.12 and Table 1.4 are provided on the Companion Website.

Example 1.1

The results of particle size analyses of four soils A, B, C and D are shown in Table 1.5. The results of limit tests on soil D are as follows:

TABLE A

Liquid limit:					
Cone penetration[①] (mm)	15.5	18.0	19.4	22.2	24.9
Water content (%)	39.3	40.8	42.1	44.6	45.6
Plastic limit:					
Water content (%)	23.9	24.3			

① 锥入度。

The fine fraction of soil C has a liquid limit[②] of $I_L = 26$ and a plasticity index[③] of $I_P = 9$.

a Determine the coefficients of uniformity and curvature for soils A, B and C.

b Allot group symbols, with main and qualifying terms to each soil.

Table 1.5 Example 1.1

Sieve	*Particle size**[④]	*Percentage smaller* Soil A	Soil B	Soil C	Soil D
63mm		100		100	
20mm		64		76	
6.3mm		39	100	65	
2mm		24	98	59	
600mm		12	90	54	
212mm		5	9	47	100
63mm		0	3	34	95
	0.020mm			23	69
	0.006mm			14	46
	0.002mm			7	31

Note: * From sedimentation test.

② 液限。

③ 塑性指数。

④ 粒径。

Solution

The particle size distribution curves are plotted in Figure 1.13. For soils A, B and C, the sizes D_{10}, D_{30} and D_{60} are read from the curves and the values of C_U and C_Z are calculated:

TABLE B

Soil	D_{10}	D_{30}	D_{60}	C_U	C_Z
A	0.47	3.5	16	34	1.6
B	0.23	0.30	0.41	1.8	0.95
C	0.003	0.042	2.4	800	0.25

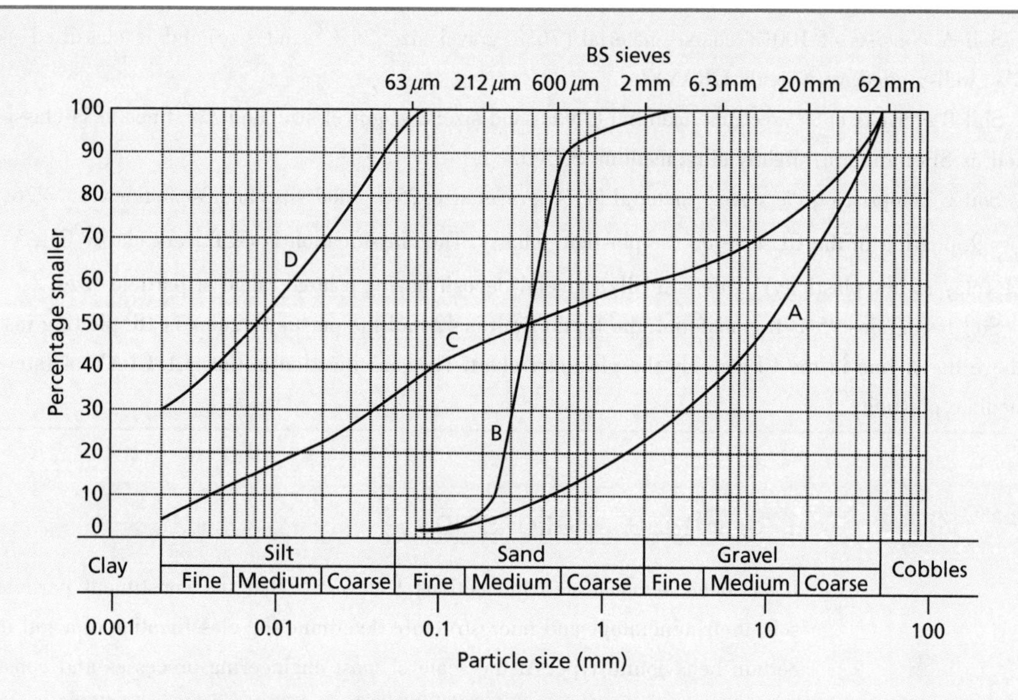

Figure 1.13 Particle size distribution curves (Example 1.1).

For soil D the liquid limit is obtained from Figure 1.14, in which fall-cone penetration is plotted against water content. The percentage water content, to the nearest integer, corresponding to a penetration of 20mm is the liquid limit, and is 42%. The plastic limit is the average of the two percentage water contents, again to the nearest integer, i. e. 24%. The plasticity index is the difference between the liquid and plastic limits, i. e. 18%.

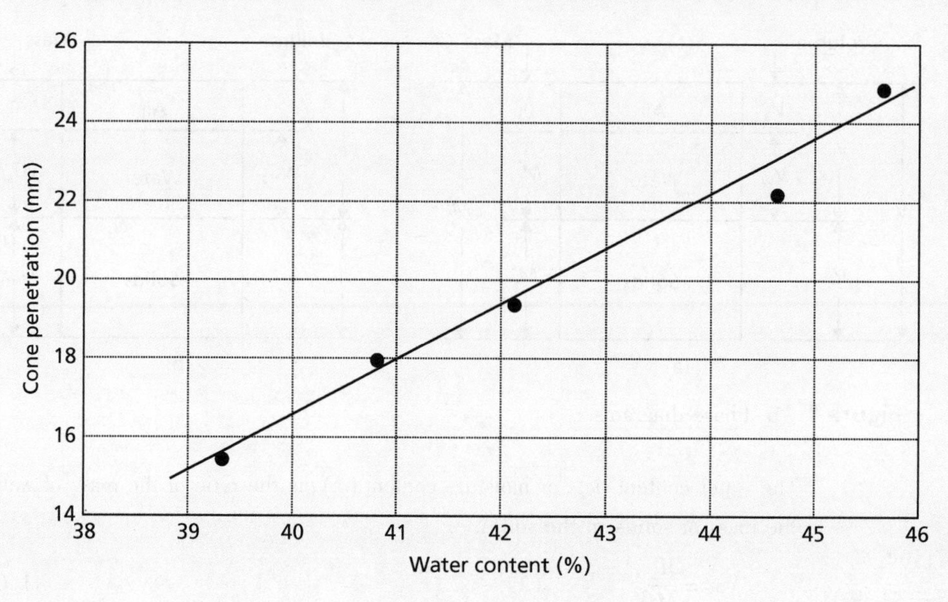

Figure 1.14 Determination of liquid limit[①] (Example 1.1).

① 液限的确定。

Soil A consists of 100% coarse material (76% gravel size; 24% sand size) and is classified as GW: well-graded, very sandy GRAVEL.

Soil B consists of 97% coarse material (95% sand size; 2% gravel size) and 3% fines. It is classified as SPu: uniform, slightly silty, medium SAND.

Soil C comprises 66% coarse material (41% gravel size; 25% sand size) and 34% fines ($w_L = 26$, $I_P = 9$, plotting in the CL zone on the plasticity chart). The classification is GCL: very clayey GRAVEL (clay of low plasticity). This is a till, a glacial deposit having a large range of particle sizes.

Soil D contains 95% fine material: the liquid limit is 42 and the plasticity index is 18, plotting just above the A-line in the CI zone on the plasticity chart. The classification is thus CI: CLAY of intermediate plasticity.

① 土的三相关系。

1.6 Phase relationships[①]

It has been demonstrated in Sections 1.1 – 1.5 that the constituent particles of soil, their mineralogy and microstructure determine the classification of a soil into a certain behavioural type. At the scale of most engineering processes and constructions, however, it is necessary to describe the soil as a continuum. Soils can be of either two-phase or three-phase composition. In a completely dry soil there are two phases, namely the solid soil particles and pore air. A fully saturated soil is also two-phase, being composed of solid soil particles and pore water. A partially saturated soil is three-phase, being composed of solid soil particles, pore water and pore air. The components of a soil can be represented by a phase diagram as shown in Figure 1.15 (a), from which the following relationships are defined.

② 体积。
③ 质量。
④ 三相图。

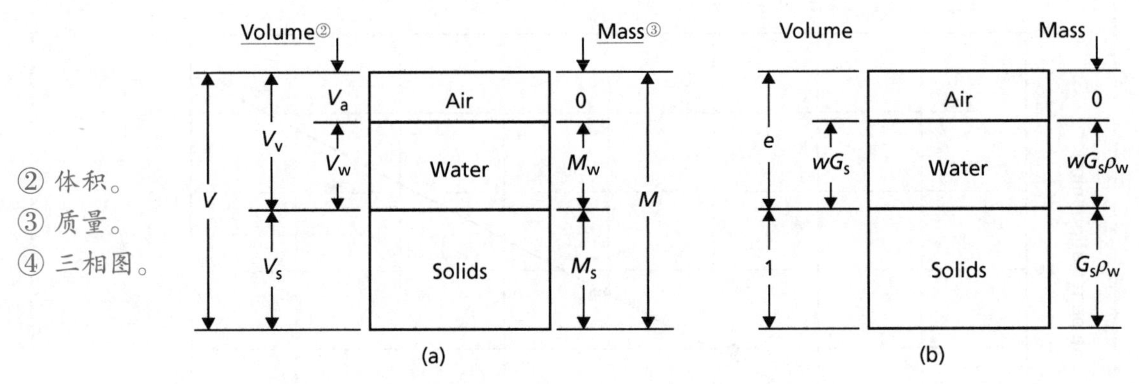

Figure 1.15 Phase diagrams.[④]

The water content (w), or moisture content (m), is the ratio of the mass of water to the mass of solids in the soil, i. e.

$$w = \frac{M_w}{M_s} \tag{1.6}$$

The water content is determined by weighing a sample of the soil and then drying the sample in an oven at a temperature of 105 – 110℃ and re-weighing.[⑤] Drying

⑤ 通过称量土样，然后在105 ~ 110℃的烤箱中烘干取出再称量，便可得出含水量。

should continue until the differences between successive weighings at four-hourly intervals are not greater than 0.1% of the original mass of the sample. A drying period of 24h is normally adequate for most soils.

The degree of saturation or **saturation ratio**① (S_r) is the ratio of the volume of water to the total volume of void space, i. e.

① 饱和度。

$$S_r = \frac{V_w}{V_v} \tag{1.7}$$

The saturation ratio can range between the limits of zero for a completely dry soil and one (or 100%) for a fully saturated soil.②

② 饱和度的范围是从完全干燥土的零到完全饱和土中的100%。

The **void ratio**③ (e) is the ratio of the volume of voids to the volume of solids, i. e.

$$e = \frac{V_v}{V_s} \tag{1.8}$$

The **porosity**④ (n) is the ratio of the volume of voids to the total volume of the soil, i. e.

③ 孔隙比。

④ 孔隙率。

$$n = \frac{V_v}{V} \tag{1.9}$$

As $V = V_v + V_s$, void ratio and porosity are interrelated as follows:

$$e = \frac{n}{1 - n} \tag{1.10}$$

$$n = \frac{e}{1 + e} \tag{1.11}$$

The **specific volume**⑤ (ν) is the total volume of soil which contains a unit volume of solids, i. e.

⑤ 比容。

$$\nu = \frac{V_v}{V_s} = 1 + e \tag{1.12}$$

The **air content**⑥ or air voids (A) is the ratio of the volume of air to the total volume of the soil, i. e.

⑥ 含气量。

$$A = \frac{V_a}{V} \tag{1.13}$$

The **bulk density**⑦ or mass density⑧ (ρ) of a soil is the ratio of the total mass to the total volume, i. e.

⑦ 体积密度。

⑧ 质量密度。

$$\rho = \frac{M}{V} \tag{1.14}$$

Convenient units for density are kg/m^3 or Mg/m^3. The density of water (1000 kg/m^3 or 1.00 Mg/m^3) is denoted by ρ_w.

The **specific gravity**⑨ of the soil particles (G_s) is given by

⑨ 相对密度。

$$G_s = \frac{M_s}{V_s\rho_w} = \frac{\rho_s}{\rho_w} \tag{1.15}$$

where ρ_s is the **particle density**⑩.

⑩ 颗粒密度。

From the definition of void ratio, if the volume of solids is 1 unit then the volume of voids is e units.⑪ The mass of solids is then $G_s\rho_w$ and, from the definition of water content, the mass of water is $wG_s\rho_w$. The volume of water is thus wG_s. These volumes and masses are represented in Figure 1.15 (b). From this figure, the fol-

⑪ 根据孔隙比的定义，如果固体的体积为单位1，则孔隙的体积为e。

lowing relationships can then be obtained.

The degree of saturation (definition in Equation 1.7) is

$$S_r = \frac{V_w}{V_v} = \frac{wG_s}{e} \tag{1.16}$$

The air content is the proportion of the total volume occupied by air, i. e.

$$A = \frac{V_a}{V} = \frac{e - wG_s}{1 + e} \tag{1.17}$$

or, from Equations 1.11 and 1.16,

$$A = n(1 - S_r) \tag{1.18}$$

From Equation 1.14, the bulk density of a soil is:

$$\rho = \frac{M}{V} = \frac{G_s(1 + w)\rho_s}{1 + e} \tag{1.19}$$

or, from Equation 1.16,

$$\rho = \frac{G_s + S_r e}{1 + e}\rho_w \tag{1.20}$$

Equation 1.20 holds true for any soil. Two special cases that commonly occur, however, are when the soil is fully saturated with either water or air. For a fully saturated soil $S_r = 1$, giving:

$$\rho_{sat} = \frac{G_s + e}{1 + e}\rho_w \tag{1.21}$$

For a completely dry soil ($S_r = 0$):

$$\rho_d = \frac{G_s}{1 + e}\rho_w \tag{1.22}$$

① 单位体积重量。

The **unit weight**[①] or weight density (γ) of a soil is the ratio of the total weight (Mg) to the total volume, i. e.

$$\gamma = \frac{Mg}{V} = \rho g$$

Multiplying Equations 1.19 and 1.20 by g then gives

$$\gamma = \frac{G_s(1 + w)}{1 + e}\gamma_w \tag{1.19a}$$

$$\gamma = \frac{G_s + S_r e}{1 + e}\gamma_w \tag{1.20a}$$

where γ_w is the unit weight of water. Convenient units are kN/m^3, the unit weight of water being $9.81 kN/m^3$ (or $10.0 kN/m^3$ in the case of sea water).

② 在砂土和砂砾中，相对密实度被用作描述原位孔隙比或试样孔隙比与土样的最大和最小孔隙比的关系。

In the case of sands and gravels the **relative density** (I_D) is used to express the relationship between the in-situ void ratio (e), or the void ratio of a sample, and the limiting values e_{max} and e_{min} representing the loosest and densest possible soil packing states respectively.[②] The relative density is defined as

$$I_D = \frac{e_{max} - e}{e_{max} - e_{min}} \tag{1.23}$$

Thus, the relative density of a soil in its densest possible state ($e = e_{min}$) is 1 (or 100%) and in its loosest possible state ($e = e_{max}$) is 0.

The maximum density is determined by compacting a sample underwater in a mould, using a circular steel tamper attached to a vibrating hammer: a 1-l mould is used for sands and a 2.3-l mould for gravels. The soil from the mould is then dried in an oven, enabling the dry density to be determined. The minimum dry density can be determined by one of the following procedures. In the case of sands, a 1-l measuring cylinder is partially filled with a dry sample of mass 1000g and the top of the cylinder closed with a rubber stopper. The minimum density is achieved by shaking and inverting the cylinder several times, the resulting volume being read from the graduations on the cylinder. In the case of gravels, and sandy gravels, a sample is poured from a height of about 0.5 m into a 2.3-l mould and the resulting dry density determined. Full details of the above tests are given in BS 1377, Part 4 (1990). Void ratio can be calculated from a value of dry density using Equation 1.22. However, the density index can be calculated directly from the maximum, minimum and in-situ values of dry density, avoiding the need to know the value of G_s (see Problem 1.5).

Example 1.2

In its natural condition, a soil sample has a mass of 2290g and a volume of $1.15 \times 10^{-3}\,m^3$. After being completely dried in an oven, the mass of the sample is 2035g. The value of G_s for the soil is 2.68. Determine the bulk density, unit weight, water content, void ratio, porosity, degree of saturation and air content.

Solution

$$\text{Bulk density, } \rho = \frac{M}{V} = \frac{2.290}{1.15 \times 10^{-3}} = 1990\,\text{kg/m}^3\ (1.99\,\text{Mg/m}^3)$$

$$\text{Unit weight, } \gamma = \frac{Mg}{V} = 1990 \times 9.8 = 19500\,\text{N/m}^3$$

$$= 19.5\,\text{kN/m}^3$$

$$\text{Water content, } w = \frac{M_w}{M_s} = \frac{2290 - 2035}{2035} = 0.125 \text{ or } 12.5\%$$

From Equation 1.19,

$$\text{Void ratio, } e = G_s(1 + w)\frac{\rho_w}{\rho} - 1$$

$$= \left(2.68 \times 1.125 \times \frac{1000}{1990}\right) - 1$$

$$= 1.52 - 1$$

$$= 0.52$$

$$\text{Porosity, } n = \frac{e}{1 + e} = \frac{0.52}{1.52} = 0.34 \text{ or } 34\%$$

$$\text{Degree of saturation, } S_r = \frac{wG_s}{e} = \frac{0.125 \times 2.68}{0.52} = 0.645 \text{ or } 64.5\%$$

$$\text{Air content, } A = n(1 - S_r) = 0.34 \times 0.355$$

$$= 0.121 \text{ or } 12.1\%$$

① 土的击实。

1.7 Soil compaction[①]

Compaction is the process of increasing the density of a soil by packing the particles closer together with a reduction in the volume of air; there is no significant change in the volume of water in the soil.[②] In the construction of fills and embankments, loose soil is typically placed in layers ranging between 75 and 450mm in thickness, each layer being compacted to a specified standard by means of rollers, vibrators or rammers. In general, the higher the degree of compaction, the higher will be the shear strength and the lower will be the compressibility of the soil[③] (see Chapters 4 and 5). An **engineered fill**[④] is one in which the soil has been selected, placed and compacted to an appropriate specification with the object of achieving a particular engineering performance, generally based on past experience. The aim is to ensure that the resulting fill possesses properties that are adequate for the function of the fill. This is in contrast to non-engineered fills, which have been placed without regard to a subsequent engineering function.

② 土的击实是通过减少土中空气体积，使颗粒间排列更紧密，从而增大土密度的过程；击实过程中，土中水的体积基本不变。

③ 通常，击实度越高，土的抗剪强度越高，渗透性越低。

④ 工程填土。

The degree of compaction of a soil is measured in terms of dry density, i. e. the mass of solids only per unit volume of soil. If the bulk density of the soil is ρ and the water content w, then from Equations 1.19 and 1.22 it is apparent that the dry density is given by

$$\rho_d = \frac{\rho}{1+w} \tag{1.24}$$

The dry density of a given soil after compaction depends on the water content and the energy supplied by the compaction equipment (referred to as the **compactive effort**[⑤]).

⑤ 击实能。

⑥ 实验室内击实。

Laboratory compaction[⑥]

The compaction characteristics of a soil can be assessed by means of standard laboratory tests. The soil is compacted in a cylindrical mould using a standard compactive effort. In the **Proctor test**[⑦], the volume of the mould is 1-1 and the soil (with all particles larger than 20mm removed) is compacted by a rammer consisting of a 2.5-kg mass falling freely through 300mm: the soil is compacted in three equal layers, each layer receiving 27 blows with the rammer. In the **modified AASHTO test**[⑧], the mould is the same as is used in the above test but the rammer consists of a 4.5-kg mass falling 450mm; the soil (with all particles larger than 20mm removed) is compacted in five layers, each layer receiving 27 blows with the rammer. If the sample contains a limited proportion of particles up to 37.5mm in size, a 2.3-1 mould should be used, each layer receiving 62 blows with either the 2.5- or 4.5-kg rammer. In the **vibrating hammer test**[⑨], the soil (with all particles larger than 37.5mm removed) is compacted in three layers in a 2.3-1 mould, using a circular tamper fitted in the vibrating hammer, each layer being compacted for a period of 60s. These tests are detailed in BS1377 – 4 (UK), EC7 – 2 (Europe) and, in

⑦ 击实试验。

⑧ 修订后的美国国家高速公路和交通运输协会标准试验。

⑨ 振动锤击试验。

the US, ASTM D698, D1557 and D7382.

After compaction using one of the three standard methods, the bulk density and water content of the soil are determined and the dry density is calculated. For a given soil the process is repeated at least five times, the water content of the sample being increased each time. Dry density is plotted against water content, and a curve of the form shown in Figure 1.16 is obtained. This curve shows that for a particular method of compaction (i. e. a given compactive effort) there is a particular value of water content, known as the **optimum water content** (w_{opt}), at which a maximum value of dry density is obtained.[1] At low values of water content, most soils tend to be stiff and are difficult to compact. As the water content is increased the soil becomes more workable, facilitating compaction and resulting in higher dry densities. At high water contents, however, the dry density decreases with increasing water content, an increasing proportion of the soil volume being occupied by water.

① 该曲线说明在某特定击实方法下（如给定的击实功），存在一个特定的含水量，在该含水量下可得到土的最大干密度，该含水量称为最优含水量（w_{opt}）

$$\rho_{d0} = \frac{G_s}{1 + wG_s}\rho_w$$

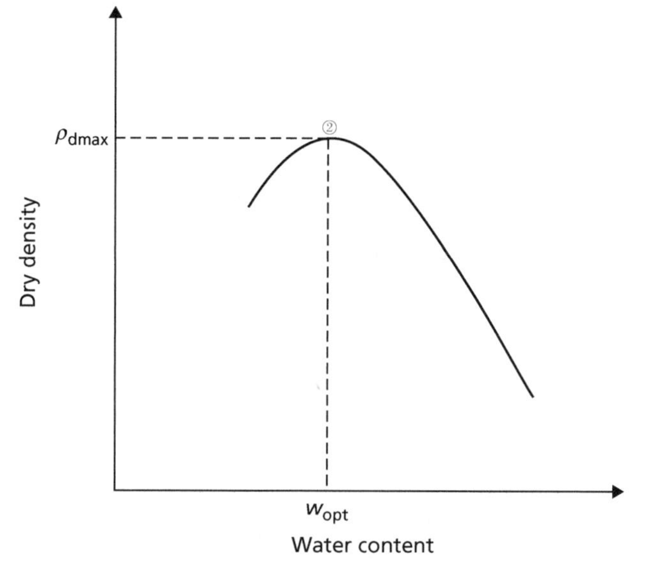

Figure 1.16 Dry density – water content relationship.[3]

② 注：最大干密度对应的最优含水量。

③ 干密度-含水量的关系图。

If all the air in a soil could be expelled by compaction, the soil would be in a state of full saturation and the dry density would be the maximum possible value for the given water content. However, this degree of compaction is unattainable in practice. The maximum possible value of dry density is referred to as the **'zero air voids' dry density**[4] (ρ_{d0}) or the saturation dry density, and can be calculated from the expression:

④ 土体无含气孔隙时的干密度。

$$\rho_{d0} = \frac{G_s}{1 + wG_s}\rho_w \tag{1.25}$$

In general, the dry density after compaction at water content w to an air content A can be calculated from the following expression, derived from Equations 1.17 and 1.22:

$$\rho_d = \frac{G_s(1 - A)}{1 + wG_s}\rho_w \tag{1.26}$$

The calculated relationship between zero air voids dry density (A = 0) and water content (for $G_s = 2.65$) is shown in Figure 1.17; the curve is referred to as the zero air voids line or the **saturation line**①. The experimental dry density – water content curve for a particular compactive effort must lie completely to the left of the saturation line. The curves relating dry density at air contents of 5% and 10% with water content are also shown in Figure 1.17, the values of dry density being calculated from Equation 1.26. These curves enable the air content at any point on the experimental dry density – water content curve to be determined by inspection.

① 完全饱和线。

For a particular soil, different dry density – water content curves are obtained for different compactive efforts. Curves representing the results of tests using the 2.5- and 4.5-kg rammers are shown in Figure 1.17. The curve for the 4.5-kg test is situated above and to the left of the curve for the 2.5-kg test. Thus, a higher compactive effort results in a higher value of maximum dry density and a lower value of optimum water content②; however, the values of air content at maximum dry density are approximately equal.

② 击实能越高，土的最大干密度越大，对应的最优含水量越小。

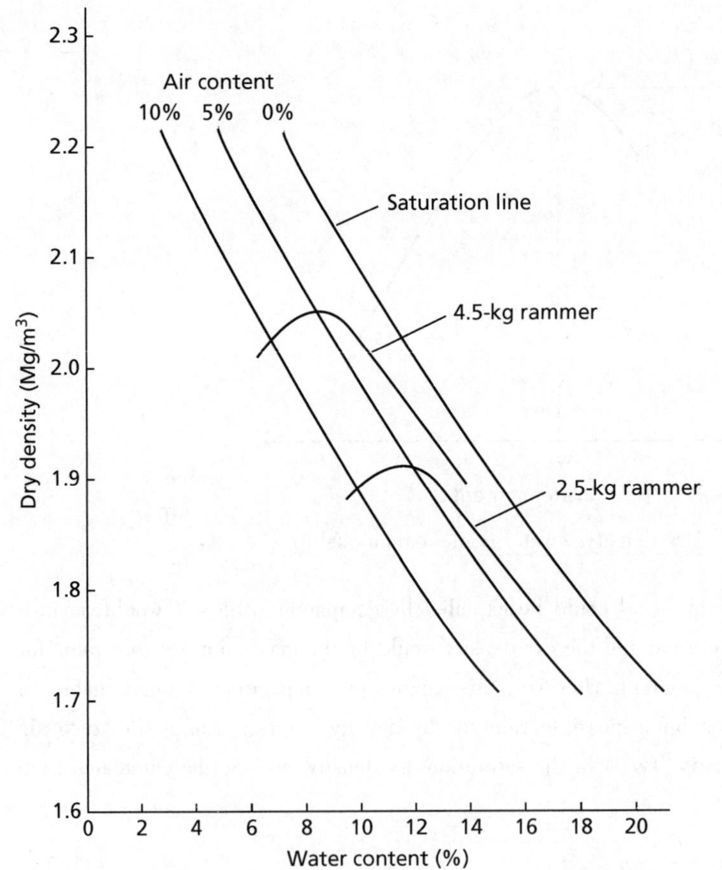

Figure 1.17 Dry density – water content curves for different compactive efforts.③

③ 不同击实能下土的干密度-含水量曲线。

The dry density – water content curves for a range of soil types using the same compactive effort (the BS 2.5-kg rammer) are shown in Figure 1.18. In general, coarse soils can be compacted to higher dry densities than fine soils.

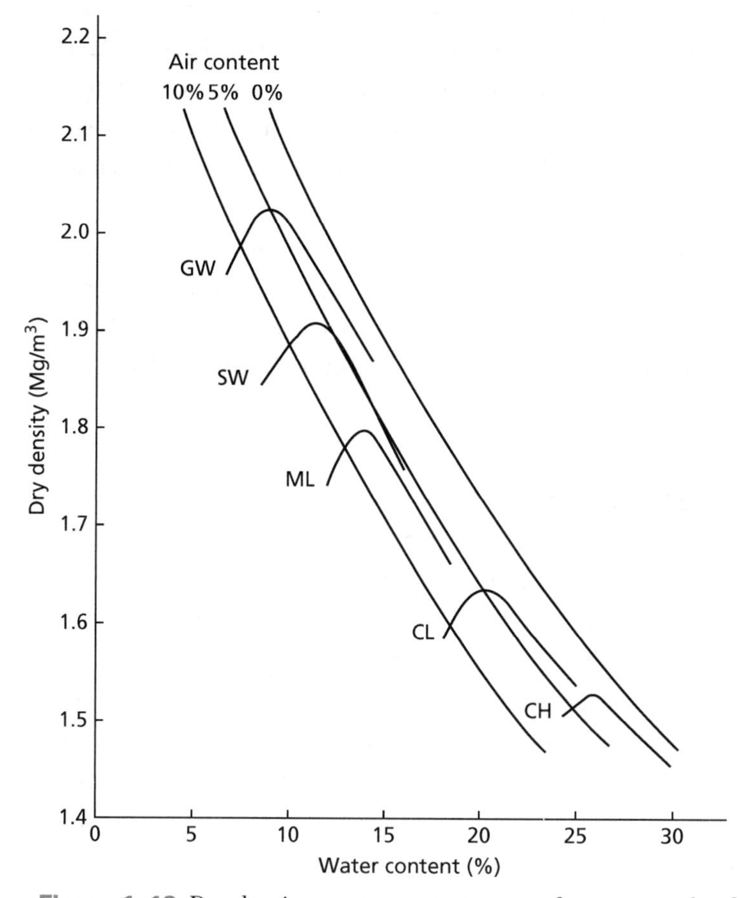

Figure 1.18 Dry density – water content curves for a range of soil types.

Field compaction[①]

① 现场压实。

The results of laboratory compaction tests are not directly applicable to field compaction because the compactive efforts in the laboratory tests are different, and are applied in a different way, from those produced by field equipment. Further, the laboratory tests are carried out only on material smaller than either 20 or 37.5mm. However, the maximum dry densities obtained in the laboratory using the 2.5- and 4.5-kg rammers cover the range of dry density normally produced by field compaction equipment.

A minimum number of passes must be made with the chosen compaction equipment to produce the required value of dry density. This number, which depends on the type and mass of the equipment and on the thickness of the soil layer, is usually within the range 3 – 12. Above a certain number of passes, no significant increase in dry density is obtained. In general, the thicker the soil layer, the heavier the equipment required to produce an adequate degree of compaction.

① 击实过程控制方法。
② 最终击实状态控制方法。
③ 土是一种由岩石风化而成的材料，土颗粒是不同尺寸的单粒组成或是黏土矿物组成。土通常由不同颗粒组成，其中黏粒会对土体的力学特性产生重要影响。
④ 土可基于颗粒级配进行描述和分类，由小颗粒(如黏粒和粉粒）组成的细粒土通常呈现为塑性行为，其可通过塑性和液性指标来定量描述。粗粒土通常没有塑性行为。
⑤ 宏观上，土是由固、水、气三相组成的连续体，它们的相对比例关系可由孔隙比(e)，含水率(w) 和饱和度(S_r) 来表示。
⑥ 土体可作为岩土工程的填料，土的击实可提高土的强度，降低土的压缩性，确定最优含水率，可通过击实试验评价土的击实特性。

There are two approaches to the achievement of a satisfactory standard of compaction in the field, known as **method compaction**① and **end-product compaction**②. In method compaction, the type and mass of equipment, the layer depth and the number of passes are specified. In the UK, these details are given, for the class of material in question, in the Specification for Highway Works (Highways Agency, 2008). In end-product compaction, the required dry density is specified: the dry density of the compacted fill must be equal to or greater than a stated percentage of the maximum dry density obtained in one of the standard laboratory compaction tests.

Field density tests can be carried out, if considered necessary, to verify the standard of compaction in earthworks, dry density or air content being calculated from measured values of bulk density and water content. A number of methods of measuring bulk density in the field are detailed in BS 1377, Part 4 (1990).

Summary

1 Soil is a particulate material formed of weathered rock. The particles may be single grains in a wide range of sizes (from boulders to silt), or clay minerals (colloidal in size). Soil is typically formed from a mixture of such particles, and the presence of clay minerals may significantly alter the mechanical properties of the soil. ③

2 Soils may be described and classified by their particle size distribution. Fine soils consisting of mainly small particles (e. g. clays and silts) typically exhibit plastic behaviour (e. g. cohesion) which may be defined by the plasticity and liquidity indices. Coarse grained soils generally do not exhibit plastic behaviour. ④

3 At the level of the macro-fabric, all soils may be idealised as a three-phase continuum, the phases being solid particles, water and air. The relative proportions of these phases are controlled by the closeness of particle packing, described by the voids ratio (e), water content (w) and saturation ratio (S_r). ⑤

4 In addition to being used in their in-situ state, soils may be used as a fill material in geotechnical constructions. Compaction of such soils increases shear strength and reduces compressibility, and is necessary to achieve optimal performance of the fill. Compaction may be quantified using the Proctor or moisture condition tests. ⑥

Problems

1.1 The results of particle size analyses and, where appropriate, limit tests on samples of four soils are given in Table 1.6. Allot group symbols and give main and qualifying terms appropriate for each soil.

Table 1.6 Problem 1.1

BS sieve	*Particle size*	*Percentage smaller*			
		Soil I	*Soil J*	*Soil K*	*Soil L*
63mm					
20mm		100			
6.3mm		94	100		
2mm		69	98		
600mm[①]		32	88	100	
212mm[②]		13	67	95	100
63mm[③]		2	37	73	99
	0.020mm		22	46	88
	0.006mm		11	25	71
	0.002mm		4	13	58
Liquid limit			Non-plastic	32	78
Plastic limit				24	31

①②③注：英文版原书中此处似有错误，应该单位为"μm"，也可分别表示为0.6mm，0.212mm，0.063mm。

1.2 A soil has a bulk density of 1.91Mg/m^3 and a water content of 9.5%. The value of G_s is 2.70. Calculate the void ratio and degree of saturation of the soil. What would be the values of density and water content if the soil were fully saturated at the same void ratio?

1.3 Calculate the dry unit weight and the saturated unit weight of a soil having a void ratio of 0.70 and a value of G_s of 2.72. Calculate also the unit weight and water content at a degree of saturation of 75%.

1.4 A soil specimen is 38mm in diameter and 76mm long, and in its natural condition weighs 168.0g. When dried completely in an oven, the specimen weighs 130.5g. The value of G_s is 2.73. What is the degree of saturation of the specimen?

1.5 The in-situ dry density of a sand is 1.72Mg/m^3. The maximum and minimum dry densities, determined by standard laboratory tests, are 1.81 and 1.54Mg/m^3, respectively. Determine the relative density of the sand.

1.6 Soil has been compacted in an embankment at a bulk density of 2.15Mg/m^3 and a water content of 12%. The value of G_s is 2.65. Calculate the dry density, void ratio, degree of saturation and air content. Would it be possible to compact the above soil at a water content of 13.5% to a dry density of 2.00Mg/m^3?

1.7 The following results were obtained from a standard compaction test on a soil:

TABLE C

Mass (*g*)	2010	2092	2114	2100	2055
Water content (%)	12.8	14.5	15.6	16.8	19.2

The value of G_s is 2.67. Plot the dry density – water content curve, and give the optimum water content and maximum dry density. Plot also the curves of zero, 5% and 10% air content, and give the value of air content at maximum dry density. The volume of the mould is 1000cm^3.

1.8 Determine the moisture condition value for the soil whose moisture condition test data are given below:

TABLE D

Number of blows	1	2	3	4	6	8	12	16	24	32	64	96	128
Penetration (*mm*)	15.0	25.2	33.0	38.1	44.7	49.7	57.4	61.0	64.8	66.2	68.2	68.8	69.7

References

ASTM D698 (2007) *Standard Test Methods for Laboratory Compaction Characteristics of Soil Using Standard Effort* (*12, 400ftlbf/ft³* (*600kNm/m³*)), American Society for Testing and Materials, West Conshohocken, PA.

ASTM D1557 (2009) *Standard Test Methods for Laboratory Compaction Characteristics of Soil Using Modified Effort* (*56, 000 ft lbf/ft³* (*2, 700kN m/m³*)), American Society for Testing and Materials, West Conshohocken, PA.

ASTM D2487 (2011) *Standard Practice for Classification of Soils for Engineering Purposes* (*Unified Soil Classification System*), American Society for Testing and Materials, West Conshohocken, PA.

ASTM D4318 (2010) *Standard Test Methods for Liquid Limit, Plastic Limit, and Plasticity Index of Soils*, American Society for Testing and Materials, West Conshohocken, PA.

ASTM D6913 – 04 (2009) *Standard Test Methods for Particle Size Distribution* (*Gradation*) *of Soils Using Sieve Analysis*, American Society for Testing and Materials, West Conshohocken, PA.

ASTM D7382 (2008) *Standard Test Methods for Determination of Maximum Dry Unit Weight and Water Content Range for Effective Compaction of Granular Soils Using a Vibrating Hammer*, American Society for Testing and Materials, West Conshohocken, PA.

Barnes, G. E. (2009) An apparatus for the plastic limit and workability of soils, *Proceedings ICE – Geotechnical Engineering*, **162** (3), 175 – 185.

British Standard 1377 (1990) *Methods of Test for Soils for Civil Engineering Purposes*, British Standards Institution, London.

British Standard 5930 (1999) *Code of Practice for Site Investigations*, British Standards Institution, London.

CEN ISO/TS 17892 (2004) *Geotechnical Investigation and Testing – Laboratory Testing of Soil*, International Organisation for Standardisation, Geneva.

Croney, D. and Croney, P. (1997) *The Performance of Road Pavements* (3rd edn), McGraw Hill, New York, NY.

Day, R. W. (2001) *Soil testing manual*, McGraw Hill, New York, NY.

EC7 – 2 (2007) *Eurocode 7: Geotechnical Design – Part 2: Ground Investigation and Testing, BS EN 1997 – 2: 2007*, British Standards Institution, London.

Highways Agency (2008) Earthworks, in *Specification for Highway Works*, HMSO, Series 600, London.

Mitchell, J. K. and Soga, K. (2005) *Fundamentals of Soil Behaviour* (3rd edn), John Wiley & Sons, New York, NY.

Oliphant, J. and Winter, M. G. (1997) Limits of use of the moisture condition apparatus, *Proceedings ICE – Transport*, **123** (1), 17 – 29.

Parsons, A. W. and Boden, J. B. (1979) *The Moisture Condition Test and its Potential Applications in Earthworks*, TRRL Report 522, Crowthorne, Berkshire.

Sivakumar, V., Glynn, D., Cairns, P. and Black, J. A. (2009) A new method of measuring plastic limit of fine

materials, *Géotechnique*, **59** (10), 813 – 823.
Waltham, A. C. (2002) *Foundations of Engineering Geology* (2nd edn), Spon Press, Abingdon, Oxfordshire.
White, D. J. (2003) PSD measurement using the single particle optical sizing (SPOS) method, *Géotechnique*, **53** (3), 317 – 326.

Further reading

Collins, K. and McGown, A. (1974) The form and function of microfabric features in a variety of natural soils, *Géotechnique*, **24** (2), 223 – 254.
Provides further information on the structures of soil particles under different depositional regimes.
Grim, R. E. (1962) *Clay Mineralogy*, McGraw-Hill, New York, NY.
Further detail about clay mineralogy in terms of its basic chemistry.
Rowe, P. W. (1972) The relevance of soil fabric to site investigation practice, *Géotechnique*, **22** (2), 195 – 300.
Presents 35 case studies demonstrating how the depositional/geological history of the soil deposits influences the selection of laboratory tests and their interpretation, and the consequences for geotechnical constructions. Also includes a number of photographs showing a range of different soil types and fabric features to aid interpretation.
Henkel, D. J. (1982) Geology, geomorphology and geotechnics, *Géotechnique*, **32** (3), 175 – 194.
Presents a series of case studies demonstrating the importance of engineering geology and geomorphological observations in geotechnical engineering.

Chapter 2

Seepage[1]

① 渗流。

Learning outcomes

After working through the material in this chapter, you should be able to:

1. Determine the permeability of soils[2] using the results of both laboratory tests and in-situ tests conducted in the field (Sections 2.1 and 2.2);
2. Understand how groundwater flows for a wide range of ground conditions, and determine seepage quantities and pore pressures within the ground (Sections 2.3 – 2.6);
3. Use computer-based tools for accurately and efficiently solving larger/more complex seepage problems (Section 2.7);
4. Assess seepage through and beneath earthen dams, and understand the design features/remedial methods which may be used to control this (Sections 2.8 – 2.10).

② 土的渗透性。

2.1 Soil water[3]

③ 土中的水。

All soils are permeable materials, water being free to flow through the interconnected pores between the solid particles.[4] It will be shown in Chapters 3 – 5 that the pressure of the pore water is one of the key parameters governing the strength and stiffness of soils. It is therefore vital that the pressure of the pore water is known both under static conditions and when pore water flow is occurring (this is known as **seepage**).

④ 所有的土都是渗透性材料，水可以在土颗粒间内部连通孔隙中自由流动。

The pressure of the pore water is measured relative to atmospheric pressure, and the level at which the pressure is atmospheric (i. e. zero) is defined as the **water table**[5] (WT) or the **phreatic surface**[6]. Below the water table the soil is assumed to be fully saturated, although it is likely that, due to the presence of small volumes of entrapped air, the degree of saturation will be marginally below 100%. The level of the water table changes according to climatic conditions, but the level can change also as a consequence of constructional operations. A **perched** water table[7] can occur locally in an **aquitard** (in which water is contained by soil of low permeability,

⑤ 地下水位。

⑥ 潜水面。

⑦ 滞水位。

① 隔水层。
② 承压水情况。

above the normal water table level) or an **aquiclude**① (where the surrounding material is impermeable). An example of a perched water table is shown schematically in Figure 2.1. **Artesian** conditions② can exist if an inclined soil layer of high permeability is confined locally by an overlying layer of low permeability; the pressure in the artesian layer is governed not by the local water table level but by a higher water table level at a distant location where the layer is unconfined.

③ 滞水位。
④ 含由高渗透性材料形成的透镜体(滞水层)。
⑤ 因毛细吸力而饱和的土。
⑥ 低渗流性材料。
⑦ 地下水位。
⑧ 完全饱和土。

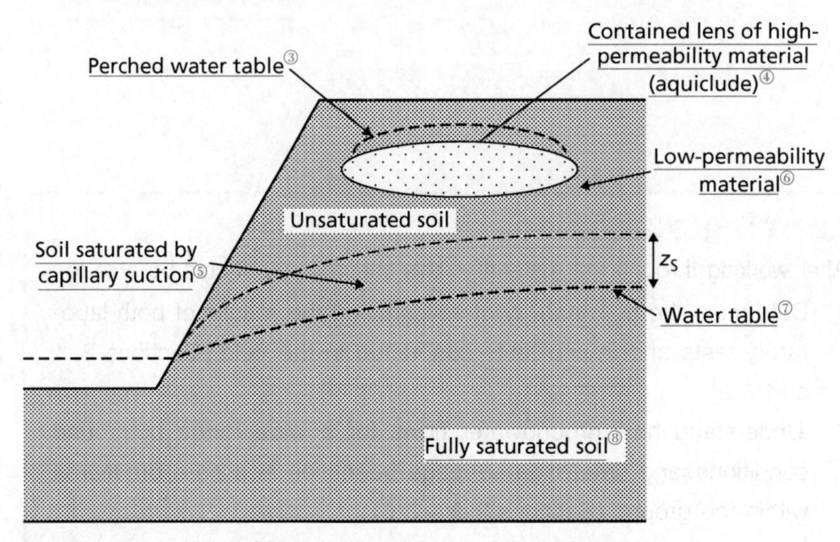

Figure 2.1 Terminology used to describe groundwater conditions.

⑨ 水力梯度。
⑩ 流速水头。

Below the water table the pore water may be static, the hydrostatic pressure depending on the depth below the water table, or may be seeping through the soil under an **hydraulic gradient**⑨: this chapter is concerned with the second case. Bernoulli's theorem applies to the pore water, but seepage velocities in soils are normally so small that velocity head⑩ can be neglected. Thus

$$h = \frac{u}{\gamma_w} + z \tag{2.1}$$

⑪ 总水头。
⑫ 位置水头。

where h is the total head⑪, u the pore water pressure, γ_w the unit weight of water (9.81kN/m^3) and z the elevation head⑫ above a chosen datum.

⑬ 在地下水位之上，土可以通过毛细水的负压力持水从而保持饱和；孔隙越小，由毛细力上升的水位越高。
⑭ 表面张力。

Above the water table, soil can remain saturated, with the pore water being held at negative pressure by capillary tension; the smaller the size of the pores, the higher the water can rise above the water table.⑬ The maximum negative pressure which can be sustained by a soil can be estimated using

$$u_e \approx -\frac{4T_s}{eD} \tag{2.2}$$

where T_s is the surface tension⑪ of the pore fluid (= 7×10^{-5} kN/m for water at 10℃), e is the voids ratio and D is the pore size. As most soils are graded, D is often taken as that at which 10% of material passes on a particle size distribution chart (i.e. D_{10}). The height of the suction zone above the water table may then be estimated by $z_s = u_c/\gamma_w$.

The capillary rise tends to be irregular due to the random pore sizes occurring in a soil. The soil can be almost completely saturated in the lower part of the capillary zone, but in general the degree of saturation decreases with height. When water percolates through the soil from the surface towards the water table, some of this water can be held by surface tension around the points of contact between particles. The negative pressure of water[①] held above the water table results in attractive forces between the particles: this attraction is referred to as soil suction[②], and is a function of pore size and water content.

① 负孔压。
② 土壤吸力。

2.2 Permeability and testing[③]

③ 渗透性与试验。

In one dimension, water flows through a fully saturated soil in accordance with Darcy's empirical law[④]:

④ 经验性的达西定律。

$$v_d = ki \quad (2.3)$$

or

$$q = v_d A = Aki$$

where q is the volume of water flowing per unit time (also termed **flow rate**[⑤]), A the cross-sectional area of soil corresponding to the flow q, k the coefficient of permeability (or **hydraulic conductivity**[⑥]), i the hydraulic gradient and v_d the discharge velocity[⑦]. The units of the coefficient of permeability are those of velocity (m/s).

⑤ 流量。
⑥ 渗透系数。
⑦ 流速。

The coefficient of permeability depends primarily on the average size of the pores, which in turn is related to the distribution of particle sizes, particle shape and soil structure.[⑧] In general, the smaller the particles, the smaller is the average size of the pores and the lower is the coefficient of permeability. The presence of a small percentage of fines in a coarse-grained soil results in a value of k significantly lower than the value for the same soil without fines. For a given soil, the coefficient of permeability is a function of void ratio. As soils become denser (i. e. their unit weight goes up), void ratio reduces. Compression of soil will therefore alter its permeability (see Chapter 4). If a soil deposit is **stratified** (layered), the permeability for flow parallel to the direction of stratification is higher than that for flow perpendicular to the direction of stratification.[⑨] A similar effect may be observed in soils with plate-like particles (e. g. clay) due to alignment of the plates along a single direction. The presence of fissures[⑩] in a clay results in a much higher value of permeability compared with that of the unfissured material, as the fissures are much larger in size than the pores of the intact material, creating preferential flow paths. This demonstrates the importance of soil fabric in understanding groundwater seepage.

⑧ 渗透性参数主要取决于孔隙的平均尺寸，其尺寸与粒径分布、颗粒形状和土体结构有关。
⑨ 若土为层状沉积，则平行成层方向的渗透速度将大于垂直于成层方向的渗透速度。
⑩ 裂隙。

The coefficient of permeability also varies with temperature, upon which the viscosity of the water[⑪] depends. If the value of k measured at 20℃ is taken as 100%, then the values at 10 and 0℃ are 77% and 56%, respectively. The coefficient of permeability can also be represented by the equation:

⑪ 水的黏滞性。

$$k = \frac{\gamma_w}{\eta_w} K$$

where γ_w is the unit weight of water, η_w the viscosity of water and K (units m^2) an absolute coefficient depending only on the characteristics of the soil skeleton.

The values of k for different types of soil are typically within the ranges shown in Table 2.1. For sands, Hazen showed that the approximate value of k is given by

$$k = 10^{-2} D_{10}^2 (\text{m/s}) \quad (2.4)$$

where D_{10} is in mm.

Table 2.1 Coefficient of permeability[①] (m/s)

1	10^{-1}	10^{-2}	10^{-3}	10^{-4}	10^{-5}	10^{-6}	10^{-7}	10^{-8}	10^{-9}	10^{-10}
					Desiccated and fissured clays					
Clean gravels[②]			Clean sands and sand – gravel mixtures[③]			Very fine sands, silts and clay-silt laminate[④]			Unfissured clays and clay-silts (>20% clay)[⑤]	

On the microscopic scale the water seeping through a soil follows a very tortuous path between the solid particles, but macroscopically the flow path (in one dimension) can be considered as a smooth line.[⑥] The average velocity at which the water flows through the soil pores is obtained by dividing the volume of water flowing per unit time by the average area of voids (A_v) on a cross-section normal to the macroscopic direction of flow: this velocity is called the seepage velocity[⑦] (v_s). Thus

$$v_s = \frac{q}{A_v}$$

The porosity of a soil is defined in terms of volume as described by Equation 1.9. However, on average, the porosity can also be expressed as

$$n = \frac{A_v}{A}$$

Hence

$$v_s = \frac{q}{nA} = \frac{v_d}{n}$$

or

$$v_s = \frac{ki}{n} \quad (2.5)$$

Alternatively, Equation 2.5 may be expressed in terms of void ratio, rather than porosity, by substituting for n using Equation 1.11.

Determination of coefficient of permeability[⑧]

Laboratory methods

The coefficient of permeability for coarse soils can be determined by means of the **constant-head** permeability test[⑨] (Figure 2.2 (a)). The soil specimen, at the appropriate density, is contained in a Perspex cylinder of cross-sectional area A and length l: the specimen rests on a coarse filter or a wire mesh. A steady vertical flow

① 渗透系数。
② 洁净卵石。
③ 洁净砂与砂卵石混合物。
④ 极细砂、粉土与黏质粉土层。
⑤ 无裂隙黏土与黏质粉土层（黏粒含量>20%）。
⑥ 在细观层面，土颗粒间水的渗流路径是非常扭曲的，但是在宏观上渗流路径（一维）可被认为是一条平滑的线。
⑦ 渗流速度。
⑧ 渗透系数的确定。
⑨ 常水头渗透试验。

of water, under a constant total head, is maintained through the soil, and the volume of water flowing per unit time (q) is measured. Tappings from the side of the cylinder enable the hydraulic gradient ($i = h/l$) to be measured. Then, from Darcy's law:

$$k = \frac{ql}{Ah}$$

A series of tests should be run, each at a different rate of flow. Prior to running the test, a vacuum is applied to the specimen to ensure that the degree of saturation under flow will be close to 100%. If a high degree of saturation is to be maintained, the water used in the test should be de-aired①.

① 无空气的。

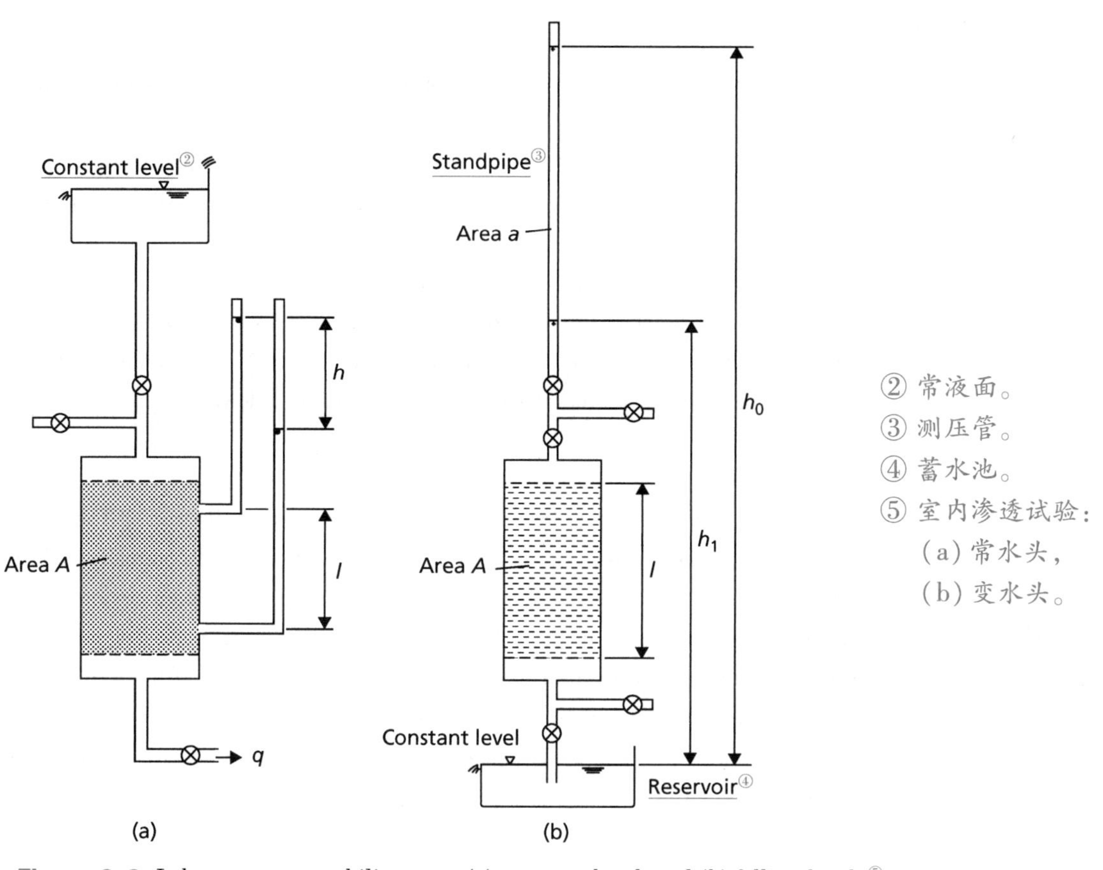

Figure 2.2 Laboratory permeability tests: (a) constant head, and (b) falling head.⑤

② 常液面。
③ 测压管。
④ 蓄水池。
⑤ 室内渗透试验:(a)常水头,(b)变水头。

For fine soils, the **falling-head test**⑥ (Figure 2.2 (b)) should be used. In the case of fine soils undisturbed specimens are normally tested (see Chapter 6), and the containing cylinder in the test may be the sampling tube itself. The length of the specimen is l and the cross-sectional area A. A coarse filter⑦ is placed at each end of the specimen, and a standpipe of internal area a is connected to the top of the cylinder. The water drains into a reservoir of constant level. The standpipe is filled with water, and a measurement is made of the time (t_1) for the water level (relative to the water level in the reservoir) to fall from h_0 to h_1. At any intermediate time t, the water level in the standpipe is given by h and its rate of change by −

⑥ 变水头试验。
⑦ 粗滤层。

$\mathrm{d}h/\mathrm{d}t$. At time t, the difference in total head between the top and bottom of the specimen is h. Then, applying Darcy's law:

$$q = Aki$$

$$-a\frac{\mathrm{d}h}{\mathrm{d}t} = Ak\frac{h}{l}$$

$$\therefore\ -a\int_{h_0}^{h_1}\frac{\mathrm{d}h}{h} = \frac{Ak}{l}\int_0^{t_1}dt \tag{2.6}$$

$$\therefore k = \frac{al}{At_1}\ln\frac{h_0}{h_1}$$

$$= 2.3\frac{al}{At_1}\log\frac{h_0}{h_1}$$

Again, precautions must be taken to ensure that the degree of saturation remains close to 100%. A series of tests should be run using different values of h_0 and h_1 and/or standpipes of different diameters.

The coefficient of permeability of fine soils can also be determined indirectly from the results of consolidation tests[①] (see Chapter 4). Standards governing the implementation of laboratory tests for permeability include BS1377 – 5 (UK), CEN ISO 17892 – 11 (Europe) and ASTM D5084 (US).

① 固结试验。

The reliability of laboratory methods depends on the extent to which the test specimens are representative of the soil mass as a whole.[②] More reliable results can generally be obtained by the in-situ methods described below.

② 室内试验方法的可靠性取决于试验试样在多大程度上可以代表原土体。

Well pumping test[③]

③ 抽水井试验。

This method is most suitable for use in homogeneous coarse soil strata.[④] The procedure involves continuous pumping at a constant rate from a well, normally at least 300mm in diameter, which penetrates to the bottom of the stratum under test. A screen or filter is placed in the bottom of the well to prevent ingress of soil particles. Perforated casing is normally required to support the sides of the well. Steady seepage is established, radially towards the well, resulting in the water table being drawn down to form a cone of depression. Water levels are observed in a number of boreholes spaced on radial lines at various distances from the well. An unconfined stratum of uniform thickness with a (relatively) impermeable lower boundary is shown in Figure 2.3 (a), the water table being below the upper surface of the stratum. A confined layer between two impermeable strata is shown in Figure 2.3 (b), the original water table being within the overlying stratum. Frequent recordings are made of the water levels in the boreholes, usually by means of an electrical dipper[⑤]. The test enables the average coefficient of permeability of the soil mass below the cone of depression to be determined.

④ 此法最适用于均匀的粗颗粒土层。

⑤ 电子水位蜂鸣器。

Analysis is based on the assumption that the hydraulic gradient at any distance r from the centre of the well is constant with depth and is equal to the slope of the water table, i. e.

$$i_r = \frac{\mathrm{d}h}{\mathrm{d}r}$$

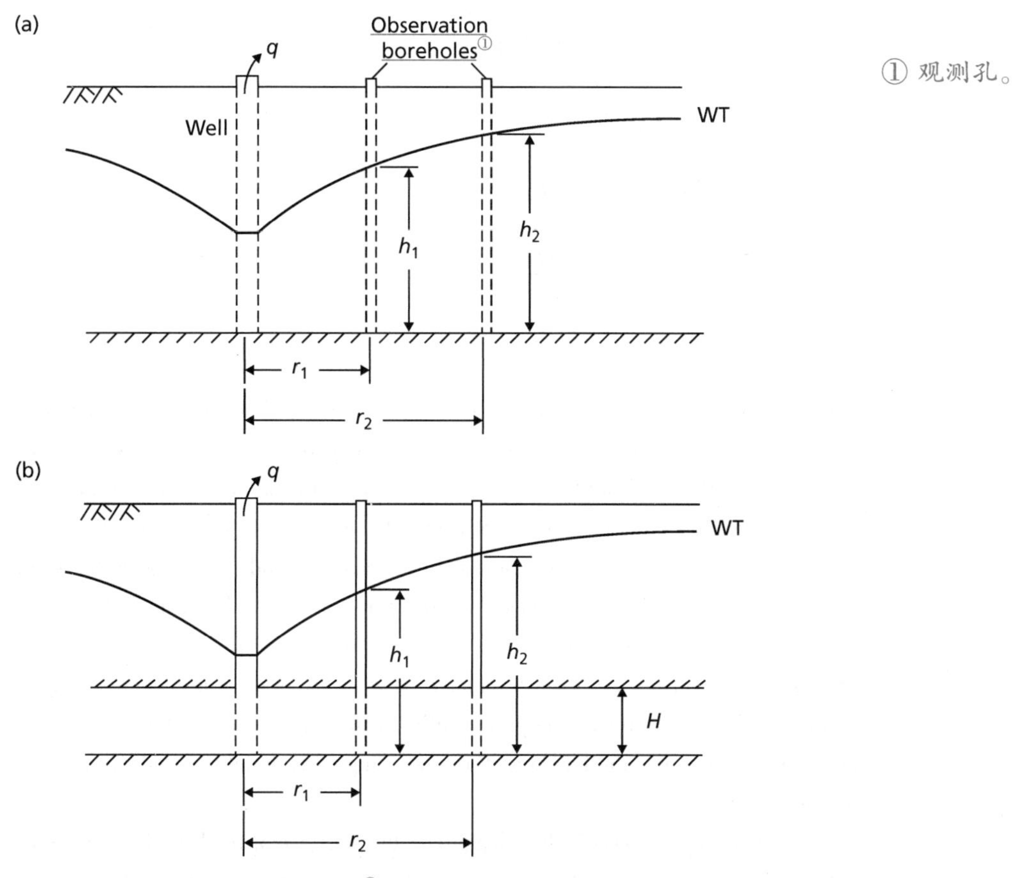

① 观测孔。

Figure 2.3 Well pumping tests[2]: (a) unconfined stratum, and (b) confined stratum.

② 抽水井试验。

where h is the height of the water table at radius r. This is known as the Dupuit assumption, and is reasonably accurate except at points close to the well.

In the case of an unconfined stratum[3] (Figure 2.3 (a)), consider two boreholes located on a radial line at distances r_1 and r_2 from the centre of the well, the respective water levels relative to the bottom of the stratum being h_1 and h_2. At distance r from the well, the area through which seepage takes place is $2\pi rh$, where r and h are variables. Then, applying Darcy's law:

③ 无承压含水层。

$$
\begin{aligned}
q &= Aki \\
q &= 2\pi rhk\frac{\mathrm{d}h}{\mathrm{d}r} \\
\therefore q\int_{r_1}^{r_2}\frac{\mathrm{d}r}{r} &= 2\pi k\int_{h_1}^{h_2}h\,\mathrm{d}h \\
\therefore q\ln\left(\frac{r_2}{r_1}\right) &= \pi k\left(h_2^2 - h_1^2\right) \\
\therefore k &= \frac{2.3q\log\left(r_2/r_1\right)}{\pi\left(h_2^2 - h_1^2\right)}
\end{aligned}
\tag{2.7}
$$

For a confined stratum[4] of thickness H (Figure 2.3 (b)) the area through which seepage takes place is $2\pi rH$, where r is variable and H is constant. Then

④ 承压含水层。

$$
\begin{aligned}
q &= 2\pi rHk\frac{\mathrm{d}h}{\mathrm{d}r} \\
\therefore q\int_{r_1}^{r_2}\frac{\mathrm{d}r}{r} &= 2\pi Hk\int_{h_1}^{h_2}\mathrm{d}h \\
\therefore q\ln\left(\frac{r_2}{r_1}\right) &= 2\pi Hk(h_2 - h_1) \\
\therefore k &= \frac{2.3q\log(r_2/r_1)}{2\pi H(h_2 - h_1)}
\end{aligned}
\tag{2.8}
$$

① 钻孔试验。

Borehole tests[①]

The general principle is that water is either introduced into or pumped out of a borehole which terminates within the stratum in question, the procedures being referred to as inflow and outflow tests, respectively. A hydraulic gradient is thus established, causing seepage either into or out of the soil mass surrounding the borehole, and the rate of flow is measured. In a constant-head test[②], the water level above the water table is maintained throughout at a given level (Figure 2.4 (a)). In a falling-head test[③], the water level is allowed to fall or rise from its initial position and the time taken for the level to change between two values is recorded (Figure 2.4 (b)). The tests indicate the permeability of the soil within a radius of only 1 – 2m from the centre of the borehole. Careful boring is essential to avoid disturbance in the soil structure.

② 常水头试验。

③ 变水头试验。

A problem in such tests is that clogging of the soil face at the bottom of the borehole tends to occur due to the deposition of sediment from the water.[④] To alleviate the problem the borehole may be extended below the bottom of the casing, as shown in Figure 2.4 (c), increasing the area through which seepage takes place. The extension may be uncased in stiff fine soils, or supported by perforated casing in coarse soils.

④ 该试验会遇到钻孔底部土面堵塞的问题，堵塞的土则是由水中沉积物的沉积造成的。

Expressions for the coefficient of permeability depend on whether the stratum is unconfined or confined, the position of the bottom of the casing within the stratum, and details of the drainage face in the soil. If the soil is anisotropic with respect to permeability and if the borehole extends below the bottom of the casing (Figure 2.4 (c)), then the horizontal permeability[⑤] tends to be measured. If, on the other hand, the casing penetrates below soil level in the bottom of the borehole (Figure 2.4 (d)), then vertical permeability[⑥] tends to be measured. General formulae can be written, with the above details being represented by an intake factor (F_i).

⑤ 水平向渗透性。

⑥ 垂直向渗透性。

For a constant-head test:

$$k = \frac{q}{F_i h_c}$$

For a falling-head test:

$$k = \frac{2.3A}{F_i(t_2 - t_1)}\log\frac{h_1}{h_2} \tag{2.9}$$

where k is the coefficient of permeability, q the rate of flow, h_c the constant head, h_1 the variable head at time t_1, h_2 the variable head at time t_2, and A the cross-sec-

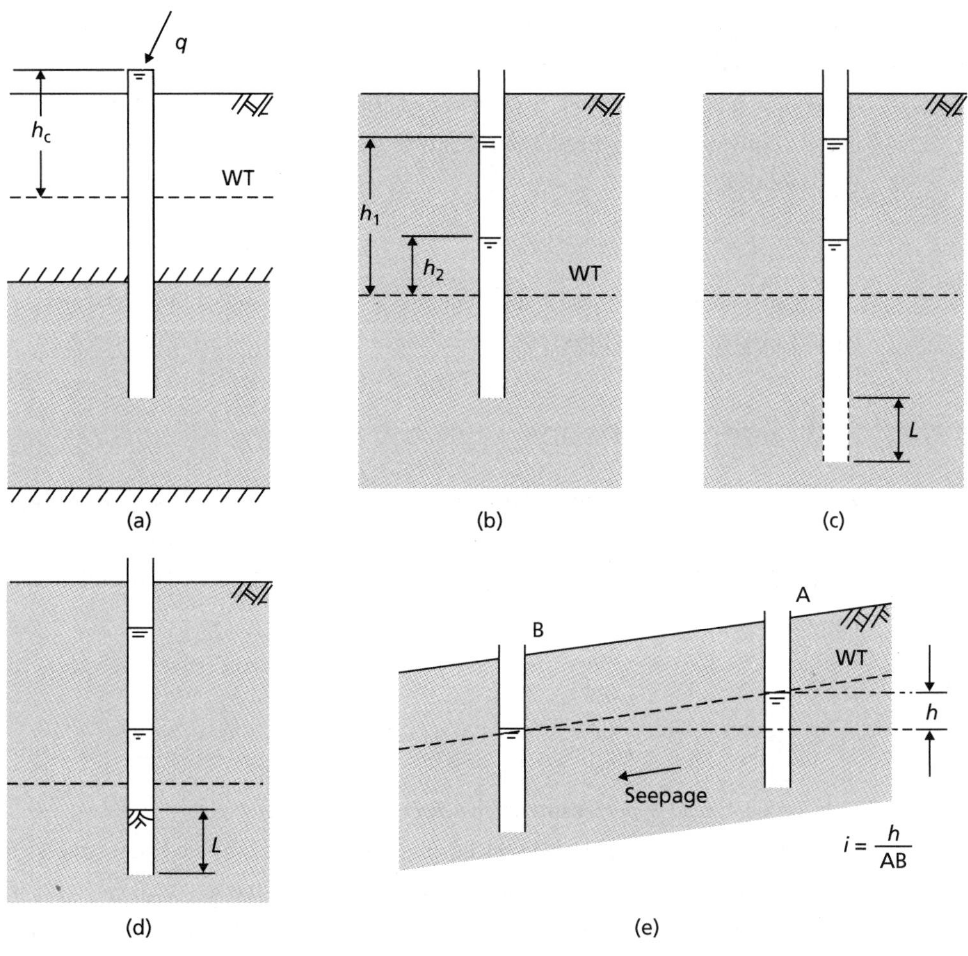

Figure 2.4 Borehole tests: (a) constant-head, (b) variable-head, (c) extension of borehole to prevent clogging, (d) measurement of vertical permeability in anisotropic soil, and (e) measurement of in-situ seepage. ①

tional area of casing or standpipe. Values of intake factor were originally published by Hvorslev (1951), and are also given in Cedergren (1989).

For the case shown in Figure 2.4 (b)

$$F_i = \frac{11R}{2}$$

where R is the radius of the inner radius of the casing②, while for Figure 2.4 (c)

$$F_i = \frac{2\pi L}{\ln(L/R)}$$

and for Figure 2.4 (d)

$$F_i = \frac{11\pi R^2}{2\pi R + 11L}$$

The coefficient of permeability for a coarse soil can also be obtained from in-situ measurements of seepage velocity③, using Equation 2.5. The method involves excavating uncased boreholes or trial pits at two points A and B (Figure 2.4 (e)), seepage taking place from A towards B. The hydraulic gradient is given by the differ-

① 钻孔测试：
(a) 常水头，
(b) 变水头，
(c) 为防堵而延长钻孔，
(d) 各向异性土的竖向渗透性测量，
(e) 原位渗流测量。

② 钻孔套管。

③ 渗流速度。

① A、B之间的水力梯度由两钻孔处的稳态水头差除以A、B之间的距离得到。

ence in the steady-state water levels in the boreholes divided by the distance AB[①]. Dye or any other suitable tracer is inserted into borehole A, and the time taken for the dye to appear in borehole B is measured. The seepage velocity is then the distance AB divided by this time. The porosity of the soil can be determined from density tests. Then

$$k = \frac{v_s n}{i}$$

Further information on the implementation of in-situ permeability tests may be found in Clayton *et al*. (1995).

② 渗流理论。

2.3 Seepage theory[②]

The general case of seepage in two dimensions will now be considered. Initially it will be assumed that the soil is homogeneous and isotropic with respect to permeability[③], the coefficient of permeability being k. In the $x-z$ plane, Darcy's law can be written in the generalised form:

③ 土体的渗透特性均匀且各向同性。

$$v_x = ki_x = -k\frac{\partial h}{\partial x} \tag{2.10a}$$

$$v_z = ki_z = -k\frac{\partial h}{\partial z} \tag{2.10b}$$

with the total head h decreasing in the directions of v_x and v_z.

An element of fully saturated soil having dimensions dx, dy and dz in the x, y and z directions, respectively, with flow taking place in the $x-z$ plane only, is shown in Figure 2.5. The components of discharge velocity of water entering the element are v_x and v_z, and the rates of change of discharge velocity in the x and z directions are $\partial v_x/\partial x$ and $\partial v_z/\partial z$, respectively. The volume of water entering the element per unit time[④] is

④ 单位时间内流入单元体内的水的体积。

$$v_x \mathrm{d}y\mathrm{d}z + v_z \mathrm{d}x\mathrm{d}y$$

and the volume of water leaving per unit time[⑤] is

⑤ 单位时间内流出单元体的水的体积。

$$\left(v_x + \frac{\partial v_x}{\partial x}\mathrm{d}x\right)\mathrm{d}y\mathrm{d}z + \left(v_z + \frac{\partial v_z}{\partial z}\mathrm{d}z\right)\mathrm{d}x\mathrm{d}y$$

If the element is undergoing no volume change and if water is assumed to be incompressible, the difference between the volume of water entering the element per unit time and the volume leaving must be zero.[⑥] Therefore

⑥ 若单元体没有体积变化，且假设水不可压缩，则单位时间内进出单元体的水量之差为零。

$$\frac{\partial v_x}{\partial x} + \frac{\partial v_z}{\partial z} = 0 \tag{2.11}$$

Equation 2.11 is the equation of continuity[⑦] in two dimensions. If, however, the volume of the element is undergoing change, the equation of continuity becomes

⑦ 连续方程。

$$\left(\frac{\partial v_x}{\partial x} + \frac{\partial v_z}{\partial z}\right)\mathrm{d}x\mathrm{d}y\mathrm{d}z = \frac{\mathrm{d}V}{\mathrm{d}t} \tag{2.12}$$

where $\mathrm{d}V/\mathrm{d}t$ is the volume change per unit time.

⑧ 势函数。

Consider, now, the function $\phi(x, z)$, called the potential function[⑧], such that

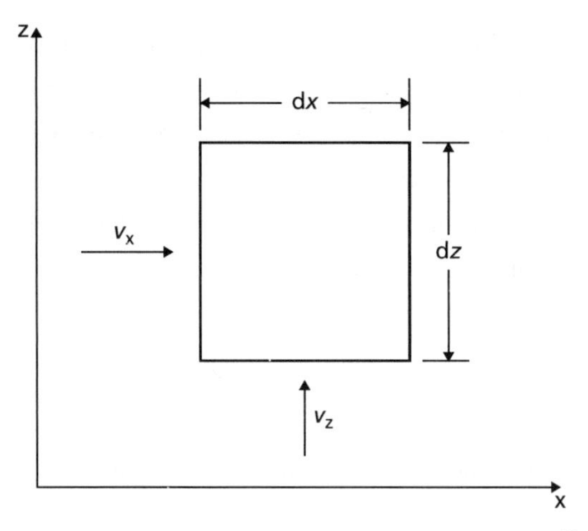

Figure 2.5 Seepage through a soil element.[①]

① 通过土单元体中的渗流。

$$\frac{\partial \phi}{\partial x} = v_x = -k\frac{\partial h}{\partial x} \tag{2.13a}$$

$$\frac{\partial \phi}{\partial z} = v_z = -k\frac{\partial h}{\partial z} \tag{2.13b}$$

From Equations 2.11 and 2.13 it is apparent that

$$\frac{\partial^2 \phi}{\partial x^2} + \frac{\partial^2 \phi}{\partial z^2} = 0 \tag{2.14}$$

i.e. the function $\phi(x, z)$ satisfies the Laplace equation.

Integrating Equation 2.13:

$$\phi(x, z) = -kh(x, z) + C$$

where C is a constant. Thus, if the function $\phi(x, z)$ is given a constant value, equal to ϕ_1 (say), it will represent a curve along which the value of total head (h_1) is constant. If the function $\phi(x, z)$ is given a series of constant values, ϕ_1, ϕ_2, ϕ_3 etc., a family of curves is specified along each of which the total head is a constant value (but a different value for each curve). Such curves are called **equipotentials**.[②]

② 因此，如果函数 $\phi(x, z)$ 给定某个值如 ϕ_1，此函数代表了总水头为 h_1 的一条曲线。如果函数 $\phi(x, z)$ 给定一系列的值如 ϕ_1、ϕ_2、ϕ_3 等，将确定总水头压力为定值（不同曲线对应不同定值）的一组曲线，这些曲线称为等势线。

A second function $\psi(x, z)$, called the flow function, is now introduced, such that

$$-\frac{\partial \psi}{\partial x} = v_z = -k\frac{\partial h}{\partial z} \tag{2.15a}$$

$$\frac{\partial \psi}{\partial z} = v_x = -k\frac{\partial h}{\partial x} \tag{2.15b}$$

It can be shown that this function also satisfies the Laplace equation.

The total differential of the function $\psi(x, z)$ is

$$\begin{aligned} d\psi &= \frac{\partial \psi}{\partial x}dx + \frac{\partial \psi}{\partial z}dz \\ &= -v_z dx + v_x dz \end{aligned}$$

If the function $\psi(x, z)$ is given a constant value ψ_1 then $d\psi = 0$ and

$$\frac{dz}{dx} = \frac{v_z}{v_x} \tag{2.16}$$

Thus, the tangent[③] at any point on the curve represented by

$$\psi(x, z) = \psi_1$$

③ 切线。

specifies the direction of the resultant discharge velocity at that point: the curve therefore represents the flow path. If the function $\psi\,(x, z)$ is given a series of constant values, ψ_1, ψ_2, ψ_3 etc., a second family of curves is specified, each representing a flow path. These curves are called **flow lines**.[①]

① 如果函数 $\psi\,(x, z)$ 给定一系列的常值如 ψ_1、ψ_2、ψ_3 等，将确定另一组曲线，这些曲线称为流线。

Referring to Figure 2.6, the flow per unit time between two flow lines for which the values of the flow function are ψ_1 and ψ_2 is given by

$$\Delta q = \int_{\psi_1}^{\psi_2} (-v_z \mathrm{d}x + v_x \mathrm{d}z)$$

$$\int_{\psi_1}^{\psi_2} \left(\frac{\partial \psi}{\partial x} \mathrm{d}x + \frac{\partial \psi}{\partial z} \mathrm{d}z \right) = \psi_2 - \psi_1$$

Thus, the flow through the 'channel' between the two flow lines is constant[②].

② 在两条流线之间流体的流速为一个常数。

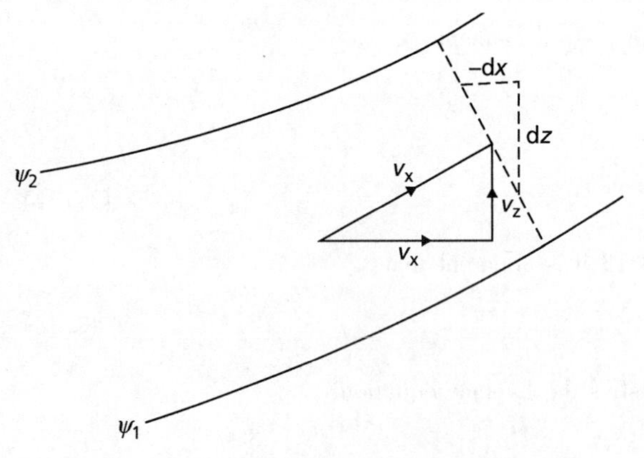

Figure 2.6 Seepage between two flow lines.[③]

③ 两条流线之间的渗流。

The total differential of the function $\phi\,(x, z)$ is

$$\mathrm{d}\phi = \frac{\partial \phi}{\partial x} \mathrm{d}x + \frac{\partial \phi}{\partial z} \mathrm{d}z$$

$$= v_x \mathrm{d}x + v_z \mathrm{d}z$$

If $\phi\,(x, z)$ is constant then $\mathrm{d}\phi = 0$ and

$$\frac{\mathrm{d}z}{\mathrm{d}x} = -\frac{v_x}{v_z} \tag{2.17}$$

Comparing Equations 2.16 and 2.17, it is apparent that the flow lines and the equipotentials intersect each other at right angles[④].

④ 很明显，流线与等势线互相正交。

Consider, now, two flow lines ψ_1 and $(\psi_1 + \Delta\psi)$ separated by the distance Δn. The flow lines are intersected orthogonally by two equipotentials ϕ_1 and $(\phi_1 + \Delta\phi)$ separated by the distance Δs, as shown in Figure 2.7. The directions s and n are inclined at angle α to the x and z axes, respectively. At point A the discharge velocity (in direction s) is v_s; the components of v_s in the x and z directions, respectively, are

$$v_x = v_s \cos \alpha$$

$$v_z = v_s \sin \alpha$$

Now

$$\frac{\partial \phi}{\partial s} = \frac{\partial \phi}{\partial x}\frac{\partial x}{\partial s} + \frac{\partial \phi}{\partial z}\frac{\partial z}{\partial s}$$

$$= v_s \cos^2 \alpha + v_s \sin^2 \alpha = v_s$$

and

$$\frac{\partial \psi}{\partial n} = \frac{\partial \psi}{\partial x}\frac{\partial x}{\partial n} + \frac{\partial \psi}{\partial z}\frac{\partial z}{\partial n}$$

$$= -v_s \sin \alpha (-\sin \alpha) + v_s \cos^2 \alpha = v_s$$

Thus

$$\frac{\partial \psi}{\partial n} = \frac{\partial \phi}{\partial s}$$

or approximately

$$\frac{\Delta \psi}{\Delta n} = \frac{\Delta \phi}{\Delta s} \tag{2.18}$$

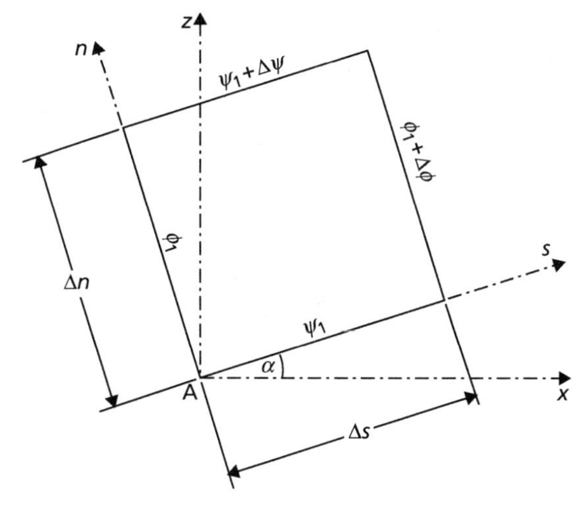

Figure 2.7 Flow lines and equipotentials.[①]

① 流线与等势线。

2.4 Flow nets[②]

② 流网。

In principle, for the solution of a practical seepage problem the functions $\phi(x, z)$ and $\psi(x, z)$ must be found for the relevant boundary conditions. The solution is represented by a family of flow lines and a family of equipotentials, constituting what is referred to as a **flow net**. Computer software based on either the finite difference or finite element methods is widely available for the solution of seepage problems. Williams *et al.* (1993) described how solutions can be obtained using the finite difference form of the Laplace equation by means of a spreadsheet. This method will be described in Section 2.7, with electronic resources available on the Companion Website to accompany the material discussed in the remainder of this chapter. Relatively simple problems can be solved by the trial-and-error sketching of the flow net, the general form of which can be deduced from consideration of the boundary conditions. Flow net sketching leads to a greater understanding of seepage principles. However, for problems in which the geometry becomes complex and there are zones of different permeability throughout the flow region, use of the finite difference method is usually necessary.

The fundamental condition to be satisfied in a flow net is that every intersection between a flow line and an equipotential must be at right angles.[①] In addition, it is convenient to construct the flow net such that $\Delta\psi$ has the same value between any two adjacent flow lines and $\Delta\phi$ has the same value between any two adjacent equipotentials. It is also convenient to make $\Delta s = \Delta n$ in Equation 2. 18, i. e. the flow lines and equipotentials form curvilinear squares throughout the flow net. Then, for any curvilinear square

① 流网中需要满足的基本条件为每条流线与每条等势线互相正交。

$$\Delta\psi = \Delta\phi$$

Now, $\Delta\psi = \Delta q$ and $\Delta\phi = k\Delta h$, therefore:

$$\Delta q = k\Delta h \tag{2.19}$$

For the entire flow net, h is the difference in total head between the first and last equipotentials, N_d the number of equipotential drops, each representing the same total head loss Δh, and N_f the number of **flow channels**[②], each carrying the same flow Δq. Then,

② 流槽。

$$\Delta h = \frac{h}{N_d} \tag{2.20}$$

and

$$q = N_f \Delta q$$

Hence, from Equation 2. 19

$$q = kh\frac{N_f}{N_d} \tag{2.21}$$

Equation 2. 21 gives the total volume of water flowing per unit time (per unit dimension in the y-direction) and is a function of the ratio N_f/N_d.

Between two adjacent equipotentials the hydraulic gradient is given by

$$i = \frac{\Delta h}{\Delta s} \tag{2.22}$$

Example of a flow net

As an illustration, the flow net for the problem detailed in Figure 2. 8 (a) will be considered. The figure shows a line of sheet piling driven 6. 00m into a stratum of soil 8. 60m thick, underlain by an impermeable stratum[③]. On one side of the piling the depth of water is 4. 50m; on the other side the depth of water (reduced by pumping) is 0. 50m. The soil has a permeability of 1.5×10^{-5}m/s.

③ 不透水层。

The first step is to consider the boundary conditions of the flow region[④] (Figure 2. 8 (b)). At every point on the boundary AB the total head is constant, so AB is an equipotential; similarly, CD is an equipotential. The datum to which total head is referred may be chosen to be at any level, but in seepage problems it is convenient to select the downstream water level as datum. Then, the total head on equipotential CD is zero after Equation 2. 1 (pressure head 0. 50m; elevation head −0. 50m) and the total head on equipotential AB is 4. 00m (pressure head 4. 50m; elevation head −0. 50m). From point B, water must flow down the upstream face

④ 第一步要考虑渗流区域的边界条件。

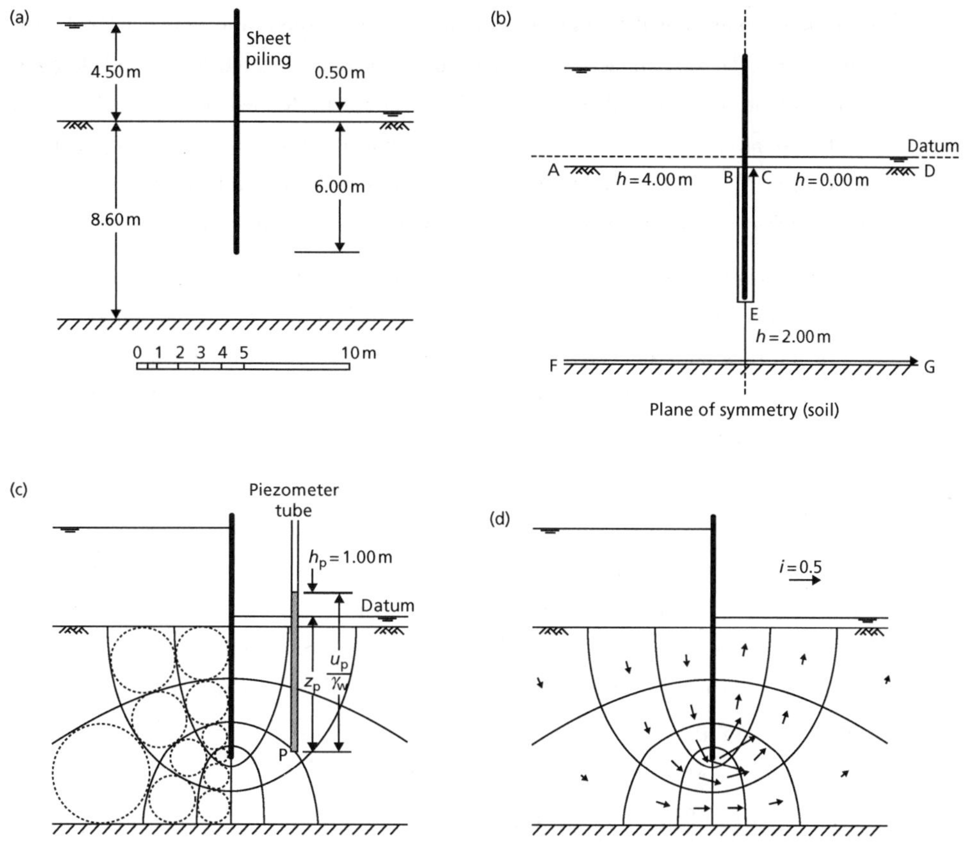

Figure 2.8 Flow net construction: (a) section, (b) boundary conditions, (c) final flow net including a check of the 'square-ness' of the curvilinear squares, and (d) hydraulic gradients inferred from flow net. [①]

① 流网的绘制：
(a) 截面，
(b) 边界条件，
(c) 最终流网图包括对曲线正方形"方度"的检查，
(d) 由流网推得的水力梯度。

BE of the piling, round the tip E and up the downstream face EC. Water from point F must flow along the impermeable surface FG. Thus, BEC and FG are flow lines. The other flow lines must lie between the extremes of BEC and FG, and the other equipotentials must lie between AB and CD. As the flow region is symmetric on either side of the sheet piling, when flowline BEC reaches point E, halfway between AB and CD (i. e. at the toe of the sheet piling), the total head must be half way between the values along AB and CD. This principle also applies to flowline FG, such that a third vertical equipotential can be drawn from point E, as shown in Figure 2.8 (b).

The number of equipotential drops should then be selected. Any number may be selected, however it is convenient to use a value of N_d which divides precisely into the total change in head through the flow region. In this example $N_d = 8$ is chosen so that each equipotential will represent a drop in head of 0.5m. The choice of N_d has a direct influence on the value of N_f. As N_d is increased, the equipotentials get closer together such that, to get a 'square' flow net, the flow channels will also have to be closer together (i. e. more flow lines will need to be plotted). This will lead to a finer net with greater detail in the distribution of seepage pressures; however, the total flow quantity will be unchanged. Figure 2. 8 (c) shows the flow net for $N_d = 8$ and $N_f = 3$. These parameters for this particular example give a 'square' flow net and a whole number of flow channels. This should be formed by trial and error: a first attempt should be made and the positions of the flow lines and equipotentials (and even N_f and N_d) should then be adjusted as necessary until a satisfactory flow net is achieved. A satisfactory flow net should satisfy the following conditions:

① 流线与等势线交汇处必为90°。

② 流网方格必须为曲线正方形。

- All intersections between flow lines and equipotentials should be at 90°. ①
- The curvilinear squares must be square② – in Figure 2. 8 (c), the 'square-ness' of the flow net has been checked by inscribing a circle within each square. The flow net is acceptable if the circle just touches the edges of the curvilinear square (i. e. there are no rectangular elements).

Due to the symmetry within the flow region, the equipotentials and flow lines may be drawn in half of the problem and then reflected about the line of symmetry (i. e. the sheet piling). In constructing a flownet it is a mistake to draw too many flow lines; typically, three to five flow channels are sufficient, depending on the geometry of the problem and the value of N_d which is most convenient.

③ 总水头损失。

In the flow net in Figure 2. 8 (c) the number of flow channels is three and the number of equipotential drops is eight; thus the ratio N_f/N_d is 0. 375. The loss in total head③ between any two adjacent equipotentials is

$$\Delta h = \frac{h}{N_d} = \frac{4.00}{8} = 0.5\text{m}$$

The total volume of water flowing under the piling per unit time per unit length of piling is given by

$$q = kh\frac{N_f}{N_d} = 1.5 \times 10^{-5} \times 4.00 \times 0.375$$
$$= 2.25 \times 10^{-5}\text{m}^3/\text{s}$$

④ 测压管。

⑤ 位置水头。

A piezometer tube④ is shown at a point P on the equipotential with total head h = 1.00m, i. e. the water level in the tube is 1.00m above the datum. The point P is at a distance $z_P = 6$m below the datum, i. e. the elevation head⑤ is $-z_P$. The pore water pressure at P can then be calculated from Bernoulli's theorem:

$$\begin{aligned} u_P &= \gamma_w [h_P - (-z_P)] \\ &= \gamma_w (h_P + z_P) \\ &= 9.81 \times (1+6) \\ &= 68.7\text{kPa} \end{aligned}$$

⑥ 最高的水力坡降（最高的渗流速度）发生在最小的方格区域，反之亦然。

The hydraulic gradient across any square in the flow net involves measuring the average dimension of the square (Equation 2. 22). The highest hydraulic gradient (and hence the highest seepage velocity) occurs across the smallest square, and vice versa.⑥ The dimension Δs has been estimated by measuring the diameter of the circles in Figure 2. 8 (c). The hydraulic gradients across each square are shown using a quiver plot in Figure 2. 8 (d) in which the length of the arrows is proportional to the magnitude of the hydraulic gradient.

Example 2.1

A river bed consists of a layer of sand 8.25m thick overlying impermeable rock; the depth of water is 2.50m. A long cofferdam 5.50m wide is formed by driving two lines of sheet piling to a depth of 6.00m below the level of the river bed, and excavation to a depth of 2.00m below bed level is carried out within the cofferdam. The water level within the cofferdam is kept at excavation level by pumping. If the flow of water into the cofferdam is 0.25m^3/h per unit length, what is the coefficient of permeability of the sand? What is the hydraulic gradient immediately below the excavated surface?

(a)

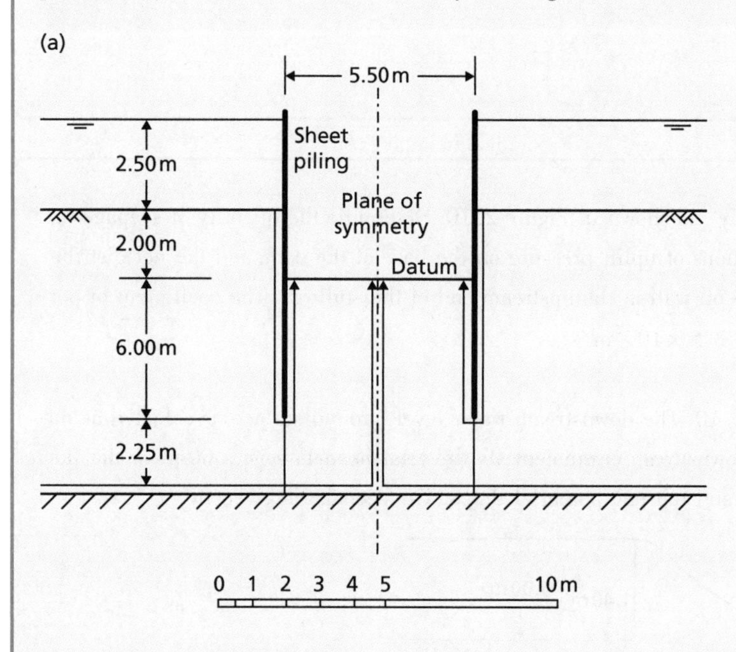

(b)

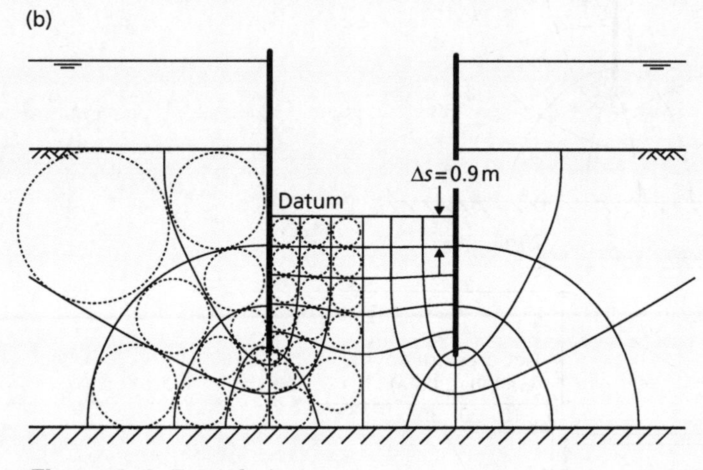

Figure 2.9 Example 2.1.

Solution

The section and boundary conditions appear in Figure 2.9 (a) and the flow net is shown in Figure 2.9 (b). In the flownet there are six flow channels (three per side) and ten equipotential drops. The total head loss[①] is 4.50m. The coefficient of permeability is given by

① 总水头损失。

$$k = \frac{q}{h\,(N_f/N_d)}$$

$$= \frac{0.25}{4.50 \times 6/10 \times 60^2} = 2.6 \times 10^{-5}\,\text{m/s}$$

The distance (Δs) between the last two equipotentials is measured as 0.9m. The required hydraulic gradient is given by

$$i = \frac{\Delta h}{\Delta s}$$

$$= \frac{4.50}{10 \times 0.9} = 0.50$$

Example 2.2

The section through a dam spillway① is shown in Figure 2.10. Determine the quantity of seepage under the dam and plot the distributions of uplift pressure on the base of the dam, and the net distribution of water pressure on the cut – off wall at the upstream end of the spillway. The coefficient of permeability of the foundation soil is 2.5×10^{-5}m/s.

Solution

The flow net is shown in Figure 2.10. The downstream water level (ground surface) is selected as datum. Between the upstream and downstream equipotentials the total head loss is 5.00m. In the flow net there are three flow channels and ten equipotential drops. The seepage is given by

① 泄洪道。

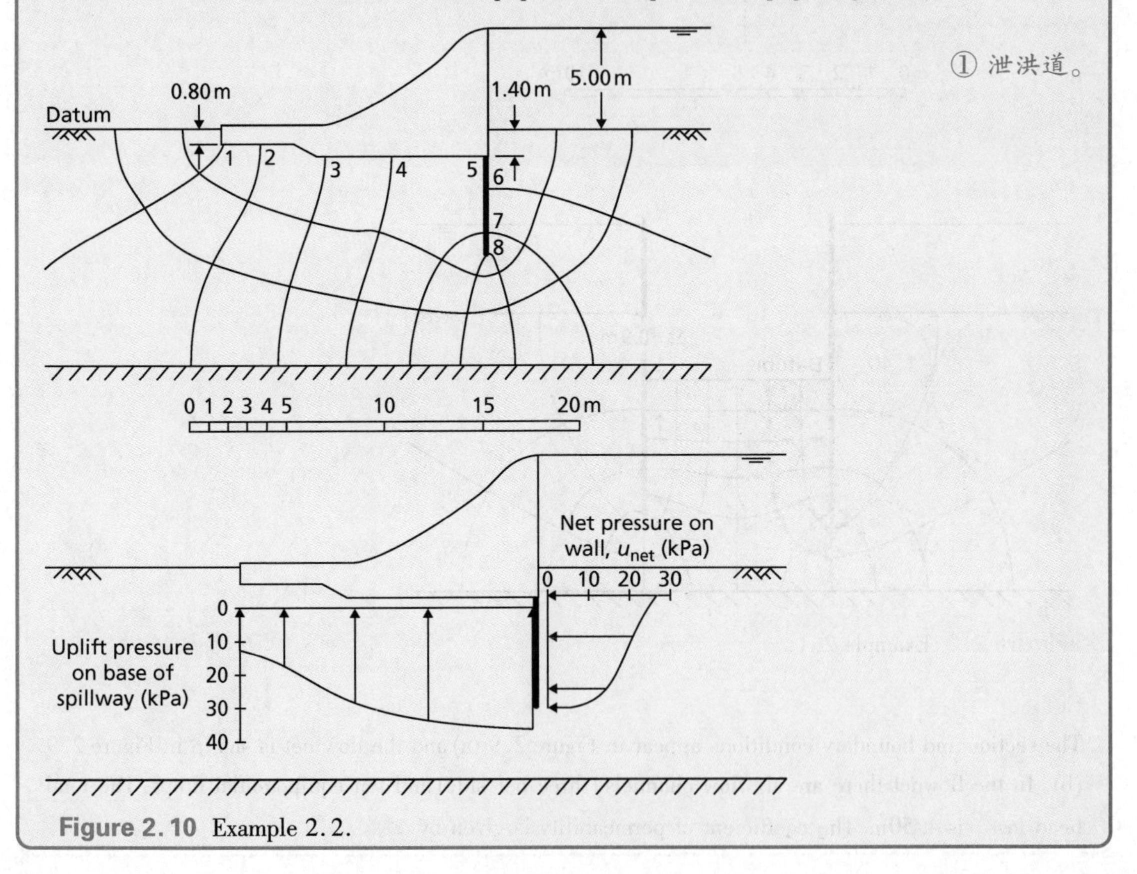

Figure 2.10 Example 2.2.

$$q = kh\frac{N_{\mathrm{f}}}{N_{\mathrm{d}}} = 2.5 \times 10^{-5} \times 5.00 \times \frac{3}{10}$$
$$= 3.75 \times 10^{-5}\,\mathrm{m^3/s}$$

This inflow rate is per metre length of the cofferdam. The pore water pressures acting on the base of the spillway are calculated at the points of intersection of the equipotentials with the base of the spillway. The total head at each point is obtained from the flow net, and the elevation head from the section. The calculations are shown in Table 2. 2, and the pressure diagram is plotted in Figure 2. 10.

The water pressures acting on the cut-off wall are calculated on both the back (h_{b}) and front (h_{f}) of the wall at the points of intersection of the equipotentials with the wall. The net pressure acting on the back face of the wall is therefore

$$u_{\mathrm{net}} = u_{\mathrm{b}} - u_{\mathrm{f}} = \left(\frac{h_{\mathrm{b}} - z}{\gamma_{\mathrm{w}}}\right) - \left(\frac{h_{\mathrm{f}} - z}{\gamma_{\mathrm{w}}}\right)$$

The calculations are shown in Table 2. 3, and the pressure diagram is plotted in Figure 2. 10. The levels (z) of points 5 – 8 in Table 2. 3 were found by scaling from the diagram.

Table 2.2 Example 2.2

Point	*h (m)*	*z (m)*	*h – z (m)*	*u = γ_w (h – z) (kPa)*
1	0.50	-0.80	1.30	12.8
2	1.00	-0.80	1.80	17.7
3	1.50	-1.40	2.90	28.4
4	2.00	-1.40	3.40	33.4
5	2.30	-1.40	3.70	36.3

Table 2.3 Example 2.2 (contd.)

Level	*z (m)*	*h_b (m)*	*u_b/γ_w (m)*	*h_f (m)*	*u_f/γ_w (m)*	*$u_b - u_b$ (kPa)*
5	-1.40	5.00	6.40	2.28	3.68	26.7
6	-3.07	4.50	7.57	2.37	5.44	20.9
7	-5.20	4.00	9.20	2.50	7.70	14.7
8	-6.00	3.50	9.50	3.00	9.00	4.9

2.5 Anisotropic soil conditions[①]

It will now be assumed that the soil, although homogeneous, is anisotropic with respect to permeability.[②] Most natural soil deposits are anisotropic, with the coefficient of permeability having a maximum value in the direction of stratification and

① 各向异性土中的渗流。

② 假定土虽然为均质材料，但渗透性为各向异性。

① 大部分天然土都是各向异性的，沿着层理方向的渗透系数达到最大值，而垂直于层理方向的渗透系数达到最小值。

a minimum value in the direction normal to that of stratification[①]; these directions are denoted by x and z, respectively, i. e.

$$k_x = k_{max} \text{ and } k_z = k_{min}$$

In this case, the generalised form of Darcy's law is

$$v_x = k_x i_x = -k_x \frac{\partial h}{\partial x} \tag{2.23a}$$

$$v_z = k_z i_z = -k_z \frac{\partial h}{\partial z} \tag{2.23b}$$

Also, in any direction s, inclined at angle α to the x direction, the coefficient of permeability is defined by the equation

$$v_s = -k_s \frac{\partial h}{\partial s}$$

Now

$$\frac{\partial h}{\partial s} = \frac{\partial h}{\partial x}\frac{\partial x}{\partial s} + \frac{\partial h}{\partial z}\frac{\partial z}{\partial s}$$

i. e.

$$\frac{v_s}{k_s} = \frac{v_x}{k_x}\cos\alpha + \frac{v_z}{k_z}\sin\alpha$$

The components of discharge velocity are also related as follows:

$$v_x = v_s \cos\alpha$$

$$v_z = v_s \sin\alpha$$

Hence

$$\frac{1}{k_s} = \frac{\cos^2\alpha}{k_x} + \frac{\sin^2\alpha}{k_z}$$

or

$$\frac{s^2}{k_s} = \frac{x^2}{k_x} + \frac{z^2}{k_z} \tag{2.24}$$

The directional variation of permeability is thus described by Equation 2. 24, which represents the ellipse shown in Figure 2. 11.

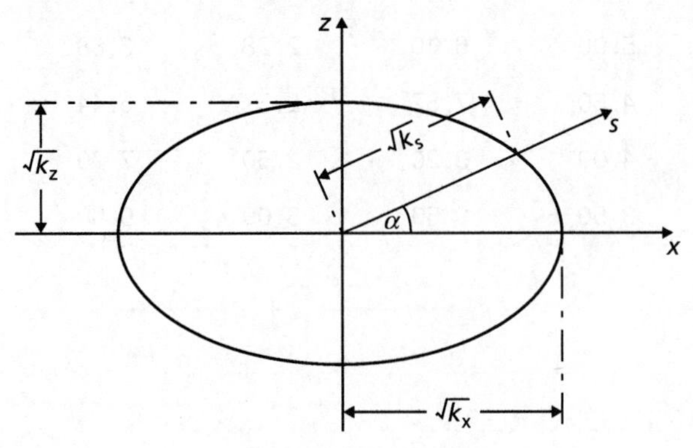

Figure 2. 11 Permeability ellipse.[②]

② 渗透椭圆。

Given the generalised form of Darcy's law (Equation 2. 23), the equation of continuity (2. 11) can be written:

$$k_x \frac{\partial^2 h}{\partial x^2} + k_z \frac{\partial^2 h}{\partial z^2} = 0 \tag{2.25}$$

or

$$\frac{\partial^2 h}{(k_z/k_x)\,\partial x^2} + \frac{\partial^2 h}{\partial z^2} = 0$$

Substituting

$$x_t = x\sqrt{\frac{k_z}{k_x}} \tag{2.26}$$

the equation of continuity becomes

$$\frac{\partial^2 h}{\partial x_t^2} + \frac{\partial^2 h}{\partial z^2} = 0 \tag{2.27}$$

which is the equation of continuity for an isotropic soil in an $x_t - z$ plane.

Thus, Equation 2.26 defines a scale factor[①] which can be applied in the x direction to transform a given anisotropic flow region into a fictitious isotropic flow region in which the Laplace equation is valid. Once the flow net (representing the solution of the Laplace equation) has been drawn for the transformed section, the flow net for the natural section can be obtained by applying the inverse of the scaling factor.[②] Essential data, however, can normally be obtained from the transformed section. The necessary transformation could also be made in the z direction.

① 比例系数。

② 一旦绘出换算截面的流网（代表拉普拉斯方程解）绘出，天然截面的流网就可以通过比例因子换算得到。

The value of coefficient of permeability applying to the transformed section, referred to as the equivalent isotropic coefficient[③], is

$$k' = \sqrt{(k_x k_z)} \tag{2.28}$$

③ 等效均质渗透系数。

A formal proof of Equation 2.28 has been given by Vreedenburgh (1936). The validity of Equation 2.28 can be demonstrated by considering an elemental flow net field through which flow is in the x direction. The flow net field is drawn to the transformed and natural scales in Figure 2.12, the transformation being in the x direction. The discharge velocity v_x can be expressed in terms of either k'(transformed section) or k_x (natural section), i. e.

$$v_x = -k'\frac{\partial h}{\partial x_t} = -k_x\frac{\partial h}{\partial x}$$

where

$$\frac{\partial h}{\partial x_t} = \frac{\partial h}{\sqrt{(k_z/k_x)}\,\partial x}$$

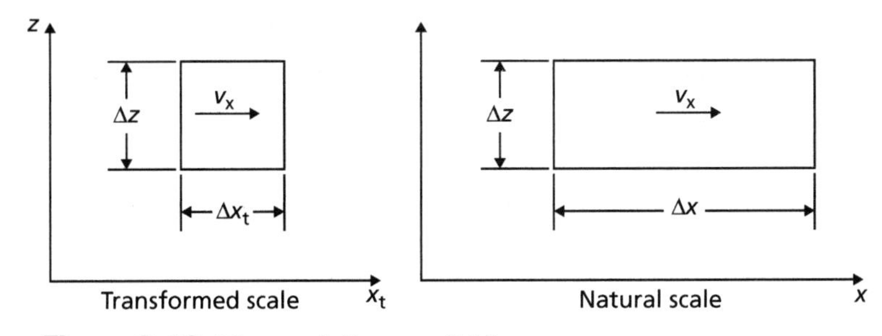

Figure 2.12 Elemental flow net field.

Thus

$$k' = k_x \sqrt{\frac{k_z}{k_x}} = \sqrt{(k_x k_z)}$$

① 非均质土中的渗流。

2.6 Non-homogeneous soil conditions[①]

Two isotropic soil layers of thicknesses H_1 and H_2 are shown in Figure 2.13, the respective coefficients of permeability being k_1 and k_2; the boundary between the layers is horizontal. (If the layers are anisotropic, k_1 and k_2 represent the equivalent isotropic coefficients for the layers.) The two layers can be approximated as a single homogeneous anisotropic layer of thickness $(H_1 + H_2)$ in which the coefficients in the directions parallel and normal to that of stratification[②] are k_x and k_z, respectively.

② 成层，层理。

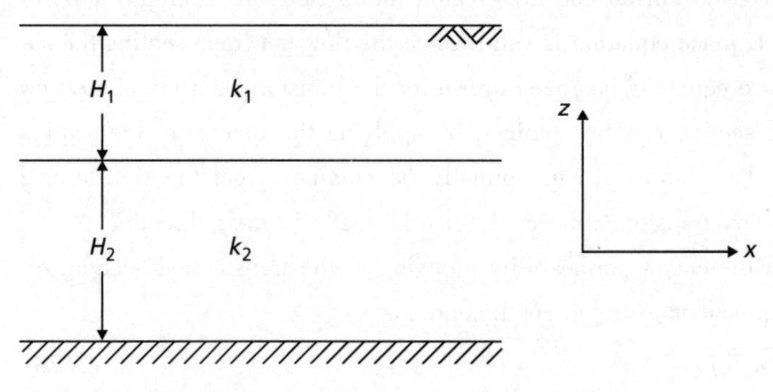

Figure 2.13 Non – homogeneous soil conditions.

For one-dimensional seepage in the horizontal direction, the equipotentials in each layer are vertical.[③] If h_1 and h_2 represent total head at any point in the respective layers, then for a common point on the boundary $h_1 = h_2$. Therefore, any vertical line through the two layers represents a common equipotential. Thus, the hydraulic gradients in the two layers, and in the equivalent single layer, are equal; the equal hydraulic gradients are denoted by i_x. The total horizontal flow per unit time is then given by

③ 对于水平方向的一维渗流，每一层的等水头线是竖直的。

$$\bar{q}_x = (H_1 + H_2)\bar{k}_x i_x = (H_1 k_1 + H_2 k_2) i_x$$

$$\therefore \bar{k}_x = \frac{H_1 k_1 + H_2 k_2}{H_1 + H_2} \tag{2.29}$$

For one-dimensional seepage in the vertical direction, the discharge velocities in each layer, and in the equivalent single layer, must be equal if the requirement of continuity is to be satisfied. Thus

$$v_z = \bar{k}_z \bar{i}_z = k_1 i_1 = k_2 i_2$$

where $\bar{i}_z$ is the average hydraulic gradient over the depth $(H_1 + H_2)$. Therefore

$$i_1 = \frac{\bar{k}_z}{k_1}\bar{i}_z \text{ and } i_2 = \frac{\bar{k}_z}{k_2}\bar{i}_z$$

Now the loss in total head over the depth $(H_1 + H_2)$ is equal to the sum of the losses in total head in the individual layers[④], i.e.

④ 在深度$(H_1 + H_2)$中发生的总水头损失等于每层水头损失之和。

$$
\begin{aligned}
\bar{i}_z(H_1+H_2) &= i_1H_1+i_2H_2 \\
&= \bar{k}_z\bar{i}_z\left(\frac{H_1}{k_1}+\frac{H_2}{k_2}\right) \\
\therefore \bar{k}_z &= \frac{H_1+H_2}{\left(\frac{H_1}{k_1}\right)+\left(\frac{H_2}{k_2}\right)}
\end{aligned} \tag{2.30}
$$

Similar expressions for $\bar{k}_x$ and $\bar{k}_z$ apply in the case of any number of soil layers. It can be shown that $\bar{k}_x$ must always be greater than $\bar{k}_z$, i. e. seepage can occur more readily in the direction parallel to stratification than in the direction perpendicular to stratification.

2.7 Numerical solution using the Finite Difference Method[①]

① 使用有限差分法的渗流数值解法。

Although flow net sketches are useful for estimating seepage-induced pore pressures and volumetric flow rates for simple problems, a great deal of practice is required to produce reliable results. As an alternative, spreadsheets (which are almost universally available to practising engineers) may be used to determine the seepage pressures more rapidly and reliably. A spreadsheet-based tool for solving a wide range of problems may be found on the Companion Website which accompanies the material in this chapter. Spreadsheets analyse seepage problems by solving Laplace's equation using the **Finite Difference Method** (FDM)[②]. The problem is first discretised into a mesh of regularly-spaced nodes[③] representing the soil in the problem. If steady-state seepage is occurring through a given region of soil with isotopic permeability k, the total head at a general node within the soil is the average of the values of head at the four connecting nodes[④], as shown in Figure 2.14:

② 有限差分法。

③ 等间距节点网格。

④ 如果在给定的各向同性且渗透系数都为 k 的土中发生稳态渗流，土中节点的总水头等于与其相连4个节点的水头平均值。

$$
h_0 = \frac{1}{4}(h_1+h_2+h_3+h_4) \tag{2.31}
$$

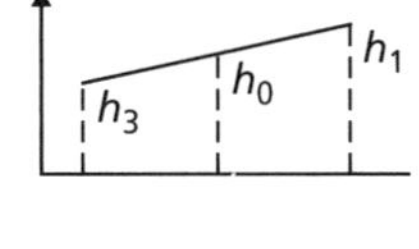

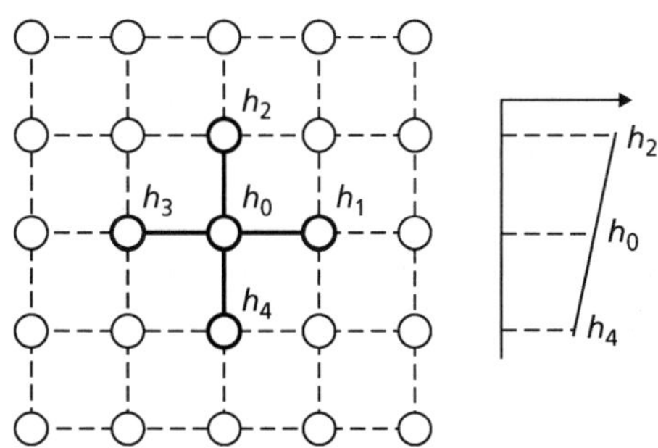

Figure 2.14 Determination of head at an FDM node.[⑤]

⑤ 有限差分法节点处水头的确定。

Modified forms of Equation 2. 31 can be derived for impervious boundaries and for determining the values of head either side of thin impermeable inclusions such as sheet piling. The analysis technique may additionally be used to study anisotropic soils at the transformed scale for soil of permeability k' as defined at the end of Section 2. 5. It is also possible to modify Equation 2. 31 to model nodes at the boundary between two soil layers of different isotropic permeabilities (k_1, k_2) as discussed in Section 2. 6.

The Companion Website contains a spreadsheet analysis tool (Seepage_CSM8. xls) that can be used for the solution of a wide range of seepage problems. Each cell is used to represent a node, with the value of the cell equal to the total head, h_0 for that node. A library of equations for h_0 for a range of different boundary conditions is included within this spreadsheet, which may be copied as necessary to build up a complete and detailed model of a wide range of problems. A more detailed description of the boundary conditions which may be used and their formulation are given in the User's Manual, which may also be found on the Companion Website. After the formulae have been copied in for the relevant nodes, the values of head are entered on the recharge and discharge boundaries and the calculations are then conducted iteratively until further iterations give a negligible change in the head distribution. This iterative process is entirely automated within the spreadsheet. On a modern computer, this should take a matter of seconds for the problems addressed in this chapter. The use of this spreadsheet will be demonstrated in the following example, which utilises all of the boundary conditions included in the spreadsheet.

Example 2.3

A deep excavation is to be made next to an existing masonry tunnel carrying underground railway lines, as shown in Figure 2. 15. The surrounding soil is layered, with isotropic permeabilities as shown in the figure. Calculate the pore water pressure distribution around the tunnel, and find also the flow-rate of water into the excavation.

Solution

Given the geometry shown in Figure 2. 15, a grid spacing of 1m is chosen in the horizontal and vertical directions, giving the nodal layout shown in the figure. The appropriate formulae are then entered in the cells representing each node, as shown in the User's Manual on the Companion Website. The datum is set at the level of the excavation. The resulting total head distribution is shown in Figure 2. 15, and by applying Equation 2. 1 the pore water pressure distribution around the tunnel can be plotted.

The flow rate of water into the excavation can be found by considering the flow between the eight adjacent nodes on the discharge boundary. Considering the nodes next to the sheet pile wall, the change in head between the last two nodes $\Delta h = 0.47$. This is repeated along the discharge boundary, and the average Δh between each set of nodes is computed. Adapting Equation 2. 19, and noting that the soil on the discharge boundary has permeability k_1, the flow rate is then given by:

$$
\begin{aligned}
q &= k_1 \sum \Delta h \\
&= 3.3 \times 10^{-9} \mathrm{m^3/s}
\end{aligned}
$$

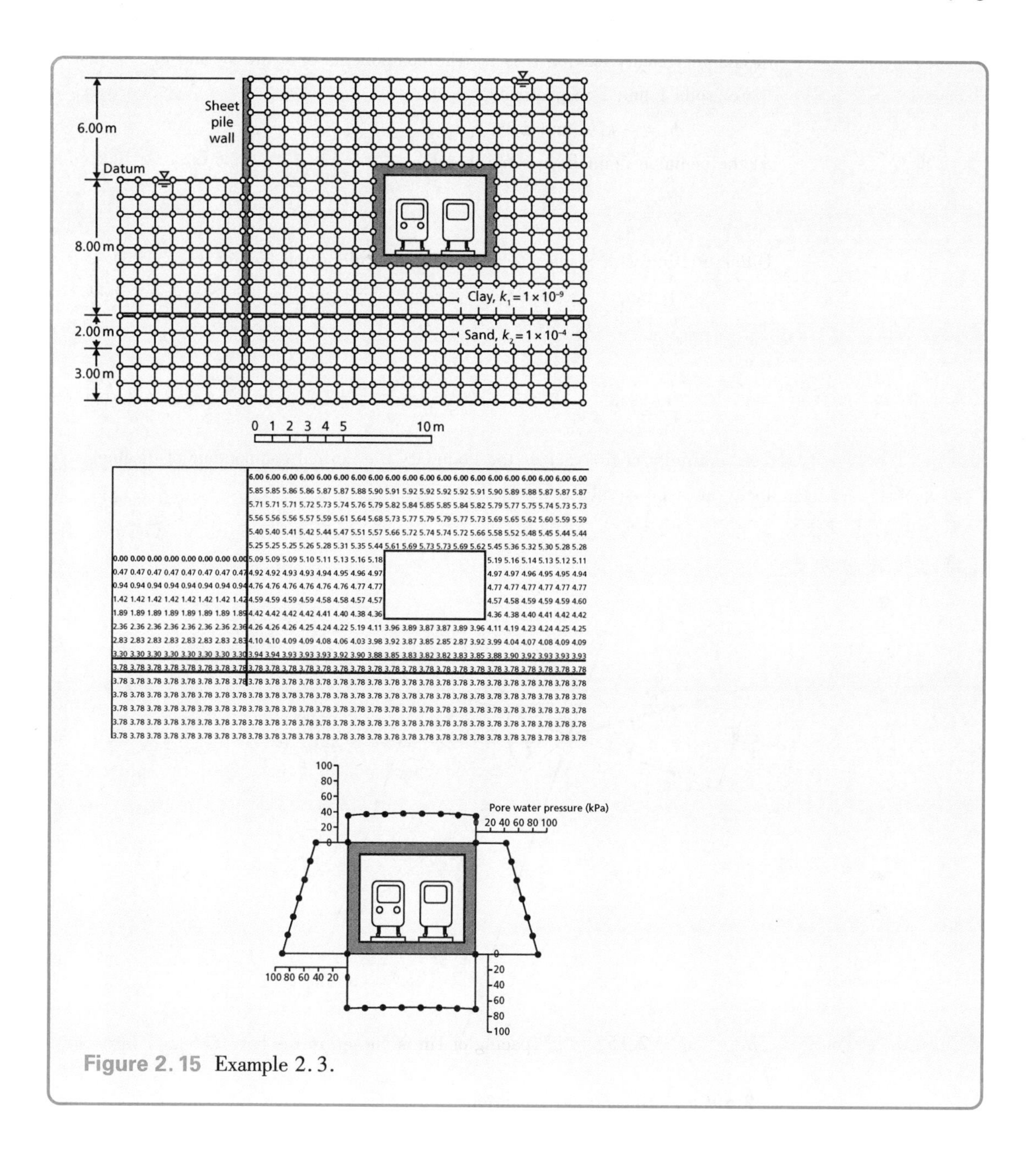

Figure 2.15 Example 2.3.

2.8 Transfer condition[①]

Consideration is now given to the condition which must be satisfied when seepage takes place diagonally across the boundary between two isotropic soils 1 and 2 having coefficients of permeability k_1 and k_2, respectively.[②] The direction of seepage approaching a point B on the boundary ABC is at angle α_1 to the normal at B, as shown in Figure 2.16; the discharge velocity approaching B is v_1. The components of v_1 along the boundary and normal to the boundary are v_{1s} and v_{1n} respectively. The direction of seepage leaving point B is at angle α_2 to the normal, as shown; the

① 转换边界上的渗流。

② 考虑不同渗透系数 k_1 和 k_2 的两种各向同性土 1 和 2 边界上，渗流以对角线穿过边界需满足的条件。

discharge velocity leaving B is v_2. The components of v_2 are v_{2s} and v_{2n}.

For soils 1 and 2 respectively

$$\phi_1 = -k_1 h_1 \text{ and } \phi_2 = -k_2 h_2$$

At the common point B, $h_1 = h_2$; therefore

$$\frac{\phi_1}{k_1} = \frac{\phi_2}{k_2}$$

Differentiating with respect to s, the direction along the boundary:

$$\frac{1}{k_1}\frac{\partial \phi_1}{\partial s} = \frac{1}{k_2}\frac{\partial \phi_2}{\partial s}$$

i. e.

$$\frac{v_{1s}}{k_1} = \frac{v_{2s}}{k_2}$$

For continuity of flow across the boundary the normal components of discharge velocity must be equal[①], i. e.

① 为了保持穿过边界渗流的连续性，流速的法向分量必须相等。

$$v_{1n} = v_{2n}$$

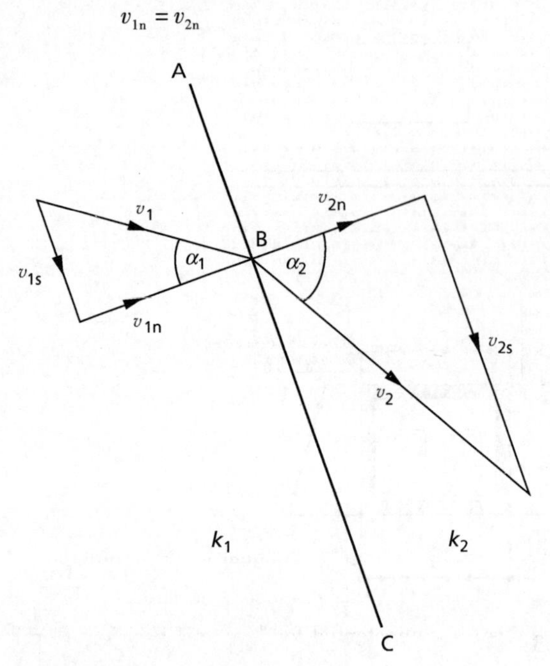

Figure 2.16 Transfer condition.

Therefore

$$\frac{1}{k_1}\frac{v_{1s}}{v_{1n}} = \frac{1}{k_2}\frac{v_{2s}}{v_{2n}}$$

Hence it follows that

$$\frac{\tan \alpha_1}{\tan \alpha_2} = \frac{k_1}{k_2} \tag{2.32}$$

Equation 2.32 specifies the change in direction of the flow line passing through point B. This equation must be satisfied on the boundary by every flow line crossing the boundary.

Equation 2.18 can be written as

$$\Delta\psi = \frac{\Delta n}{\Delta s}\Delta\phi$$

i. e.

$$\Delta q = \frac{\Delta n}{\Delta s}k\Delta\phi$$

If Δq and Δh are each to have the same values on both sides of the boundary, then

$$\left(\frac{\Delta n}{\Delta s}\right)_1 k_1 = \left(\frac{\Delta n}{\Delta s}\right)_2 k_2$$

and it is clear that curvilinear squares are possible only in one soil. If

$$\left(\frac{\Delta n}{\Delta s}\right)_1 = 1$$

then

$$\left(\frac{\Delta n}{\Delta s}\right)_2 = \frac{k_1}{k_2} \tag{2.33}$$

If the permeability ratio is less than $\frac{1}{10}$, it is unlikely that the part of the flow net in the soil of higher permeability needs to be considered.

2.9 Seepage through embankment dams[①]

① 堤坝中的渗流。

This problem is an example of unconfined seepage, one boundary of the flow region being a phreatic surface[②] on which the pressure is atmospheric[③]. In section, the phreatic surface constitutes the top flow line and its position must be estimated before the flow net can be drawn.

② 自由水面。

③ 大气压。

Consider the case of a homogeneous isotropic embankment dam on an impermeable foundation, as shown in Figure 2. 17. The impermeable boundary BA is a flow line, and CD is the required top flow line. At every point on the upstream slope BC the total head is constant (u/γ_w and z varying from point to point but their sum remaining constant); therefore, BC is an equipotential. If the downstream water level is taken as datum, then the total head on equipotential BC is equal to h, the difference between the upstream and downstream water levels. The discharge surface AD, for the case shown in Figure 2. 17 only, is the equipotential for zero total head. At every point on the top flow line the pressure is zero (atmospheric), so total head is equal to elevation head and there must be equal vertical intervals Δz between the points of intersection between successive equipotentials and the top flow line.[④]

④ 流线顶部每一点的压力都为 0，所以总水头等于位置水头，并且必须等于连续等势线与顶端流线的交点之间的竖向间距 Δz。

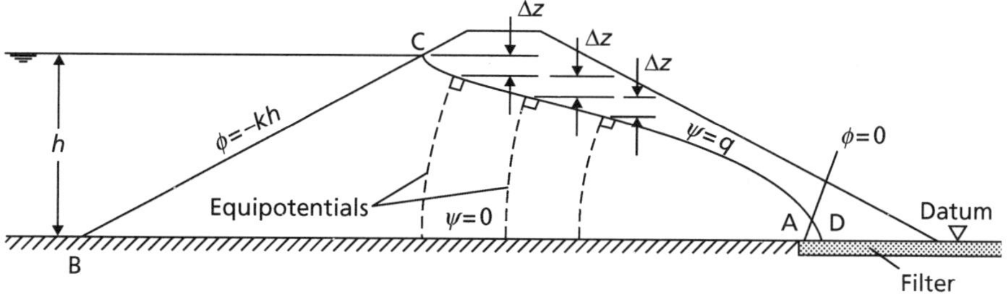

Figure 2. 17 Homogeneous embankment dam section.[⑤]

⑤ 均质土石坝截面。

① 反滤层。
② 溢出面。
③ 逐步侵蚀。
④ 下游坡面。
⑤ 在这两个例子中，溢出面投影 AD 线既不是流线，也不是等势线，因为流速在 AD 上即有法向分量也有切向分量。

A suitable filter① must always be constructed at the discharge surface② in an embankment dam. The function of the filter is to keep the seepage entirely within the dam; water seeping out onto the downstream slope would result in the gradual erosion③ of the slope. The consequences of such erosion can be severe. In 1976, a leak was noticed close to one of the abutments of the Teton Dam in Idaho, USA. Seepage was subsequently observed through the downstream face④. Within two hours of this occurring the dam failed catastrophically, causing extensive flooding, as shown in Figure 2.18. The direct and indirect costs of this failure were estimated at close to $1 billion. A horizontal under-filter is shown in Figure 2.17. Other possible forms of filter are illustrated in Figures 2.22 (a) and (b); in these two cases the discharge surface AD is neither a flow line nor an equipotential since there are components of discharge velocity both normal and tangential to AD⑤.

⑥ 提顿坝垮塌，1976。

Figure 2.18 Failure of the Teton Dam, 1976⑥ (photo courtesy of the Bureau of Reclamation).

The boundary conditions of the flow region ABCD in Figure 2.17 can be written as follows:

Equipotential BC: $\phi = -kh$

Equipotential AD: $\phi = 0$

Flow line CD: $\psi = q$ (also, $\phi = -kz$)

Flow line BA: $\psi = 0$

⑦ 保角变换。
⑧ 复变函数理论。

The conformal transformation⑦ $r = w^2$

Complex variable theory⑧ can be used to obtain a solution to the embankment dam problem. Let the complex number $w = \phi + i\psi$ be an analytic function of $r = x + iz$. Consider the function

$$r = w^2$$

Thus

$$(x + iz) = (\phi + i\psi)^2 = (\phi^2 + 2i\phi\psi - \psi^2)$$

Equating real and imaginary parts:

$$x = \phi^2 - \psi^2 \tag{2.34}$$

$$z = 2\phi\psi \tag{2.35}$$

Equations 2.34 and 2.35 govern the transformation of points between the r and w planes.

Consider the transformation of the straight lines $\psi = n$, where $n = 0, 1, 2, 3$ (Figure 2.19 (a)). From Equation 2.35

$$\phi = \frac{z}{2n}$$

and Equation 2.34 becomes

$$x = \frac{z^2}{4n^2} - n^2 \tag{2.36}$$

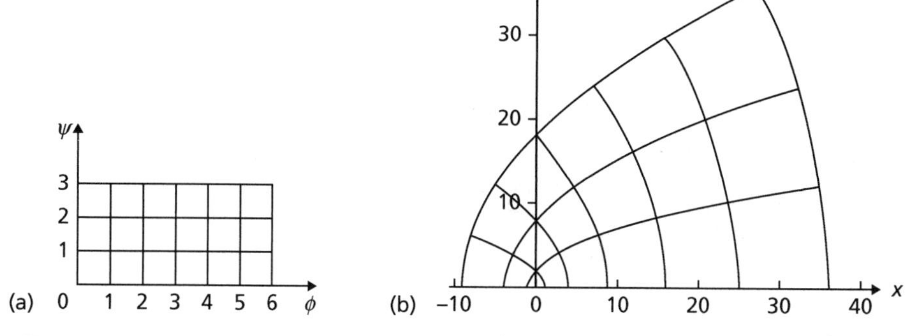

Figure 2.19 Conformal transformation $r = w^2$: (a) w plane, and (b) r plane.[①]

① 保角变换 $r = w^2$：(a) w 面，(b) r 面。

Equation 2.36 represents a family of confocal parabolas. For positive values of z the parabolas for the specified values of n are plotted in Figure 2.19 (b).

Consider also the transformation of the straight lines $\phi = m$, where $m = 0, 1, 2, \ldots, 6$ (Figure 2.19 (a)). From Equation 2.35

$$\psi = \frac{z}{2m}$$

and Equation 2.34 becomes

$$x = m^2 - \frac{z^2}{4m^2} \tag{2.37}$$

Equation 2.37 represents a family of confocal parabolas conjugate with the parabolas represented by Equation 2.36. For positive values of z the parabolas for the specified values of m are plotted in Figure 2.19 (b). The two families of parabolas satisfy the requirements of a flow net.

Application to embankment dam sections[②]

② 在堤坝截面的应用。

The flow region in the w plane satisfying the boundary conditions for the section (Figure 2.17) is shown in Figure 2.20 (a). In this case, the transformation function

$$r = Cw^2$$

will be used, where C is a constant. Equations 2.34 and 2.35 then become

$$x = C(\phi^2 - \psi^2)$$
$$z = 2C\phi\psi$$

The equation of the top flow line can be derived by substituting the conditions

$$\psi = q$$
$$\phi = -kz$$

Thus

$$z = -2Ckzq$$
$$\therefore C = -\frac{1}{2kq}$$

Hence

$$x = -\frac{1}{2kq}(k^2z^2 - q^2)$$
$$x = \frac{1}{2}\left(\frac{q}{k} - \frac{k}{q}z^2\right) \tag{2.38}$$

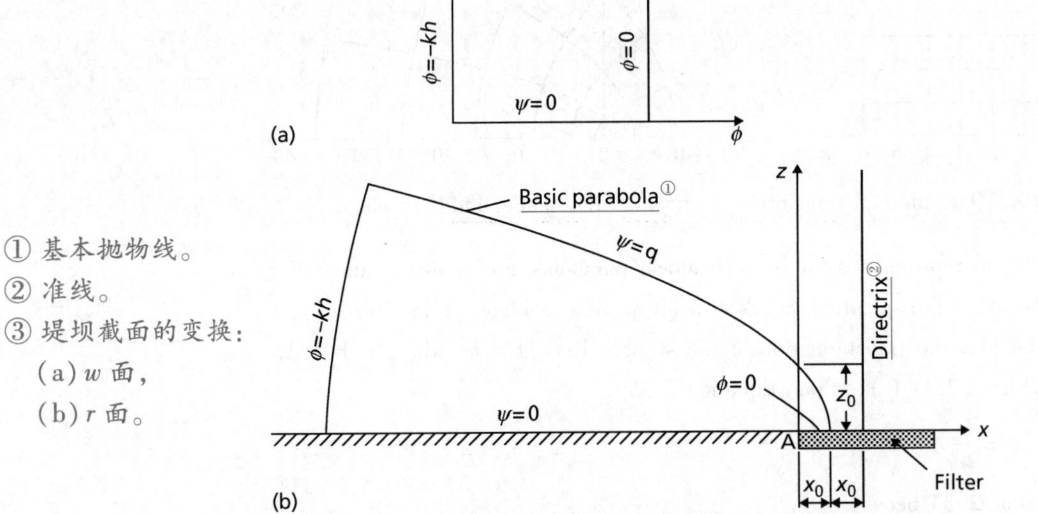

① 基本抛物线。
② 准线。
③ 堤坝截面的变换：(a) w 面，(b) r 面。

Figure 2.20 Transformation for embankment dam section: (a) w plane, and (b) r plane.[3]

④ 基本抛物线。

The curve represented by Equation 2.38 is referred to as Kozeny's basic parabola[4] and is shown in Figure 2.20 (b), the origin and focus both being at A.

When $z = 0$, the value of x is given by

$$x_0 = \frac{q}{2k} \tag{2.39}$$
$$\therefore q = 2kx_0$$

⑤ $2x_0$ 为基本抛物线准线的间距。

where $2x_0$ is the directrix distance of the basic parabola[5]. When $x = 0$, the value of z is given by

$$z_0 = \frac{q}{k} = 2x_0$$

Substituting Equation 2.39 in Equation 2.38 yields

$$x = x_0 - \frac{z^2}{4x_0} \tag{2.40}$$

The basic parabola can be drawn using Equation 2.40, provided the coordinates of one point on the parabola are known initially.

An inconsistency arises due to the fact that the conformal transformation of the straight line $\phi = -kh$ (representing the upstream equipotential) is a parabola, whereas the upstream equipotential in the embankment dam section is the upstream slope. Based on an extensive study of the problem, Casagrande (1940) recommended that the initial point of the basic parabola should be taken at G (Figure 2.21) where GC = 0.3HC. The coordinates of point G, substituted in Equation 2.40, enable the value of x_0 to be determined; the basic parabola can then be plotted. The top flow line must intersect the upstream slope at right angles; a correction CJ must therefore be made (using personal judgement) to the basic parabola. The flow net can then be completed as shown in Figure 2.21.

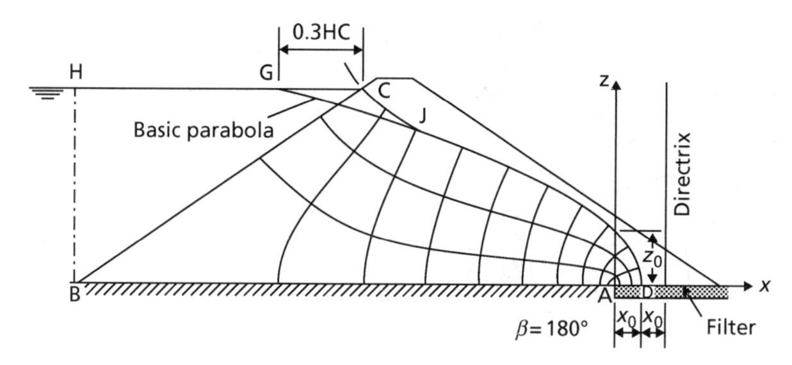

Figure 2.21 Flow net for embankment dam section.[①]

① 堤坝截面流网。

If the discharge surface AD is not horizontal, as in the cases shown in Figure 2.22, a further correction KD to the basic parabola is required. The angle β is used to describe the direction of the discharge surface relative to AB. The correction can be made with the aid of values of the ratio MD/MA = $\Delta a/a$, given by Casagrande for the range of values of β (Table 2.4).

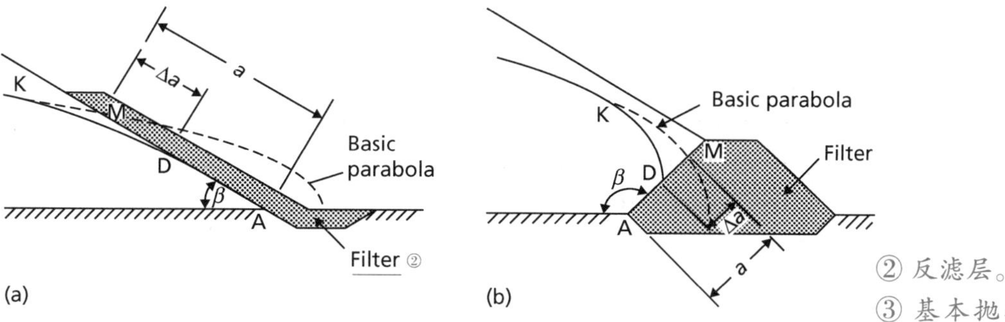

Figure 2.22 Downstream correction to basic parabola.[③]

② 反滤层。

③ 基本抛物线的下游校正。

Table 2.4 Downstream correction to basic parabola. Reproduced from A. Casagrande (1940) ' Seepage through dams', in *Contributions to Soil Mechanics 1925 – 1940*, by permission of the Boston Society of Civil Engineers

β	30°	60°	90°	120°	150°	180°
$\Delta a/a$	(0.36)	0.32	0.26	0.18	0.10	0

Example 2.4

A homogeneous anisotropic embankment dam section is detailed in Figure 2.23 (a), the coeffcients of permeability in the x and z directions being 4.5×10^{-8} and 1.6×10^{-8} m/s, respectively. Construct the flow net and determine the quantity of seepage through the dam. What is the pore water pressure at point P?

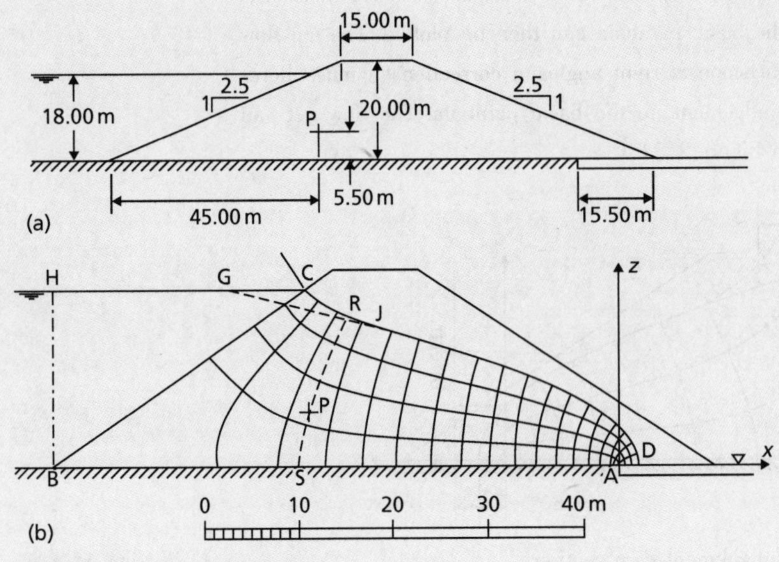

Figure 2.23 Example 2.4.

Solution

The scale factor for transformation in the x direction is

$$\sqrt{\frac{k_z}{k_x}} = \sqrt{\frac{1.6}{4.5}} = 0.60$$

The equivalent isotropic permeability is

$$k' = \sqrt{(k_x k_z)}$$

$$= \sqrt{(4.5 \times 1.6)} \times 10^{-8} = 2.7 \times 10^{-8} \text{ m/s}$$

The section is drawn to the transformed scale in Figure 2.23 (b). The focus of the basic parabola is at point A. The basic parabola passes through point G such that

$$GC = 0.3HC = 0.3 \times 27.00 = 8.10\text{m}$$

i.e. the coordinates of G are

$$x = -40.80, z = +18.00$$

Substituting these coordinates in Equation 2.40:

$$-40.80 = x_0 - \frac{18.00^2}{4x_0}$$

Hence

$$x_0 = 1.90\text{m}$$

Using Equation 2.40, the coordinates of a number of points on the basic parabola are now calculated:

TABLE E

x	1.90	0	−5.00	−10.00	−20.00	−30.00
z	0	3.80	7.24	9.51	12.90	15.57

The basic parabola is plotted in Figure 2.23 (b). The upstream correction is made (JC) and the flow net completed, ensuring that there are equal vertical intervals between the points of intersection of successive equipotentials with the top flow line. In the flow net there are four flow channels and 18 equipotential drops. Hence, the quantity of seepage (per unit length) is

$$q = k'h\frac{N_f}{N_d}$$

$$= 2.7 \times 10^{-8} \times 18 \times \frac{4}{18} = 1.1 \times 10^{-7}\text{m}^3/\text{s}$$

The quantity of seepage can also be determined from Equation 2.39 (without the necessity of drawing the flow net):

$$q = 2k'x_0$$

$$= 2 \times 2.7 \times 10^{-8} \times 1.90 = 1.0 \times 10^{-7}\text{m}^3/\text{s}$$

To determine the pore water pressure at P, Level AD is first selected as datum. An equipotential RS is drawn through point P (transformed position). By inspection, the total head at P is 15.60m. At P the elevation head is 5.50m, so the pressure head is 10.10m and the pore water pressure is

$$u_P = 9.81 \times 10.10 = 99\text{kPa}$$

Alternatively, the pressure head at P is given directly by the vertical distance of P below the point of intersection (R) of equipotential RS with the top flow line.

Seepage control in embankment dams[①]

The design of an embankment dam section and, where possible, the choice of soils are aimed at reducing or eliminating the detrimental effects of seeping water.[②] Where high hydraulic gradients exist there is a possibility that the seeping water may cause internal erosion[③] within the dam, especially if the soil is poorly compacted. Erosion can work its way back into the embankment, creating voids in the form of channels or 'pipes', and thus impairing the stability of the dam. This form of erosion is referred to as **piping**[④].

A section with a central core of low permeability, aimed at reducing the volume of seepage, is shown in Figure 2.24 (a). Practically all the total head is lost in the core, and if the core is narrow, high hydraulic gradients will result. There is a parti-

① 堤坝中的渗流控制。
② 堤坝截面设计及填土类型选择（如果可能）的目的在于减少或消除渗流的不利影响。
③ 内部侵蚀。
④ 管涌。

cular danger of erosion at the boundary between the core and the adjacent soil (of higher permeability) under a high exit gradient from the core. Protection against this danger can be given by means of a 'chimney' drain[1] (Figure 2. 24 (a)) at the downstream boundary of the core. The drain, designed as a filter (see Section 2. 10) to provide a barrier to soil particles from the core, also serves as an interceptor, keeping the downstream slope in an unsaturated state.

① 竖向排水系统。

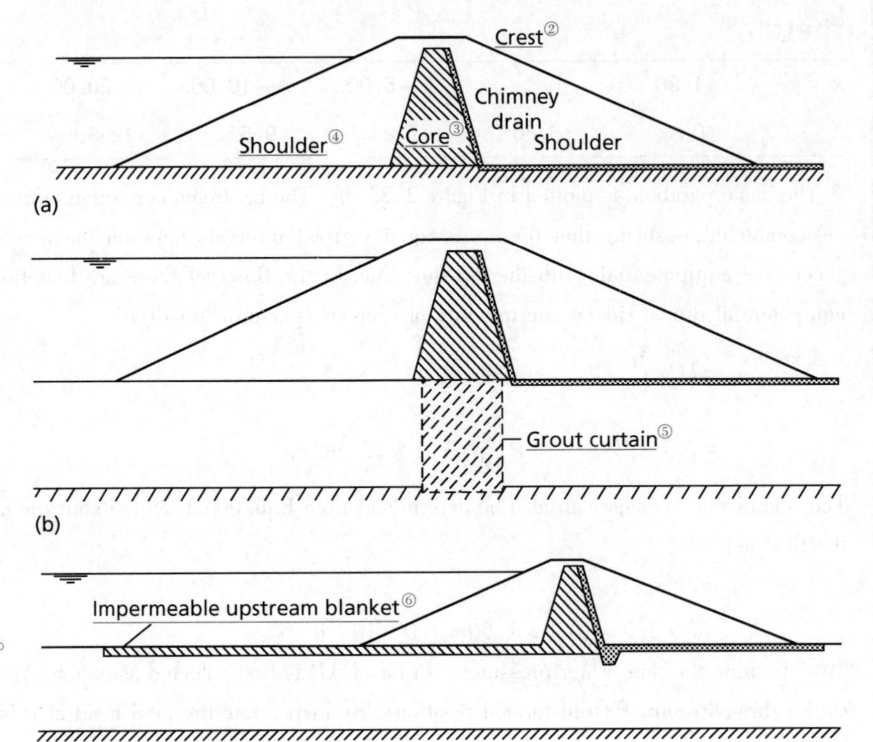

Figure 2. 24 (a) Central core and chimney drain, (b) grout curtain, and (c) impermeable upstream blanket.

② 坝顶。
③ 芯墙。
④ 坝肩。
⑤ 注浆帷幕。
⑥ 上游不透水卷材。

Most embankment dam sections are non-homogeneous owing to zones of different soil types, making the construction of the flow net more difficult. The basic parabola construction for the top flow line applies only to homogeneous sections, but the condition that there must be equal vertical distances between the points of intersection of equipotentials with the top flow line applies equally to a non-homogeneous section. The transfer condition (Equation 2. 32) must be satisfied at all zone boundaries. In the case of a section with a central core of low permeability, the application of Equation 2. 32 means that the lower the permeability ratio, the lower the position of the top flow line in the downstream zone (in the absence of a chimney drain).

If the foundation soil is more permeable than the dam, the control of **underseepage** is essential.[7] Underseepage can be virtually eliminated by means of an 'impermeable' cut-off[8] such as a grout curtain[9] (Figure 2. 24 (b)). Another form of cut-off is the concrete diaphragm wall (see Section 11. 10). Any measure designed to

⑦ 如果基础的渗透性高于坝体，则坝底渗流的控制非常重要。
⑧ 不透水截断。
⑨ 注浆帷幕。

lengthen the seepage path, such as an impermeable upstream blanket (Figure 2.24 (c)), will result in a partial reduction in underseepage.

An excellent treatment of seepage control is given by Cedergren (1989).

Example 2.5

Consider the concrete dam spillway from Example 2.2 (Figure 2.10). Determine the effect of the length of the cut-off wall on the reduction in seepage flow beneath the spillway. Determine also the reduction in seepage flow due to an impermeable upstream blanket of length L_b, and compare the efficacy of the two methods of seepage control.

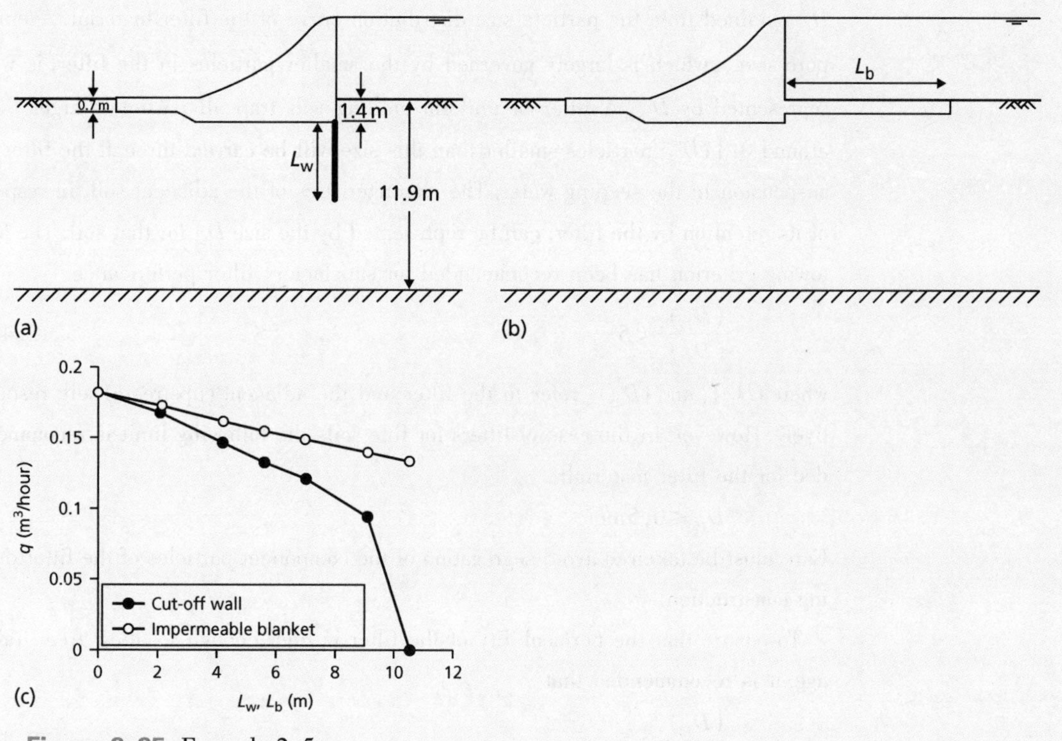

Figure 2.25 Example 2.5.

Solution

The seepage flow can be determined using the spreadsheet tool on the Companion Website. Repeating the calculations for different lengths, L_w, of cut-off wall (Figure 2.25 (a)) and separately for different lengths of impermeable blanket, L_b, as shown in Figure 2.25 (b), the results shown in the figure can be determined. By comparing the two methods, it can be seen that for thin layers such as in this problem, cut-off walls are generally more effective at reducing seepage and typically require a lower volume of material (and therefore lower cost) to achieve the same reduction in seepage flow.

2.10 Filter design[①]

① 反滤层设计。

Filters used to control seepage must satisfy certain fundamental requirements. The pores must be small enough to prevent particles from being carried in from the

① 孔隙必须足够小，防止邻近的土颗粒被水排出，同时必须保证渗透性足够大，保证自由水可以从滤层排出。

② 滤层不必为完全饱和。

③ 平均孔隙尺寸。

adjacent soil.[①] The permeability must simultaneously be high enough to ensure the free drainage of water entering the filter. The capacity of a filter should be such that it does not become fully saturated.[②] In the case of an embankment dam, a filter placed downstream from the core should be capable of controlling and sealing any leak which develops through the core as a result of internal erosion. The filter must also remain stable under the abnormally high hydraulic gradient which is liable to develop adjacent to such a leak.

Based on extensive laboratory tests by Sherard *et al.* (1984a, 1984b) and on design experience, it has been shown that filter performance can be related to the size D_{15} obtained from the particle size distribution curve of the filter material. Average pore size[③], which is largely governed by the smaller particles in the filter, is well represented by D_{15}. A filter of uniform grading will trap all particles larger than around $0.11D_{15}$; particles smaller than this size will be carried through the filter in suspension in the seeping water. The characteristics of the adjacent soil, in respect of its retention by the filter, can be represented by the size D_{85} for that soil. The following criterion has been recommended for satisfactory filter performance:

$$\frac{(D_{15})_f}{(D_{85})_s} < 5 \qquad (2.41)$$

where $(D_{15})_f$ and $(D_{85})_s$ refer to the filter and the adjacent (upstream) soil, respectively. However, in the case of filters for fine soils the following limit is recommended for the filter material:

$$D_{15} \leqslant 0.5\text{mm}$$

Care must be taken to avoid segregation of the component particles of the filter during construction.

To ensure that the permeability of the filter is high enough to allow free drainage, it is recommended that

$$\frac{(D_{15})_f}{(D_{15})_s} > 5 \qquad (2.42)$$

Graded filters comprising two (or more) layers with different gradings can also be used, the finer layer being on the upstream side. The above criterion (Equation 2.41) would also be applied to the component layers of the filter.

Summary

1 The permeability of soil is strongly affected by the size of the voids. As a result, the permeability of fine soils may be many orders of magnitude smaller than those for coarse soils. Falling head tests are commonly used to measure permeability in fine soils, and constant head tests are used for coarse soils. These may be conducted in the laboratory on undisturbed samples removed from the ground (see Chapter 6) or in-situ.

2 Groundwater will flow wherever a hydraulic gradient exists. For flow in two-dimensions, a flow net may be used to determine the distribution of total head, pore pressure and seepage quantity. This technique can also account for soils which are layered or anisotropic in permeability, both of which parameters significantly affect seepage.
3 Complex or large seepage problems, for which flow net sketching would prove impractical, can be accurately and efficiently solved using the finite difference method. A spreadsheet implementation of this method is available on the Companion Website.
4 Seepage through earth dams is more complex as it is unconfined. The conformal transformation provides a straightforward and efficient method for determining flow quantities and developing the flow net for such a case. Seepage through and beneath earth dams, which may affect their stability, may be controlled using a range of techniques, including a low permeability core, cut-off walls, or impermeable blankets. The efficacy of these methods may be determined using the techniques outlined in the chapter.

Problems

2.1 In a falling-head permeability test the initial head of 1.00m dropped to 0.35m in 3h, the diameter of the standpipe being 5mm. The soil specimen was 200mm long by 100mm in diameter. Calculate the coefficient of permeability of the soil.

2.2 The section through part of a cofferdam is shown in Figure 2.26, the coefficient of permeability of the soil being 2.0×10^{-6} m/s. Draw the flow net and determine the quantity of seepage.

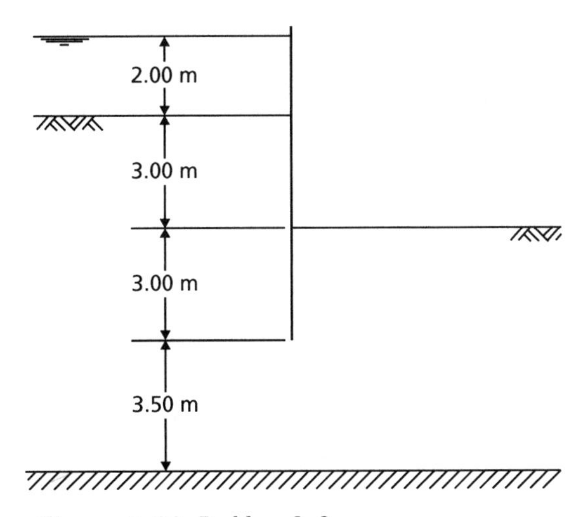

Figure 2.26 Problem 2.2.

2.3 The section through a long cofferdam is shown in Figure 2.27, the coefficient of permeability of the soil being 4.0×10^{-7} m/s. Draw the flow net and determine the quantity of seepage entering the cofferdam.

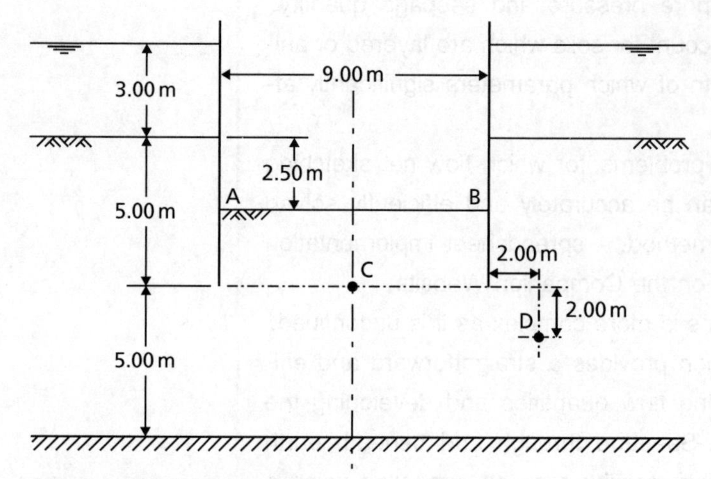

Figure 2.27 Problem 2.3.

2.4 The section through a sheet pile wall along a tidal estuary is given in Figure 2.28. At low tide the depth of water in front of the wall is 4.00m; the water table behind the wall lags 2.50m behind tidal level. Plot the net distribution of water pressure on the piling.

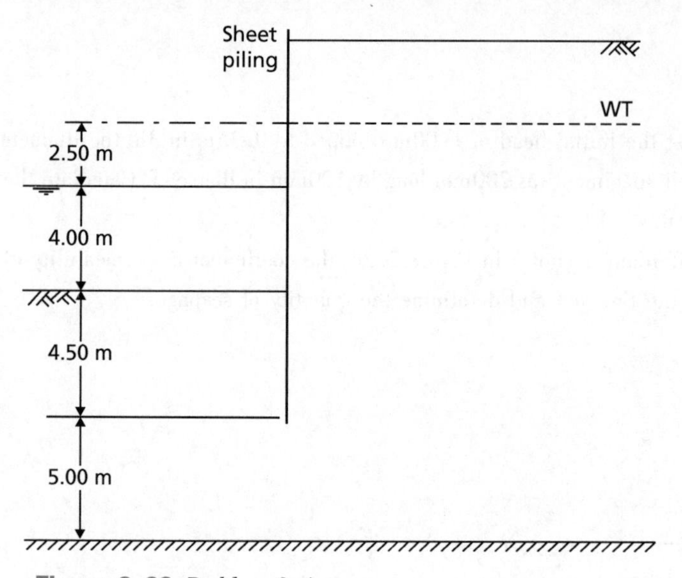

Figure 2.28 Problem 2.4.

2.5 Details of an excavation adjacent to a canal are shown in Figure 2.29. Determine the quantity of seepage into the excavation if the coefficient of permeability is 4.5×10^{-5} m/s.

2.6 The dam shown in section in Figure 2.30 is located on anisotropic soil. The coefficients of permeability in the x and z directions are 5.0×10^{-7} and 1.8×10^{-7} m/s, respectively. Determine the quantity of seepage under the dam.

2.7 Determine the quantity of seepage under the dam shown in section in Figure 2.31. Both layers of soil are isotropic, the coefficients of permeability of the upper and lower layers being 2.0×10^{-6} and $1.6 \times$

10^{-5} m/s, respectively.

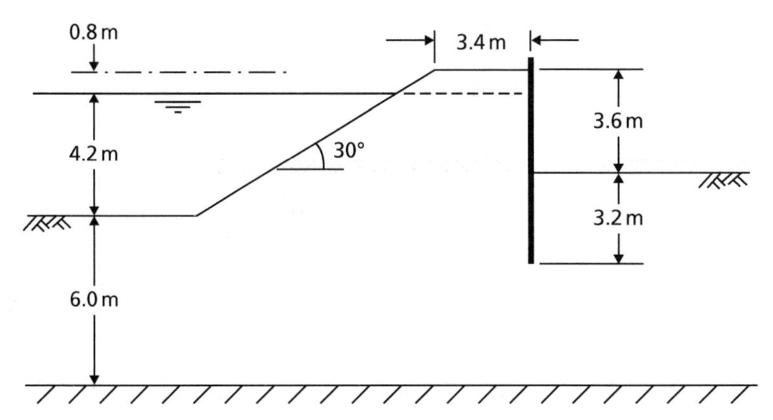

Figure 2.29 Problem 2.5.

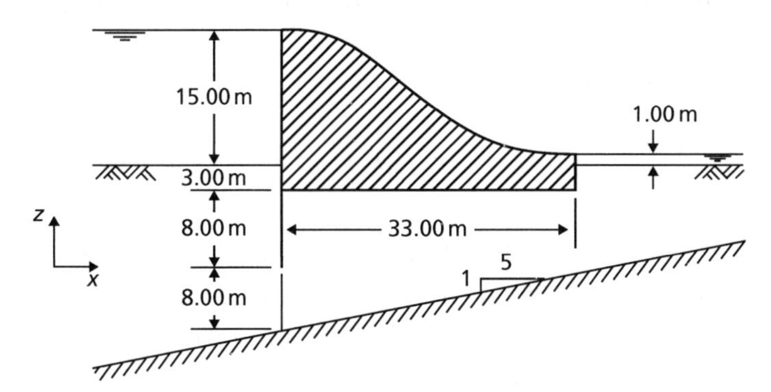

Figure 2.30 Problem 2.6.

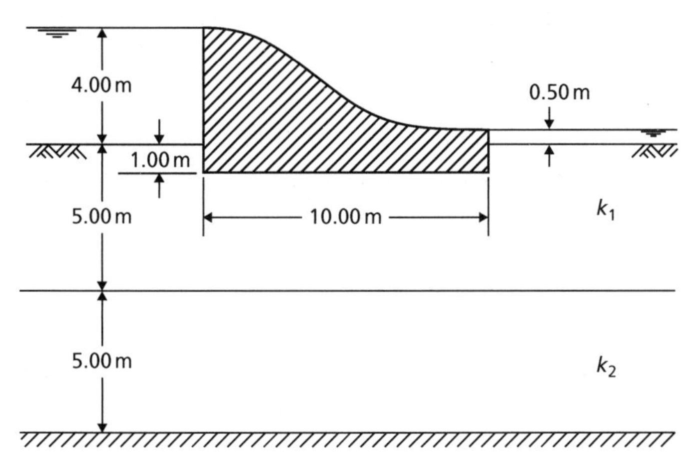

Figure 2.31 Problem 2.7.

2.8 An embankment dam is shown in section in Figure 2.32, the coefficients of permeability in the horizontal and vertical directions being 7.5×10^{-6} and 2.7×10^{-6} m/s, respectively. Construct the top flow line and determine the quantity of seepage through the dam.

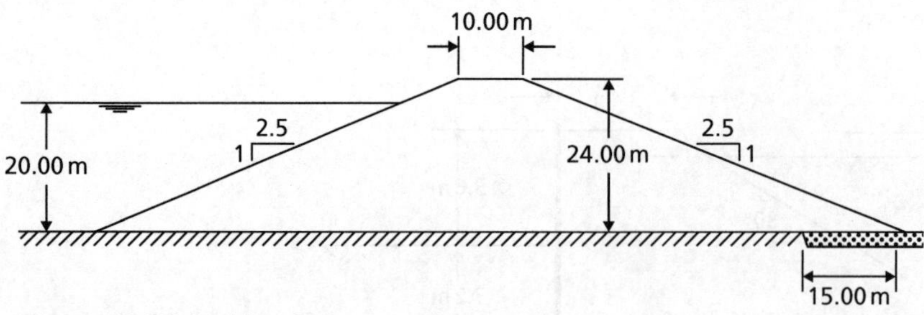

Figure 2.32 Problem 2.8.

References

ASTM D5084 (2010) *Standard Test Methods for Measurement of Hydraulic Conductivity of Saturated Porous Materials Using a Flexible Wall Permeameter*, American Society for Testing and Materials, West Conshohocken, PA.

British Standard 1377 (1990) *Methods of Test for Soils for Civil Engineering Purposes*, British Standards Institution, London.

CEN ISO/TS 17892 (2004) *Geotechnical Investigation and Testing – Laboratory Testing of Soil*, International Organisation for Standardisation, Geneva.

Casagrande, A. (1940) Seepage through dams, in *Contributions to Soil Mechanics* 1925 – 1940, Boston Society of Civil Engineers, Boston, MA, pp. 295 – 336.

Cedergren, H. R. (1989) *Seepage, Drainage and Flow Nets* (3rd edn), John Wiley & Sons, New York, NY.

Clayton, C. R. I., Matthews, M. C. and Simons, N. E. (1995) *Site Investigation* (2nd edn), Blackwell, London.

Hvorslev, M. J. (1951) *Time Lag and Soil Permeability in Ground-Water Observations*, Bulletin No. 36, Waterways Experimental Station, US Corps of Engineers, Vicksburg, MS.

Sherard, J. L., Dunnigan, L. P. and Talbot, J. R. (1984a) Basic properties of sand and gravel filters, *Journal of the ASCE*, **110** (GT6), 684 – 700.

Sherard, J. L., Dunnigan, L. P. and Talbot, J. R. (1984b) Filters for silts and clays, *Journal of the ASCE*, **110** (GT6), 701 – 718.

Vreedenburgh, C. G. F. (1936) On the steady flow of water percolating through soils with homogeneousanisotropic permeability, in *Proceedings of the 1st International Conference on SMFE, Cambridge, MA*, Vol. 1.

Williams, B. P., Smyrell, A. G. and Lewis, P. J. (1993) Flownet diagrams – the use of finite differences and a spreadsheet to determine potential heads, *Ground Engineering*, **25** (5), 32 – 38.

Further reading

Cedergren, H. R. (1989) *Seepage, Drainage and Flow Nets* (3rd edn), John Wiley & Sons, New York, NY.

This is still the definitive text on seepage, particularly regarding flow net construction. The book also includes case histories showing the application of flow net techniques to real problems.

Preene, M., Roberts, T. O. L., Powrie, W. and Dyer, M. R. (2000) *Groundwater Control – Design and Practice*, CIRIA Publication C515, CIRIA, London.

This text covers groundwater control in more detail (it has only been touched on here). Also a valuable source of practical guidance.

Chapter 3

Effective stress[①]

① 有效应力。

Learning outcomes

After working through the material in this chapter, you should be able to:

1 Understand how total stress[②], pore water pressure[③] and effective stress are related and the importance of effective stress in soil mechanics (Sections 3.1, 3.2 and 3.4);

2 Determine the effective stress state within the ground, both under hydrostatic[④] conditions and when seepage[⑤] is occurring (Sections 3.3 and 3.6);

3 Describe the phenomenon of liquefaction[⑥], and determine the hydraulic conditions[⑦] within the groundwater under which liquefaction will occur (Section 3.7).

② 总应力。
③ 孔隙水压力。
④ 静水压力。
⑤ 渗流。
⑥ 液化。
⑦ 水力条件。

3.1 Introduction

A soil can be visualised as a skeleton of solid particles enclosing continuous voids which contain water and/or air.[⑧] For the range of stresses usually encountered in practice, the individual solid particles and water can be considered incompressible; air, on the other hand, is highly compressible[⑨]. The volume of the soil skeleton as a whole can change due to rearrangement of the soil particles[⑩] into new positions, mainly by rolling and sliding, with a corresponding change in the forces acting between particles. The actual compressibility of the soil skeleton will depend on the structural arrangement of the solid particles, i.e. the void ratio[⑪], e. In a fully saturated soil, since water is considered to be incompressible, a reduction[⑫] in volume is possible only if some of the water can escape from the voids. In a dry or a partially saturated soil a reduction in volume is always possible, due to compression of the air in the voids, provided there is scope for particle rearrangement (i.e. the soil is not already in its densest possible state, $e > e_{min}$).

Shear stress can be resisted only by the skeleton of solid particles, by means of reaction forces developed at the interparticle contacts.[⑬] Normal stress[⑭] may similarly

⑧ 土体可以被视为由土颗粒组成的土骨架，以及颗粒间充满水和（或）空气的连续孔隙组成。
⑨ 单个土颗粒和水视为不可压缩，另一方面，空气视为可高度压缩。
⑩ 土颗粒重分布。
⑪ 孔隙比。
⑫ 减少。
⑬ 剪应力只能由土颗粒组成的骨架通过颗粒间的接触力承担。
⑭ 正应力。

① 颗粒间接触力。

be resisted by the soil skeleton through an increase in the interparticle forces[①]. If the soil is fully saturated, the water filling the voids can also withstand normal stress by an increase in pore water pressure.

② 有效应力原理。

3.2 The principle of effective stress[②]

③ 有效应力原理只适用于完全饱和土。
④ 总法向应力。
⑤ 孔隙水压力。
⑥ 有效正应力。
⑦ 作用面上的有效正应力，代表只通过土骨架传递的应力。
⑧ 物理模型。

The importance of the forces transmitted through the soil skeleton from particle to particle was recognised by Terzaghi (1943), who presented his **Principle of Effective Stress**, an intuitive relationship based on experimental data. The principle applies only to fully saturated soils[③], and relates the following three stresses:

1 the **total normal stress**[④] (σ) on a plane within the soil mass, being the force per unit area transmitted in a normal direction across the plane, imagining the soil to be a solid (single-phase) material;

2 the **pore water pressure**[⑤] (u), being the pressure of the water filling the void space between the solid particles;

3 the **effective normal stress**[⑥] (σ') on the plane, representing the stress transmitted through the soil skeleton only[⑦] (i. e. due to interparticle forces).

The relationship is:

$$\sigma = \sigma' + u \tag{3.1}$$

The principle can be represented by the following physical model[⑧]. Consider a 'plane' XX in a fully saturated soil, passing through points of interparticle contact only, as shown in Figure 3.1. The wavy plane XX is really indistinguishable from a true plane on the mass scale due to the relatively small size of individual soil particles. A normal force P applied over an area A may be resisted partly by interparticle forces and partly by the pressure in the pore water. The interparticle forces are very random in both magnitude and direction throughout the soil mass, but at every point of contact on the wavy plane may be split into components normal and tangential[⑨] to the direction of the true plane to which XX approximates; the normal and tangential components are N' and T, respectively. Then, the effective normal stress is approximated as the sum of all the components N' within the area A, divided by the area A, i. e.

⑨ 起伏不平的截面上的任一点的接触力可分解为法向分量和切向分量。

$$\sigma' = \frac{\sum N'}{A} \tag{3.2}$$

The total normal stress is given by

$$\sigma = \frac{P}{A} \tag{3.3}$$ [⑩]

⑩ 此公式表明有效应力为整个截面面积上的平均接触应力，而不是平均接触力。

If point contact is assumed between the particles, the pore water pressure will act on the plane over the entire area A. Then, for equilibrium in the direction normal to XX

$$P = \sum N' + uA$$

or

$$\frac{P}{A} = \frac{\sum N'}{A} + u$$

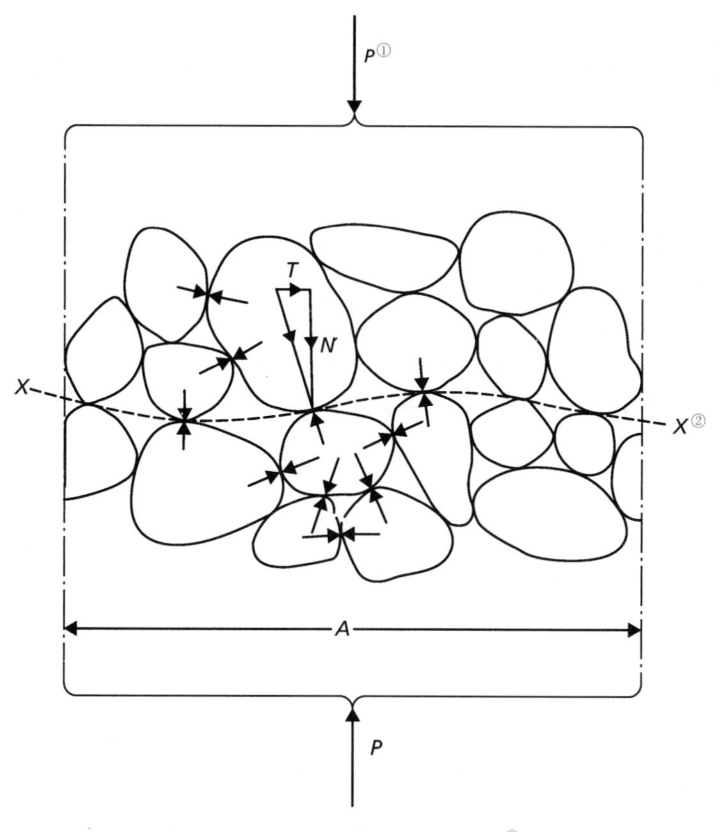

① 外力。
② 假想接触面。
③ 有效应力示意图。

Figure 3.1 Interpretation of effective stress. ③

i. e.

$$\sigma = \sigma' + u$$

The pore water pressure which acts equally in every direction will act on the entire surface of any particle, but is assumed not to change the volume of the particle (i. e. the soil particles themselves are incompressible); also, the pore water pressure does not cause particles to be pressed together. The error involved in assuming point contact between particles is negligible in soils, the actual contact area a normally being between 1% and 3% of the cross-sectional area A. ④ It should be understood that σ' does not represent the true contact stress between two particles, which would be the random but very much higher stress N'/a. ⑤

④ 颗粒间为点接触的假设，引起的误差在土体中是可以忽略的，实际的接触面积 a 一般占横截面面积 A 的 1% ~3%。
⑤ 需要明白，有效应力 σ' 不是真实两个颗粒间的真实接触应力，真实的接触应力是随机的且为很高的应力值 N'/a。

Effective vertical stress due to self-weight of soil⑥

Consider a soil mass having a horizontal surface and with the water table at surface level. The total vertical stress (i. e. the total normal stress on a horizontal plane) σ_v at depth z is equal to the weight of all material (solids + water) per unit area above that depth, i. e.

$$\sigma_v = \gamma_{sat} z$$

The pore water pressure at any depth will be hydrostatic since the void space between the solid particles is continuous⑦, so at depth z

⑥ 土体自重产生的竖向有效应力。
⑦ 因为颗粒间的孔隙是自由连通的，所以任一深度处的孔隙水压力为静水压力。

$$u = \gamma_w z$$

Hence, from Equation 3.1 the effective vertical stress at depth z in this case will be

$$\sigma'_v = \sigma_v - u$$
$$= (\gamma_{sat} - \gamma_w) z$$

Example 3.1

A layer of saturated clay 4m thick is overlain by sand 5m deep, the water table being 3m below the surface, as shown in Figure 3.2. The saturated unit weights of the clay and sand are 19 and 20kN/m³, respectively; above the water table the (dry) unit weight of the sand is 17kN/m³. Plot the values of total vertical stress and effective vertical stress against depth. If sand to a height of 1m above the water table is saturated with capillary water, how are the above stresses affected?

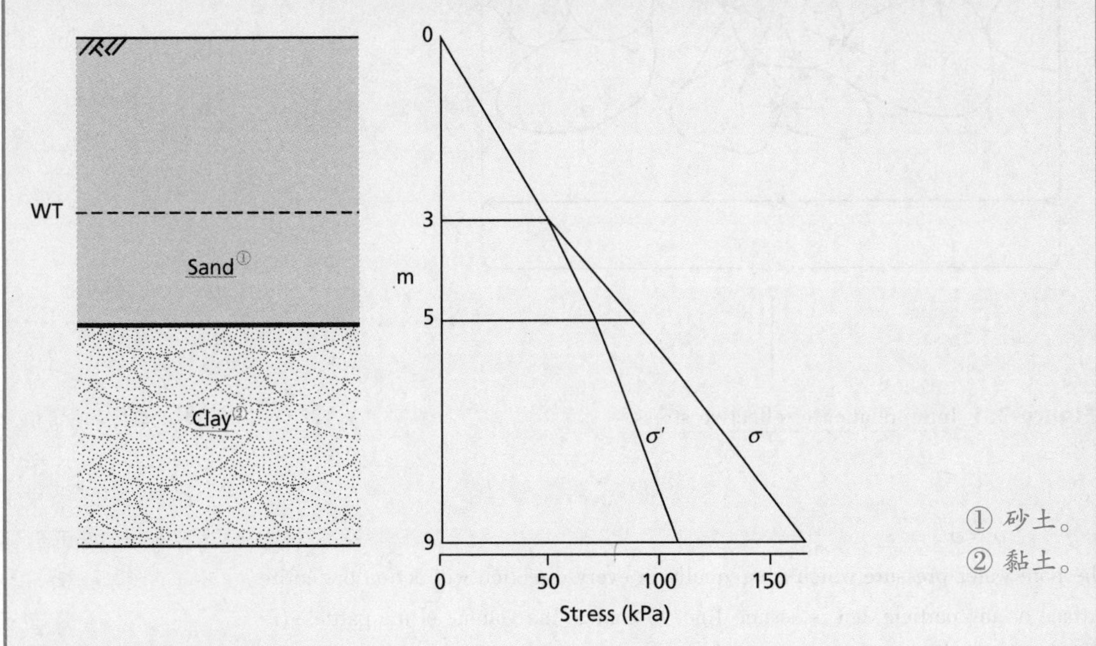

Figure 3.2 Example 3.1.

Solution

The total vertical stress is the weight of all material (solids + water) per unit area above the depth in question. Pore water pressure is the hydrostatic pressure corresponding to the depth below the water table. The effective vertical stress is the difference between the total vertical stress and the pore water pressure at the same depth. The stresses need only be calculated at depths where there is a change in unit weight (Table 3.1).

Table 3.1 Example 3.1

Depth (m)	*σ_v (kPa)*		*u (kPa)*	*$\sigma'_v = \sigma_v - u$ (kPa)*
3	3 × 17	= 51.0	0	51.0
5	(3 × 17) + (2 × 20)	= 91.0	2 × 9.81 = 19.6	71.4
9	(3 × 17) + (2 × 20) + (4 × 19)	= 167.0	6 × 9.81 = 58.9	108.1

In all cases the stresses would normally be rounded off to the nearest whole number. The stresses are plotted against depth in Figure 3.2.

From Section 2.1, the water table[①] is the level at which pore water pressure is atmospheric (i.e. $u = 0$). Above the water table, water is held under negative pressure and, even if the soil is saturated above the water table, does not contribute to hydrostatic pressure below the water table. The only effect of the 1-m capillary rise, therefore, is to increase the total unit weight of the sand between 2 and 3m depth from 17 to 20kN/m^3, an increase of 3kN/m^3. Both total and effective vertical stresses below 3m depth are therefore increased by the constant amount $3 \times 1 = 3.0$kPa, pore water pressures being unchanged.[②]

3.3 Numerical solution using the Finite Difference Method[③]

The calculation of the effective stress profile within the ground (i.e. variation with depth) under hydrostatic groundwater conditions is amenable to analysis using spreadsheets. In this way, the repeated calculations required to define the total stress pore pressure and effective stress over a fine depth scale can be automated for rapid analysis.

The soil body in question is divided into thin slices, i.e. a one-dimensional finite difference grid[④]. The total stress due to each slice is then calculated as the product of its thickness and saturated or dry unit weight as appropriate. The total stress at any depth is then found as the sum of the total stress increments due to each of the slices above that depth. By subdividing the problem in this way, it is straightforward to tackle even problems with many layers of different soils. Because the pore fluid is continuous through the interconnected pore structure, the pore water pressure at any depth can be found from the depth below the water table (hydrostatic distribution[⑤]), suitably modified to account for artesian or seepage-induced pressure[⑥] (see Chapter 2) as necessary. The effective stress at any depth is then found by a simple application of Equation 3.1.

A spreadsheet tool for undertaking such analysis, Stress_CSM8.xls, may be found on the Companion Website. This is able to address a wide range of problems including up to ten layers of soil with different unit weights, a variable water table (either below or above the ground surface), surcharge loading at the ground surface, and confined layers under artesian fluid pressure (see Section 2.1).

① 水位线。
② (应理解该算例)毛细水上升 1m 的效果就只是指2～3m 深度砂土的重度由 17kN/m^3 变为 20kN/m^3，增长了 3kN/m^3。因此 3m 以下土体的总应力和有效应力均提高 3kPa，但孔隙水压力不变。
③ 使用有限差分法的数值解法。
④ 一维差分网格。
⑤ 静水压力分布。
⑥ 承压水或渗流引起的孔压。

3.4 Response of effective stress to a change in total stress[⑦]

As an illustration of how effective stress responds to a change in total stress, consider the case of a fully saturated soil subject to an increase in total vertical stress $\Delta\sigma$ and in which the lateral strain is zero[⑧], volume change being entirely due to deformation of the soil in the vertical direction. This condition may be assumed in

⑦ 有效应力对于总应力变化的响应。
⑧ 无侧向应变。

① （大面积堆载条件）比所研究土层厚度大。
② 静孔隙水压力。
③ 由于孔隙水需要时间通过渗流排出，因此总应力增加后，孔隙水压力立即增加超过其静态值。
④ 超静孔隙水压力。
⑤ 水力梯度。
⑥ 消散。
⑦ 排水条件。
⑧ 不排水条件。
⑨ 需要注意"排水"一词不意味着所有水已流出土的孔隙，而是意味着孔隙水中没有由于应力增加（超静水压）导致的压力。在整个孔隙水消散过程中土体保持完全饱和状态。

practice when there is a change in total vertical stress over an area which is large compared with the thickness of the soil layer in question①.

It is assumed initially that the pore water pressure is constant at a value governed by a constant position of the water table. This initial value is called the **static pore water pressure**② (u_s). When the total vertical stress is increased, the solid particles immediately try to take up new positions closer together. However, if water is incompressible and the soil is laterally confined, no such particle rearrangement, and therefore no increase in the interparticle forces, is possible unless some of the pore water can escape. Since it takes time for the pore water to escape by seepage, the pore water pressure is increased above the static value immediately after the increase in total stress takes place.③ The component of pore water pressure above the static value is known as the **excess pore water pressure**④ (u_e). This increase in pore water pressure will be equal to the increase in total vertical stress, i. e. the increase in total vertical stress is carried initially entirely by the pore water ($u_e = \Delta\sigma$). Note that if the lateral strain were not zero, some degree of particle rearrangement would be possible, resulting in an immediate increase in effective vertical stress, and the increase in pore water pressure would be less than the increase in total vertical stress by Terzaghi's Principle.

The increase in pore water pressure causes a hydraulic pressure gradient⑤, resulting in transient flow of pore water (i. e. seepage, see Chapter 2) towards a free-draining boundary of the soil layer. This flow or drainage will continue until the pore water pressure again becomes equal to the value governed by the position of the water table, i. e. until it returns to its static value. It is possible, however, that the position of the water table will have changed during the time necessary for drainage to take place, so that the datum against which excess pore water pressure is measured will have changed. In such cases, the excess pore water pressure should be expressed with reference to the static value governed by the new water table position. At any time during drainage, the overall pore water pressure (u) is equal to the sum of the static and excess components, i. e.

$$u = u_s + u_e \tag{3.4}$$

The reduction of excess pore water pressure as drainage takes place is described as **dissipation**⑥, and when this has been completed (i. e. when $u_e = 0$ and $u = u_s$) the soil is said to be in the **drained** condition⑦. Prior to dissipation, with the excess pore water pressure at its initial value, the soil is said to be in the **undrained** condition⑧. It should be noted that the term 'drained' does not mean that all of the water has flowed out of the soil pores; it means that there is no stress-induced (excess) pressure in the pore water. The soil remains fully saturated throughout the process of dissipation.⑨

As drainage of pore water takes place the solid particles become free to take up new positions, with a resulting increase in the interparticle forces. In other words, as the excess pore water pressure dissipates, the effective vertical stress increases, accompanied by a corresponding reduction in volume. When dissipation of excess pore water pressure is complete, the increment of total vertical stress will be carried entirely by the soil skeleton. The time taken for drainage to be completed depends on the permeability[①] of the soil. In soils of low permeability, drainage will be slow; in soils of high permeability, drainage will be rapid. The whole process is referred to as **consolidation**[②]. With deformation taking place in one direction only (vertical as described here), consolidation is described as one-dimensional. This process will be described in greater detail in Chapter 4.

① 渗透性。

② 固结。

When a soil is subject to a reduction in total normal stress the scope for volume increase is limited, because particle rearrangement due to total stress increase is largely irreversible[③]. As a result of increase in the interparticle forces there will be small elastic strains (normally ignored) in the solid particles, especially around the contact areas, and if clay mineral particles are present in the soil they may experience bending. In addition, the adsorbed water surrounding clay mineral particles will experience recoverable compression due to increases in interparticle forces, especially if there is face-to-face orientation of the particles[④]. When a decrease in total normal stress takes place in a soil there will thus be a tendency for the soil skeleton to expand to a limited extent, especially so in soils containing an appreciable proportion of clay mineral particles. As a result, the pore water pressure will initially be reduced and the excess pore water pressure will be negative[⑤]. The pore water pressure will gradually increase to the static value, flow taking place into the soil, accompanied by a corresponding reduction in effective normal stress and increase in volume. This process is known as **swelling**[⑥].

③ 不可恢复。

④ 颗粒面-面接触。

⑤ 超孔隙水压力为负值。

⑥ 回弹。

Under seepage (as opposed to static) conditions, the excess pore water pressure due to a change in total stress is the value above or below the **steady-state seepage pore water pressure**[⑦] (u_{ss}), which is determined, at the point in question, from the appropriate flow net[⑧] (see Chapter 2).

⑦ 稳态渗流孔隙水压力。

⑧ 流网。

Consolidation analogy[⑨]

⑨ 固结的模拟模型。

The mechanics of the one-dimensional consolidation[⑩] process can be represented by means of a simple analogy. Figure 3.3 (a) shows a spring inside a cylinder filled with water, and a piston, filled with a valve, on top of the spring. It is assumed that there can be no leakage between the piston and the cylinder, and no friction. The spring represents the compressible soil skeleton, the water in the cylinder the pore water, and the bore diameter of the valve the permeability of the soil. The cylinder itself simulates the condition of no lateral strain in the soil. [⑪]

⑩ 一维固结。

⑪ 弹簧代表可压缩的土骨架，圆柱中的水代表孔隙水，孔径代表土的渗透性，圆柱本身模拟土的无侧向应变条件。

Suppose a load is now placed on the piston with the valve closed, as in Figure 3.3(b). Assuming water to be incompressible, the piston will not move as long as the valve is closed, with the result that no load can be transmitted to the spring; the load will be carried by the water, the increase in pressure in the water being equal to the load divided by the piston area. This situation with the valve closed corresponds to the undrained condition in the soil.

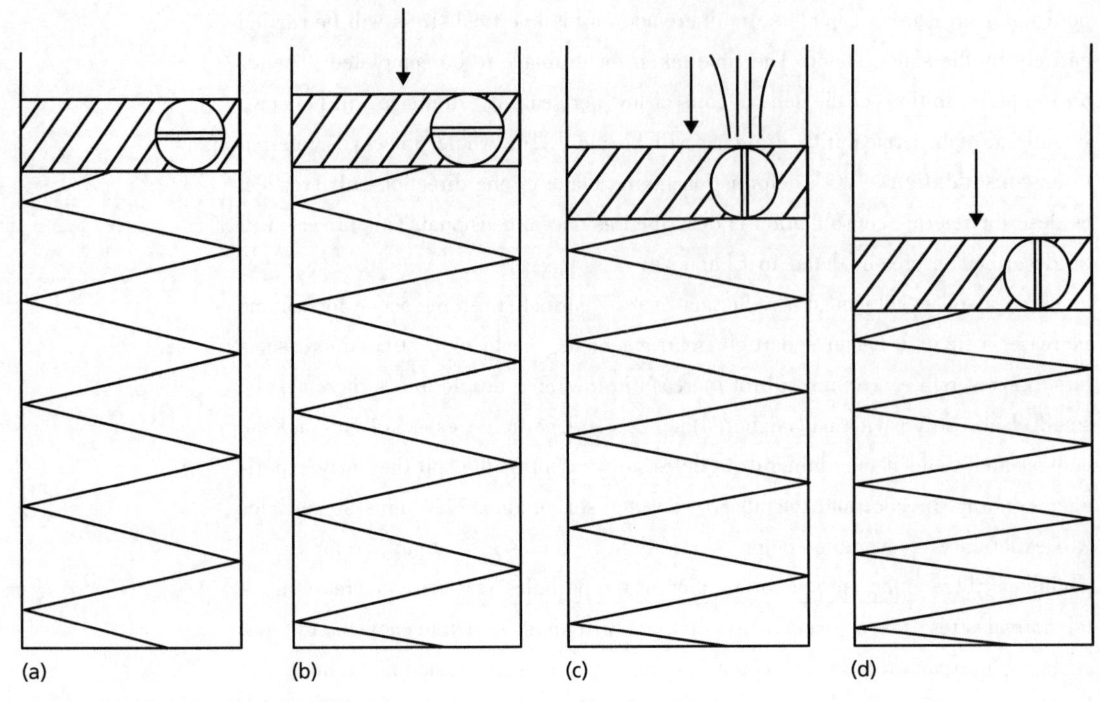

Figure 3.3 Consolidation analogy.

If the valve is now opened, water will be forced out through the valve at a rate governed by the bore diameter. This will allow the piston to move and the spring to be compressed as load is gradually transferred to it. This situation is shown in Figure 3.3 (c). At any time, the increase in load on the spring will be directly proportional to the reduction in pressure in the water. Eventually, as shown in Figure 3.3 (d), all the load will be carried by the spring and the piston will come to rest, this corresponding to the drained condition in the soil. At any time, the load carried by the spring represents the effective normal stress in the soil, the pressure of the water in the cylinder represents the pore water pressure, and the load on the piston represents the total normal stress. The movement of the piston represents the change in volume of the soil, and is governed by the compressibility of the spring (the equivalent of the compressibility of the soil skeleton①, see Chapter 4). The piston and spring analogy represents only an element of soil, since the stress conditions vary from point to point throughout a soil mass.②

① 土骨架的压缩性。
② 由于土体中的应力条件处处不同，活塞-弹簧模拟模型只能代表土中一点(即土单元)。

Example 3.2

A 5-m depth of sand overlies a 6-m thick layer of clay, the water table being at the surface; the permeability of the clay is very low. The saturated unit weight of the sand is 19kN/m^3 and that of the clay is 20kN/m^3. A 4-m depth of fill material of unit weight 20kN/m^3 is placed on the surface over an extensive area. Determine the effective vertical stress at the centre of the clay layer (a) immediately after the fill has been placed, assuming this to take place rapidly, and (b) many years after the fill has been placed.

Solution

The soil profile is shown in Figure 3.4. Since the fill covers an extensive area, it can be assumed that the condition of zero lateral strain applies. As the permeability of the clay is very low, dissipation of excess pore water pressure will be very slow; immediately after the rapid placement of the fill, no appreciable dissipation will have taken place. The initial stresses and pore water pressure at the centre of the clay layer are

$$\sigma_v{}^{①} = (5 \times 19) + (3 \times 20) = 155\text{kPa}$$

$$u_s{}^{②} = 8 \times 9.81 = 78\text{kPa}$$

$$\sigma'_v{}^{③} = \sigma_v - u_s = 77\text{kPa}$$

Many years after placement of the fill, dissipation of excess pore water pressure should be essentially complete such that the increment of total stress from the fillis entirely carried by the soil skeleton (effective stress). The effective vertical stress at the centre of the clay layer is then

$$\sigma'_v{}^{④} = 77 + (4 \times 20) = 157\text{kPa}$$

Immediately after the fill has been placed, the total vertical stress at the centre of the clay increases by 80kPa due to the weight of the fill. Since the clay is saturated and there is no lateral strain, there will be a corresponding increase in pore water pressure of $u_e = 80$kPa (the initial excess pore water pressure). The static pore water pressure was calculated previously as $u_s = 78$kPa. Immediately after placement, the pore water pressure therefore increases from 78 to 158kPa, and then during subsequent consolidation gradually decreases again to 78kPa, accompanied by the gradual increase of effective vertical stress from 77 to 157kPa.

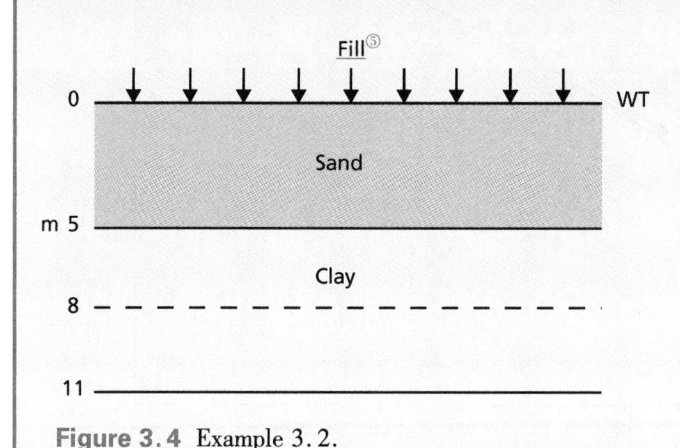

Figure 3.4 Example 3.2.

① 初始竖向总应力。
② 孔隙水压力。
③ 竖向有效应力。
④ 最终竖向有效应力。
⑤ 填土。

3.5 Effective stress in partially saturated soils⑥

⑥ 非饱和土的有效应力。

In the case of partially saturated soils, part of the void space is occupied by water and part by air. The pore water pressure (u_w) must always be less than the pore air pressure (u_a) due to surface tension.⑦ Unless the degree of saturation is close to unity, the pore air will form continuous channels through the soil and the pore water will be concentrated in the regions around the interparticle contacts. The bound-

⑦ 由于表面张力的存在，孔隙水压力(u_w)总是小于孔隙气压力(u_a)。

aries between pore water and pore air will be in the form of menisci whose radii will depend on the size of the pore spaces within the soil.[①] Part of any wavy plane through the soil will therefore pass part through water, and part through air.

① 孔隙气与孔隙水之间的界面边界为新月状，其半径取决于土内孔隙空间的尺寸。

Bishop (1959) proposed the following effective stress equation for partially saturated soils:

$$\sigma = \sigma' + u_a - \chi(u_a - u_w) \tag{3.5}$$ [②]

② Bishop 非饱和土有效应力方程。

where χ is a parameter, to be determined experimentally, related primarily to the degree of saturation of the soil. The term $(\sigma' - u_a)$ is also described as the **net stress**[③], while $(u_a - u_w)$ is a measure of the suction in the soil[④].

③ 净应力。

④ 土体吸力。

Vanapalli and Fredlund (2000) conducted a series of laboratory triaxial tests (this test is described in Chapter 5) on five different soils, and found that

$$\chi = (S_r)^{\kappa} \tag{3.6}$$

where κ is a fitting parameter that is predominantly a function of plasticity index (I_P) as shown in Figure 3.5. Note that for non-plastic (coarse) soils[⑤], $\kappa = 1$. For a fully saturated soil ($S_r = 1$), $\chi = 1$; and for a completely dry soil ($S_r = 0$), $\chi = 0$. Equation 3.5 thus reduces to Equation 3.1 when $S_r = 1$. The value of χ is also influenced, to a lesser extent, by the soil structure[⑥] and the way the particular degree of saturation[⑦] was brought about.

⑤ 非塑性土（或粗粒土）。

⑥ 土体结构。

⑦ 达到某一饱和度的方式。

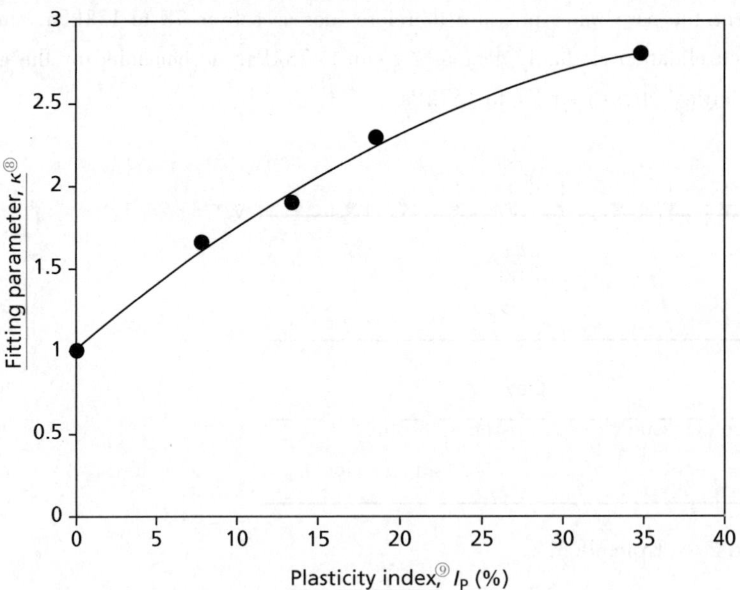

Figure 3.5 Relationship of fitting parameter κ to soil plasticity[⑩] (re-plotted after Vanapalli and Fredlund, 2000).

⑧ 拟合参数。

⑨ 塑性指数。

⑩ 拟合参数与土体塑性指数的关系图。

A physical model may be considered in which the parameter χ is interpreted as the average proportion of any cross-section which passes through water. Then, across a given section of gross area A (Figure 3.6), total force is given by the equation

$$\sigma A = \sigma' A + u_w \chi A + u_a(1 - \chi)A \tag{3.7}$$

which leads to Equation 3.5.

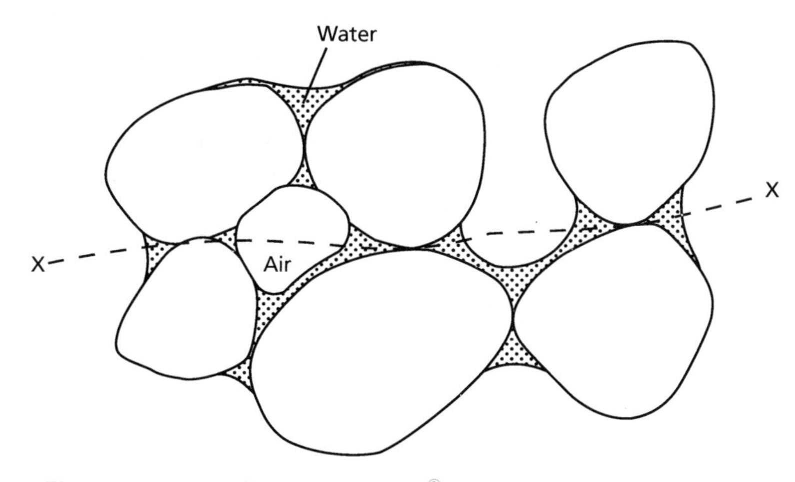

Figure 3.6 Partially saturated soil.[①]

① 非饱和土。

If the degree of saturation of the soil is close to unity, it is likely that the pore air will exist in the form of bubbles within the pore water and it is possible to draw a wavy plane through pore water only. The soil can then be considered as a fully saturated soil, but with the pore water having some degree of compressibility due to the presence of the air bubbles[②]. This is reasonable for most common applications in temperate climates (such as the UK). Equation 3.1 may then represent effective stress with sufficient accuracy for most practical purposes. A notable application where it is of greater importance to understand the behaviour of partially saturated soil is the stability of slopes under seasonal groundwater changes[③].

② 气泡。

③ 季节性水位变化。

3.6 Influence of seepage on effective stress[④]

④ 渗流对有效应力的影响。

When water is seeping through the pores of a soil, total head[⑤] is dissipated as viscous friction producing a frictional drag, acting in the direction of flow, on the solid particles. A transfer of energy thus takes place from the water to the solid particles, and the force corresponding to this energy transfer is called **seepage force**[⑥]. Seepage force acts on the particles of a soil in addition to gravitational force, and the combination of the forces on a soil mass due to gravity and seeping water is called the **resultant body force**[⑦]. It is the resultant body force that governs the effective normal stress on a plane within a soil mass through which seepage is taking place.[⑧]

⑤ 总水头。

⑥ 渗流力。

⑦ 体力合力。

⑧ 体力合力控制着土体内某一渗流平面上的有效正应力。

Consider a point in a soil mass where the direction of seepage is at angle θ below the horizontal. A square element ABCD of dimension L (unit dimension normal to the page) is centred at the above point with sides parallel and normal to the direction of seepage, as shown in Figure 3.7 (a) – i. e. the square element can be considered as a flow net element (curvilinear square). Let the drop in total head between the sides AD and BC be Δh. Consider the pore water pressures on the boundaries of the element, taking the value of pore water pressure at point A as

u_A. The difference in pore water pressure between A and D is due only to the difference in elevation head[①] between A and D, the total head[②] being the same at A and D. However, the difference in pore water pressure between A and either B or C is due to the difference in elevation head and the difference in total head between A and either B or C. If the total head at A is h_A and the elevation head at the same point is z_A, applying Equation 2.1 gives:

① 位置(高程)水头。
② 总水头。

$$u_A = [h_A - z_A]\gamma_w$$
$$u_B = [(h_A - \Delta h) - (z_A - L\sin\theta)]\gamma_w$$
$$u_C = [(h_A - \Delta h) - (z_A - L\sin\theta - L\cos\theta)]\gamma_w$$
$$u_D = [(h_A) - (z_A - L\cos\theta)]\gamma_w$$

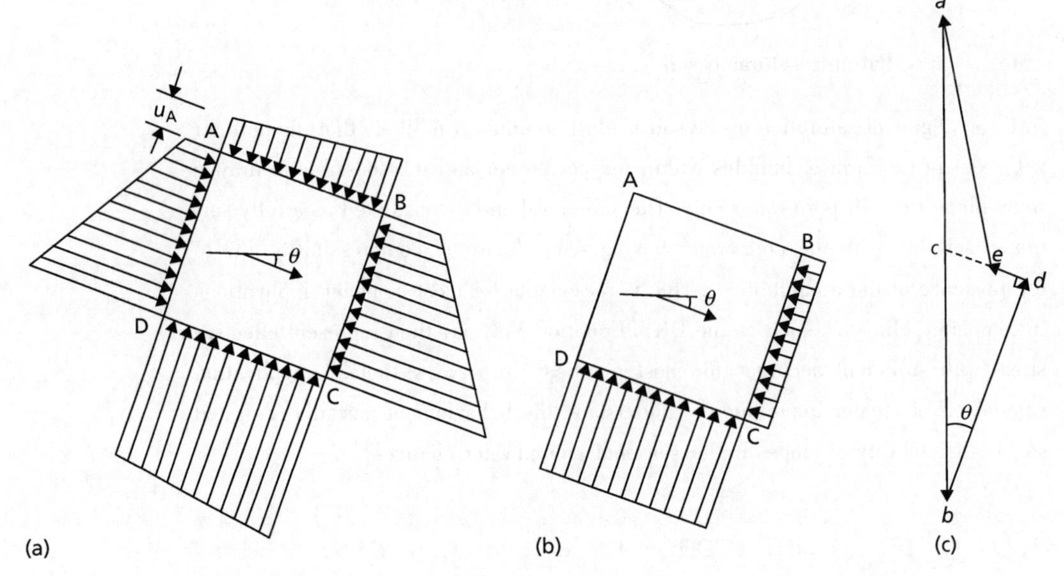

Figure 3.7 Forces under seepage conditions[③] (reproduced after D. W. Taylor (1948) *Fundamentals of Soil Mechanics*, © John Wiley & Sons Inc., by permission).

③ 渗流条件下的力。

The following pressure differences can now be established:

$$u_B - u_A = u_C - u_D = [-\Delta h + L\sin\theta]\gamma_w$$
$$u_D - u_A = u_C - u_B = [L\cos\theta]\gamma_w$$

These values are plotted in Figure 3.7 (b), giving the distribution diagrams of net pressure across the element in directions parallel and normal to the direction of flow as shown.

Therefore, the force on BC due to pore water pressure acting on the boundaries of the element, called the boundary water force[④], is given by

④ 边界水压力。

$$\gamma_w(-\Delta h + L\sin\theta)L$$

or

$$\gamma_w L^2\sin\theta - \Delta h\gamma_w L$$

and the boundary water force on CD by

$$\gamma_w L^2\cos\theta$$

If there were no seepage, i.e. if the pore water were static, the value of Δh would be zero, the forces on BC and CD would be $\gamma_w L^2 \sin\theta$ and $\gamma_w L^2 \cos\theta$, respectively, and their resultant would be $\gamma_w L^2$ acting in the vertical direction. The force $\Delta h\ \gamma_w L$ represents the only difference between the static and seepage cases and is therefore the seepage force (J), acting in the direction of flow (in this case normal to BC).

Now, the average hydraulic gradient[①] across the element is given by

$$i = \frac{\Delta h}{L}$$

hence,

$$J = \Delta h \gamma_w L = \frac{\Delta h}{L}\gamma_w L^2 = i\gamma_w L^2$$

or

$$J = i\gamma_w V \tag{3.8}$$

where V is the volume of the soil element.

The seepage pressure (j) is defined as the seepage force per unit volume[②], i. e.

$$j = i\gamma_w \tag{3.9}$$

It should be noted that j (and hence J) depends only on the value of the hydraulic gradient.

All the forces, both gravitational and forces due to seeping water, acting on the element ABCD, may be represented in the vector diagram shown in Figure 3.7 (c). The magnitude of the forces shown are summarised below.

Total weight of the element $= \gamma_{sat} L^2 =$ vector *ab*
Boundary water force[③] on CD $= \gamma_w L^2 \cos\theta =$ vector *bd*
Boundary water force on BC $= \gamma_w L^2 \sin\theta - \Delta h \gamma_w L =$ vector *de*
Resultant body force = vector *ea*

The resultant body force can be obtained as the vector summation of $ab + bd + de$, as shown in Figure 3.7 (c). The seepage force $J = \Delta h \gamma_w L$ is represented by the dashed line in Figure 3.7 (c), i. e. vector *ce*. Considering the triangle cbd:

$$bc = \gamma_w L^2$$

$$ac = ab - bc = (\gamma_{sat} - \gamma_w) L^2 = \gamma' L^2$$

Applying the cosine rule to triangle *ace*, and recognising that angle $ace = 90 + \theta$ degrees, it can be shown that the magnitude of the resultant body force (length of $|ea|$) is given by:

$$|ea| = \sqrt{(\gamma' L^2)^2 + (\Delta h \gamma_w L)^2 + (2\gamma' L^3 \Delta h \gamma_w \sin\theta)} \tag{3.10}$$

This resultant body force acts at an angle to the vertical described by angle *bae*. Applying the sine rule to triangle *ace* gives:

$$\angle bae = \sin^{-1}\left(\frac{\Delta h \gamma_w L}{ea}\cos\theta\right) \tag{3.11}$$

Only the resultant body force contributes to effective stress.[④] A component of seepage force acting vertically upwards will therefore reduce a vertical effective stress component from the static value. A component of seepage force acting verti-

① 平均水力梯度。

② 渗透压力定义为单位体积土中的渗透力。

③ 边界水压力。

④ 只有体力合力对有效应力才有贡献。

cally downwards will increase a vertical effective stress component from the static value.[1]

① 竖直向上的渗流力分量将减小竖向有效应力，竖直向下的渗流力分量将增大竖向有效应力。

3.7 Liquefaction

Seepage-induced liquefaction[2]

② 渗流导致的液化。

Consider the special case of seepage vertically upwards ($\theta = -90°$). The vector *ce* in Figure 3.7 (c) would then be vertically upwards, and if the hydraulic gradient were high enough the resultant body force would be zero. The value of hydraulic gradient corresponding to zero resultant body force is called the **critical hydraulic gradient**[3] (i_{cr}). Substituting $|ea| = 0$ and $\theta = -90°$ into Equation 3.10 gives

③ 临界水力梯度。

$$0 = (\gamma' L^2)^2 + (\Delta h \gamma_w L)^2 - 2(\gamma' L^3 \Delta h \gamma_w)$$

$$= (\gamma' L^2 - \Delta h \gamma_w L)^2$$

$$\therefore i_{cr} = \frac{\Delta h}{L} = \frac{\gamma'}{\gamma_w}$$

Therefore

$$i_{cr} = \frac{\gamma'}{\gamma_w} = \frac{G_s - 1}{1 + e} \qquad (3.12)$$

The ratio γ'/γ_w, and hence the critical hydraulic gradient, is approximately 1.0 for most soils.

When the hydraulic gradient is i_{cr}, the effective normal stress on any plane will be zero, gravitational forces having been cancelled out by upward seepage forces.[4] In the case of sands, the contact forces between particles will be zero and the soil will have no strength.[5] The soil is then said to be **liquefied**[6], and if the critical gradient is exceeded the surface will appear to be 'boiling'[7] as the particles are moved around in the upward flow of water. It should be realised that 'quicksand' is not a special type of soil, but simply sand through which there is an upward flow of water under a hydraulic gradient equal to or exceeding i_{cr}. In the case of clays, liquefaction may not necessarily result when the hydraulic gradient reaches the critical value given by Equation 3.12.

④ 当水力梯度为临界水力梯度（i_{cr}）时，在任意平面上的有效正应力将为零，重力已被渗流力抵消。

⑤ 此时，砂土颗粒无接触力，强度为零。

⑥ 液化的。

⑦ 砂沸。

Conditions adjacent to sheet piling[8]

⑧ 悬臂式板桩附近的渗流情况。

High upward hydraulic gradients may be experienced in the soil adjacent to the downstream face of a sheet pile wall. Figure 3.8 shows part of the flow net for seepage under a sheet pile wall, the embedded length[9] on the downstream side being *d*. A mass of soil adjacent to the piling may become unstable and be unable to support the wall. Model tests have shown that failure is likely to occur within a soil mass of approximate dimensions $d \times d/2$ in section (ABCD in Figure 3.8). Failure first shows in the form of a rise or **heave**[10] at the surface, associated with an expansion of the soil which results in an increase in permeability. This in turn leads to increased flow, surface 'boiling' in the case of sands, and complete failure of the wall.

⑨ 插入深度。

⑩ 隆起。

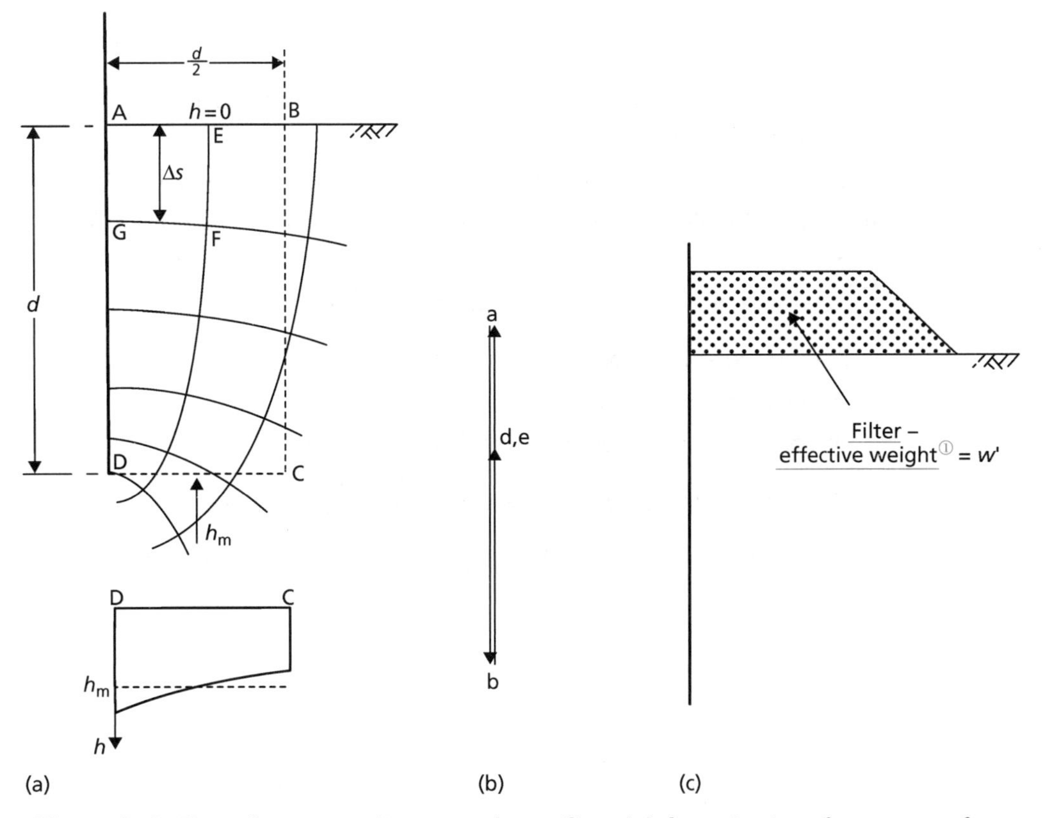

Figure 3.8 Upward seepage adjacent to sheet piling: (a) determination of parameters from flow net, (b) force diagram, and (c) use of a filter to suppress heave.②

① 反滤层的有效重力。

② 悬臂式板桩附近的向上渗流：(a) 由流网确定参数，(b) 受力图，(c) 应用滤层防隆起。

The variation of total head on the lower boundary CD of the soil mass can be obtained from the flow net equipotentials, or directly from the results of analysis using the FDM, such as the spreadsheet analysis tool described in Chapter 2. For the purpose of analysis, however, it is usually sufficient to determine the average total head h_m by inspection. The total head on the upper boundary AB is zero. The average hydraulic gradient is given by

$$i_m = \frac{h_m}{d}$$

Since failure due to heaving may be expected when the hydraulic gradient becomes i_{cr}, the factor of safety (F) against heaving may be expressed as

$$F = \frac{i_{cr}}{i_m} \tag{3.13}$$

In the case of sands, a factor of safety can also be obtained with respect to 'boiling' at the surface. The exit hydraulic gradient③ (i_e) can be determined by measuring the dimension Δs of the flow net field AEFG adjacent to the piling:

③ 渗流溢出处的水力梯度。

$$i_e = \frac{\Delta h}{\Delta s}$$

where Δh is the drop in total head between equipotentials④ GF and AE. Then, the factor of safety is

④ 等势线。

$$F = \frac{i_{cr}}{i_e} \tag{3.14}$$

There is unlikely to be any appreciable difference between the values of F given by Equations 3.13 and 3.14.

The sheet pile wall problem shown in Figure 3.8 can also be used to illustrate the graphical method for determining seepage forces outlined in Section 3.6:

Total weight of mass ABCD = vector $ab = \frac{1}{2}\gamma_{sat}d^2$

Average total head on CD = h_m

Elevation head on CD = $-d$

Average pore water pressure on CD = $(h_m + d)\gamma_w$

Boundary water force on CD = vector $bd = \frac{d}{2}(h_m + d)\gamma_w$

Boundary water force on BC = vector $de = 0$ (as seepage is vertically upwards).

Resultant body force of ABCD = $ab - bd - de$

$$= \frac{1}{2}\gamma_{sat}d^2 - \frac{d}{2}(h_m + d)\gamma_w - 0$$

$$= \frac{1}{2}(\gamma' + \gamma_w)d^2 - \frac{1}{2}(h_m d + d^2)\gamma_w$$

$$= \frac{1}{2}\gamma' d^2 - \frac{1}{2}h_m\gamma_w d$$

The resultant body force will be zero, leading to heave, when

$$\frac{1}{2}h_m\gamma_w d = \frac{1}{2}\gamma' d^2$$

The factor of safety can then be expressed as

$$F = \frac{\frac{1}{2}\gamma' d^2}{\frac{1}{2}h_m\gamma_w d} = \frac{\gamma' d}{h_m\gamma_w} = \frac{i_{cr}}{i_m}$$

① 抗隆起安全系数。

If the factor of safety against heave[①] is considered inadequate, the embedded length d may be increased or a surcharge load in the form of a filter may be placed on the surface AB, the filter being designed to prevent entry of soil particles, following the recommendations of Section 2.10. Such a filter is shown in Figure 3.8 (c). If the effective weight of the filter per unit area is w', then the factor of safety becomes

$$F = \frac{\gamma' d + w'}{h_m\gamma_w}$$

Example 3.3

The flow net for seepage under a sheet pile wall is shown in Figure 3.9 (a), the saturated unit weight of the soil being 20kN/m^3. Determine the values of effective vertical stress at points A and B.

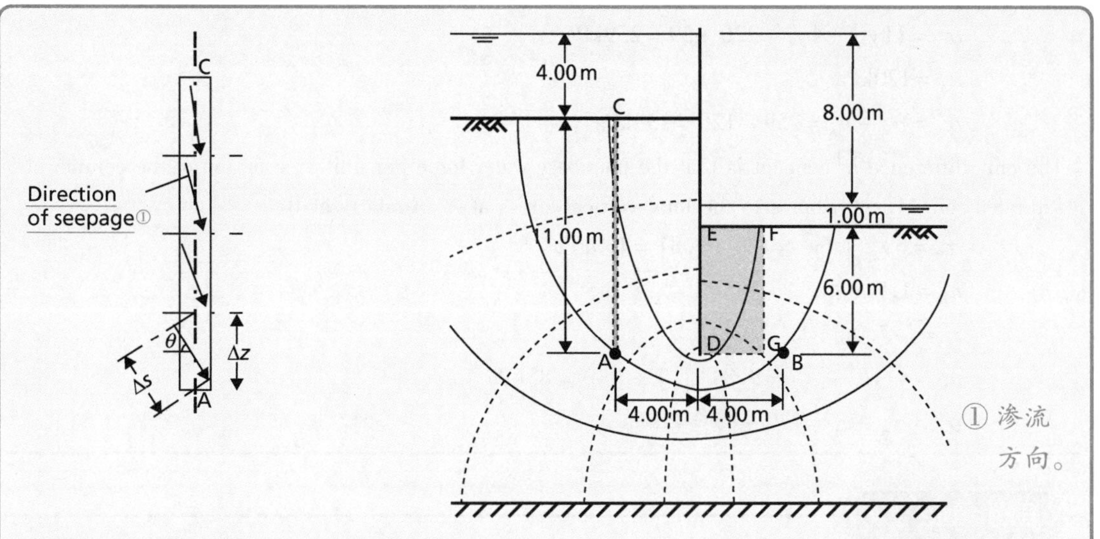

① 渗流方向。

Figure 3.9 Example 3.3 and 3.4.

Solution

First, consider the column of saturated soil of unit area between A and the soil surface at C. The total weight of the column is $11\gamma_{sat}$(220kN). Due to the change in level of the equipotentials across the column, the boundary water forces on the sides of the column will not be equal, although in this case the difference will be small. There is thus a net horizontal boundary water force on the column. However, as the effective vertical stress is to be calculated, only the vertical component of the result-ant body force is required and the net horizontal boundary water force need not be considered. The vertical component of the boundary water force on the top surface of the column is due only to the depth of water above C, and is $4\gamma_w$ (39kN). The boundary water force on the bottom surface of the column must be determined; in this example the FDM spreadsheet described in Section 2.7 is used as the required values of total head can be directly read from the spreadsheet, though this may equally be determined from a hand-drawn flow net. The calculated total head distribution may be found in Seepage_CSM8.xls on the Companion Website.

Total head at A, $h_A = 5.2$m

Elevation head at A, $zh_A = -7.0$m

Pore water pressure at A, $u_A = \gamma_w (h_A - z_A) = 9.81\ (5.2 + 7.0) =$ 120kPa

i.e. boundary water force on bottom surface = 120kN

Net vertical boundary water force = 120 − 39 = 81kN

Total weight of the column = 220kN

Vertical component of resultant body force = 220 − 81 = 139kN

i.e. effective vertical stress at A = 139kPa. ②

② 注：此类题目一般先基于流网求总水头，再求位置水头，最后反算孔隙水压力。

It should be realised that the same result would be obtained by the direct application of the effective stress equation, the total vertical stress at A being the weight of saturated soil and water, per unit area, above A. Thus

$$\sigma_A = 11\gamma_{sat} + 4\gamma_w = 220 + 39 = 259\text{kPa}$$

$$u_A = 120\text{kPa}$$

$$\sigma'_A = \sigma_A - u_A = 259 - 120 = 139\text{kPa}$$

The only difference in concept is that the boundary water force per unit area on top of the column of saturated soil AC contributes to the total vertical stress at A. Similarly at B

$$\sigma_B = 6\gamma_{sat} + 1\gamma_w = 120 + 9.81 = 130\text{kPa}$$

$$h_B = 1.7\text{m}$$

$$z_B = -7.0\text{m}$$

$$u_B = \gamma_w (h_B - z_B) = 9.81\,(1.7 + 7.0) = 85\text{kPa}$$

$$\sigma'_B = \sigma_B - u_B = 130 - 85 = 45\text{kPa}$$

Example 3.4

Using the flow net in Figure 3.9 (a), determine the factor of safety against failure by heaving adjacent to the downstream face of the piling. The saturated unit weight of the soil is 20kN/m^3.

Solution

The stability of the soil mass DEFG in Figure 3.9 (a), 6m by 3m in section, will be analysed. By inspection of the flow net or from Seepage_CSM8.xls on the Companion Website, the average value of total head on the base DG is given by

$$h_m = 2.6\text{m}$$

The average hydraulic gradient between DG and the soil surface EF is

$$i_m = \frac{2.6}{6} = 0.43$$

Critical hydraulic gradient, $i_{cr} = \dfrac{\gamma'}{\gamma_w} = \dfrac{10.2}{9.8} = 1.04$

Factor of safety, $F = \dfrac{i_{cr}}{i_m} = \dfrac{1.04}{0.43} = 2.4$

Dynamic/seismic liquefaction[1]

In the previous examples, seepage-induced or **static liquefaction**[2] of soil has been discussed – that is to say, situations where the effective stress within the soil is reduced to zero as a result of high pore water pressures due to seepage. Pore water pressure may also be increased due to dynamic loading of soil. As soil is sheared cyclically it has a tendency to contract[3], reducing the void ratio *e*. If this shearing and resulting contraction happens rapidly, then there may not be sufficient time for the pore water to escape from the voids, such that the reduction in volume will lead to an increase in pore water pressure due to the incompressibility of water.[4]

Consider a uniform layer of fully saturated soil with the water table at the surface. The total stress at any depth *z* within the soil is

$$\sigma_v = \gamma_{sat} z$$

① 动力/地震液化。
② 静态液化。
③ 体积缩小。
④ 如果剪切导致收缩迅速发生，没有足够时间让孔隙水消散，那么由于水的不可压缩性，土体体积的减少将导致孔隙水压力上升。

The soil will liquefy at this depth when the effective stress becomes zero. By Terzaghi's Principle (Equation 3.1) this will occur when $u = \sigma_v$. From Equation 3.4, the pore water pressure u is made up of two components: hydrostatic pressure[①] u_s (present initially before the soil is loaded) and an excess component u_e (which is induced by the dynamic load[②]). Therefore, the critical excess pore water pressure at the onset of liquefaction (u_{eL}) is given by

① 静水压力。

② 动荷载。

$$u = \sigma_v$$

$$u_s + u_{eL} = \gamma_{sat} z$$

$$\gamma_w z + u_{eL} = \gamma_{sat} z$$

$$u_{eL} = \gamma' z \tag{3.15}$$

i. e. for soil to liquefy, the excess pore water pressure must be equal to the initial effective stress in the ground (prior to application of the dynamic load). Furthermore, considering a datum at the surface of the soil, from Equation 2.1

$$u = \gamma_w (h + z)$$

$$\therefore h = \frac{(\gamma_{sat} - \gamma_w)}{\gamma_w} z$$

$$\frac{h}{z} = \frac{\gamma'}{\gamma_w}$$

This demonstrates that there will be a positive hydraulic gradient h/z between the soil at depth z and the surface (i. e. vertically upwards), when liquefaction has been achieved.[③] This is the same as the critical hydraulic gradient defined by Equation 3.12 for seepage-induced liquefaction.

③ 这说明：液化时，z 深度处的土与地面间将存在正的（即垂直向上的）水力梯度 h/z。

Excess pore water pressure rise due to volumetric contraction may be induced by vibrating loads from direct sources on or in the soil, or may be induced due to cyclic ground motion during an earthquake. An example of the former case is a shallow foundation for a piece of machinery such as a power station turbine. In this case, the cyclic straining (and therefore volumetric contraction and induced excess pore pressure) in the soil will generally decrease with distance from the source. From Equation 3.15, it can be seen that the amount of excess pore water pressure required to initiate liquefaction increases with depth[④]. As a result, any liquefaction will be concentrated towards the surface of the ground, close to the source.

④ 由式（3.15）可知，随深度增加，发生液化需要的超静水压力越高。

In the case of an earthquake, ground motion is induced as a result of powerful stress waves which are transmitted from within the Earth's crust (i. e. far beneath the soil). As a result, liquefaction may extend to much greater depths. Combining Equations 3.12 and 3.15

$$u_{eL} = i_{cr} \gamma_w z = \frac{\gamma_w (G_s - 1)}{1 + e} z \tag{3.16}$$

The soil towards the surface is often at a lower density (higher e) than the soil beneath.[⑤] Combined with the shallow depth z, it is clear that, under earthquake shaking, liquefaction will start at the ground surface and move downwards as shaking continues, requiring larger excess pore water pressures at depth to liquefy the deeper layers. It is also clear from Equation 3.16 that looser soils at high e will require

⑤ 地表土体通常比其下土体的密度更低（孔隙比更高）。

lower excess pore water pressures to cause liquefaction.[①] Soils with high voids ratio also have a higher potential for densification[②] when shaken (towards e_{min}, the densest possible state), so loose soils are particularly vulnerable to liquefaction. Indeed, strong earthquakes may fully liquefy layers of loose soil many metres thick.

① 式(3.16)表明，高孔隙比的松散土发生液化的超静孔压较低。

② 变密实。

It will be demonstrated in Chapter 5 that the shear strength of a soil (which resists applied loads due to foundations and other geotechnical constructions) is proportional to the effective stresses within the ground.[③] It is therefore clear that the occurrence of liquefaction ($\sigma_v' = 0$) can lead to significant damage to structures – an example, observed during the 1964 Niigata earthquake in Japan, is shown in Figure 3.10.

③ 如第5章阐述土体的抗剪强度与土中有效应力成正比。

Figure 3.10 Foundation failure due to liquefaction, 1964 Niigata earthquake, Japan.[④]

④ 液化导致的基础破坏，1964年日本新潟地震。

Summary

1 Total stress is used to define the applied stresses on an element of soil (both due to external applied loads and due to self weight). Soils support total stresses through a combination of effective stress due to interparticle contact and pore water pressure in the voids. This is known as Terzaghi's Principle[⑤] (Equation 3.1).

⑤ 太沙基原理。

2 Under hydrostatic conditions, the effective stress state at any depth within the ground can be found from knowledge of the unit weight of the soil layers and the location of the water table. If seepage is occurring, a flow net or finite difference mesh can be used to determine the pore water pressures at any point within the ground, with effective stress subsequently being found using Terzaghi's Principle.

3 A consequence of Terzaghi's Principle is that if there is significant excess pore water pressure developed in the ground, the soil skeleton may become unloaded (zero effective stress). This condition is known as liquefaction and may occur due to seepage, or due to dynamic external loads which cause the soil to contract rapidly. Seepage-induced liquefaction can lead to uplift or boiling of soil along the downstream face of a sheet piled excavation, and subsequent failure of the excavation. ①

① 渗流产生的液化可导致沿板桩支护开挖的基坑下游侧的土体发生上浮或砂沸，从而使基坑发生破坏。

Problems

3.1 A river is 2m deep. The river bed consists of a depth of sand of saturated unit weight $20kN/m^3$. What is the effective vertical stress 5m below the top of the sand?

3.2 The North Sea is 200m deep. The sea bed consists of a depth of sand of saturated unit weight $20kN/m^3$. What is the effective vertical stress 5m below the top of the sand? Compare your answer to the value found in 3.1 – how does the water level above the ground surface affect the stresses within the ground?

3.3 A layer of clay 4m thick lies between two layers of sand each 4m thick, the top of the upper layer of sand being ground level. The water table is 2m below ground level but the lower layer of sand is under artesian pressure, the piezometric surface being 4m above ground level. The saturated unit weight of the clay is $20kN/m^3$ and that of the sand $19kN/m^3$; above the water table the unit weight of the sand is $16.5kN/m^3$. Calculate the effective vertical stresses at the top and bottom of the clay layer.

3.4 In a deposit of fine sand the water table is 3.5m below the surface, but sand to a height of 1.0m above the water table is saturated by capillary water; above this height the sand may be assumed to be dry. The saturated and dry unit weights, respectively, are 20 and $16kN/m^3$. Calculate the effective vertical stress in the sand 8m below the surface.

3.5 A layer of sand extends from ground level to a depth of 9 m and overlies a layer of clay, of very low permeability, 6m thick. The water table is 6m below the surface of the sand. The saturated unit weight of the sand is $19kN/m^3$ and that of the clay $20kN/m^3$; the unit weight of the sand above the water table is $16kN/m^3$. Over a short period of time the water table rises by 3m, and is expected to remain permanently at this new level. Determine the effective vertical stress at depths of 8 and 12m below ground level (a) immediately after the rise of the water table, and (b) several years after the rise of the water table.

3.6 An element of soil with sides horizontal and vertical measures 1m in each direction. Water is seeping through the element in a direction inclined upwards at 30° above the horizontal under a hydraulic gradient of 0.35. The saturated unit weight of the soil is 21kN/m^3. Draw a force diagram to scale showing the following: total and effective weights, resultant boundary water force, seepage force. What is the magnitude and direction of the resultant body force?

3.7 For the seepage situations shown in Figure 3.11, determine the effective normal stress on plane XX in each case, (a) by considering pore water pressure and (b) by considering seepage pressure. The saturated unit weight of the soil is 20kN/m^3.

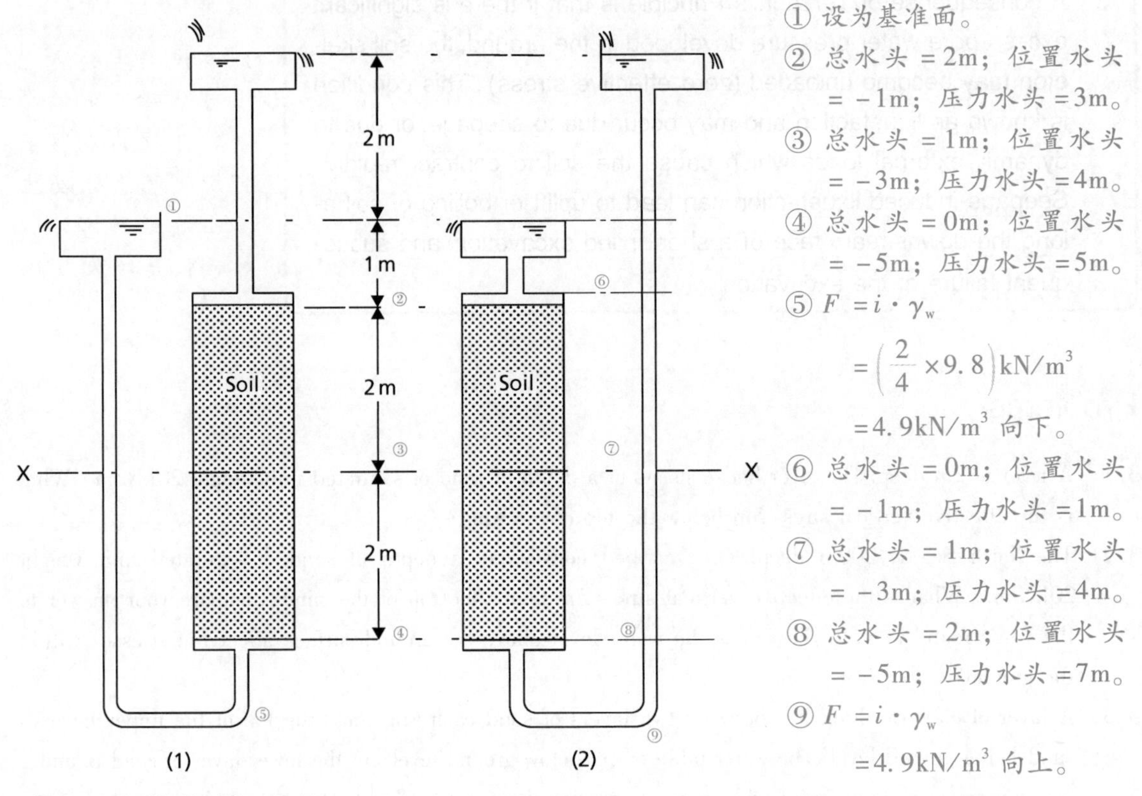

Figure 3.11 Problem 3.7.

3.8 The section through a long cofferdam is shown in Figure 2.27, the saturated unit weight of the soil being 20kN/m^3. Determine the factor of safety against 'boiling' at the surface AB, and the values of effective vertical stress at C and D.

3.9 The section through part of a cofferdam is shown in Figure 2.26, the saturated unit weight of the soil being 19.5kN/m^3. Determine the factor of safety against heave failure in the excavation adjacent to the sheet piling. What depth of filter (unit weight 21kN/m^3) would be required to ensure a factor of safety of 3.0?

References

Bishop, A. W. (1959) The principle of effective stress, *Tekniche Ukeblad*, **39**, 4 – 16.

Taylor, D. W. (1948) *Fundamentals of Soil Mechanics*, John Wiley & Sons, New York, NY.

Terzaghi, K. (1943) *Theoretical Soil Mechanics*, John Wiley & Sons, New York, NY.

Vanapalli, S. K. and Fredlund, D. G. (2000) Comparison of different procedures to predict unsaturated soil shear strength, *Proceedings of Geo-Denver* 2000, *ASCE Geotechnical Special Publication*, **99**, 195 – 209.

Further reading

Rojas, E. (2008a) Equivalent Stress Equation for Unsaturated Soils. I: Equivalent Stress, *International Journal of Geomechanics*, **8** (5), 285 – 290.

Rojas, E. (2008b) Equivalent Stress Equation for Unsaturated Soils. II: Solid-Porous Model, *International Journal of Geomechanics*, **8** (5), 291 – 299.

These companion papers describe in detail the development of a strength model for unsaturated soil, aiming to address one of the key questions in geotechnical engineering for which a satisfactory and widely accepted answer is still unclear.

Chapter 4

Consolidation[1]

① 固结。

Learning outcomes

After working through the material in this chapter, you should be able to:

1. Understand the behaviour of soil during consolidation (drainage of pore water pressure), and determine the mechanical properties which characterise this behaviour from laboratory testing (Sections 4.1 –4.2 and 4.6 –4.7);
2. Calculate ground settlements as a function of time due to consolidation both analytically and using computer-based tools for more complex problems (Sections 4.3 –4.5 and 4.8 –4.9);
3. Design a remedial scheme of vertical drains to speed-up consolidation and meet specified performance criteria (Section 4.10).

4.1 Introduction

As explained in Chapter 3, **consolidation** is the gradual reduction in volume of a fully saturated soil of low permeability due to change of effective stress.[2] This may be as a result of drainage of some of the pore water, the process continuing until the excess pore water pressure set up by an increase in total stress has completely dissipated; consolidation may also occur due to a reduction in pore water pressure, e. g. from groundwater pumping or well abstraction (see Example 4.5). The simplest case is that of one-dimensional consolidation, in which the stress increment is applied in one direction only (usually vertical) with a condition of zero lateral strain being implicit. The process of **swelling**[3], the reverse of consolidation, is the gradual increase in volume of a soil under negative excess pore water pressure.

② 低渗透性的完全饱和土由于有效应力改变而产生体积逐渐减少的现象称为固结。

③ 回弹，膨胀。

Consolidation settlement[4] is the vertical displacement of the soil surface corresponding to the volume change at any stage of the consolidation process. Consolidation settlement will result, for example, if a structure (imposing additional total stress) is built over a layer of saturated clay, or if the water table is lowered permanently in a stratum overlying a clay layer. On the other hand, if an excavation (re-

④ 固结沉降。

duction in total stress) is made in a saturated clay, **heave** (upward displacement) will result in the bottom of the excavation due to swelling of the clay.[①] In cases in which significant lateral strain takes place there will be an immediate settlement due to deformation of the soil under undrained conditions, in addition to consolidation settlement. The determination of immediate settlement will be discussed further in Chapter 8. This chapter is concerned with the prediction of both the magnitude and the rate of consolidation settlement under one-dimensional conditions (i. e. where the soil deforms only in the vertical direction). This is extended to the case when the soil can strain laterally (such as beneath a foundation) in Section 8.7.

① 另一方面，如果在饱和黏土中开挖（总应力减少），坑底会由于土的回弹而发生隆起（向上位移）。

The progress of consolidation in-situ can be monitored by installing piezometers to record the change in pore water pressure with time (these are described in Chapter 6). The magnitude of settlement can be measured by recording the levels of suitable reference points on a structure or in the ground: precise levelling is essential, working from a benchmark which is not subject to even the slightest settlement. Every opportunity should be taken of obtaining settlement data in the field, as it is only through such measurements that the adequacy of theoretical methods can be assessed.

4.2 The oedometer test[②]

The characteristics of a soil during one-dimensional consolidation or swelling can be determined by means of the oedometer test. Figure 4.1 shows diagrammatically a cross-section through an oedometer. The test specimen is in the form of a disc of soil, held inside a metal ring and lying between two porous stones. The upper porous stone, which can move inside the ring with a small clearance, is fixed below a

② 固结试验。
③ 荷载。
④ 透水石。
⑤ 试样。
⑥ 侧限环刀。
⑦ 固结仪。
⑧ 试验仪器。

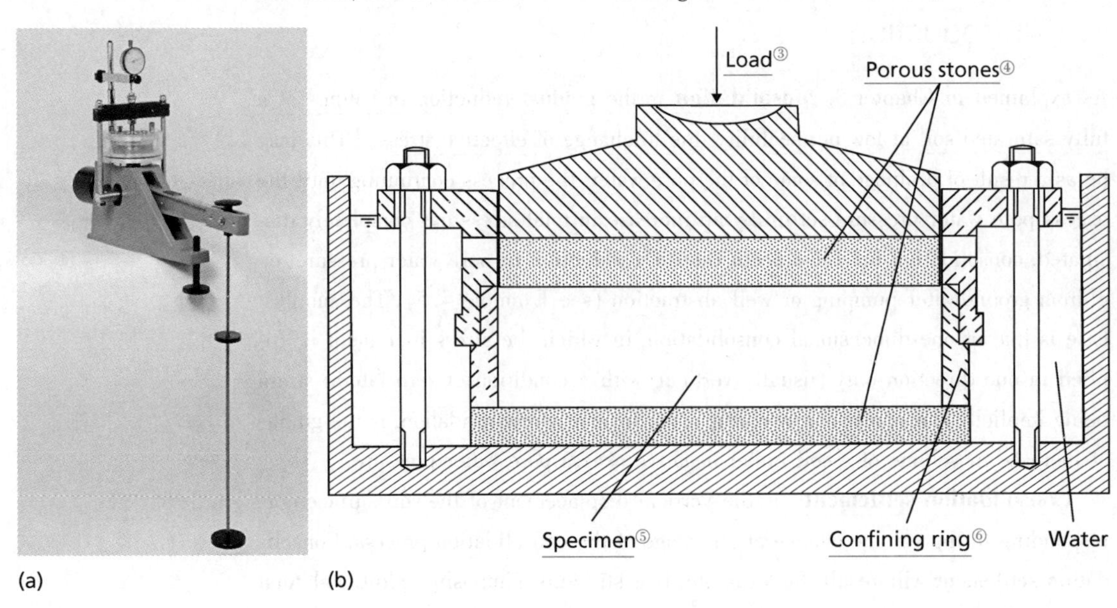

Figure 4.1 The oedometer[⑦]: (a) test apparatus[⑧], (b) test arrangement (image courtesy of Impact Test Equipment Ltd.).

metal loading cap through which pressure can be applied to the specimen. The whole assembly sits in an open cell of water to which the pore water in the specimen has free access. The ring confining the specimen may be either fixed (clamped to the body of the cell) or floating (being free to move vertically); the inside of the ring should have a smooth polished surface to reduce side friction. The confining ring imposes a condition of zero lateral strain on the specimen. The compression of the specimen under pressure is measured by means of a dial gauge or electronic displacement transducer operating on the loading cap.

The test procedure has been standardised in CEN ISO/TS17892 – 5 (Europe) and ASTM D2435 (US), though BS 1377, Part 5 remains current in the UK, which specifies that the oedometer shall be of the fixed ring type. The initial pressure (total stress) applied will depend on the type of soil; following this, a sequence of pressures is applied to the specimen, each being double the previous value. Each pressure is normally maintained for a period of 24h (in exceptional cases a period of 48h may be required), compression readings being observed at suitable intervals during this period. At the end of the increment period, when the excess pore water pressure has completely dissipated, the applied total stress equals the effective vertical stress in the specimen. The results are presented by plotting the thickness (or percentage change in thickness) of the specimen or the void ratio at the end of each increment period against the corresponding effective stress. The effective stress may be plotted to either a natural or a logarithmic scale, though the latter is normally adopted due to the reduction in volume change in a given increment as total stress increases. If desired, the expansion of the specimen can additionally be measured under successive decreases in applied pressure to observe the swelling behaviour. However, even if the swelling characteristics of the soil are not required, the expansion of the specimen due to the removal of the final pressure should be measured.

Eurocode 7, Part 2 recommends that a minimum of two tests are conducted in a given soil stratum; this value should be doubled if there is considerable discrepancy in the measured compressibility, especially if there is little or no previous experience relating to the soil in question.

The void ratio at the end of each increment period can be calculated from the displacement readings and either the water content or the dry weight of the specimen at the end of the test.[①] Referring to the phase diagram in Figure 4.2, the two methods of calculation are as follows:

1 Water content measured at end of test $= w_1$

Void ratio at end of test $= e_1 = w_1 G_s$ (assuming $S_r = 100\%$)

Thickness of specimen at start of test $= H_0$

Change in thickness during test $= \Delta H$

Void ratio at start of test $= e_0 = e_1 + \Delta e$ where

$$\frac{\Delta e}{\Delta H} = \frac{1 + e_0}{H_0} \tag{4.1}$$

In the same way Δe can be calculated up to the end of any increment period.

① 每一加载段结束时的孔隙比可以通过所测位移和试样最终含水量或试样最终干密度计算求得。

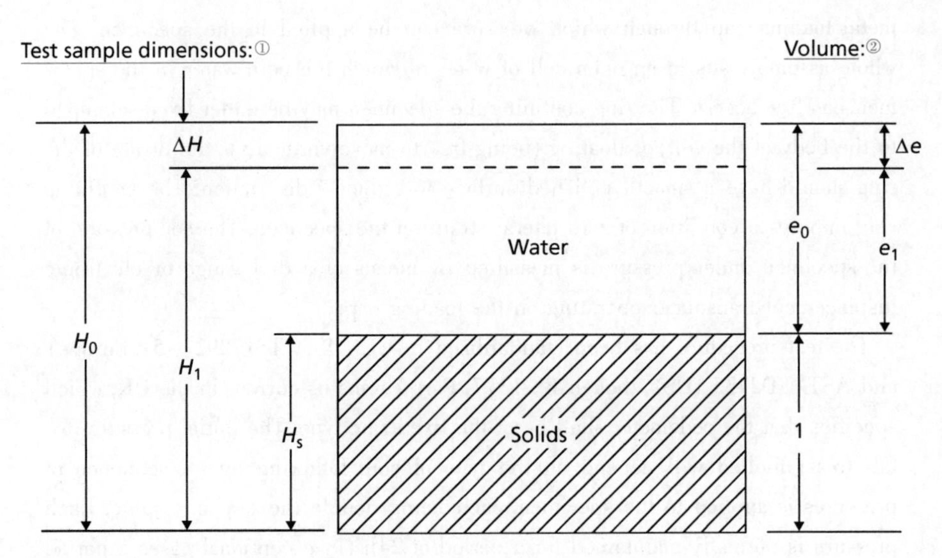

Figure 4.2 Phase diagram③.

① 试样尺寸。
② 体积。
③ 饱和土的水-土两相图。

2 Dry weight measured at end of test = M_s (i. e. mass of solids)

Thickness at end of any increment period = H_1

Area of specimen = A

Equivalent thickness of solids = $H_s = M_s / A G_s \rho w$

Void ratio,

$$e_1 = \frac{H_1 - H_s}{H_s} = \frac{H_1}{H_s} - 1 \tag{4.2}$$

Stress history

④ 孔隙比与有效应力的关系取决于土的应力历史。
⑤ 正常固结。
⑥ 超固结。
⑦ 超固结比。

The relationship between void ratio and effective stress depends on the **stress history**④ of the soil. If the present effective stress is the maximum to which the soil has ever been subjected, the clay is said to be **normally consolidated**⑤. If, on the other hand, the effective stress at some time in the past has been greater than the present value, the soil is said to be **overconsolidated**⑥. The maximum value of effective stress in the past divided by the present value is defined as the **overconsolidation ratio**⑦ (OCR). A normally consolidated soil thus has an overconsolidation ratio of unity; an overconsolidated soil has an overconsolidation ratio greater than unity. The overconsolidation ratio can never be less than one.

Most soils will initially be formed by sedimentation of particles, which leads to gradual consolidation under increasing self-weight. Under these conditions, the effective stresses within the soil will be constantly increasing as deposition continues, and the soil will therefore be normally consolidated. Seabed or riverbed soils are common examples of soils which are naturally in a normally consolidated state (or close to it). Overconsolidation is usually the result of geological factors – for example, the erosion of overburden (due to glacier motion, wind, wave or ocean currents), the melting of ice sheets (and therefore reduction in stress) after glaciation, or permanent rise of the water table. Overconsolidation may also occur due to man-made processes: for example, the demolition of an old structure to redevelop the land will remove the total stresses that were applied by the building's foundations causing heave, such that, for the redevelopment, the soil will initially be overconsolidated.

Compressibility characteristics[1]

Typical plots of void ratio (e) after consolidation against effective stress (σ') for a saturated soil are shown in Figure 4.3, the plots showing an initial compression followed by unloading and recompression. The shapes of the curves are related to the stress history of the soil. The $e - \log \sigma'$ relationship for a normally consolidated soil is linear (or nearly so), and is called the **virgin (one-dimensional) compression line**[2] (1DCL). During compression along this line, permanent (irreversible) changes in soil structure continuously take place and the soil does not revert to the original structure during expansion.[3] If a soil is overconsolidated, its state will be represented by a point on the expansion or recompression parts of the $e - \log \sigma'$ plot. The changes in soil structure along this line are almost wholly recoverable as shown in Figure 4.3.

① 压缩特性。
② 一维初始压缩曲线。
③ 在初始(一维)压缩曲线上，土体结构随着加压产生永久的(不可逆的)改变，同时在卸载回弹时土体也不能再回到原来的结构状态。
④ 初始压缩曲线。
⑤ 再压缩曲线。
⑥ 卸载-再加载曲线。
⑦ 回弹（膨胀)。
⑧ 孔隙比-有效应力关系图。

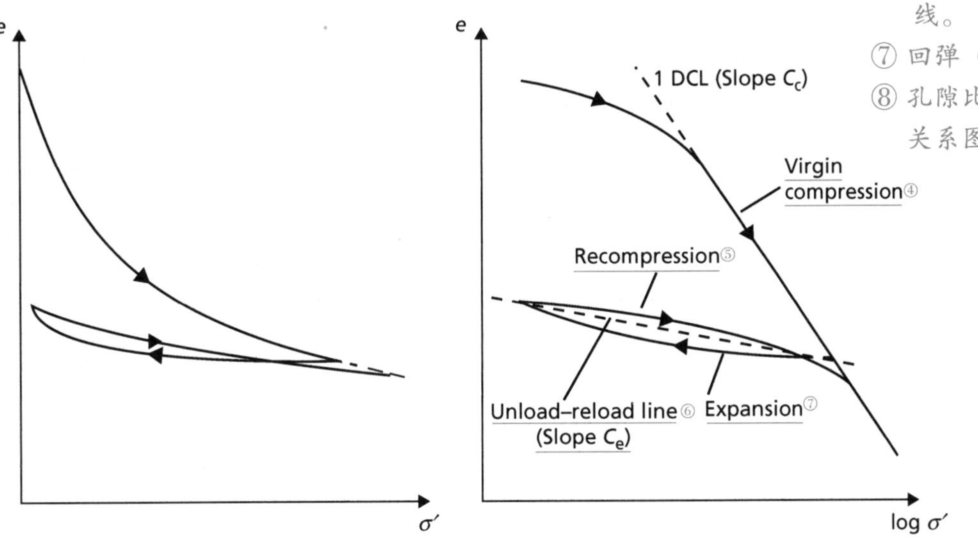

Figure 4.3 Void ratio – effective stress relationship[8].

The recompression curve ultimately rejoins the virgin compression line: further compression then occurs along the virgin line. The plots show that a soil in the overconsolidated state will be much less compressible than that in a normally consolidated state.

The compressibility of the soil can be quantified by one of the following coefficients.

1 The **coefficient of volume compressibility**[9] (m_v), defined as the volume change per unit volume per unit increase in effective stress (i. e. ratio of volu-

⑨ 体积压缩系数。

metric strain to applied stress). The units of m_v are the inverse of stiffness (m^2/MN). The volume change may be expressed in terms of either void ratio or specimen thickness. If, for an increase in effective stress from σ'_0 to σ'_1, the void ratio decreases from e_0 to e_1, then

① 注：体积压缩系数定义为在单位体积中单位有效应力增量等效的体积改变量。

$$m_v = \frac{1}{1+e_0}\left(\frac{e_0 - e_1}{\sigma'_1 - \sigma'_0}\right) \tag{4.3}$$ [①]

$$m_v = \frac{1}{H_0}\left(\frac{H_0 - H_1}{\sigma'_1 - \sigma'_0}\right) \tag{4.4}$$

The value of m_v for a particular soil is not constant but depends on the stress range over which it is calculated, as this parameter appears in the denominator of Equations 4.3 and 4.4. Most test standards specify a single value of the coefficient m_v calculated for a stress increment of $100 kN/m^2$ in excess of the in-situ vertical effective stress of the soil sample at the depth it was sampled from (also termed the effective **overburden pressure**[②]), although the coefficient may also be calculated, if required, for any other stress range, selected to represent the expected stress changes due to a particular geotechnical construction.

② 上覆应力。

2 The **constrained modulus**[③] (also called one-dimensional elastic modulus) E'_{oed} is the reciprocal of m_v (i.e. having units of stiffness, $MN/m^2 = MPa$) where:

③ 压缩模量。

$$E'_{oed} = \frac{1}{m_v} \tag{4.5}$$

3 The **compression index**[④] (C_c) is the slope of the 1DCL, which is a straight line on the $e - \log \sigma'$ plot, and is dimensionless. For any two points on the linear portion of the plot

④ 压缩指数。

$$C_c = \frac{e_0 - e_1}{\log(\sigma'_1/\sigma'_0)} \tag{4.6}$$

The expansion part of the $e - \log \sigma'$ plot can be approximated to a straight line, the slope of which is referred to as the **expansion index**[⑤] C_e (also called swell-back index). The expansion index is usually many times less than the compression index (as indicated in Figure 4.3).

⑤ 回弹指数。

It should be noted that although C_c and C_e represent negative gradients on the $e - \log \sigma'$ plot, their value is always given as positive (i.e. they represent the magnitude of the gradients).

Preconsolidation pressure

Casagrande (1936) proposed an empirical construction to obtain, from the $e - \log \sigma'$ curve for an overconsolidated soil, the maximum effective vertical stress that has acted on the soil in the past, referred to as the **pre-consolidation pressure**[⑥] (σ'_{max}). This parameter may be used to determine the in-situ OCR for the soil tested:

⑥ 先期固结压力。

$$OCR = \frac{\sigma'_{max}}{\sigma'_{v0}} \tag{4.7}$$

where σ'_{v0} is the in-situ vertical effective stress of the soil sample at the depth it was sampled from (effective overburden pressure), which may be calculated using the methods outlined in Chapter 3.

Figure 4.4 shows a typical $e - \log \sigma'$ curve for a specimen of soil which is initially overconsolidated. The initial curve (AB) and subsequent transition to a linear compression (BC) indicates that the soil is undergoing recompression in the oedometer, having at some stage in its history undergone swelling. Swelling of the soil in-situ may, for example, have been due to melting of ice sheets, erosion of overburden, or a rise in water table level. The construction for estimating the preconsolidation pressure consists of the following steps:

1 Produce back the straight-line part (BC) of the curve;[①]
2 Determine the point (D) of maximum curvature on the recompression part (AB) of the curve;[②]
3 Draw a horizontal line through D;[③]
4 Draw the tangent to the curve at D and bisect the angle between the tangent and the horizontal through D;[④]
5 The vertical through the point of intersection of the bisector and CB produced gives the approximate value of the preconsolidation pressure.[⑤]

① 将曲线中的直线段 *BC* 向后延长。

② 确定曲线再压缩段 *AB* 上的最大曲率点 *D*。

③ 过点 *D* 作水平线。

④ 过 *D* 点作曲线的切线，并作直线平分过 *D* 点的切线与水平线所夹的角。

⑤ 过角平分线与 *CB* 段延长线的交点作垂直于水平线的直线，该直线与 log σ' 轴的交点的值即为先期固结压力的估值。

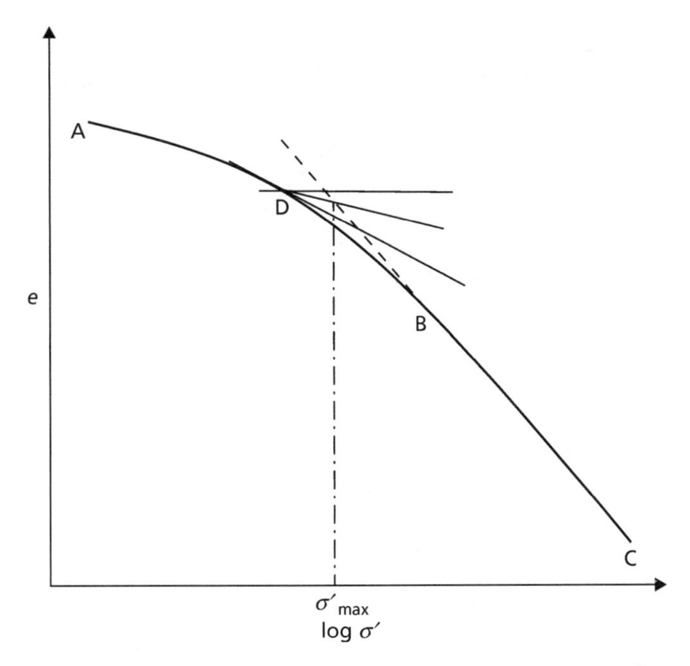

Figure 4.4 Determination of preconsolidation pressure.[⑥]

⑥ 先期固结压力的确定。

Whenever possible, the preconsolidation pressure for an overconsolidated clay should not be exceeded in construction. Compression will not usually be great if the effective vertical stress remains below σ'_{max}, as the soil will be always on the unload – reload part of the compression curve. Only if σ'_{max} is exceeded will compression be large. This is the key principle behind **preloading**[⑦], which is a technique used to reduce the compressibility of soils to make them more suitable for use in foundations; this is discussed in Section 4.11.

⑦ 预加荷载。

In-situ $e-\log\sigma'$ curve[1]

① 原位 e- log σ′ 曲线。
② 扰动。
③ 原位土的初始压缩曲线的斜率稍大于室内试验初始压缩曲线的斜率。

Due to the effects of sampling (see Chapter 6) and test preparation, the specimen in an oedometer test will be slightly disturbed.[2] It has been shown that an increase in the degree of specimen disturbance results in a slight decrease in the slope of the virgin compression line. It can therefore be expected that the slope of the line representing virgin compression of the in-situ soil will be slightly greater than the slope of the virgin line obtained in a laboratory test.[3]

No appreciable error will be involved in taking the in-situ void ratio as being equal to the void ratio (e_0) at the start of the laboratory test. Schmertmann (1953) pointed out that the laboratory virgin line may be expected to intersect the in-situ virgin line at a void ratio of approximately 0.42 times the initial void ratio. Thus the in-situ virgin line can be taken as the line EF in Figure 4.5 where the coordinates of E are $\log\sigma'_{max}$ and e_0, and F is the point on the laboratory virgin line at a void ratio of $0.42e_0$.

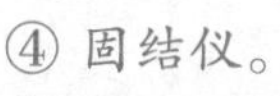

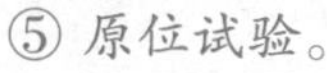

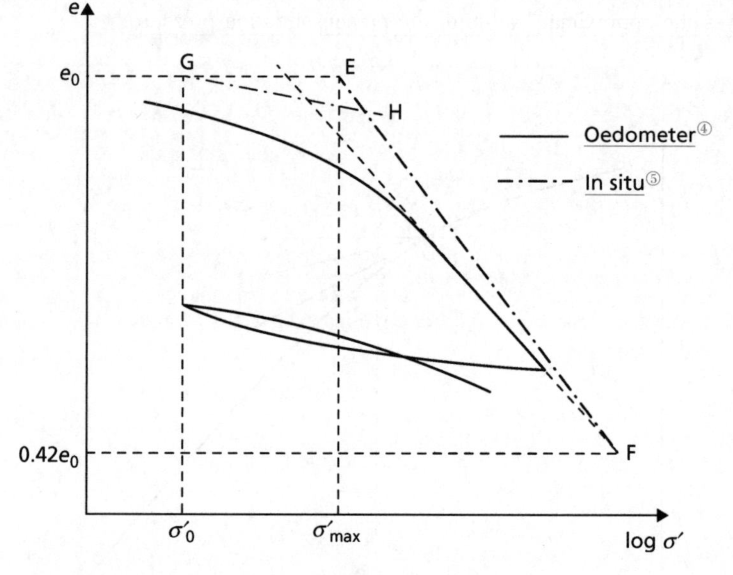

Figure 4.5 In-situ $e-\log\sigma'$ curve.

In the case of overconsolidated clays, the in-situ condition is represented by the point (G) having coordinates σ'_0 and e_0, where σ'_{max} is the present effective overburden pressure. The in-situ recompression curve can be approximated to the straight line GH parallel to the mean slope of the laboratory recompression curve.

Example 4.1

The following compression readings were obtained in an oedometer test on a specimen of saturated clay ($G_s = 2.73$):

TABLE F

Pressure (kPa)	0	54	107	214	429	858	1716	3432	0
Dial gauge after 24h (mm)	5.000	4.747	4.493	4.108	3.449	2.608	1.676	0.737	1.480

The initial thickness of the specimen was 19.0mm, and at the end of the test the water content was 19.8%. Plot the e-log σ' curve and determine the preconsolidation pressure. Determine the values of m_v for the stress increments 100 – 200 and 1000 – 1500kPa. What is the value of C_c for the latter increment?

Solution

$$\text{Void ratio at end of test} = e_1 = w_1 G_s = 0.198 \times 2.73 = 0.541$$

$$\text{Void ratio at end of test} = e_0 = e_1 + \Delta e$$

Now

$$\frac{\Delta e}{\Delta H} = \frac{1 + e_0}{H_0} = \frac{1 + e_1 + \Delta e}{H_0}$$

i. e.

$$\frac{\Delta e}{3.520} = \frac{1.541 + \Delta e}{19.0}$$

$$\Delta e = 0.350$$

$$e_0 = 0.541 + 0.350 = 0.891$$

In general, the relationship between Δe and ΔH is given by

$$\frac{\Delta e}{\Delta H} = \frac{1.891}{19.0}$$

i. e. $\Delta e = 0.0996\ \Delta H$, and can be used to obtain the void ratio at the end of each increment period (see Table 4.1). The e-log σ' curve using these values is shown in Figure 4.6. Using Casagrande's construction, the value of the preconsolidation pressure is 325kPa.

Table 4.1 Example 4.1

Pressure (kPa)	*ΔH (mm)*	*Δe*	*e*
0	0	0	0.891
54	0.253	0.025	0.866
107	0.507	0.050	0.841
214	0.892	0.089	0.802
429	1.551	0.154	0.737
858	2.392	0.238	0.653
1716	3.324	0.331	0.560
3432	4.263	0.424	0.467
0	3.520	0.350	0.541

$$m_v = \frac{1}{1+e_0} \cdot \frac{e_0 - e_1}{\sigma_1' - \sigma_0'}$$

For $\sigma_0' = 100\text{kPa}$ and $\sigma_1' = 200\text{kPa}$

$e_0 = 0.845$ and $e_1 = 0.808$

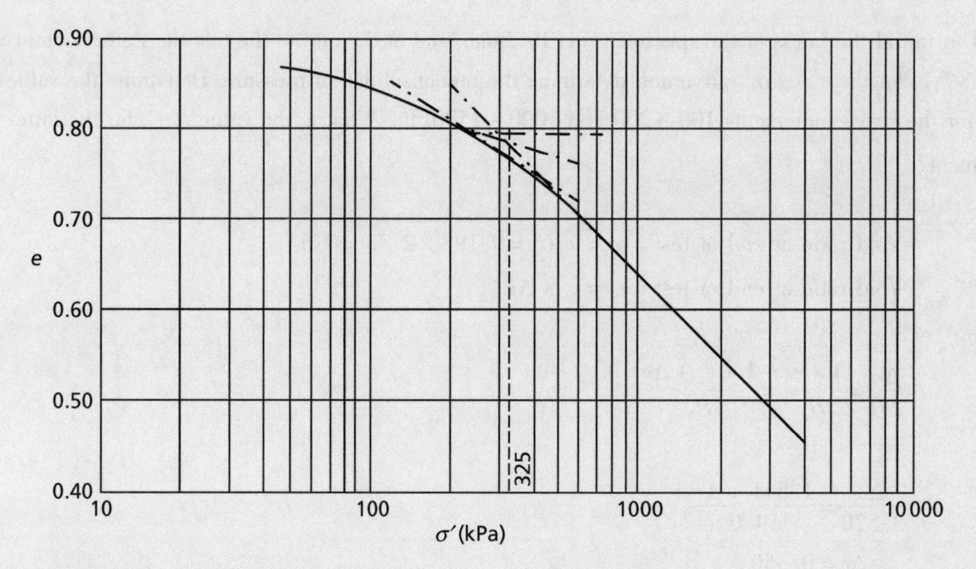

Figure 4.6 Example 4.1.

and therefore

$$m_v = \frac{1}{1.845} \times \frac{0.037}{100} = 2.0 \times 10^{-4}\text{m}^2/\text{kN} = 0.20\text{m}^2/\text{MN}$$

For $\sigma_0' = 1000\text{kPa}$ and $\sigma_1' = 1500\text{kPa}$

$e_0 = 0.632$ and $e_1 = 0.577$

and therefore

$$m_v = \frac{1}{1.845} \times \frac{0.055}{500} = 6.7 \times 10^{-5}\text{m}^2/\text{kN} = 0.07\text{m}^2/\text{MN}$$

and

$$C_C = \frac{0.632 - 0.557}{\log(1500/1000)} = \frac{0.055}{0.176} = 0.31$$

Note that C_c will be the same for any stress range on the linear part of the $e - \log\sigma'$ curve; m_v will vary according to the stress range, even for ranges on the linear part of the curve.

① 固结沉降。

4.3 Consolidation settlement[①]

In order to estimate one-dimensional consolidation settlement, the value of either the coefficient of volume compressibility or the compression index is required. Consider a layer of saturated soil of thickness H. Due to construction, the total vertical stress in an elemental layer of thickness dz at depth z is increased by a **surcharge**

pressure[1] $\Delta\sigma$ (Figure 4.7). It is assumed that the condition of zero lateral strain applies within the soil layer. This is an appropriate assumption if the surcharge is applied over a wide area. After the completion of consolidation, an equal increase $\Delta\sigma'$ in effective vertical stress will have taken place corresponding to a stress increase from σ_0' to σ_1' and a reduction in void ratio from e_0 to e_1 on the e-σ' curve. The reduction in volume per unit volume of soil can be written in terms of void ratio:

$$\frac{\Delta V}{V_0}=\frac{e_0-e_1}{1+e_0}$$

Since the lateral strain is zero, the reduction in volume per unit volume is equal to the reduction in thickness per unit thickness,[2] i. e. the settlement per unit depth. Therefore, by proportion, the settlement of the layer of thickness dz will be given by

$$\begin{aligned} \mathrm{d}s_{\mathrm{oed}} &= \frac{e_0-e_1}{1+e_0}\mathrm{d}z \\ &= \left(\frac{e_0-e_1}{\sigma_1'-\sigma_0'}\right)\left(\frac{\sigma_1'-\sigma_0'}{1+e_0}\right)\mathrm{d}z \\ &= m_{\mathrm{v}}\Delta\sigma'\mathrm{d}z \end{aligned}$$

where s_{oed} = consolidation settlement[3] (one dimensional).

The settlement of the layer of thickness H is given by integrating the incremental change ds_{oed} over the height of the layer

$$s_{\mathrm{oed}}=\int_0^H m_{\mathrm{v}}\Delta\sigma'\mathrm{d}z \tag{4.8}$$

If m_{v} and $\Delta\sigma'$ are constant with depth, then

$$s_{\mathrm{oed}}=m_{\mathrm{v}}\Delta\sigma' H \tag{4.9}$$

Substituting Equation 4.3 for m_{v}, Equation 4.9 may also be written as

$$s_{\mathrm{oed}}=\frac{e_0-e_1}{1+e_0}H \tag{4.10}$$

or, in the case of a normally consolidated soil, by substituting Equation 4.6, Equation 4.9 is rewritten as

$$s_{\mathrm{oed}}=\frac{C_{\mathrm{c}}\log(\sigma_1'/\sigma_0')}{1+e_0}H \tag{4.11}$$

In order to take into account the variation of m_{v} and/or $\Delta\sigma'$ with depth (i. e. non-uniform soil), the graphical procedure shown in Figure 4.8 can be used to determine s_{oed}. The variations of initial effective vertical stress (σ_0') and effective vertical stress increment ($\Delta\sigma'$) over the depth of the layer are represented in Figure 4.8 (a); the variation of m_{v} is represented in Figure 4.8 (b). The curve in Figure 4.8 (c) represents the variation with depth of the dimensionless product $m_{\mathrm{v}}\Delta\sigma'$, and the area under this curve is the settlement of the layer. Alternatively, the layer can be divided into a suitable number of sub-layers and the product $m_{\mathrm{v}}\Delta\sigma'$ evaluated at the centre of each sub-layer: each product $m_{\mathrm{v}}\Delta\sigma'$ is then multiplied by the appropriate sub-layer thickness to give the sub-layer settlement. The settlement of the whole layer is equal to the sum of the sub-layer settlements. The sub-layer technique may also be used to analyse layered soil profiles where the soil layers have very different compressional characteristics.

① 堆载。

② 由于侧向应变为零，单位体积的体积减少量与单位厚度的厚度减少量数值上相等。

③ 固结沉降。

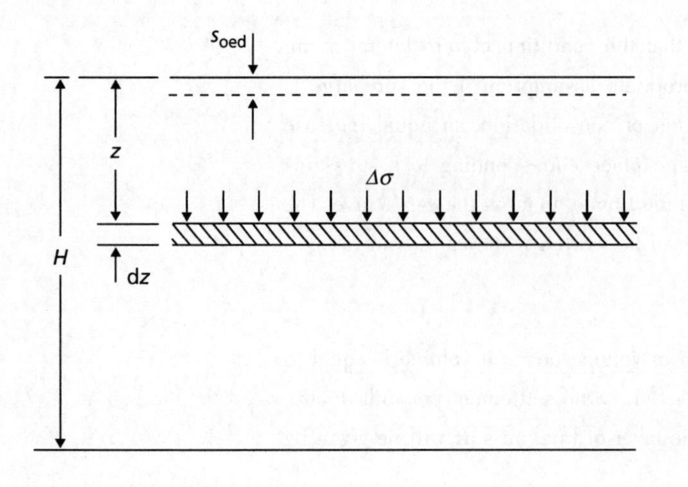

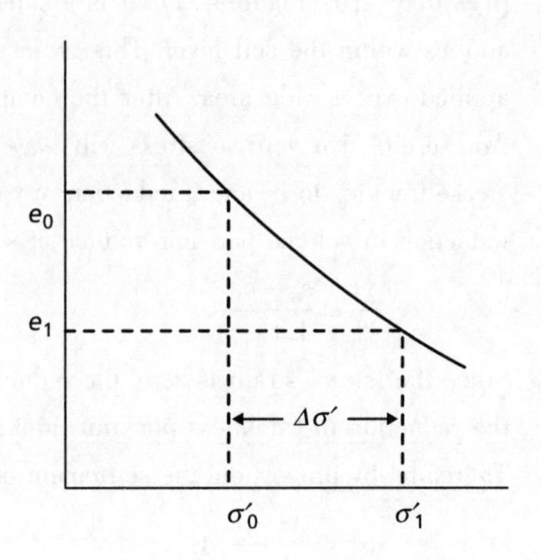

Figure 4.7 Consolidation settlement.

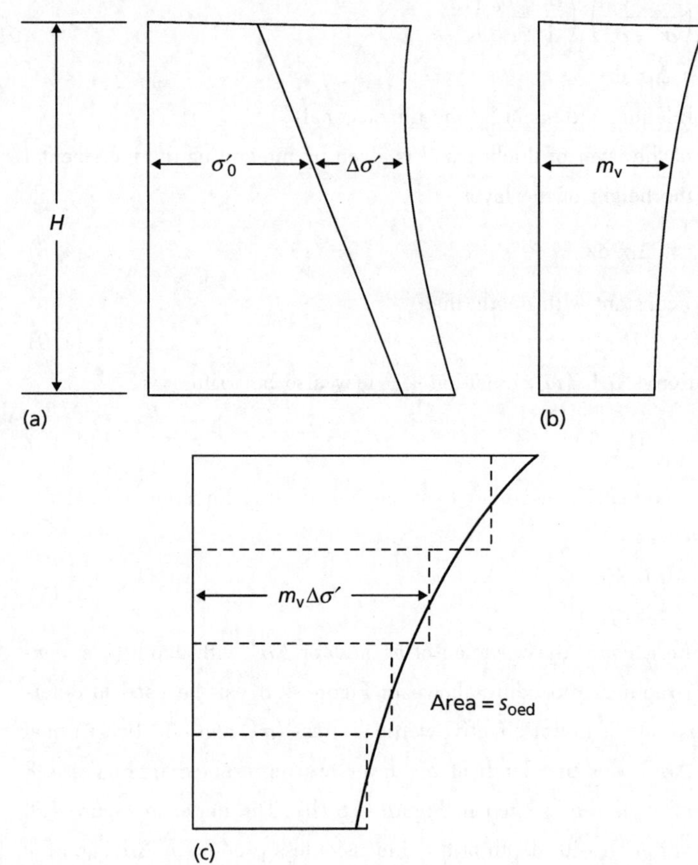

Figure 4.8 Consolidation settlement: graphical procedure.[①]

① 固结沉降：图解步骤。

Example 4.2

A long embankment 30m wide is to be built on layered ground as shown in Figure 4.9. The net vertical pressure applied by the embankment (assumed to be uniformly distributed) is 90kPa. The soil profile

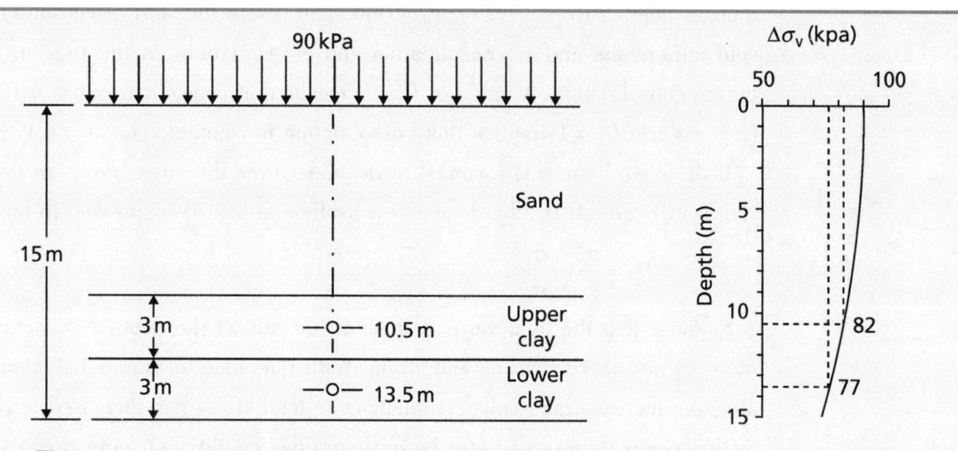

Figure 4.9 Example 4.2.

and stress distribution beneath the centre of the embankment are as shown in Figure 4.9 (the methods used to determine such a distribution are outlined in Section 8.5). The value of m_v for the upper clay is $0.35\text{m}^2/\text{MN}$, and for the lower clay $m_v = 0.13\text{m}^2/\text{MN}$. The permeabilities of the clays are 10^{-10} m/s and 10^{-11} m/s for the upper and lower soils respectively. Determine the final settlement under the centre of the embankment due to consolidation.

Solution

The clay layers are thin relative to the width of the applied surcharge from the embankment, therefore it can be assumed that consolidation is approximately one-dimensional. Considering the stresses shown in Figure 4.9, it will be sufficiently accurate to treat each clay layer as a single sub-layer. For one-dimensional consolidation $\Delta\sigma' = \Delta\sigma$, with the stress increments at the middle of each layer indicated in Figure 4.9. Equation 4.9 is then applied to determine the settlement of each of the layers, with H being the layer thickness (= 3m), and these are combined to give the total consolidation settlement. The calculations are shown in Table 4.2.

Table 4.2 Example 4.2

Sublayer	*z (m)*	*Δσ′ (kPa)*	*m_v(m^2/MN)*	*H (m)*	*s_{oed} (mm)*
1 (Upper clay)	10.5	82	0.35	3	86.1
2 (Lower clay)	13.5	77	0.13	3	30.0
					116.1

4.4 Degree of consolidation

For an element of soil at a particular depth z in a layer of soil, the progress of the consolidation process under a particular total stress increment can be expressed in terms of void ratio as follows:

$$U_v = \frac{e_0 - e}{e_0 - e_1}$$

where U_v is defined as the **degree of consolidation**[1], at a particular instant of

① 固结度。

time, at depth z ($0 \leqslant U_v \leqslant 1$); e_0 = void ratio before the start of consolidation; e_1 = void ratio at the end of consolidation and e = void ratio, at the time in question, during consolidation. A value of $U_v = 0$ means that consolidation has not yet begun (i. e. $e = e_0$); $U_v = 1$ implies that consolidation is complete (i. e. $e = e_1$)

If the $e-\sigma'$ curve is assumed to be linear over the stress range in question, as shown in Figure 4.10, the degree of consolidation can be expressed in terms of σ':

$$U_v = \frac{\sigma' - \sigma_0'}{\sigma_1' - \sigma_0'}$$

Suppose that the total vertical stress in the soil at the depth z is increased from σ_0 to σ_1 and there is no lateral strain. With reference to Figure 4.11, immediately after the increase takes place, although the total stress has increased to σ_1, the effective vertical stress will still be σ_0' (undrained conditions); only after the completion of consolidation will the effective stress become σ_1' (drained conditions). During consolidation, the increase in effective vertical stress is numerically equal to the decrease in excess pore water pressure.[①] If σ' and u_e are, respectively, the values of effective stress and excess pore water pressure at any time during the consolidation process and if u_i is the initial excess pore water pressure (i. e. the value immediately after the increase in total stress), then, referring to Figure 4.10:

① 固结过程中，竖向有效应力的增加量与超静孔隙水压力的减少量在数值上相等。

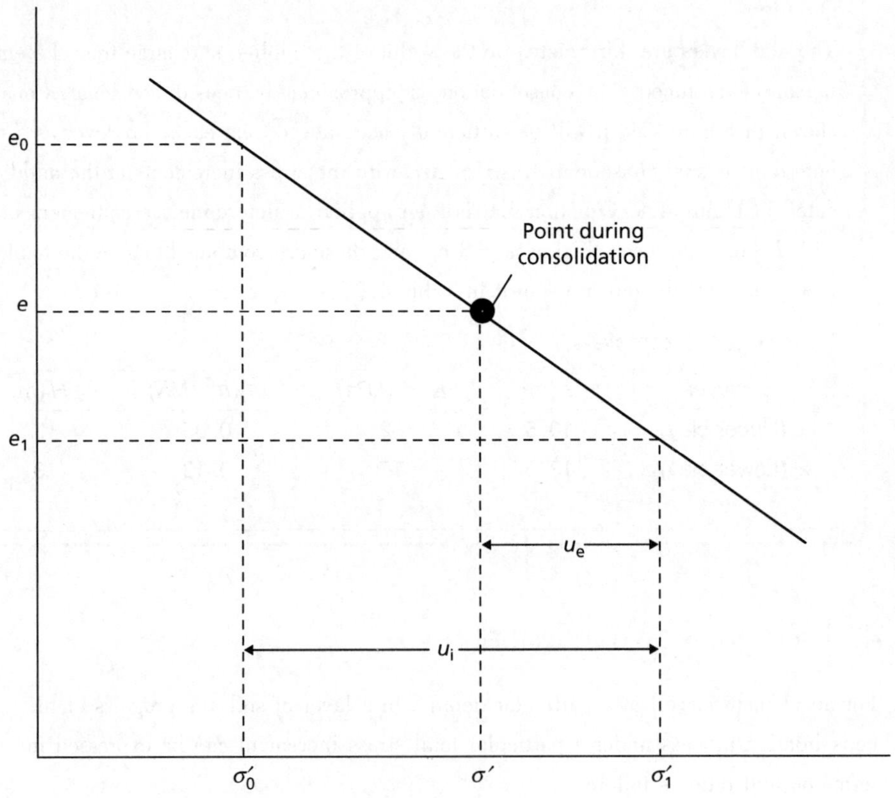

Figure 4.10 Assumed linear $e-\sigma'$ relationship[②].

② 假设的 e-σ' 线性关系。

$$\sigma_1' = \sigma_0' + u_i = \sigma' + u_e$$

The degree of consolidation can then be expressed as

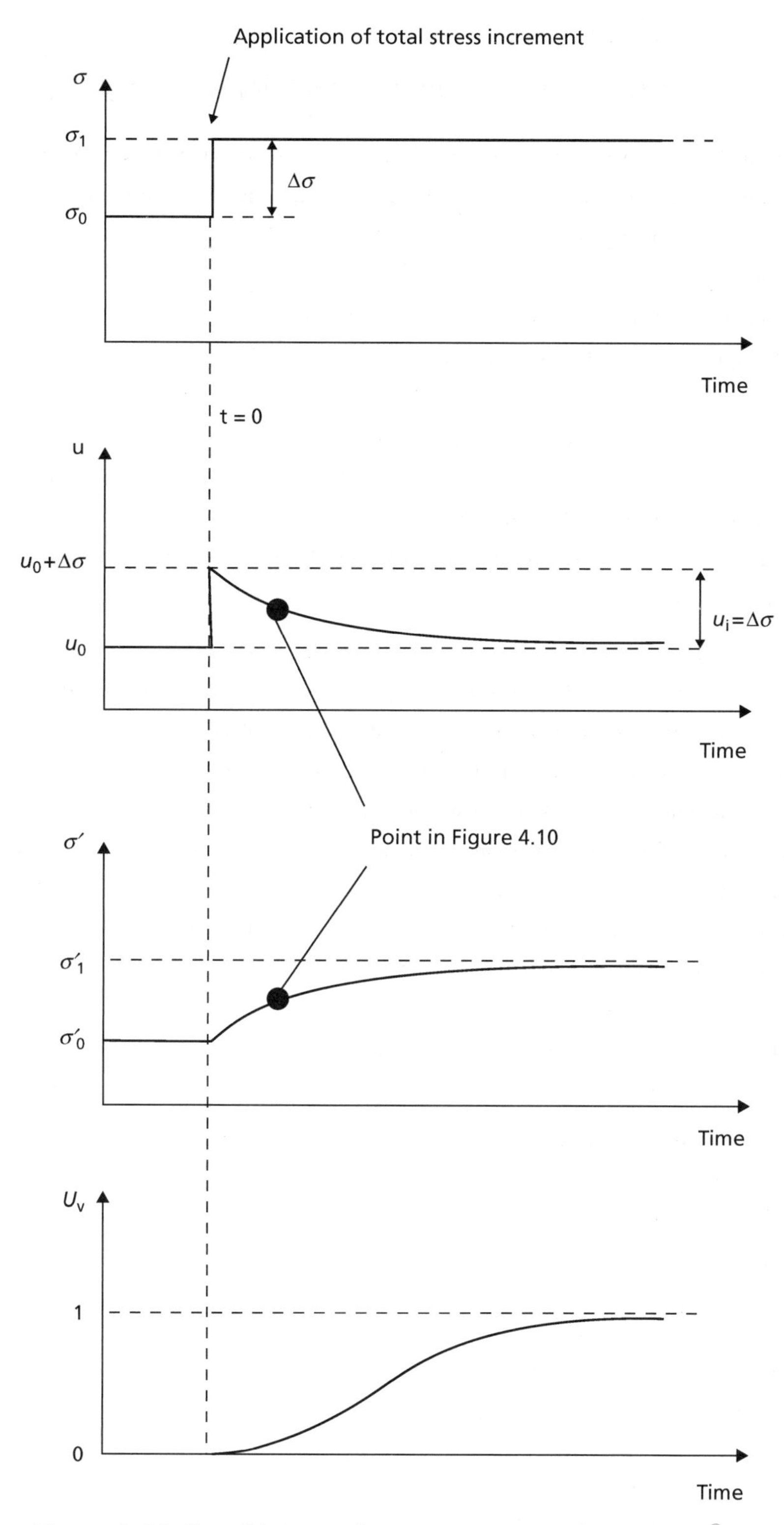

Figure 4.11 Consolidation under an increase in total stress $\Delta\sigma$.[①]

① 总应力增长量为$\Delta\sigma$时的固结。

$$U_v = \frac{u_i - u_e}{u_i} = 1 - \frac{u_e}{u_i} \tag{4.12}$$

4.5 Terzaghi's theory of one-dimensional consolidation[1]

① 太沙基一维固结理论。

Terzaghi (1943) developed an analytical model for determining the degree of consolidation within soil at any time, t. The assumptions[2] made in the theory are:

1 The soil is homogeneous.[3]
2 The soil is fully saturated.[4]
3 The solid particles and water are incompressible.[5]
4 Compression and flow are one-dimensional (vertical).[6]
5 Strains are small.[7]
6 Darcy's law is valid at all hydraulic gradients.[8]
7 The coefficient of permeability and the coefficient of volume compressibility remain constant throughout the process.[9]
8 There is a unique relationship, independent of time, between void ratio and effective stress.[10]

② 假设。
③ 土是均质的。
④ 土处于完全饱和状态。
⑤ 土颗粒与水是不可压缩的。
⑥ 压缩与渗流都是一维的（竖向）。
⑦ 应变很小。
⑧ 达西定律适用于所有水力梯度。
⑨ 渗透系数与体积压缩系数在整个过程中保持不变。
⑩ 孔隙比与有效应力之间存在一种与时间无关的唯一关系。

Regarding assumption 6, there is evidence of deviation from Darcy's law at low hydraulic gradients. Regarding assumption 7, the coefficient of permeability decreases as the void ratio decreases during consolidation (Al-Tabbaa and Wood, 1987). The coefficient of volume compressibility also decreases during consolidation since the $e-\sigma'$ relationship is non-linear. However, for small stress increments assumption 7 is reasonable. The main limitations of Terzaghi's theory (apart from its one-dimensional nature) arise from assumption 8. Experimental results show that the relationship between void ratio and effective stress is not independent of time.

The theory relates the following three quantities.

1 the excess pore water pressure (u_e);
2 the depth (z) below the top of the soil layer;
3 the time (t) from the instantaneous application of a total stress increment.

Consider an element having dimensions dx, dy and dz within a soil layer of thickness $2d$, as shown in Figure 4.12. An increment of total vertical stress $\Delta\sigma$ is applied to the element.

The flow velocity through the element is given by Darcy's law as

$$v_z = ki_z = -k\frac{\partial h}{\partial z}$$

Since at a fixed position z any change in total head[11] (h) is due only to a change in pore water pressure:

⑪ 总水头。

$$v_z = -\frac{k}{\gamma_w}\frac{\partial u_e}{\partial z}$$

The condition of continuity (Equation 2.12) can therefore be expressed as

$$-\frac{k}{\gamma_w}\frac{\partial^2 u_e}{\partial z^2}\mathrm{d}x\mathrm{d}y\mathrm{d}z = \frac{\mathrm{d}V}{\mathrm{d}t} \tag{4.13}$$

The rate of volume change can be expressed in terms of m_v:

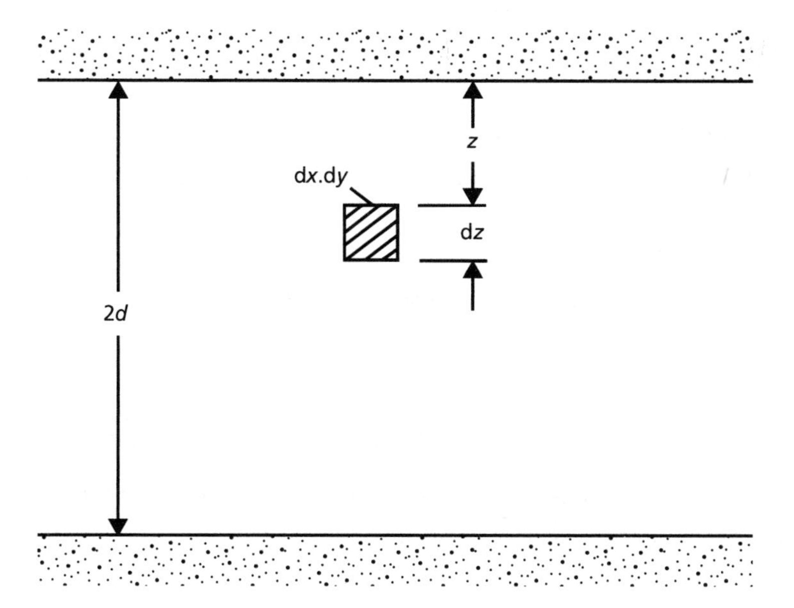

Figure 4.12 Element within a consolidating layer of soil.[①]

① 土层中正在固结的单元体。

$$\frac{\mathrm{d}V}{\mathrm{d}t}=m_{\mathrm{v}}\frac{\partial\sigma'}{\partial t}\mathrm{d}x\mathrm{d}y\mathrm{d}z$$

The total stress increment is gradually transferred to the soil skeleton, increasing effective stress, as the excess pore water pressure decreases. Hence the rate of volume change can be expressed as

$$\frac{\mathrm{d}V}{\mathrm{d}t}=-m_{\mathrm{v}}\frac{\partial u_{\mathrm{e}}}{\partial t}\mathrm{d}x\mathrm{d}y\mathrm{d}z \tag{4.14}$$

Combining Equations 4.13 and 4.14,

$$m_{\mathrm{v}}\frac{\partial u_{\mathrm{e}}}{\partial t}=\frac{k}{\gamma_{\mathrm{w}}}\frac{\partial^2 u_{\mathrm{e}}}{\partial z^2}$$

or

$$\frac{\partial u_{\mathrm{e}}}{\partial t}=c_{\mathrm{v}}\frac{\partial^2 u_{\mathrm{e}}}{\partial z^2} \tag{4.15}$$

This is the differential equation of consolidation,[②] in which

② 固结微分方程。

$$c_{\mathrm{v}}=\frac{k}{m_{\mathrm{v}}\gamma_{\mathrm{w}}} \tag{4.16}$$

c_{v} being defined as the **coefficient of consolidation**[③], with a suitable unit being m^2/year. Since k and m_{v} are assumed as constants (assumption 7), c_{v} is constant during consolidation.

③ 固结系数。

Solution of the consolidation equation[④]

④ 固结方程的求解。

The total stress increment is assumed to be applied instantaneously, as in Figure 4.11. At zero time, therefore, the increment will be carried entirely by the pore water, i. e. the initial value of excess pore water pressure (u_{i}) is equal to $\Delta\sigma$ and the initial condition is

$$u_e = u_i \text{ for } 0 \leqslant z \leqslant 2d \text{ when } t = 0$$

The upper and lower boundaries of the soil layer are assumed to be free-draining, the permeability of the soil adjacent to each boundary being very high compared to that of the soil.[①] Water therefore drains from the centre of the soil element to the upper and lower boundaries simultaneously so that the drainage path length is d if the soil is of thickness $2d$. Thus the boundary conditions at any time after the application of $\Delta\sigma$ are

① 土层的上下表面都假设为自由排水，邻近土层界面的土比该土层的渗透性高很多。

$$u_e = 0 \text{ for } z = 0 \text{ and } z = 2d \text{ when } t > 0$$

The solution for the excess pore water pressure at depth z after time t from Equation 4.15 is

$$u_e = \sum_{n=1}^{n=\infty} \left(\frac{1}{d} \int_0^{2d} u_i \sin \frac{n\pi z}{2d} \mathrm{d}z \right) \left(\sin \frac{n\pi z}{2d} \right) \exp \left(- \frac{n^2 \pi^2 c_v t}{4d^2} \right) \tag{4.17}$$

where u_i = initial excess pore water pressure, in general a function of z. For the particular case in which u_i is constant throughout the clay layer:

$$u_e = \sum_{n=1}^{n=\infty} \frac{2u_i}{n\pi} (1 - \cos n\pi) \left(\sin \frac{n\pi z}{2d} \right) \exp \left(- \frac{n^2 \pi^2 c_v t}{4d^2} \right) \tag{4.18}$$

When n is even, $(1 - \cos n\pi) = 0$, and when n is odd, $(1 - \cos n\pi) = 2$. Only odd values of n are therefore relevant, and it is convenient to make the substitutions

$$n = 2m + 1$$

and

$$M = \frac{\pi}{2}(2m + 1)$$

It is also convenient to substitute

$$T_v = \frac{c_v t}{d^2} \tag{4.19}$$

a dimensionless number called the **time factor**[②]. Equation 4.18 then becomes

② 时间因子。

$$u_e = \sum_{m=0}^{m=\infty} \frac{2u_i}{M} \left(\sin \frac{Mz}{d} \right) \exp(-M^2 T_v) \tag{4.20}$$

The progress of consolidation can be shown by plotting a series of curves of u_e against z for different values of t. Such curves are called **isochrones**,[③] and their form will depend on the initial distribution of excess pore water pressure and the drainage conditions at the boundaries of the soil layer. A layer for which both the upper and lower boundaries are free-draining is described as an **open layer** (sometimes called double drainage);[④] a layer for which only one boundary is free-draining is a **half-closed layer**[⑤] (sometimes called single drainage). Examples of isochrones are shown in Figure 4.13. In part (a) of the figure the initial distribution of u_i is constant and for an open layer of thickness $2d$ the isochrones are symmetrical about the centre line. The upper half of this diagram also represents the case of a half-closed layer of thickness d. The slope of an isochrone at any depth gives the hydraulic gradient and also indicates the direction of flow.[⑥] In parts (b) and (c) of the figure, with a triangular distribution of u_i, the direction of flow changes over certain parts of the layer. In part

③ 瞬压曲线。

④ 双面排水。

⑤ 单面排水。

⑥ 在任意深度瞬压曲线的斜率即为水力梯度，同时也指明了孔隙水流动的方向。

(c) the lower boundary is impermeable, and for a time swelling takes place in the lower part of the layer.

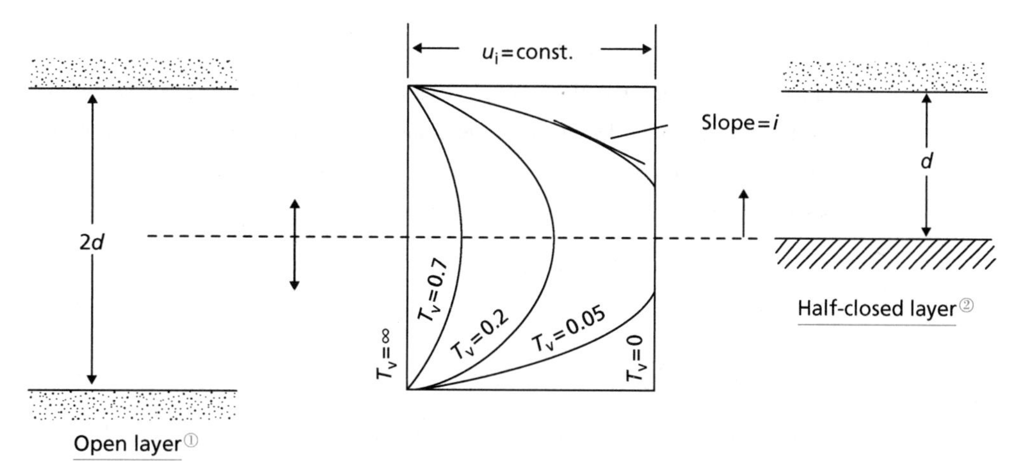

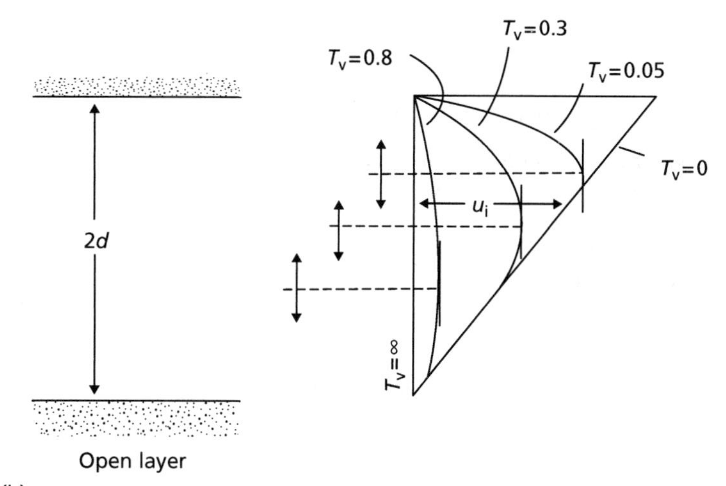

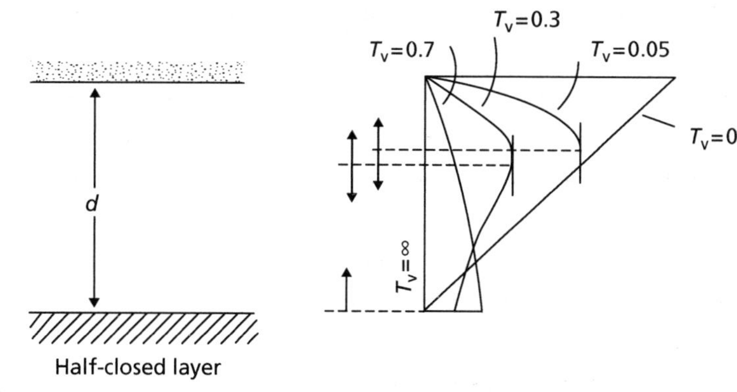

Figure 4.13 Isochrones.③

① 双面排水。
② 单面排水。
③ 瞬压曲线。

The degree of consolidation at depth z and time t can be obtained by substituting the value of u_e(Equation 4.20) in Equation 4.12, giving

$$U_v = 1 - \sum_{m=0}^{m=\infty} \frac{2}{M}\left(\sin\frac{Mz}{d}\right)\exp(-M^2T_v) \tag{4.21}$$

① 平均固结度。
② 最终沉降。

In practical problems it is the average degree of consolidation[①] (U_v) over the depth of the layer as a whole that is of interest, the consolidation settlement at time t being given by the product of U_v and the final settlement.[②] The average degree of consolidation at time t for constant u_i is given by

$$U_v = 1 - \frac{(1/2d)\int_0^{2d} u_e \mathrm{d}z}{u_i}$$

$$= 1 - \sum_{m=0}^{m=\infty} \frac{2}{M^2}\exp(M^2T_v) \tag{4.22}$$

The relationship between U_v and T_v given by Equation 4.22 is represented by curve 1 in Figure 4.14. Equation 4.22 can be represented almost exactly by the following empirical equations:

$$T_v = \begin{cases} \frac{\pi}{4}U_v^2 & U_v < 0.60 \\ -0.933\log(1 - U_v) - 0.085 & U_v > 0.60 \end{cases} \tag{4.23}$$

If u_i is not constant, the average degree of consolidation is given by

$$U_v = 1 - \frac{\int_0^{2d} u_e \mathrm{d}z}{\int_0^{2d} u_i \mathrm{d}z} \tag{4.24}$$

where

$$\int_0^{2d} u_e \mathrm{d}z = \text{area under isochrone at the time in question}$$

and

$$\int_0^{2d} u_i \mathrm{d}z = \text{area under initial isochrone}$$

(For a half-closed layer, the limits of integration are 0 and d in the above equations.)

The initial variation of excess pore water pressure in a clay layer can usually be approximated in practice to a linear distribution. Curves 1, 2 and 3 in Figure 4.14 represent the solution of the consolidation equation for the cases shown in Figure 4.15, where

i Represents the initial conditions u_i in the oedometer test, and also the field case where the height of the water table has been changed (water table raised = positive u_i; water table lowered = negative u_i);

ii Represents virgin (normal) consolidation;

iii Represents approximately the field condition when a surface loading is applied (e.g. placement of surcharge or construction of foundation).

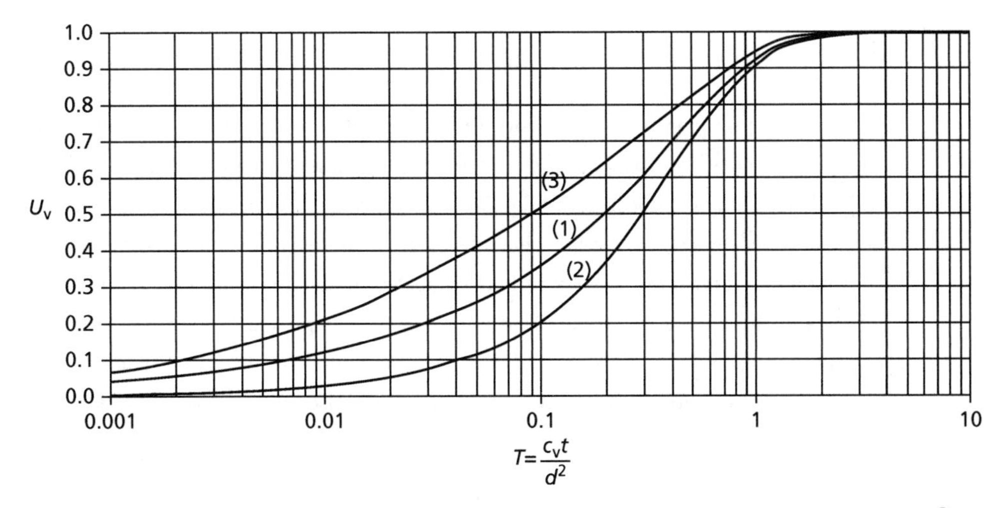

Figure 4.14 Relationships between average degree of consolidation and time factor.[①]

① 平均固结度与时间因子的关系。

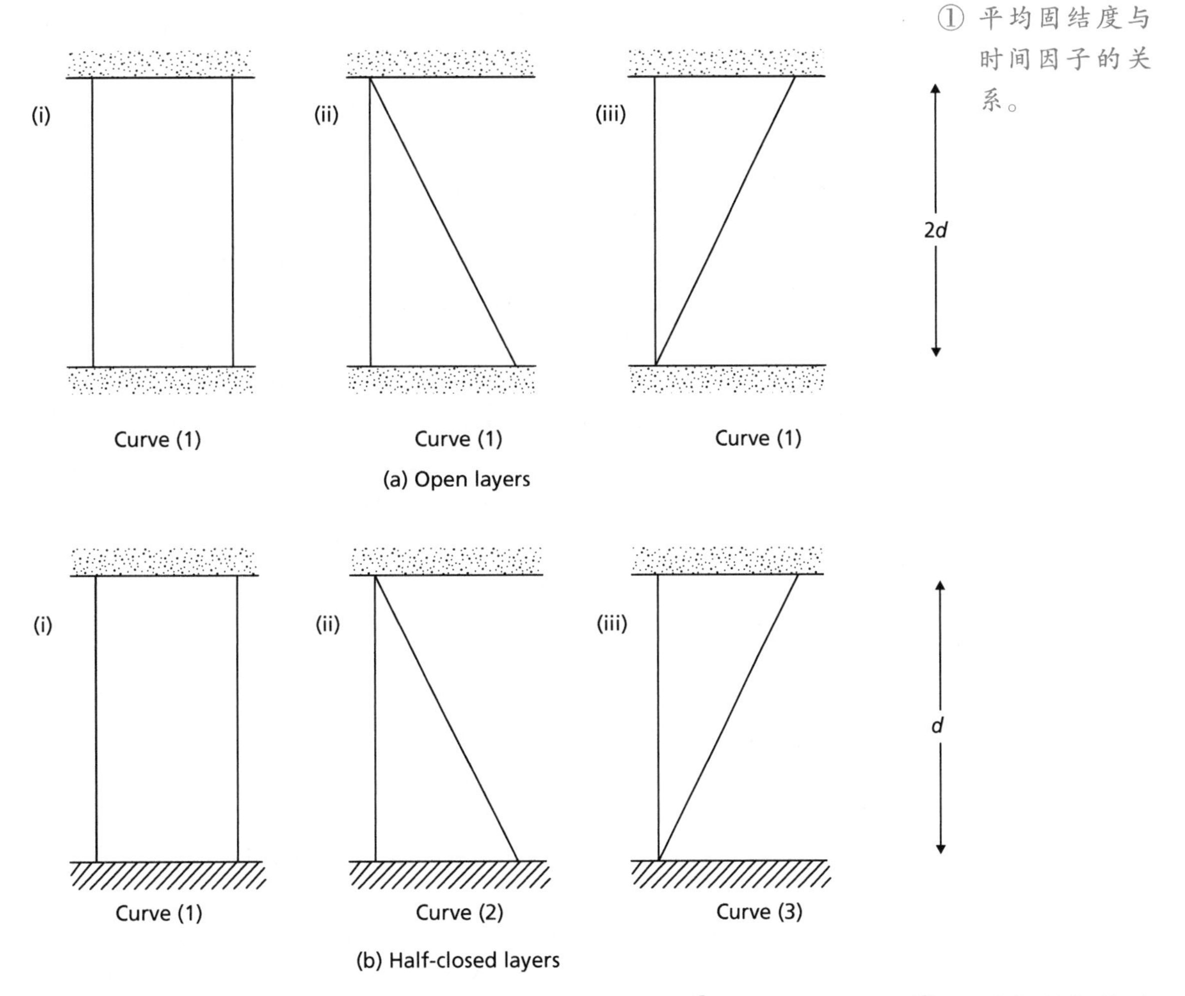

Figure 4.15 Initial variations of excess pore water pressure.[②]

② 不同初始超静孔隙水压力的分布。

① 固结系数的确定。

4.6 Determination of coefficient of consolidation[①]

In order to apply consolidation theory in practice, it is necessary to determine the value of the coefficient of consolidation for use in either Equation 4.23 or Figure 4.14. The value of c_v for a particular pressure increment in the oedometer test can be determined by comparing the characteristics of the experimental and theoretical consolidation curves, the procedure being referred to as curve-fitting.[②] The characteristics of the curves are brought out clearly if time is plotted to a square root or a logarithmic scale. It should be noted that once the value of c_v has been determined, the coefficient of permeability can be calculated from Equation 4.16, the oedometer test being a useful method for obtaining the permeability of finegrained soils.

② 曲线拟合。

③ 对数时间法（Casagrande 法）。

The log time method (due to Casagrande)[③]

The forms of the experimental and theoretical curves are shown in Figure 4.16. The experimental curve is obtained by plotting the dial gauge readings in the oedometer test against the time in minutes, plotted on a logarithmic axis. The theoretical curve (inset) is given as the plot of the average degree of consolidation against the logarithm of the time factor. The theoretical curve consists of three parts: an initial curve which approximates closely to a parabolic[④] relationship, a part which is linear and a final curve to which the horizontal axis is an asymptote at $U_v = 1.0$ (or 100%). In the experimental curve, the point corresponding to $U_v = 0$ can be determined by using the fact that the initial part of the curve represents an approximately parabolic relationship between compression and time. Two points on the curve are selected (A and B in Figure 4.16) for which the values of t are in the ratio of 1:4, and the vertical distance between them, ζ, is measured. In Figure 4.16, point A is shown at one minute and point B at four minutes. An equal distance of ζ above point A fixes the dial gauge reading (a_s) corresponding to $U_v = 0$. As a check, the procedure should be repeated using different pairs of points. The point corresponding to $U_v = 0$ will not generally correspond to the point (a_0) representing the initial dial gauge reading, the difference being due mainly to the compression of small quantities of air in the soil, the degree of saturation being marginally below 100%: this compression is called **initial compression**.[⑤] The final part of the experimental curve is linear but not horizontal, and the point (a_{100}) corresponding to $U_v = 100\%$ is taken as the intersection of the two linear parts of the curve. The compression between the a_s and a_{100} points is called **primary consolidation**[⑥], and represents that part of the process accounted for by Terzaghi's theory. Beyond the point of intersection, compression of the soil continues at a very slow rate for an indefinite period of time and is called **secondary compression**[⑦] (see Section 4.7). The point a_f in Figure 4.16 represents the final dial gauge reading before a subsequent total stress increment is applied.

④ 抛物线。

⑤ 初始压缩（瞬时压缩）。

⑥ 主固结。

⑦ 次压缩（次固结）。

The point corresponding to $U_v = 50\%$ can be located midway between the a_s and

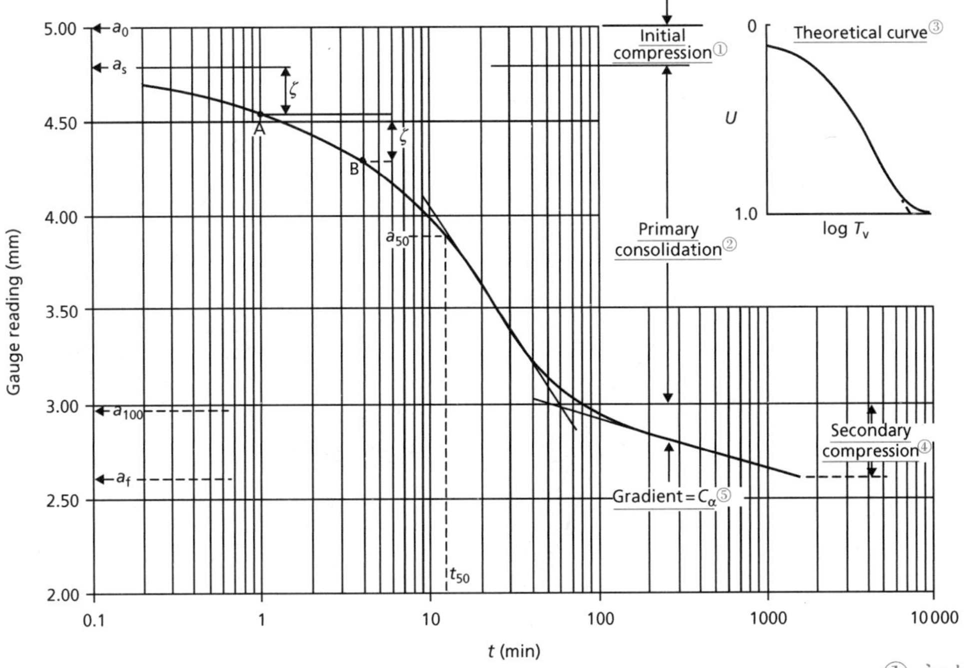

Figure 4.16 The log time method.

a_{100} points, and the corresponding time t_{50} obtained. The value of T_v corresponding to $U_v = 50\%$ is 0.196[⑥] (Equation 4.22 or Figure 4.14, curve 1), and the coefficient of consolidation is given by

$$c_v = \frac{0.196d^2}{t_{50}} \tag{4.25}$$

the value of d being taken as half the average thickness of the specimen for the particular pressure increment, due to the two-way drainage in the oedometer cell (to the top and bottom). If the average temperature of the soil in-situ is known and differs from the average test temperature, a correction should be applied to the value of c_v, correction factors being given in test standards (see Section 4.2).

The root time method (due to Taylor)[⑦]

Figure 4.17 shows the forms of the experimental and theoretical curves, the dial gauge readings being plotted against the square root of time in minutes and the average degree of consolidation against the square root of time factor respectively. The theoretical curve is linear up to approximately 60% consolidation, and at 90% consolidation the abscissa (AC) is 1.15 times the abscissa (AB) of the extrapolation of the linear part of the curve.[⑧] This characteristic is used to determine the point on the experimental curve corresponding to $U_v = 90\%$.[⑨]

The experimental curve usually consists of a short curved section representing initial compression, a linear part and a second curve. The point (D) corresponding to $U_v = 0$ is obtained by extrapolating the linear part of the curve to the ordinate at

① 初始压缩（瞬时压缩）。
② 主固结。
③ 理论曲线。
④ 次压缩（次固结）。
⑤ 梯度。
⑥ 平均固结度为50%对应的时间因子为0.196。
⑦ 时间平方根法（Taylor法）。
⑧ 理论固结曲线在固结度为60%前呈线性，在固结度为90%时，过90%固结度的水平线与固结曲线交点组成的直线段AC的长度是固结曲线直线段的外延段与水平线交点组成的直线段AB的长度的1.5倍。
⑨ 此特征用来确定试验曲线中对应固结度为90%的点。

① 初始压缩（瞬时压缩）。
② 理论曲线。
③ 主固结。
④ 次固结。

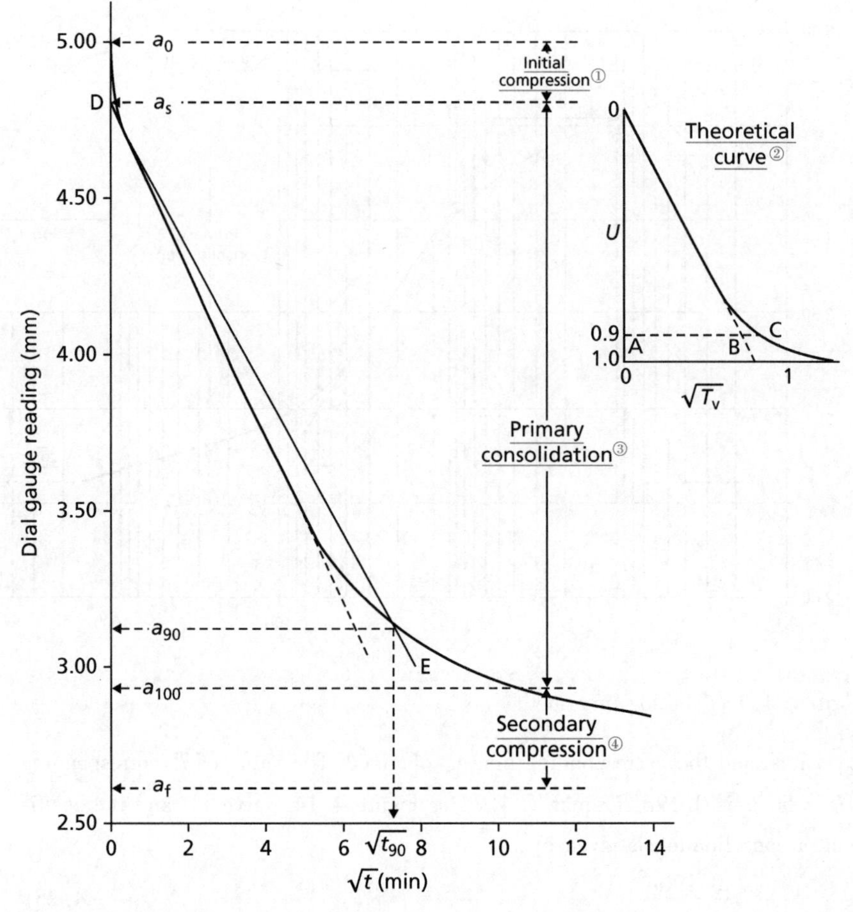

Figure 4. 17 The root time method.

zero time. A straight line (DE) is then drawn having abscissae 1. 15 times the corresponding abscissae on the linear part of the experimental curve. The intersection of the line DE with the experimental curve locates the point (a_{90}) corresponding to $U_v = 90\%$ and the corresponding value $\sqrt{t_{90}}$ can be obtained. The value of T_v corresponding to $U_v = 90\%$ is 0. 848⑤ (Equation 4. 22 or Figure 4. 14, curve 1), and the coefficient of consolidation is given by

⑤ 平均固结度为90%对应的时间因子为0. 848。

$$c_v = \frac{0.848d^2}{t_{90}} \tag{4.26}$$

If required, the point (a_{100}) on the experimental curve corresponding to $U_v = 100\%$, the limit of primary consolidation, can be obtained by proportion. As in the log time plot, the curve extends beyond the 100% point into the secondary compression range. The root time method requires compression readings covering a much shorter period of time compared with the log time method, which requires the accurate definition of the second linear part of the curve well into the secondary compression range. On the other hand, a straight-line portion is not always obtained on the root time plot, and in such cases the log time method should be used.

Other methods of determining c_v have been proposed by Naylor and Doran (1948), Scott (1961) and Cour (1971).

Compression ratios[1]

[1] 压缩比。

The relative magnitudes of the initial compression, the compression due to primary consolidation and the secondary compression can be expressed by the following ratios (refer to Figures 4.16 and 4.17).

$$\text{Initial compression ratio: } r_0 = \frac{a_0 - a_s}{a_0 - a_f} \tag{4.27}$$

$$\text{Primary compression ratio (log time): } r_p = \frac{a_s - a_{100}}{a_0 - a_f} \tag{4.28}$$

$$\text{Primary compression ratio (root time): } r_p = \frac{10\,(a_s - a_{90})}{9\,(a_0 - a_f)} \tag{4.29}$$

$$\text{Secondary compression ratio: } r_s = 1 - (r_0 + r_p) \tag{4.30}$$

In-situ value of c_v[2]

[2] 原位固结系数值。

Settlement observations have indicated that the rates of settlement of full-scale structures are generally much greater than those predicted using values of c_v obtained from oedometer tests on small specimens (e.g. 75mm diameter × 20mm thick). Rowe (1968) has shown that such discrepancies are due to the influence of the soil macro-fabric on drainage behaviour. Features such as laminations, layers of silt and fine sand, silt-filled fissures, organic inclusions and root-holes, if they reach a major permeable stratum, have the effect of increasing the overall permeability of the soil mass. In general, the macro-fabric of a field soil is not represented accurately in a small oedometer specimen, and the permeability of such a specimen will be lower than the mass permeability in the field.[3]

[3] 室内试样的渗透性将比场地的渗透性小。

In cases where fabric effects are significant, more realistic values of c_v can be obtained by means of the hydraulic oedometer developed by Rowe and Barden (1966) and manufactured for a range of specimen sizes. Specimens 250mm in diameter by 100mm in thick are considered large enough to represent the natural macro-fabric of most clays: values of c_v obtained from tests on specimens of this size have been shown to be consistent with observed settlement rates.

Details of a hydraulic oedometer are shown in Figure 4.18. Vertical pressure is applied to the specimen by means of water pressure acting across a convoluted rubber jack. The system used to apply the pressure must be capable of compensating for pressure changes due to leakage and specimen volume change. Compression of the specimen is measured by means of a central spindle passing through a sealed housing in the top plate of the oedometer. Drainage from the specimen can be either vertical or radial, and pore water pressure can be measured during the test The apparatus can also be used for flow tests, from which the coefficient of permeability can be determined directly (see Section 2.2).

Piezometers installed into the ground (see Chapter 6) can be used for the in-situ

① 橡胶千斤顶。
② 固结压力施加。
③ 轴。
④ 排水阀。
⑤ 透水板。
⑥ 试样。
⑦ 孔隙水压力表。
⑧ 液压固结仪。

determination of c_v, but the method requires the use of three-dimensional consolidation theory. The most satisfactory procedure is to maintain a constant head at the piezometer tip (above or below the ambient pore water pressure in the soil) and measure the rate of flow into or out of the system. If the rate of flow is measured at various times, the value of c_v (and of the coefficient of permeability k) can be deduced. Details have been given by Gibson (1966, 1970) and Wilkinson (1968).

Another method of determining c_v is to combine laboratory values of m_v (which from experience are known to be more reliable than laboratory values of c_v) with in-situ measurements of k, using Equation 4.16.

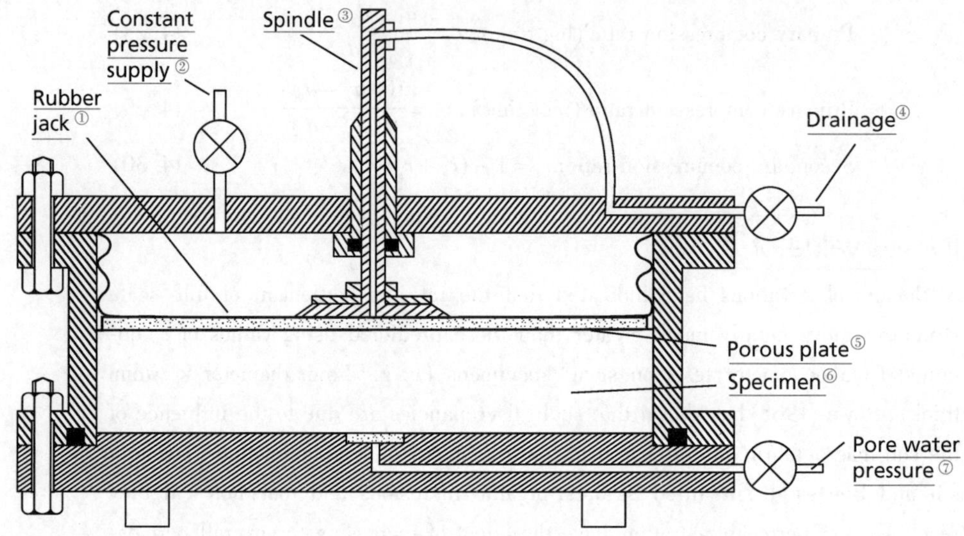

Figure 4.18 Hydraulic oedometer. ⑧

Example 4.3

The following compression readings were taken during an oedometer test on a saturated clay specimen ($G_s = 2.73$) when the applied pressure was increased from 214 to 429kPa:

TABLE G

Time (min)	0	$\frac{1}{4}$	$\frac{1}{2}$	1	$2\frac{1}{4}$	4	9	16	25
Gauge (mm)	5.00	4.67	4.62	4.53	4.41	4.28	4.01	3.75	3.49
Time (min)	36	49	64	81	100	200	400	1440	
Gauge (mm)	3.28	3.15	3.06	3.00	2.96	2.84	2.76	2.61	

After 1440min, the thickness of the specimen was 13.60mm and the water content was 35.9%. Determine the coefficient of consolidation from both the log time and the root time methods and the values of the three compression ratios. Determine also the value of the coefficient of permeability.

Solution

Total change in thickness during increment = 5.00 − 2.61 = 2.39mm

$$\text{Average thickness during increment} = 13.60 + \frac{2.39}{2} = 14.80\text{mm}$$

$$\text{Length of drainage path } d = \frac{14.80}{2} = 7.40\text{mm}$$

From the log time plot (data shown in Figure 4.16),

$$t_{50} = 12.5\text{min}$$

$$c_v = \frac{0.196d^2}{t_{50}} = \frac{0.196 \times 7.40^2}{12.5} \times \frac{1440 \times 365}{10^6} = 0.45\text{m}^2/\text{year}$$

$$r_0 = \frac{5.00 - 4.79}{5.00 - 2.61} = 0.088$$

$$r_p = \frac{4.79 - 2.98}{5.00 - 2.61} = 0.757$$

$$r_s = 1 - (0.088 + 0.757) = 0.155$$

From the root time plot (data shown in Figure 4.17) $\sqrt{t_{90}} = 7.30$, and therefore

$$t_{90} = 53.3\text{min}$$

$$c_v = \frac{0.848d^2}{t_{90}} = \frac{0.848 \times 7.40^2}{53.3} \times \frac{1440 \times 365}{10^6} = 0.46\text{m}^2/\text{year}$$

$$r_0 = \frac{5.00 - 4.81}{5.00 - 2.61} = 0.080$$

$$r_p = \frac{10\,(4.81 - 3.12)}{9\,(5.00 - 2.61)} = 0.785$$

$$r_s = 1 - (0.080 + 0.785) = 0.135$$

In order to determine the permeability, the value of m_v must be calculated.

Final void ratio, $e_1 = w_1 G_s = 0.359 \times 2.73 = 0.98$

Initial void ratio, $e_0 = e_1 + \Delta e$

Now

$$\frac{\Delta e}{\Delta H} = \frac{1 + e_0}{H_0}$$

i. e.

$$\frac{\Delta e}{2.39} = \frac{1.98 + \Delta e}{15.99}$$

Therefore

$$\Delta e = 0.35 \text{ and } e_0 = 1.33$$

Now

$$m_v = \frac{1}{1 + e_0} \cdot \frac{e_0 - e_1}{\sigma_1' - \sigma_0'}$$

$$= \frac{1}{2.33} \times \frac{0.35}{215} = 7.0 \times 10^{-4}\text{m}^2/\text{kN}$$

$$= 0.70\text{m}^2/\text{MN}$$

Coefficient of permeability:

$$k = c_v m_v \gamma_w$$

$$= \frac{0.45 \times 0.70 \times 9.8}{60 \times 1440 \times 365 \times 10^3}$$

$$= 1.0 \times 10^{-10}\text{m/s}$$

① 次固结。

4.7 Secondary compression[①]

In the Terzaghi theory, it is implied by assumption 8 that a change in void ratio is due entirely to a change in effective stress brought about by the dissipation of excess pore water pressure, with permeability alone governing the time dependency of the process. However, experimental results show that compression does not cease when the excess pore water pressure has dissipated to zero but continues at a gradually decreasing rate under constant effective stress (see Figure 4.16). Secondary compression is thought to be due to the gradual readjustment of fine-grained particles into a more stable configuration following the structural disturbance[②] caused by the decrease in void ratio, especially if the soil is laterally confined. An additional factor is the gradual lateral displacements which take place in thick layers subjected to shear stresses. The rate of secondary compression is thought to be controlled by the highly viscous film of adsorbed water surrounding clay mineral particles in the soil.[③] A very slow viscous flow of adsorbed water takes place from the zones of film contact, allowing the solid particles to move closer together. The viscosity of the film increases as the particles move closer, resulting in a decrease in the rate of compression of the soil. It is presumed that primary consolidation and secondary compression proceed simultaneously from the time of loading.[④]

② 结构性扰动。

③ 土中黏土矿物周围吸附的高黏滞性水膜。

④ 通常假定从加载开始，主固结与次固结是同时进行的。

⑤ 天然土的次固结特性。

Table 4.3 Secondary compression characteristics of natural soils[⑤] (after Mitchell and Soga, 2005)

Soil type	C_α / C_c
Sands (low fines content)	0.01 – 0.03
Clays and silts	0.03 – 0.08
Organic soils	0.05 – 0.10

The rate of secondary compression in the oedometer test can be defined by the slope (C_α) of the final part of the compression – log time curve (Figure 4.16). Mitchell and Soga (2005) collated data on C_α for a range of natural soils, normalising the data by C_c. These data are summarised in Table 4.3, from which it can be seen that the secondary compression is typically between 1% and 10% of the primary compression, depending on soil type.

The magnitude of secondary compression occurring over a given time is generally greater in normally consolidated clays than in overconsolidated clays.[⑥] In overconsolidated clays strains are mainly elastic, but in normally consolidated clays significant plastic strains occur. For certain highly plastic clays and organic clays, the secondary compression part of the compression – log time curve may completely mask the primary consolidation part. For a particular soil the magnitude of secondary compression over a given time, as a percentage of the total compression, increases as the ratio of pressure increment to initial pressure decreases; the magnitude of

⑥ 在给定时间内，正常固结黏土中的次固结压缩变形比超固结黏土的大。

secondary compression also increases as the thickness of the oedometer specimen decreases and as temperature increases. Thus, the secondary compression characteristics of an oedometer specimen cannot normally be extrapolated to the full-scale situation.

In a small number of normally consolidated clays it has been found that secondary compression forms the greater part of the total compression under applied pressure. Bjerrum (1967) showed that such clays have gradually developed a reserve resistance against further compression as a result of the considerable decrease in void ratio which has occurred, under constant effective stress, over the hundreds or thousands of years since deposition. These clays, although normally consolidated, exhibit a quasi-preconsolidation pressure.[①] It has been shown that, provided any additional applied pressure is less than approximately 50% of the difference between the quasi-preconsolidation pressure and the effective overburden pressure, the resultant settlement will be relatively small.

① 准先期固结压力。

4.8 Numerical solution using the Finite Difference Method[②]

The one-dimensional consolidation equation can be solved numerically by the method of finite differences. The method has the advantage that any pattern of initial excess pore water pressure can be adopted, and it is possible to consider problems in which the load is applied gradually over a period of time. The errors associated with the method are negligible, and the solution is easily programmed for the computer. A spreadsheet tool, Consolidation_CSM8. xls, which implements the FDM for solution of consolidation problems, as outlined in this section, is provided on the Companion Website.

② 使用有限差分法的数值解法。

The method is based on a depth – time grid (or mesh) as shown in Figure 4. 19. This differs from the FDM described in Chapter 2, where the two-dimensional geometry meant that only the steady-state pore water pressures[③] could be determined straightforwardly. Due to the simpler one-dimensional nature of the consolidation process, the second dimension of the finite difference grid can be used to determine the evolution of pore water pressure with time.

③ 稳态孔隙水压力。

The depth of the consolidating soil layer is divided into m equal parts of thickness Δz, and any specified period of time is divided into n equal intervals Δt. Any point on the grid can be identified by the subscripts i and j, the depth position of the point being denoted by i ($0 \leqslant i \leqslant m$) and the elapsed time by j ($0 \leqslant j \leqslant n$). The value of excess pore water pressure at any depth after any time is therefore denoted by $u_{i,j}$. (In this section the subscript e is dropped from the symbol for excess pore water pressure, i. e. u represents u_e as defined in Section 3. 4). The following finite difference approximations can be derived from Taylor's theorem:

$$\frac{\partial u}{\partial t} = \frac{1}{\Delta t}(u_{i,j+1} - u_{i,j})$$

$$\frac{\partial^2 u}{\partial z^2}=\frac{1}{(\Delta z)^2}(u_{i-1,j}+u_{i+1,j}-2u_{i,j})$$

Substituting these values in Equation 4.15 yields the finite difference approximation of the one-dimensional consolidation equation:

$$u_{i,j+1}=u_{i,j}+\frac{c_v\Delta t}{(\Delta z)^2}(u_{i-1,j}+u_{i+1,j}-2u_{i,j}) \tag{4.31}$$

It is convenient to write

$$\beta=\frac{c_v\Delta t}{(\Delta z)^2} \tag{4.32}$$

this term being called the operator of Equation 4.31. It has been shown that for convergence the value of the operator must not exceed 1/2.[①] The errors due to neglecting higher-order derivatives[②] in Taylor's theorem are reduced to a minimum when the value of the operator is 1/6.

① 为了收敛，该算子的值一定不能超过1/2。
② 高阶分量。

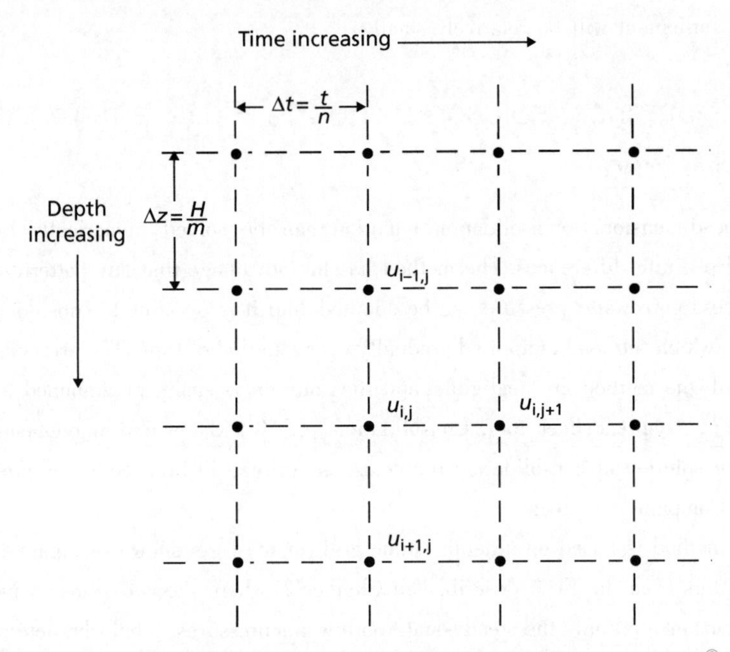

Figure 4.19 One-dimensional depth-time Finite Difference mesh.[③]

③ 一维深度-时间有限差分网格。

It is usual to specify the number of equal parts m into which the depth of the layer is to be divided, and as the value of β is limited a restriction is thus placed on the value of Δt. For any specified period of time t in the case of an open layer:

$$T_v=\frac{c_v(n\Delta t)}{\left(\frac{1}{2}m\Delta z\right)^2}$$

$$=4\frac{n}{m^2}\beta \tag{4.33}$$

In the case of a half-closed layer the denominator becomes $(m\Delta z)^2$ and

$$T_v=\frac{n}{m^2}\beta \tag{4.34}$$

A value of n must therefore be chosen such that the value of β in Equation 4.33 or

4.34 does not exceed 1/2.

Equation 4.31 does not apply to points on an impermeable boundary. There can be no flow across an impermeable boundary, a condition represented by the equation:

$$\frac{\partial u}{\partial z}=0$$

which can be represented by the finite difference approximation:

$$\frac{1}{2\Delta z}(u_{i-1,j}-u_{i+1,j})=0$$

the impermeable boundary being at a depth position denoted by subscript i, i. e.

$$u_{i-1,j}=u_{i+1,j}$$

For all points on an impermeable boundary, Equation 4.31 therefore becomes

$$u_{i+1,j}=u_{i,j}+\frac{c_v\Delta t}{(\Delta z)^2}(2u_{i-1,j}-2u_{i,j}) \tag{4.35}$$

The degree of consolidation at any time t can be obtained by determining the areas under the initial isochrone and the isochrone at time t as in Equation 4.24. An implementation of this methodology may be found within Consolidation_CSM8.xls on the Companion Website.

Example 4.4

A half-closed clay layer (free-draining at the upper boundary) is 10m thick and the value of c_v is 7.9m^2/year. The initial distribution of excess pore water pressure is as follows:

TABLE H

Depth (m)	0	2	4	6	8	10
Pressure (kPa)	60	54	41	29	19	15

Obtain the values of excess pore water pressure after consolidation has been in progress for 1 year.

Solution

The layer is half-closed, and therefore $d=10\text{m}$. For $t=1$ year,

$$T_v=\frac{c_v t}{d^2}=\frac{7.9\times 1}{10^2}=0.079$$

Table 4.4 Example 4.4

i	*j* 0	1	2	3	4	5	6	7	8	9	10
0	0	0	0	0	0	0	0	0	0	0	0
1	54.0	40.6	32.6	27.3	23.5	20.7	18.5	16.7	15.3	14.1	13.1
2	41.0	41.2	38.7	35.7	32.9	30.4	28.2	26.3	24.6	23.2	21.9
3	29.0	29.4	29.9	30.0	29.6	29.0	28.3	27.5	26.7	26.0	25.3
4	19.0	20.2	21.3	22.4	23.3	24.0	24.5	24.9	25.1	25.2	25.2
5	15.0	16.6	18.0	19.4	20.6	21.7	22.6	23.4	24.0	24.4	24.7

The layer is divided into five equal parts, i. e. $m = 5$. Now

$$T_v = \frac{n}{m^2}\beta$$

Therefore

$$n\beta = 0.079 \times 5^5 = 1.98 \text{ (say 2.0)}$$

(This makes the actual value of $T_v = 0.080$ and $t = 1.01$ years.) The value of n will be taken as 10 (i. e. $\Delta t = 1/10$ year), making $\beta = 0.2$. The finite difference equation then becomes

$$u_{i,j+1} = u_{i,j} + 0.2\,(u_{i-1,j} + u_{i+1,j} - 2u_{i,j})$$

but on the impermeable boundary:

$$u_{i,j+1} = u_{i,j} + 0.2\,(2u_{i-1,j} - 2u_{i,j})$$

On the permeable boundary, $u = 0$ for all values of t, assuming the initial pressure of 60kPa instantaneously becomes zero.

The calculations are set out in Table 4.4. The use of the spreadsheet analysis tool Consolidation_CSM8.xls to analyse this example, and the other worked examples in this chapter, may be found on the Companion Website.

4.9 Correction for construction period[1]

[1] 施工期的修正。

In practice, structural loads are applied to the soil not instantaneously but over a period of time. Initially there is usually a reduction in net load due to excavation, resulting in swelling: settlement will not begin until the applied load exceeds the weight of the excavated soil. Terzaghi (1943) proposed an empirical method of correcting the instantaneous time – settlement curve to allow for the construction period.

The net load (P') is the gross load less the weight of soil excavated, and the effective construction period (t_c) is measured from the time when P' is zero. It is assumed that the net load is applied uniformly over the time t_c (Figure 4.20) and that the degree of consolidation at time t_c is the same as if the load P' had been acting as a constant load for the period $1/2t_c$. Thus the settlement at any time during the construction period is equal to that occurring for instantaneous loading at half that time; however, since the load then acting is not the total load, the value of settlement so obtained must be reduced in the proportion of that load to the total load. This procedure is shown graphically in Figure 4.20.

For the period subsequent to the completion of construction, the settlement curve will be the instantaneous curve offset by half the effective construction period. Thus at any time after the end of construction, the corrected time corresponding to any value of settlement is equal to the time from the start of loading less half the effective construction period. After a long period of time, the magnitude of settlement is not appreciably affected by the construction time.

Alternatively, a numerical solution (Section 4.8) can be used, successive increments of excess pore water pressure being applied over the construction period.

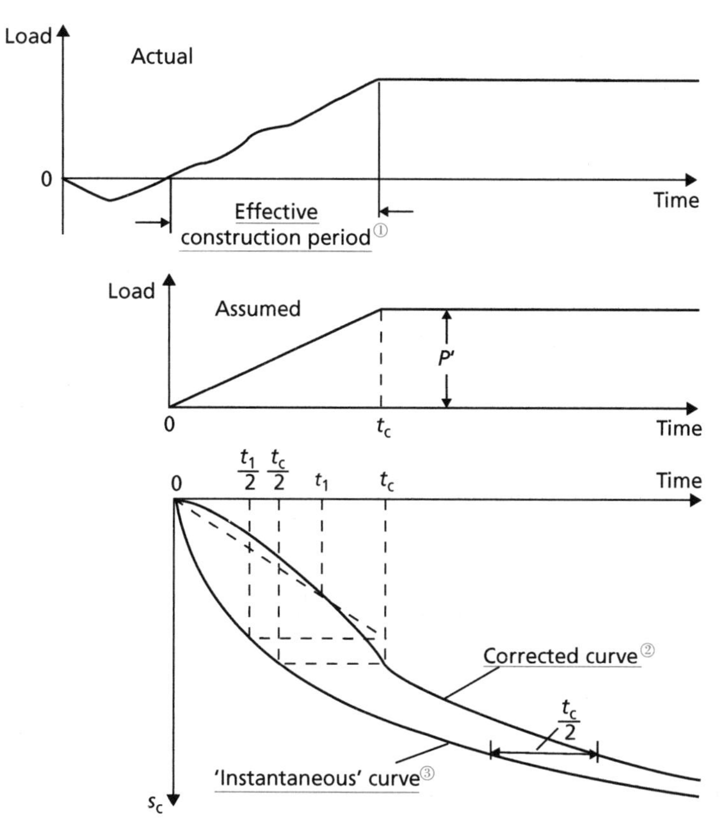

Figure 4.20 Correction for construction period.④

① 有效施工期。
② 修正曲线。
③ 瞬时一次加载曲线。
④ 施工期的修正。

Example 4.5

A layer of clay 8m thick lies between two layers of sand as shown in Figure 4.21. The upper sand layer extends from ground level to a depth of 4m, the water table being at a depth of 2m. The lower sand layer is under artesian pressure, the piezometric level being 6m above ground level. For the clay, $m_v = 0.94\text{m}^2/\text{MN}$ and $c_v = 1.4\text{m}^2/\text{year}$. As a result of pumping from the artesian layer, the piezometric level falls by 3m over a period of two years. Draw the time – settlement curve due to consolidation of the clay for a period of five years from the start of pumping.

Solution

In this case, consolidation is due only to the change in pore water pressure at the lower boundary of the clay: there is no change in total vertical stress. The effective vertical stress remains unchanged at the top of the clay layer, but will be increased by $3\gamma_w$ at the bottom of the layer due to the decrease in pore water pressure in the adjacent artesian layer. The distribution of $\Delta\sigma'$ is shown in Figure 4.21. The problem is one-dimensional, since the increase in effective vertical stress is the same over the entire area in question. In calculating the consolidation settlement, it is necessary to consider only the value of $\Delta\sigma'$ at the centre of the layer. Note that in order to obtain the value of m_v it would have been necessary to calculate the initial and final values of effective vertical stress in the clay.

At the centre of the clay layer, $\Delta\sigma' = 1.5\gamma_w = 14.7\text{kPa}$. The final consolidation settlement is given by

$$s_{oed} = m_v \Delta\sigma' H$$

$$= 0.94 \times 14.7 \times 8 = 110\text{mm}$$

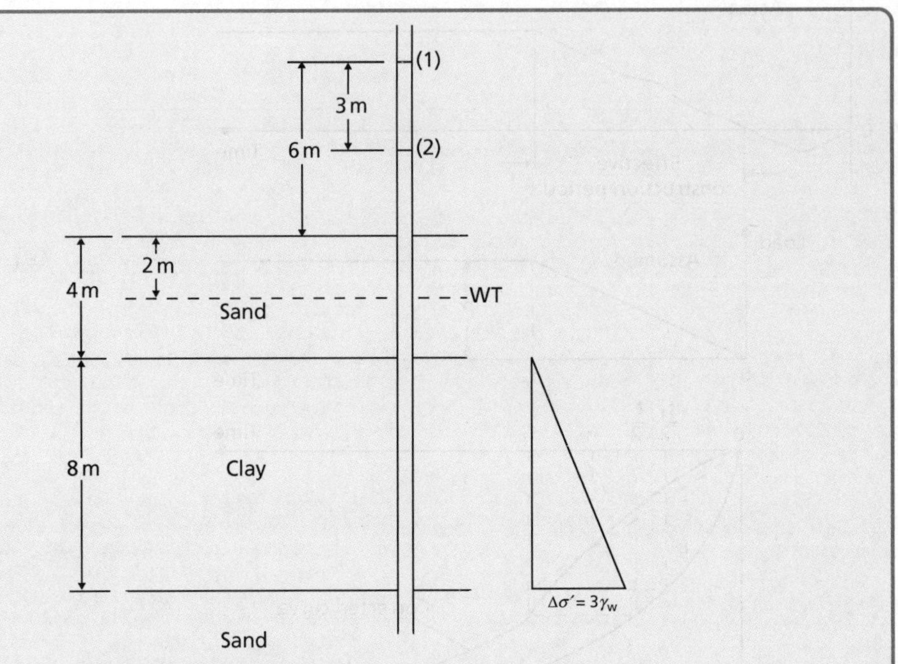

Figure 4.21 Example 4.5.

The clay layer is open and two-way drainage can occur due to the high permeability of the sand above and below the clay: therefore $d = 4\text{m}$. For $t = 5$ years,

$$T_v = \frac{c_v t}{d^2}$$

$$= \frac{1.4 \times 5}{4^2}$$

$$= 0.437$$

From curve 1 of Figure 4.14, the corresponding value of U_v is 0.73. To obtain the time – settlement relationship, a series of values of U_v is selected up to 0.73 and the corresponding times calculated from the time factor equation (4.19): the corresponding values of consolidation settlement (s_c) are given by the product of U_v and s_{oed} (see Table 4.5). The plot of s_c against t gives the 'instantane-ous' curve. Terzaghi's method of correction for the two-year period over which pumping takes place is then carried out as shown in Figure 4.22.

Table 4.5 Example 4.5

U_v	T_v	t (*years*)	s_c (*mm*)
0.10	0.008	0.09	11
0.20	0.031	0.35	22
0.30	0.070	0.79	33
0.40	0.126	1.42	44
0.50	0.196	2.21	55
0.60	0.285	3.22	66
0.73	0.437	5.00	80

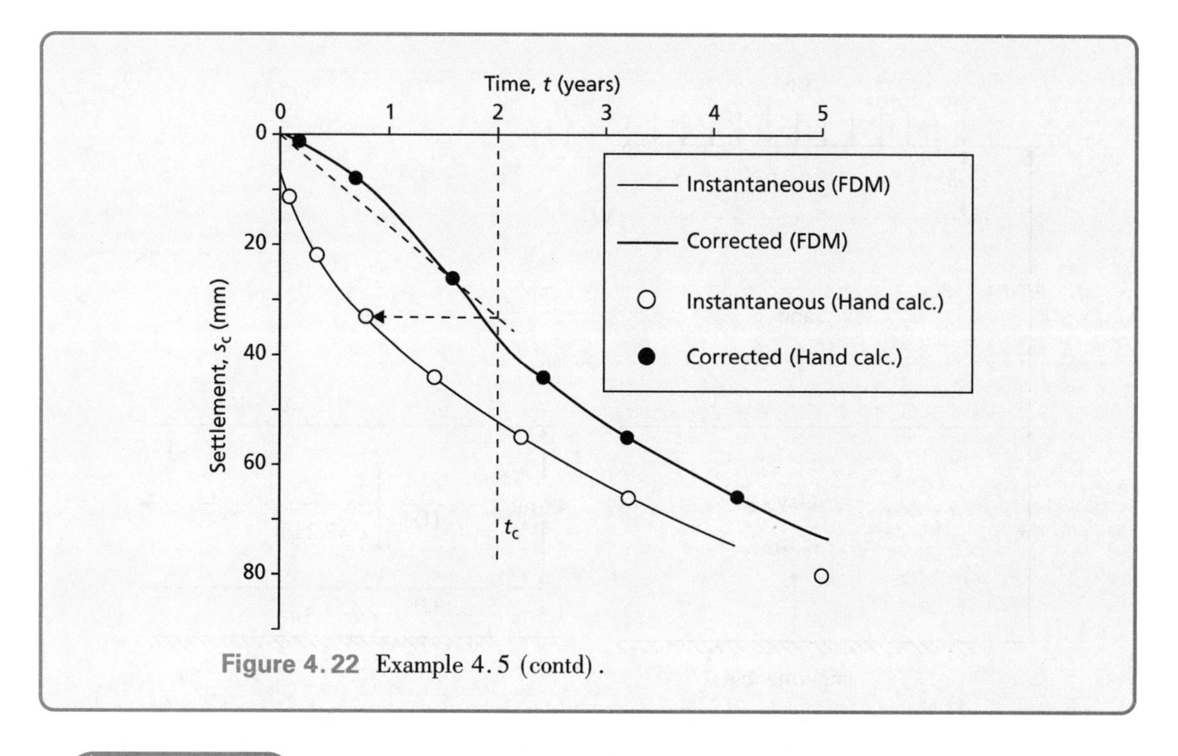

Figure 4.22 Example 4.5 (contd).

Example 4.6

An 8m depth of sand overlies a 6m layer of clay, below which is an impermeable stratum (Figure 4.23); the water table is 2m below the surface of the sand. Over a period of one year, a 3-m depth of fill (unit weight 20kN/m^3) is to be placed on the surface over an extensive area. The saturated unit weight of the sand is 19kN/m^3 and that of the clay is 20kN/m^3; above the water table the unit weight of the sand is 17kN/m^3. For the clay, the relationship between void ratio and effective stress (units kPa) can be represented by the equation

$$e = 0.88 - 0.32\log\left(\frac{\sigma'}{100}\right)$$

and the coefficient of consolidation is $1.26\text{m}^2/\text{year}$.

a Calculate the final settlement of the area due to consolidation of the clay and the settlement after a period of three years from the start of fill placement.

b If a very thin layer of sand, freely draining, existed 1.5m above the bottom of the clay layer (Figure 4.23 (b)), what would be the values of the final and three-year settlements?

Solution

a Since the fill covers a wide area, the problem can be considered to be one-dimensional. The consolidation settlement will be calculated in terms of C_c, considering the clay layer as a whole, and therefore the initial and final values of effective vertical stress at the centre of the clay layer are required.

$$\sigma_0' = (17 \times 2) + (9.2 \times 6) + (10.2 \times 3) = 119.8\text{kPa}$$

$$e_0 = 0.88 - 0.32\log 1.198 = 0.88 - 0.025 = 0.855$$

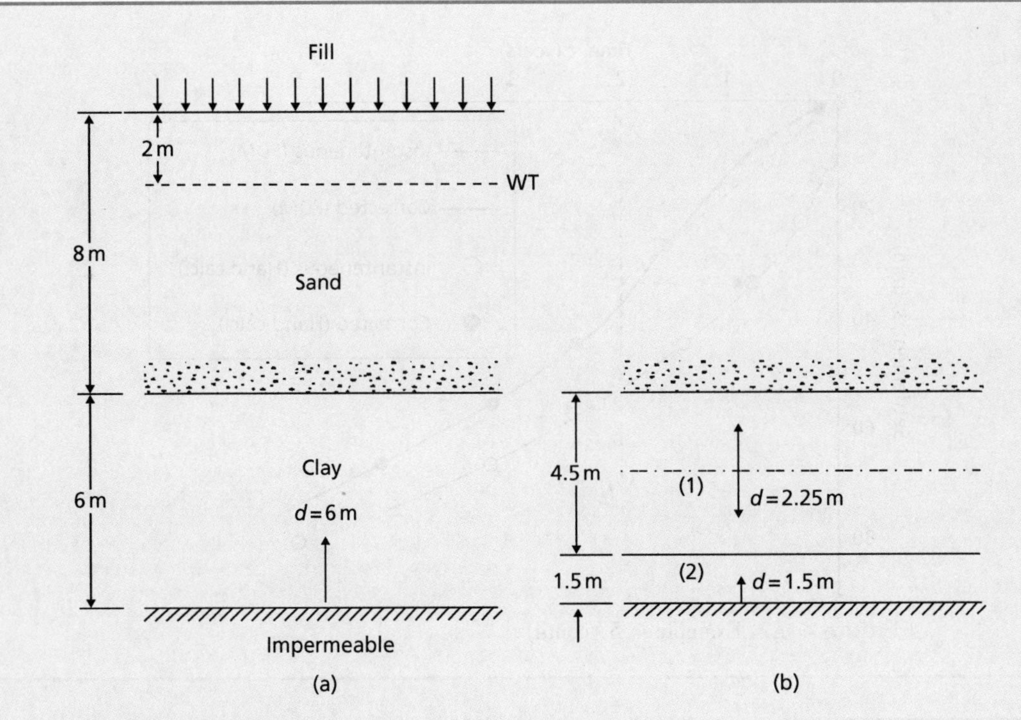

Figure 4.23 Example 4.6.

$$\sigma_1' = 119.8 + (3 \times 20) = 179.8\text{kPa}$$

$$\log\left(\frac{179.8}{119.8}\right) = 0.176$$

The final settlement is calculated from Equation 4.11:

$$s_{\text{oed}} = \frac{0.32 \times 0.176 \times 6000}{1.855} = 182\text{mm}$$

In the calculation of the degree of consolidation three years after the start of fill placement, the corrected value of time to allow for the one-year filling period is

$$t = 3 - \frac{1}{2} = 2.5 \text{ years}$$

The layer is half-closed, and therefore $d = 6\text{m}$. Then

$$T_v = \frac{c_v t}{d^2} = \frac{1.26 \times 2.5}{6^2}$$

$$= 0.0875$$

From curve 1 of Figure 4.14, $U_v = 0.335$. Settlement after three years:

$$s_c = 0.335 \times 182 = 61\text{mm}$$

b The final settlement will still be 182mm (ignoring the thickness of the drainage layer); only the rate of settlement will be affected. From the point of view of drainage there is now an open layer of thickness 4.5m ($d = 2.25\text{m}$) above a half-closed layer of thickness 1.5m ($d = 1.5\text{m}$): these layers are numbered 1 and 2, respectively.

By proportion

$$T_{v1} = 0.0875 \times \frac{6^2}{2.25^2} = 0.622$$

$$\therefore U_1 = 0.825$$

and

$$T_{v2} = 0.0875 \times \frac{6^2}{1.5^2} = 1.40$$

$$\therefore U_2 = 0.97$$

Now for each layer, $s_c = U_v s_{oed}$, which is proportional to $U_v H$. Hence if $\bar{U}$ is the overall degree of consolidation for the two layers combined:

$$4.5U_1 + 1.5U_2 = 6.0\bar{U}$$

i.e. $(4.5 \times 0.825) + (1.5 \times 0.97) = 6.0\bar{U}$.

Hence $\bar{U} = 0.86$ and the 3-year settlement is

$$s_c = 0.86 \times 182 = 157\text{mm}$$

4.10 Vertical drains

The slow rate of consolidation in saturated clays of low permeability may be accelerated by means of vertical drains which shorten the drainage path within the clay. Consolidation is then due mainly to horizontal radial drainage, resulting in the faster dissipation of excess pore water pressure and consequently more rapid settlement; vertical drainage becomes of minor importance. In theory the final magnitude of consolidation settlement is the same, only the rate of settlement being affected.

In the case of an embankment constructed over a highly compressible clay layer (Figure 4.24), vertical drains installed in the clay would enable the embankment to be brought into service much sooner. A degree of consolidation of the order of 80% would be desirable at the end of construction, to keep in-service settlements to an acceptable level. Any advantages, of course, must be set against the additional cost of the installation.

The traditional method of installing vertical drains was by driving boreholes through the clay layer and backfilling with a suitably graded sand, typical diameters being 200 – 400mm to depths of over 30m. Prefabricated drains are now generally used, and tend to be more economic than backfilled drains for a given area of treatment. One type of drain (often referred to as a '**sandwick**') consists of a filter stocking, usually of woven polypropylene, filled with sand. Compressed air is used to ensure that the stocking is completely filled with sand. This type of drain, a typical diameter being 65mm, is very flexible and is generally unaffected by lateral soil displacement, the possibility of necking being virtually eliminated. The drains are installed either by insertion into pre-bored holes or, more commonly, by placing them inside a mandrel or casing which is then driven or vibrated into the ground.

Another type of prefabricated drain is the **band drain**, consisting of a flat plastic core indented with drainage channels, surrounded by a layer of filter fabric. The fabric must have sufficient strength to prevent it from being squeezed into the channels, and the mesh size must be small enough to prevent the passage of soil particles which could clog the channels. Typical dimensions of a band drain are 100 × 4mm, and in design the equivalent diameter is assumed to be the perimeter divided by π. Band drains are installed by placing them inside a steel mandrel which is either pushed, driven or vibrated into the ground. An anchor is attached to the lower end of the drain to keep it in position as the mandrel is withdrawn. The anchor also prevents soil from entering the mandrel during installation.

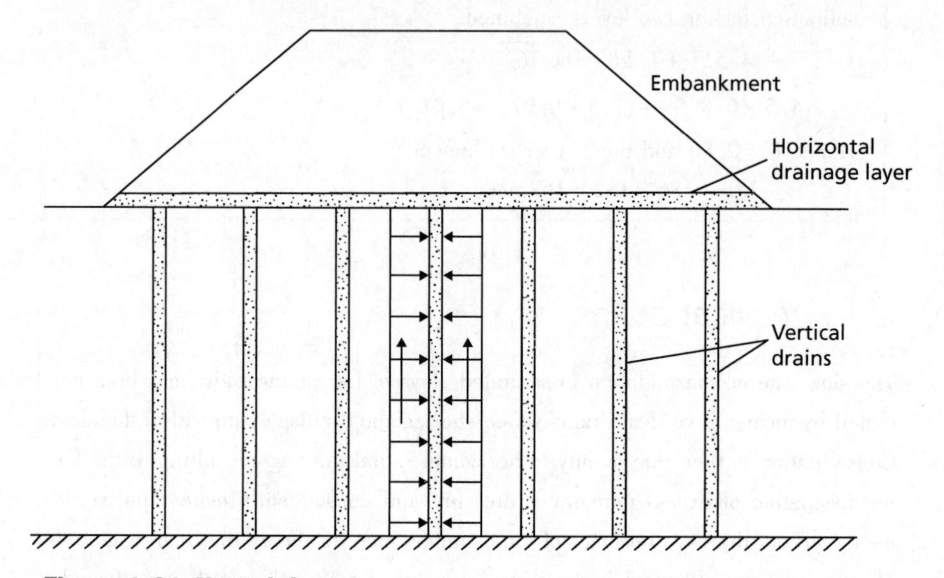

Figure 4.24 Vertical drains.

Drains are normally installed in either a square or a triangular pattern. As the object is to reduce the length of the drainage path, the spacing of the drains is the most important design consideration. The spacing must obviously be less than the thickness of the clay layer for the consolidation rate to be improved; there is therefore no point in using vertical drains in relatively thin layers. It is essential for a successful design that the coefficients of consolidation in both the horizontal and the vertical directions (c_h and c_v, respectively) are known as accurately as possible. In particular, the accuracy of c_h is the most crucial factor in design, being more important than the effect of simplifying assumptions in the theory used. The ratio c_h/c_v is normally between 1 and 2; the higher the ratio, the more advantageous a drain installation will be. A design complication in the case of large-diameter sand drains is that the column of sand tends to act as a weak pile (see Chapter 9), reducing the vertical stress increment imposed on the clay layer by an unknown degree, resulting in lower excess pore water pressure and therefore reduced consolidation settlement. This effect is minimal in the case of prefabricated drains because of their flexibility.

Vertical drains may not be effective in overconsolidated clays if the vertical stress after consolidation remains less than the preconsolidation pressure. Indeed, disturbance of overconsolidated clay during drain installation might even result in increased final consolidation settlement. It should be realised that the rate of secondary compression cannot be controlled by vertical drains.

In polar coordinates, the three-dimensional form of the consolidation equation, with different soil properties in the horizontal and vertical directions, is

$$\frac{\partial u_e}{\partial t} = c_h \left(\frac{\partial^2 u_e}{\partial r^2} + \frac{1}{r} \frac{\partial u_e}{\partial r} \right) + c_v \frac{\partial^2 u_e}{\partial z^2} \tag{4.36}$$

The vertical prismatic blocks of soil which are drained by and surround each drain are replaced by cylindrical blocks, of radius R, having the same cross-sectional area (Figure 4.25). The solution to Equation 4.36 can be written in two parts:

$$U_v = f(T_v)$$

and

$$U_r = f(T_r)$$

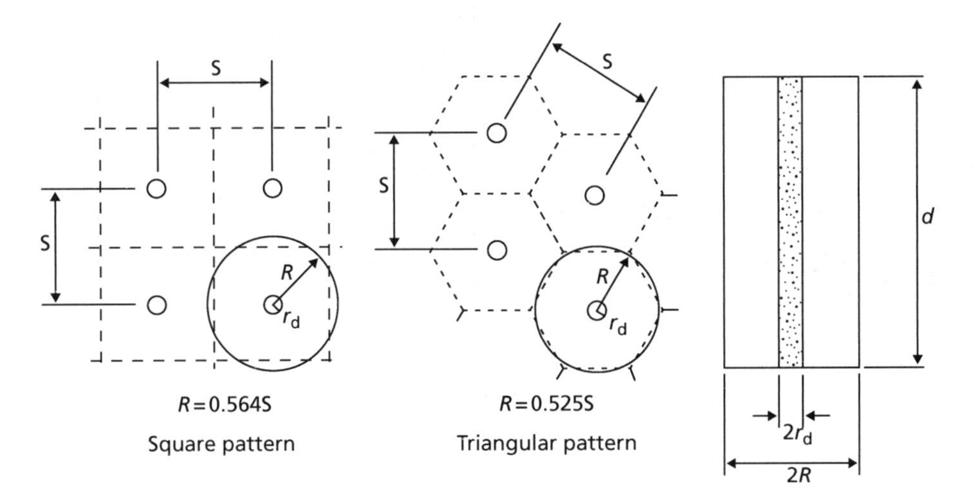

Figure 4.25 Cylindrical blocks.

where U_v = average degree of consolidation due to vertical drainage only; U_r = average degree of consolidation due to horizontal (radial) drainage only;

$$T_v = \frac{c_v t}{d^2} \tag{4.37}$$

= time factor for consolidation due to vertical drainage only

$$T_r = \frac{c_h t}{4R^2} \tag{4.38}$$

= time factor for consolidation due to radial drainage only

The expression for T_r confirms the fact that the closer the spacing of the drains, the quicker consolidation due to radial drainage proceeds.① The solution of the consolidation equation for radial drainage only may be found in Barron (1948); a simplified version which is appropriate for design is given by Hansbo (1981) as

$$U_r = 1 - e^{-\frac{8T_r}{\mu}} \tag{4.39}$$

① T_r 的表达式表明，排水间距越小，径向排水所提高的固结速度越快。

where

$$\mu=\frac{n^2}{n^2-1}\left(\ln n-\frac{3}{4}+\frac{1}{n^2}+\frac{1}{n^4}\right)\approx\ln n-\frac{3}{4} \tag{4.40}$$

In Equation 4.40, $n=R/r_d$, R is the radius of the equivalent cylindrical block and r_d the radius of the drain. The consolidation curves given by Equation 4.39 for various values of n are plotted in Figure 4.26. Some vertical drainage will continue to occur even if vertical drains have been installed and it can also be shown that

$$(1-U)=(1-U_v)(1-U_r) \tag{4.41}$$

where U is the average degree of consolidation under combined vertical and radial drainage.

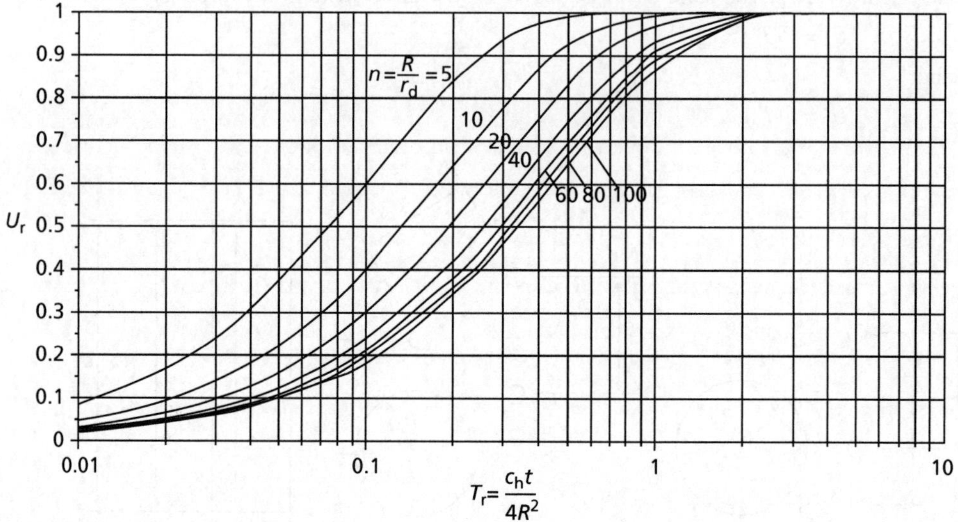

Figure 4.26 Relationships between average degree of consolidation and time factor for radial drainage.[①]

① 径向排水条件下平均固结度与时间因数的关系。

Installation effects[②]

② 施工效应。

The values of the soil properties for the soil immediately surrounding the drains may be significantly reduced due to remoulding during installation, especially if boring is used, an effect known as **smear**.[③] The smear effect can be taken into account either by assuming a reduced value of c_h or by using a reduced drain diameter in Equations 4.39 – 4.41. Alternatively, if the extent and permeability (k_s) of the smeared material are known or can be estimated, the expression for μ in Equation 4.40 can be modified after Hansbo (1979) as follows:

③ 涂抹（效应）。

$$\mu\approx\ln\frac{n}{S}+\frac{k}{k_s}\ln S-\frac{3}{4} \tag{4.42}$$

Example 4.7

An embankment is to be constructed over a layer of clay 10m thick, with an impermeable lower boundary. Construction of the embankment will increase the total vertical stress in the clay layer by 65kPa. For the clay, $c_v=4.7\text{m}^2/\text{year}$, $c_h=7.9\text{m}^2/\text{year}$ and $m_v=0.25\text{m}^2/\text{MN}$. The design requirement is that all but 25mm of the settlement due to consolidation of the clay layer will have taken place after 6 months. Determine the spacing, in a square pattern, of 400-mm diameter sand drains to achieve the above requirement.

Solution

$$\text{Final settlement} = m_v \Delta\sigma' H = 0.25 \times 65 \times 10$$
$$= 162\text{mm}$$

For $t = 6$ months,

$$U = \frac{162 - 25}{162} = 0.85$$

For vertical drainage only, the layer is half-closed, and therefore $d = 10$m.

$$T_v = \frac{c_v t}{d^2} = \frac{4.7 \times 0.5}{10^2} = 0.0235$$

From curve 1 of Figure 4.14, or using the spreadsheet tool on the Companion Website $U_v = 0.17$.

For radial drainage the diameter of the sand drains is 0.4m, i.e. $r_d = 0.2$m.

Radius of cylindrical block:

$$R = nr_d = 0.2n$$

$$T_r = \frac{c_h t}{4R^2} = \frac{7.9 \times 0.5}{4 \times 0.2^2 \times n^2} = \frac{24.7}{n^2}$$

i.e.

$$U_r = 1 - e^{-\left[\frac{8 \times 24.7}{n^2(\ln n - 0.75)}\right]}$$

Now $(1 - U) = (1 - U_v)(1 - U_r)$, and therefore

$$0.15 = 0.83(1 - U_r)$$

$$U_r = 0.82$$

The value of n for $U_r = 0.82$ may then be found by evaluating U_r at different values of n using Equations 4.39 and 4.40 and interpolating the value of n at which $U_r = 0.82$. Alternatively a simple 'Goal seek' or optimisation routine in a standard spreadsheet may be used to solve the equations iteratively. For the first of these methods, Figure 4.27 plots the value of U_r against n, from which it may

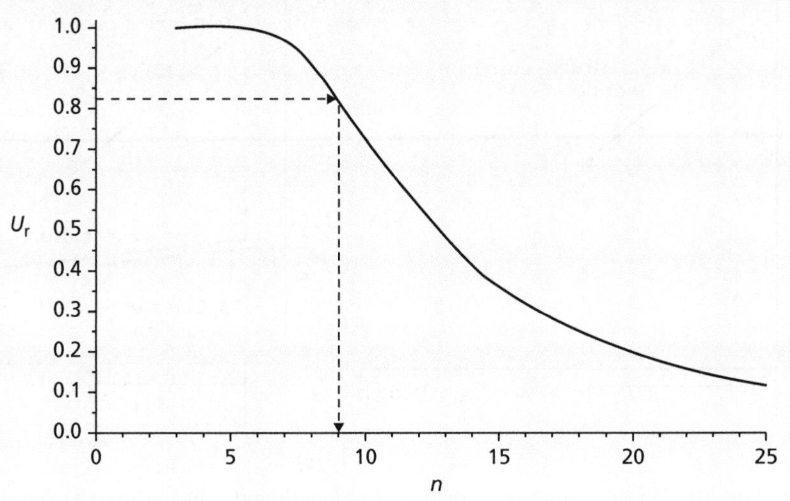

Figure 4.27 Example 4.7.

be seen that $n = 9$. It should be noted that this process is greatly shortened by the use of a spreadsheet to perform the calculations. Therefore

$$R = 0.2 \times 9 = 1.8\text{m}$$

Spacing of drains in a square pattern is given by

$$S = \frac{R}{0.564} = \frac{1.8}{0.564} = 3.2\text{m}$$

① 预压。

4.11 Pre-loading[①]

If construction is to be undertaken on very compressible ground (i. e. soil with high m_v), it is likely that the consolidation settlements which will occur will be too large to be tolerable by the construction. If the compressible material is close to the surface and of limited depth, one solution would be to excavate the material and replace it with properly compacted fill (see Section 1.7). If the compressible material is of significant thickness, this may prove prohibitively expensive. An alternative solution in such a case would be to pre-load the ground for an appropriate period by applying a surcharge, normally in the form of a temporary embankment, and allowing most of the settlement to take place before the final construction is built.

② 建筑基础。

③ 预压的应用：(a) 高压缩性土上的基础建造，(b) 预压后的基础建造。

An example for a normally consolidated soil is shown in Figure 4.28. In Figure 4.28 (a), a foundation, applying a pressure of q_f is built without applying any pre-

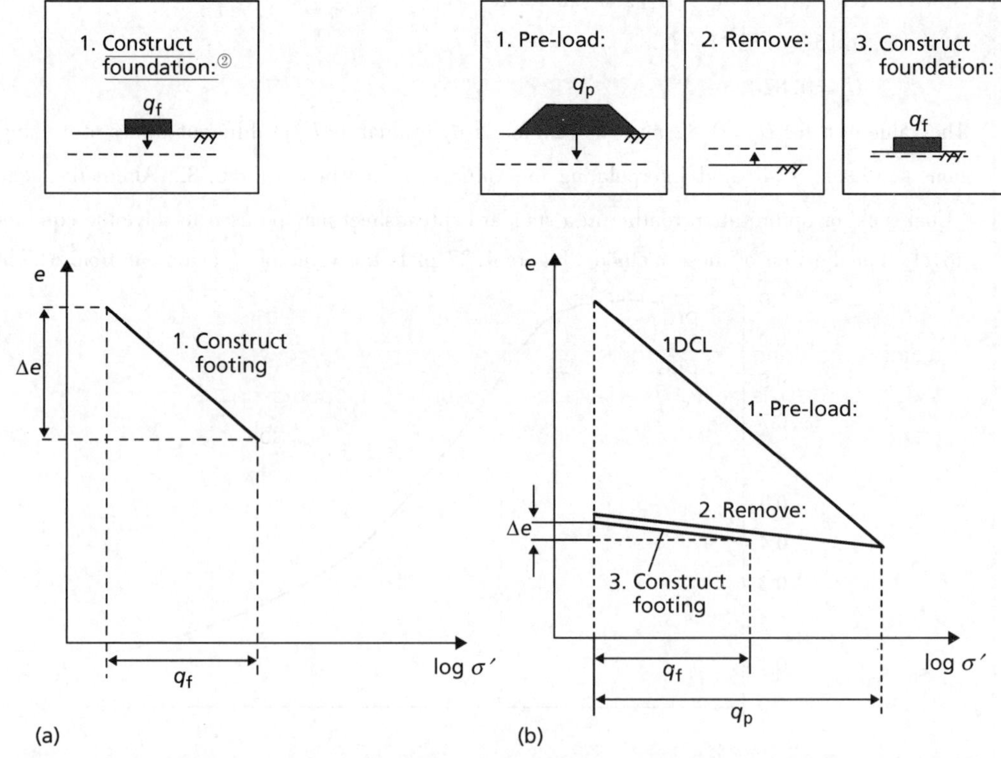

Figure 4.28 Application of pre-loading: (a) foundation construction on highly compressible soil, (b) foundation constructed following pre-loading.[③]

load; consolidation will proceed along the virgin compression line and the settlement will be large. In Figure 4. 28 (b) a pre-load is first applied by constructing a temporary embankment, applying a pressure q_p, which induces significant volumetric change (and hence settlement) along the virgin compression line. Once consolidation is complete, the pre-load is removed and the soil swells along an unload-reload line. The foundation is then built and, provided the pressure applied is less than the pre-load pressure ($q_f < q_p$), the soil will remain on the unload – reload line and the settlement in this phase (which the foundation sees) will be small.

The pre-load may have to be applied in increments to avoid undrained shear failure of the supporting soil (see Chapters 5 and 8). As consolidation proceeds, the shear strength of the soil increases, allowing further load increments to be applied.① The rate of dissipation of excess pore water pressure can be monitored by installing piezometers (Chapter 6), thus providing a means of controlling the rate of loading. It should be appreciated that differential settlement② will occur if non-uniform soil③ conditions exist and the ground surface may need re-levelling at the end of the pre-loading period. One of the principal disadvantages of pre-loading is the need to wait for consolidation to be complete under the pre-load before construction can begin on the foundation. One solution to this is to combine pre-loading with vertical drains (Section 4. 10) to speed up the pre-load stage.

① 随着固结，土的抗剪强度增加，可承受进一步的荷载增量。

② 不均匀沉降。

③ 非均质土。

The principle of pre-loading can also be applied to accelerate the settlement of embankments by surcharging with an additional depth of fill. At the end of an appropriate period, surcharge is removed down to the final (formation) level.

Summary

1 Due to the finite permeability of soils, changes in applied (total) stress or fluctuations in groundwater level do not immediately lead to corresponding increments in effective stress. Under such conditions, the volume of the soil will change with time. Dissipation of positive excess pore pressures (due to loading or lowering of the water table) leads to compression which is largely irrecoverable. Dissipation of negative excess pore pressures (unloading or raising of the water table) leads to swelling which is recoverable and of much lower magnitude than initial compression. The compression behaviour is quantified mathematically by the virgin compression and unload/re-load lines in $e - \log\sigma'$ space. The compressibility (m_v), coefficient of consolidation (c_v) and (indirectly) the permeability (k) may be determined within the oedometer.

2 The final amount of settlement due to one-dimensional consolidation can be straightforwardly determined using the compressibility (m_v) and the known changes in total stress or pore pressure condi-

tions induced by construction processes. A more detailed description of settlement (or heave) with time may be found either analytically or using the Finite Difference Method. A spreadsheet implementation of this method is available on the Companion Website.

3 For soils of low permeability (e. g. fine-grained soils) it may be necessary in construction to speed up the process of consolidation. This may be achieved by adding vertical drains. An appropriate drain specification and layout can be determined using standard solutions for radial pore water drainage to meet a specified level of performance (e. g. U_r% consolidation by time t).

Problems

4.1 In an oedometer test on a specimen of saturated clay ($G_s = 2.72$), the applied pressure was increased from 107 to 214kPa and the following compression readings recorded:

TABLE I

Time (min)	0	$\frac{1}{4}$	$\frac{1}{2}$	1	$2\frac{1}{4}$	4	$6\frac{1}{4}$	9	16
Gauge (mm)	7.82	7.42	7.32	7.21	6.99	6.78	6.61	6.49	6.37
Time (min)	25	36	49	64	81	100	300	1440	
Gauge (mm)	6.29	6.24	6.21	6.18	6.16	6.15	6.10	6.02	

After 1440min, the thickness of the specimen was 15.30mm and the water content was 23.2%. Determine the values of the coefficient of consolidation and the compression ratios from (a) the root time plot and (b) the log time plot. Determine also the values of the coefficient of volume compressibility and the coefficient of permeability.

4.2 In an oedometer test, a specimen of saturated clay 19mm thick reaches 50% consolidation in 20min. How long would it take a layer of this clay 5m thick to reach the same degree of consolidation under the same stress and drainage conditions? How long would it take the layer to reach 30% consolidation?

4.3 The following results were obtained from an oedometer test on a specimen of saturated clay:

TABLE J

Pressure (kPa)	27	54	107	214	429	214	107	54
Void ratio	1.243	1.217	1.144	1.068	0.994	1.001	1.012	1.024

A layer of this clay 8m thick lies below a 4m depth of sand, the water table being at the surface. The saturated unit weight for both soils is 19kN/m^3. A 4-m depth of fill of unit weight 21kN/m^3 is placed on the sand over an extensive area. Determine the final settlement due to consolidation of the clay. If the fill were to be removed some time after the completion of consolidation, what heave would eventually take place due to swelling of the clay?

4.4 Assuming the fill in Problem 4.3 is dumped very rapidly, what would be the value of excess pore water pressure at the centre of the clay layer after a period of three years? The layer is open and the value of c_v is $2.4m^2/year$.

4.5 A 10-m depth of sand overlies an 8-m layer of clay, below which is a further depth of sand. For the clay, $m_v = 0.83m^2/MN$ and $c_v = 4.4m^2/year$. The water table is at surface level but is to be lowered permanently by 4m, the initial lowering taking place over a period of 40 weeks. Calculate the final settlement due to consolidation of the clay, assuming no change in the weight of the sand, and the settlement two years after the start of lowering.

4.6 An open clay layer is 6m thick, the value of c_v being $1.0m^2/year$. The initial distribution of excess pore water pressure varies linearly from 60kPa at the top of the layer to zero at the bottom. Using the finite difference approximation of the one-dimensional consolidation equation, plot the isochrone after consolidation has been in progress for a period of three years, and from the isochrone determine the average degree of consolidation in the layer.

4.7 A half-closed clay layer is 8m thick and it can be assumed that $c_v = c_h$. Vertical sand drains 300mm in diameter, spaced at 3m centres in a square pattern, are to be used to increase the rate of consolidation of the clay under the increased vertical stress due to the construction of an embankment. Without sand drains, the degree of consolidation at the time the embankment is due to come into use has been calculated as 25%. What degree of consolidation would be reached with the sand drains at the same time?

4.8 A layer of saturated clay is 10m thick, the lower boundary being impermeable; an embankment is to be constructed above the clay. Determine the time required for 90% consolidation of the clay layer. If 300-mm diameter sand drains at 4m centres in a square pattern were installed in the clay, in what time would the same overall degree of consolidation be reached? The coefficients of consolidation in the vertical and horizontal directions, respectively, are 9.6 and $14.0m^2/year$.

References

Al-Tabbaa, A. and Wood, D. M. (1987) Some measurements of the permeability of kaolin, *Géotechnique*, **37** (4), 499 – 503.

ASTM D2435 (2011) *Standard Test Methods for One-Dimensional Consolidation Properties of Soils Using Incremental Loading*, American Society for Testing and Materials, West Conshohocken, PA.

Barron, R. A. (1948) Consolidation of fine grained soils by drain wells, *Transactions of the ASCE*, **113**, 718 – 742.

Bjerrum, L. (1967) Engineering geology of Norwegian normally-consolidated marine clays as related to settlement of buildings, *Géotechnique*, **17** (2), 83 – 118.

British Standard 1377 (1990) *Methods of Test for Soils for Civil Engineering Purposes*, British Standards Institution, London.

Casagrande, A. (1936) Determination of the preconsolidation load and its practical significance, in *Proceedings of the International Conference on SMFE, Harvard University, Cambridge, MA*, Vol. III, pp. 60 – 64.

CEN ISO/TS 17892 (2004) *Geotechnical Investigation and Testing – Laboratory Testing of Soil*, International Organisation for Standardisation, Geneva.

Cour, F. R. (1971) Inflection point method for computing c_v, Technical Note, *Journal of the ASCE*, **97** (SM5), 827 – 831.

EC7 – 2 (2007) *Eurocode 7: Geotechnical design – Part 2: Ground Investigation and Testing, BS EN 1997 – 2:*

2007, British Standards Institution, London.

Gibson, R. E. (1966) A note on the constant head test to measure soil permeability *in-situ*, *Géotechnique*, **16** (3), 256 – 259.

Gibson, R. E. (1970) An extension to the theory of the constant head *in-situ* permeability test, *Géotechnique*, **20** (2), 193 – 197.

Hansbo, S. (1979) Consolidation of clay by band-shaped prefabricated drains, *Ground Engineering*, **12** (5), 16 – 25.

Hansbo, S. (1981) Consolidation of fine-grained soils by prefabricated drains, in *Proceedings of the 10th International Conference on SMFE, Stockholm*, Vol. III, pp. 677 – 682.

Mitchell, J. K. and Soga, K. (2005) *Fundamentals of Soil Behaviour* (3rd edn), John Wiley & Sons, New York, NY.

Naylor, A. H. and Doran, I. G. (1948) Precise determination of primary consolidation, in *Proceedings of the 2nd International Conference on SMFE, Rotterdam*, Vol. 1, pp. 34 – 40.

Rowe, P. W. (1968) The influence of geological features of clay deposits on the design and performance of sand drains, in *Proceedings ICE* (Suppl. Vol.), Paper 70585.

Rowe, P. W. and Barden, L. (1966) A new consolidation cell, *Géotechnique*, **16** (4), 162 – 170.

Schmertmann, J. H. (1953) Estimating the true consolidation behaviour of clay from laboratory test results, *Proceedings ASCE*, **79**, 1 – 26.

Scott, R. F. (1961) New method of consolidation coefficient evaluation, *Journal of the ASCE*, **87**, No. SM1.

Taylor, D. W. (1948) *Fundamentals of Soil Mechanics*, John Wiley & Sons, New York, NY.

Terzaghi, K. (1943) *Theoretical Soil Mechanics*, John Wiley & Sons, New York, NY.

Wilkinson, W. B. (1968) Constant head *in-situ* permeability tests in clay strata, *Géotechnique*, **18** (2), 172 – 194.

Further reading

Burland, J. B. (1990) On the compressibility and shear strength of natural clays, *Géotechnique*, **40** (3), 329 – 378.

This paper describes in detail the role of depositional structure on the initial compressibility of natural clays (rather than those reconstituted in the laboratory). It contains a large amount of experimental data, and is therefore useful for reference.

McGown, A. and Hughes, F. H. (1981) Practical aspects of the design and installation of deep vertical drains, *Géotechnique*, **31** (1), 3 – 17.

This paper discusses practical aspects related to the use of vertical drains, which were introduced in Section 4.10.

Chapter 5

Soil behaviour in shear[①]

① 土的剪切特性。

Learning outcomes

After working through the material in this chapter, you should be able to:

1. Understand how soil may be modelled as a continuum, and how its mechanical behaviour (strength and stiffness)[②] may be adequately described using elastic and plastic material (constitutive) models (Section 5.1 – 5.3);
2. Understand the method of operation of standard laboratory testing apparatus and derive strength and stiffness properties of soil from these tests for use in subsequent geotechnical analyses (Section 5.4);
3. Appreciate the different strength characteristics of coarse and fine-grained soils and derive material parameters to model these (Sections 5.5 – 5.6 and Section 5.8);
4. Understand the critical state concept[③] and its important role in coupling strength and volumetric behaviour in soil (Section 5.7);
5. Use simple empirical correlations to estimate strength properties of soil based on the results of index tests (see Chapter 1) and appreciate how these may be used to support the results from laboratory tests (Section 5.9).

② 力学特性(强度和刚度)。

③ 临界状态的概念。

5.1 An introduction to continuum mechanics[④]

④ 连续介质力学简介。

This chapter is concerned with the resistance of soil to failure in shear, a knowledge of which is required in the analysis of the stability of soil masses and, therefore, for the design of geotechnical structures. Many problems can be treated by analysis in two dimensions, i. e. where only the stresses and displacements in a single plane need to be considered. This simplification will be used initially in this chapter while the framework for the constitutive behaviour of soil is described.

An element of soil in the field will typically by subjected to total normal stresses in the vertical (z) and horizontal (x) directions due to the self-weight of the soil and any applied external loading (e. g. from a foundation). The latter may also induce an applied shear stress which additionally acts on the element. The total normal stresses and shear stresses in the x- and z- directions on an element of soil are shown in Figure 5. 1 (a), the stresses being positive in magnitude as shown; the stresses vary across the element. The rates of change of the normal stresses in the x- and z- directions are $\partial\sigma_x/\partial x$ and $\partial\sigma_z/\partial z$ respectively; the rates of change of the shear stresses are $\partial\tau_{xz}/\partial x$ and $\partial\tau_{xz}/\partial z$. Every such element within a soil mass must be in static equilibrium.① By equating moments② about the centre point of the element, and neglecting higher-order differentials, it is apparent that $\tau_{xz} = \tau_{zx}$. By equating forces in the x-and z-directions the following equations are obtained:

① 土体中每一个单元体都必须处于静力平衡状态。

② 力矩。

$$\frac{\partial\sigma_x}{\partial x} + \frac{\partial\tau_{xz}}{\partial z} = 0 \tag{5.1a}$$

$$\frac{\partial\tau_{xz}}{\partial x} + \frac{\partial\sigma_z}{\partial z} - \gamma = 0 \tag{5.1b}$$

These are the **equations of equilibrium**③ in two dimensions in terms of total stresses; for dry soils, the body force (or unit weight) $\gamma = \gamma_{dry}$, while for saturated soil, $\gamma = \gamma_{sat}$. Equation 5. 1 can also be written in terms of effective stress. From Terzaghi's Principle (Equation 3. 1) the effective body forces will be 0 and $\gamma' = \gamma - \gamma_w$ in the x-and z-directions respectively. Furthermore, if seepage is taking place with hydraulic gradients of i_x and i_z in the x-and z-directions, then there will be additional body forces due to seepage (see Section 3. 6) of $i_x\gamma_w$ and $i_z\gamma_w$ in the x- and z- directions, i. e.:

③ 平衡方程组。

$$\frac{\partial\sigma_x'}{\partial x} + \frac{\partial\tau_{zx}'}{\partial z} - i_x\gamma_w = 0 \tag{5.2a}$$

$$\frac{\partial\tau_{xz}'}{\partial x} + \frac{\partial\tau_z'}{\partial z} - (\gamma' + i_z\gamma_w) = 0 \tag{5.2b}$$

The effective stress components are shown in Figure 5. 1 (b).

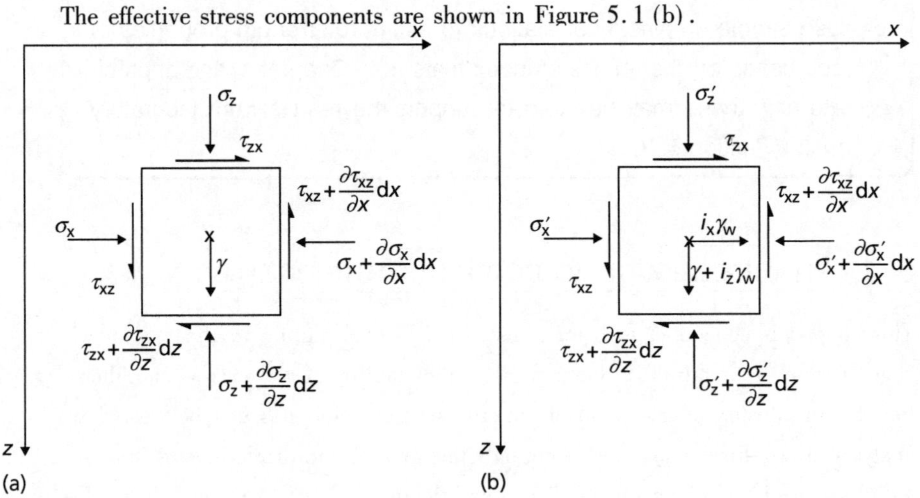

④ 土单元中的二维应力状态：(a) 总应力，(b) 有效应力。

Figure 5. 1 Two-dimensional state of stress in an element of soil: (a) total stresses, (b) effective stresses.④

The effective stress components are shown in Figure 5. 1(b).

Due to the applied loading, points within the soil mass will be displaced relative to the axes and to one another, as shown in Figure 5. 2. If the components of displacement in the x- and z-directions are denoted by u and w, respectively, then the normal strains in these directions (ε_x and ε_z, respectively) are given by

$$\varepsilon_x = \frac{\partial u}{\partial x}, \varepsilon_z = \frac{\partial w}{\partial z}$$

and the shear strain by

$$\gamma_{xz} = \frac{\partial u}{\partial z} + \frac{\partial w}{\partial x}$$

However, these strains are not independent; they must be compatible with each other if the soil mass as a whole is to remain continuous.[①] This requirement leads to the following relationship, known as the **equation of compatibility** in two dimensions:

① 然而，这些应变并不是独立的，若土体保持整体连续性，则应变必须相互协调。

$$\frac{\partial^2 \varepsilon_x}{\partial z^2} + \frac{\partial^2 \varepsilon_z}{\partial x^2} - \frac{\partial \gamma_{xz}}{\partial x \partial z} = 0 \tag{5.3}$$

The rigorous solution of a particular problem requires that the equations of equilibrium and compatibility are satisfied for the given boundary conditions (i. e. applied loads and known displacement conditions) at all points within a soil mass; an appropriate stress- strain relationship is also required to link the two equations. Equations 5. 1 – 5. 3, being independent of material properties, can be applied to soil with any stress – strain relationship (also termed a **constitutive model**).[②] In general, soils are non- homogeneous, exhibit **anisotropy**[③] (i. e. have different values of a given property in different directions) and have nonlinear stress – strain relationships which are dependent on stress history (see Section 4. 2) and the particular stress path followed. This can make solution difficult.

② 本构模型。

③ 各向异性。

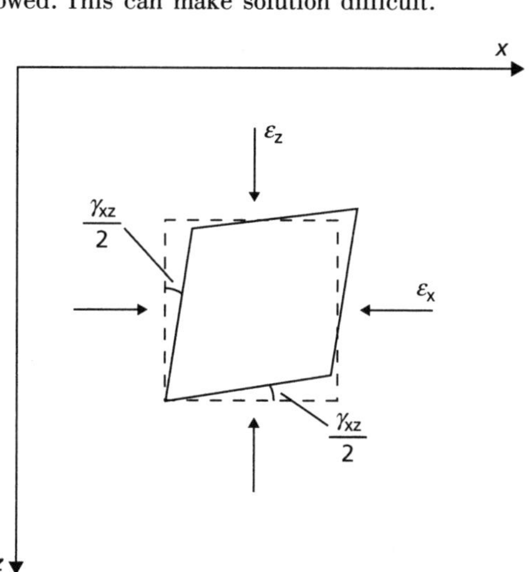

Figure 5. 2 Two- dimensional induced state of strain in an element of soil, due to stresses shown in Figure 5. 1.[④]

④ 由图 5.1 中应力状态引起的土单元体中的二维应变状态。

In analysis therefore, an appropriate idealisation of the stress – strain relation-

① 理想弹塑性模型。
② 理想刚塑性模型。
③ 应变硬化的弹塑性模型。
④ 屈服应力的增大是应变硬化的特征之一。
⑤ (a) 土的典型应力-应变关系，(b) 理想弹塑性模型，(c) 理想刚塑性模型，(d) 应变硬化与应变软化的弹塑性模型。

ship is employed to simplify computation. One such idealisation is shown by the dotted lines in Figure 5.3 (a), linearly elastic behaviour (i. e. Hooke's Law) being assumed between O and Y′ (the assumed yield point) followed by unrestricted plastic strain (or flow) Y′P at constant stress. This idealisation, which is shown separately in Figure 5.3 (b), is known as the **elastic – perfectly plastic model**[①] of material behaviour. If only the collapse condition (soil failure) in a practical problem is of interest, then the elastic phase can be omitted and the **rigid – perfectly plastic model**[②], shown in Figure 5.3 (c), may be used. A third idealisation is the **elastic – strain hardening plastic model**[③], shown in Figure 5.3 (d), in which plastic strain beyond the yield point necessitates further stress increase, i. e. the soil hardens or strengthens as it strains. If unloading and reloading were to take place subsequent to yielding in the strain hardening model, as shown by the dotted line Y″U in Figure 5.3 (d), there would be a new yield point Y″ at a stress level higher than that at the first yield point Y′. An increase in yield stress is a characteristic of strain hardening[④]. No such increase takes place in the case of perfectly plastic (i. e. non-hardening) behaviour, the stress at Y″ being equal to that at Y′ as

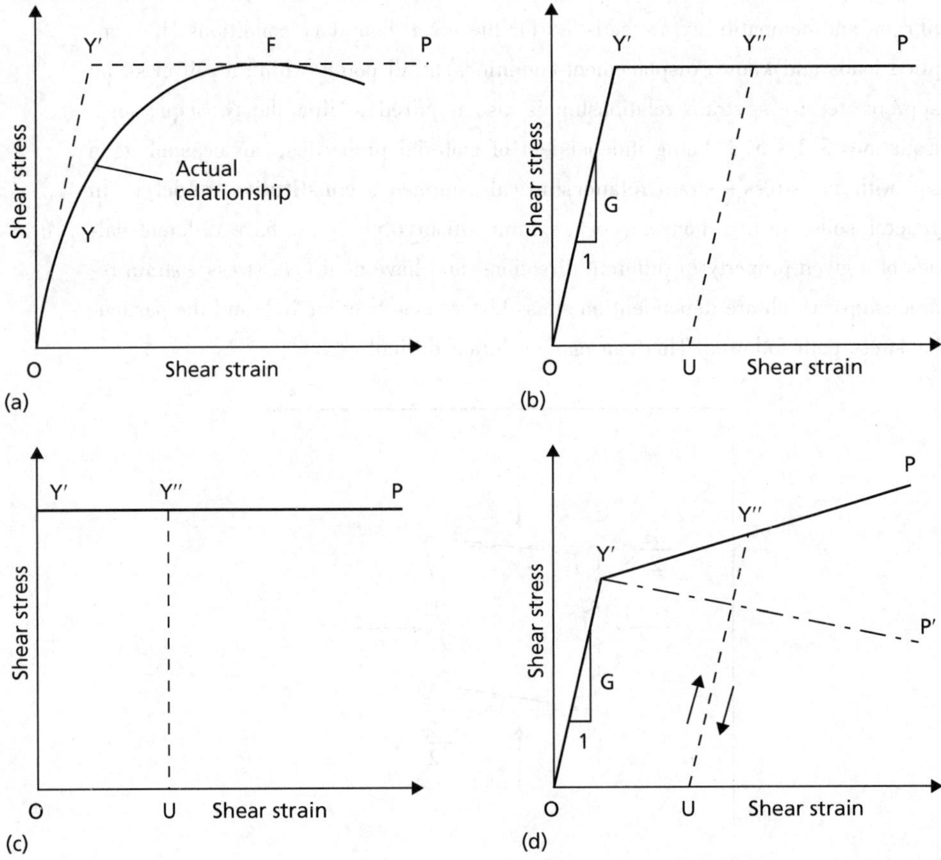

Figure 5.3 (a) Typical stress – strain relationship for soil, (b) elastic – perfectly plastic model, (c) rigid – perfectly plastic model, and (d) strain hardening and strain softening elastic – plastic models.[⑤]

shown in Figures 5.3 (b) and (c). A further idealisation is the **elastic – strain softening plastic model**[1], represented by OY′P′in Figure 5.3 (d), in which the plastic strain beyond the yield point is accompanied by stress decrease or softening of the material.

① 应变软化的弹塑性模型。

In plasticity theory (Hill, 1950; Calladine, 2000) the characteristics of yielding, hardening and flow[2] are considered; these are described by a yield function, a hardening law and a flow rule, respectively. The yield function is written in terms of stress components or principal stresses, and defines the yield point as a function of current effective stresses and stress history. The **Mohr – Coulomb criterion**[3], which will be described later in Section 5.3, is one possible (simple) yield function if perfectly plastic behaviour is assumed. The hardening law represents the relationship between the increase in yield stress and the corresponding plastic strain components, i.e. defining the gradient of Y′P or Y′P′ in Figure 5.3 (d). The flow rule specifies the relative (i.e. not absolute) magnitudes of the plastic strain components during yielding under a particular state of stress. The remainder of this book will consider simple elastic – perfectly plastic material models for soil, as shown in Figure 5.3 (b), in which elastic behaviour is isotropic (Section 5.2) and plastic behaviour is defined by the Mohr – Coulomb criterion (Section 5.3).

② 屈服、硬化和流动。

③ 莫尔-库仑准则。

5.2 Simple models of soil elasticity[4]

④ 土体弹性简化模型。

Linear elasticity[5]

⑤ 线弹性。

The initial region of soil behaviour, prior to plastic collapse (yield) of the soil, may be modeled using an elastic constitutive model.[6] The simplest model is (isotropic) linear elasticity in which shear strain is directly proportional to the applied shear stress. Such a model is shown by the initial parts of the stress – strain relationships in Figure 5.3 (a, b, d), in which the relationship is a straight line, the gradient of which is the **shear modulus**[7], G, i.e.

⑥ 土的力学特性的最初阶段，在土体发生塑性屈服前，可以用弹性本构模型来描述。

$$\tau_{xz} = \tau_{zx} = G\gamma_{xz} \tag{5.4}$$

⑦ 剪切模量。

For a general 2-D element of soil, as shown in Figures 5.1 and 5.2, loading may not just be due to an applied shear stress τ_{xz}, but also by normal stress components σ_x and σ_z. For a linearly elastic constitutive model, the relationship between stress and strain is given by Hooke's Law, in which

$$\varepsilon_x = \frac{1}{E}(\sigma_x' - \nu\sigma_z') \tag{5.5a}$$

$$\varepsilon_z = \frac{1}{E}(\sigma_z' - \nu\sigma_x') \tag{5.5b}$$

where E is the **Young's Modulus**[8] (= normal stress/normal strain) and ν is the **Poisson's ratio**[9] of the soil. While the soil remains elastic, determination of the soil response (strain) to applied stresses requires knowledge only of the elastic properties of the soil, defined by G, E and ν. For an **isotropically**[10] elastic material (i.e. uniform behaviour in all directions), it can further be shown that the three

⑧ 杨氏模量。

⑨ 泊松比。

⑩ 各向同性。

elastic material constants are related by

$$G=\frac{E}{2(1+\nu)} \tag{5.6}$$

It is therefore only necessary to know any two of the elastic properties; the third can always be found using Equation 5.6. It is preferable in soil mechanics to use ν and G as the two properties. From Equation 5.4 it can be seen that the behaviour of soil under pure shear is independent of the normal stresses and therefore is not influenced by the pore water (water cannot carry shearing stresses[1]). G may therefore be measured for soil which is either fully **drained**[2] (e.g. after consolidation is completed) or under an **undrained**[3] condition (before consolidation has begun), with both values being the same. E, on the other hand, is dependent on the normal stresses in the soil (Equation 5.5), and therefore is influenced by pore water. To determine response under immediate and long-term loading[4] it would be necessary to know two values of E, but only one of G.

① 水不能承受剪应力。
② 排水。
③ 不排水。
④ 瞬长期加载。

Poisson's ratio, which is defined as the ratio of strains in the two perpendicular directions under uniaxial loading ($\nu=\varepsilon_x/\varepsilon_z$ under applied loads σ_z', $\sigma_x'=0$), is also dependent on the drainage conditions. For fully or partially drained conditions, $\nu<0.5$, and is normally between 0.2 and 0.4 for most soils under fully drained conditions. Under undrained conditions, the soil is incompressible (no pore-water drainage has yet occurred). The volumetric strain of an element of linearly elastic material under normal stresses σ_x and σ_z is given by

$$\frac{\Delta V}{V}=\varepsilon_x+\varepsilon_z{}^{⑤}=\frac{1-2\nu}{E}(\sigma_x'+\sigma_z')$$

⑤ 注：此处应改为 $=\varepsilon_x+\varepsilon_z+\varepsilon_y$ $=\frac{1}{E}(\sigma_x'-\nu\sigma_z')$ $+\frac{1}{E}(\sigma_z'-\nu\sigma_x')$ $+\frac{1}{E}(-\nu\sigma_x'-\nu\sigma_z')$

where V is the volume of the soil element. Therefore, for undrained conditions $\Delta V/V=0$ (no change in volume), hence $\nu=\nu_u=0.5$. This is true for all soils provided conditions are completely undrained.

For given applied stress components σ_x, σ_z and τ_{xz}, the resulting strains may be found by solving Equations 5.4 and 5.5 simultaneously. Alternatively, the equations may be written in matrix form

$$\begin{bmatrix}\varepsilon_x\\ \varepsilon_z\\ \gamma_{xz}\end{bmatrix}=\frac{1}{2G(1+\nu)}\begin{bmatrix}1 & -\nu & 0\\ -\nu & 1 & 0\\ 0 & 0 & 2(1+\nu)\end{bmatrix}\begin{bmatrix}\sigma_x'\\ \sigma_z'\\ \tau_{xz}\end{bmatrix} \tag{5.7}$$

The elastic constants G, E and ν can further be related to the constrained modulus (E'_{oed}) described in Section 4.2. In the oedometer test, soil strains in the z-direction under drained conditions, but the lateral strains are zero[6]. From the 3-D version of Equation 5.7 which will be introduced later (Equation 5.29), the zero lateral strain condition gives:

⑥ 在固结试验中，排水条件下土在z方向发生应变，但是侧向应变为零。

$$\varepsilon_z=\frac{1}{E'}\left(\frac{1-\nu'-2\nu'^2}{1-\nu'}\right)\sigma_z'$$

where E' and ν' are the Young's modulus and Poisson's ratio for fully drained conditions. Then, from the definition of E'_{oed} (Equations 4.3 and 4.5):

$$E'_{oed}=\frac{\sigma_z'}{\varepsilon_z}=\frac{E'(1-\nu')}{(1-\nu'-2\nu'^2)}$$

$$\therefore E' = E'_{oed}\frac{(1+\nu')(1-2\nu')}{1-\nu'} \quad (5.8)$$

Substituting Equation 5.6

$$2G(1+\nu') = E'_{oed}\frac{(1+\nu')(1-2\nu')}{1-\nu'}$$

$$\therefore G = E'_{oed}\frac{(1-2\nu')}{2(1-\nu')} \quad (5.9)$$

Therefore, the results of oedometer tests can be used to define the shear modulus.

Non-linear elasticity[①]

In reality, the shear modulus of soil is not a material constant, but is a highly non-linear function of shear strain and effective confining stress,[②] as shown in Figure 5.4 (a). At very small values of strain, the shear modulus is a maximum (defined as G_0). The value of G_0 is independent of strain, but increases with increasing effective stress. As a result, G_0 generally increases with depth within soil masses.[③] If the shear modulus is normalised by G_0 to remove the stress dependence, a single non-linear curve of G/G_0 versus shear strain is obtained (Figure 5.4 (b)). Atkinson (2000) has suggested that this relationship may be approximated by Equation 5.10:

$$\frac{G}{G_0} = \frac{1-(\gamma_p/\gamma)^B}{1-(\gamma_p/\gamma_0)^B} \leqslant 1.0 \quad (5.10)$$

where γ is the shear strain, γ_0 defines the maximum strain at which the small-strain stiffness G_0 is still applicable (typically around 0.001% strain),[④] γ_p defines the strain at which the soil becomes plastic (typically around 1% strain) and B defines the shape of the curve between $G/G_0 = 0$ and 1, typically being between 0.1 – 0.5 depending on the soil type. This relationship is shown in Figure 5.4 (b).

For most common geotechnical structures, the operative levels of strain will mean that the shear modulus $G < G_0$. Common strain ranges are shown in Figure 5.4 (c), and these may be used to estimate an appropriate linearised value of G for a given problem from the non-linear relationship. The full nonlinear $G-\gamma$ relationship may be determined by:

1 undertaking triaxial testing[⑤] (described later in Section 5.4) on modern machines with small-strain sample measurements;[⑥] this equipment is now available in most soil testing laboratories;
2 determining the value of G_0 using seismic wave techniques[⑦] (either in a triaxial cell using specialist bender elements,[⑧] or in-situ as described in Chapters 6 and 7) and combining this with a normalised G/G_0 versus γ relationship (e.g. Equation 5.10; see Atkinson, 2000 for further details).

Of these methods, the second is usually the cheapest and quickest to implement in practice. In principle, the value of G can also be estimated from the curve relating principal stress difference and axial strain in an undrained triaxial test[⑨] (this will be described in Section 5.4). Without small-strain sample measurements, howev-

① 非线性弹性。
② 在实际中，土的剪切模量并不是一个材料常数，而是一个高度非线性且由剪应变与有效围压决定的函数。
③ 最大剪切模量 G_0 的值与应变无关，但随着有效应力的增加而增加，因此 G_0 将随着土体埋深的增加而增加。
④ γ_0 为小应变刚度 G_0 适用范围内的最大应变（通常为0.001%的剪应变）。
⑤ 三轴试验。
⑥ 小应变测量。
⑦ 波动测试。
⑧ 弯曲元。
⑨ 原则上，可以通过不排水三轴试验中建立的主应力差与轴向应变的关系曲线来估算 G 的值。

er, the data are only likely to be available for $\gamma > 0.1\%$ (see Figure 5.4 (d)). Because of the effects of sampling disturbance (see Chapter 6), it can be preferable to determine G (or E) from the results of in-situ rather than laboratory tests.

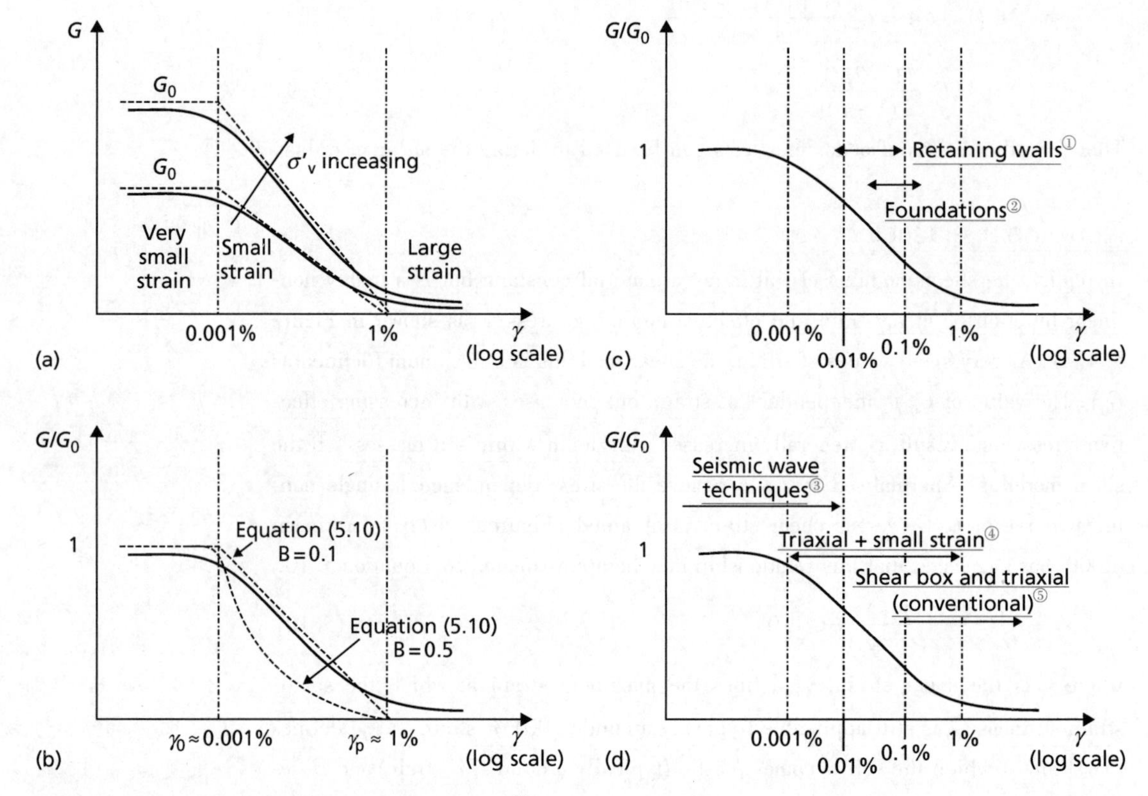

Figure 5.4 Non-linear soil shear modulus.⑥

① 挡土墙。
② 基础工程。
③ 波动测试。
④ 小应变测量的三轴试验。
⑤ 传统的剪切盒试验或三轴试验。
⑥ 土体剪切模量的非线性。
⑦ 土体塑性简化模型。
⑧ 摩擦性的。
⑨ 滑动面是粗糙的。
⑩ 摩擦系数。

5.3 Simple models of soil plasticity⑦

Soil as a frictional material

The use of elastic models alone (as described previously) implies that the soil is infinitely strong. If at a point on any plane within a soil mass the shear stress becomes equal to the shear strength of the soil, then failure will occur at that point. Coulomb originally proposed that the limiting strength of soils was frictional⑧, imagining that if slip (plastic failure) occurred along any plane within an element of closely packed particles (soil), then the slip plane would be rough⑨ due to all of the individual particle-to-particle contacts. Friction is commonly described by:

$$T = \mu N$$

where T is the limiting frictional force, N is the normal force acting perpendicular to the slip plane and μ is the coefficient of friction.⑩ This is shown in Figure 5.5 (a). In an element of soil, it is more useful to use shear stress and normal stress instead of T and N

$$\tau_f = (\tan\phi)\sigma$$

where tan ϕ is equivalent to the coefficient of friction, which is an intrinsic material property related to the roughness of the shear plane (i. e. the shape, size and angularity of the soil particles). The frictional relationship in terms of stresses is shown in Figure 5.5 (b).

While Coulomb's frictional model represented loosely packed particle arrangements well, if the particles are arranged in a dense packing then additional initial interlocking① between the particles can cause the frictional resistance τ_f to be higher than that predicted considering friction alone. If the normal stress is increased, it can become high enough that the contact forces between the individual particles cause particle breakage, which reduces the degree of interlocking and makes slip easier.② At high normal stresses, therefore, the interlocking effect disappears and the material behaviour becomes purely frictional again. This is also shown in Figure 5.5 (b).

① 额外的初始咬合作用。

② 如果法向应力增加，颗粒间的接触力就会变得足够高而引起颗粒破碎，最终导致颗粒间咬合度下降，更容易产生滑动。

In accordance with the principle that shear stress in a soil can be resisted only by the skeleton of solid particles and not the pore water, the shear strength (τ_f) of a soil at a point on a particular plane is normally expressed as a function of effective normal stress (σ') rather than total stress.

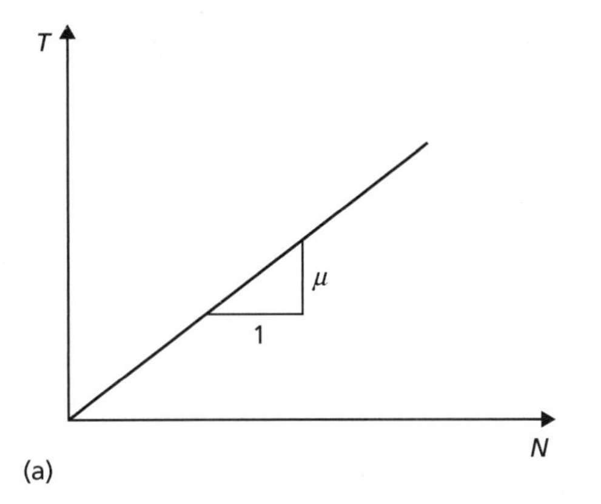

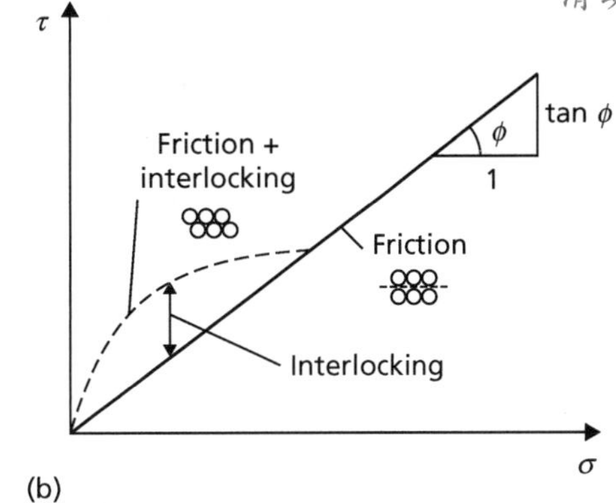

Figure 5.5 (a) Frictional strength along a plane of slip,③ (b) strength of an assembly of particles along a plane of slip.④

③ 沿一个滑动面的摩擦强度。

④ 沿一个滑动面的颗粒集合体强度。

The Mohr – Coulomb model⑤

⑤ 莫尔-库仑模型。

As described in Section 5.1, the state of stress in an element of soil is defined in terms of the normal and shear stresses applied to the boundaries of the soil element. States of stress in two dimensions can be represented on a plot of shear stress (τ) against effective normal stress (σ'). The stress state for a 2-D element of soil can be represented either by a pair of points with coordinates (σ_z', τ_{zx}) and (σ_x', τ_{xz}), or by a **Mohr circle**⑥ defined by the effective **principal stresses**⑦ σ_1' and σ_3', as shown in Figure 5.6. The stress points at either end of a diameter through a Mohr circle at an angle of 2θ to the horizontal represent the stress conditions on a plane

⑥ 莫尔圆。

⑦ 主应力。

at an angle of θ to the minor principal stress. The circle therefore represents the stress states on all possible planes within the soil element. The principal stress components alone are enough to fully describe the position and size of the Mohr circle and so are often used to describe the stress state, as it reduces the number of stress variables from three (σ_x', σ_z', τ_{zx}) to two (σ_1', σ_3'). When the element of soil reaches failure, the circle will just touch the failure envelope, at a single point.[①] The failure envelope is defined by the frictional model described above; however, it can be difficult to deal with the non-linear part of the envelope associated with interlocking,[②] so that it is common practice to approximate the failure envelope by a straight line described by:

$$\tau_f = c' + \sigma' \tan\phi' \tag{5.11}$$

where c' and ϕ' are shear strength parameters referred to as the **cohesion inter-**

① 当土单元体破坏时，莫尔圆将在破坏点处刚好与破坏包络线接触。

② 因颗粒咬合导致的破坏包络线的非线性。

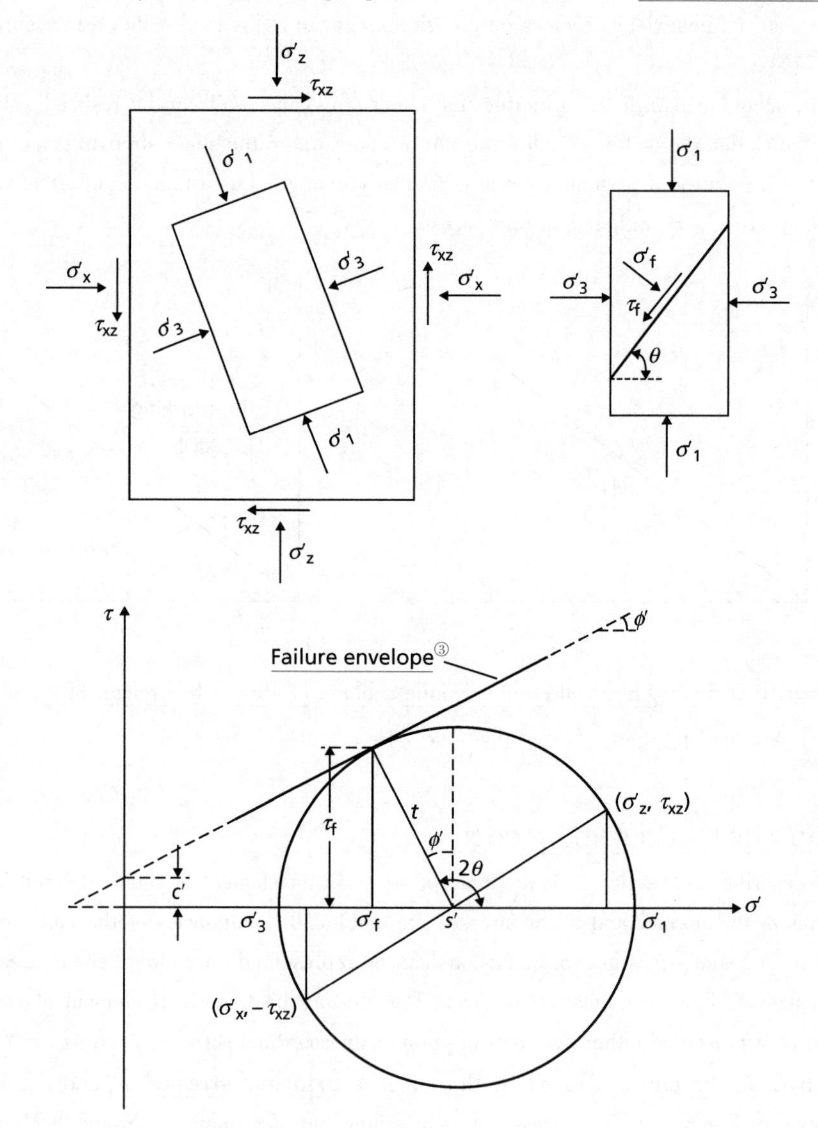

Figure 5.6 Mohr-Coulomb failure criterion.[④]

③ 破坏包络线。

④ 莫尔-库仑破坏准则。

cept[①] and the **angle of shearing resistance**[②], respectively. Failure will thus occur at any point in the soil where a critical combination of shear stress and effective normal stress develops. It should be appreciated that c' and ϕ' are simply mathematical constants defining a linear relationship between shear strength and effective normal stress. Shearing resistance is developed mechanically due to inter- particle contact forces and friction, as described above; therefore, if effective normal stress is zero then shearing resistance must also be zero (unless there is cementation or some other bonding between the particles) and the value of c' would be zero. This point is crucial to the interpretation of shear strength parameters (described in Section 5.4).

① 黏聚力截距。
② 内摩擦角。

A state of stress represented by a stress point that plots above the failure envelope, or by a Mohr circle, part of which lies above the envelope, is impossible.

With reference to the general case with $c' > 0$ shown in Figure 5.6, the relationship between the shear strength parameters and the effective principal stresses at failure at a particular point can be deduced, compressive stress being taken as positive. The coordinates of the tangent point are τ_f and σ_f' where

$$\tau_f = \frac{1}{2}(\sigma_1' - \sigma_3')\sin 2\theta \tag{5.12}$$

$$\sigma_f' = \frac{1}{2}(\sigma_1' + \sigma_3') + \frac{1}{2}(\sigma_1' - \sigma_3')\cos 2\theta \tag{5.13}$$

and θ is the theoretical angle between the minor principal plane[③] and the plane of failure. It is apparent that $2\theta = 90° + \phi'$, such that

③ 最小主应力面。

$$\theta = 45° + \frac{\phi'}{2} \tag{5.14}$$

Now

$$\sin\phi' = \frac{\frac{1}{2}(\sigma_1' - \sigma_3')}{c'\cot\phi' + \frac{1}{2}(\sigma_1' + \sigma_3')}$$

Therefore

$$(\sigma_1' - \sigma_3') = (\sigma_1' + \sigma_3')\sin\phi' + 2c'\cos\phi' \tag{5.15a}$$

or

$$\sigma_1' = \sigma_3'\tan^2\left(45° + \frac{\phi'}{2}\right) + 2c'\tan\left(45° + \frac{\phi'}{2}\right) \tag{5.15b}$$

Equation 5.15 is referred to as the Mohr – Coulomb failure criterion, defining the relationship between principal stresses at failure for given material properties c' and ϕ'.

For a given state of stress it is apparent that, because $\sigma_1' = \sigma_1 - u$ and $\sigma_3' = \sigma_3 - u$, the Mohr circles for total and effective stresses have the same diameter but their centres are separated by the corresponding pore water pressure u, as shown in Figure 5.7. Similarly, total and effective stress points are separated by the value of u.

Effect of drainage conditions on shear strength[④]

④ 排水条件对抗剪强度的影响。

The shear strength of a soil under undrained conditions is different from that under

① 有效应力的莫尔圆。
② 总应力的莫尔圆。

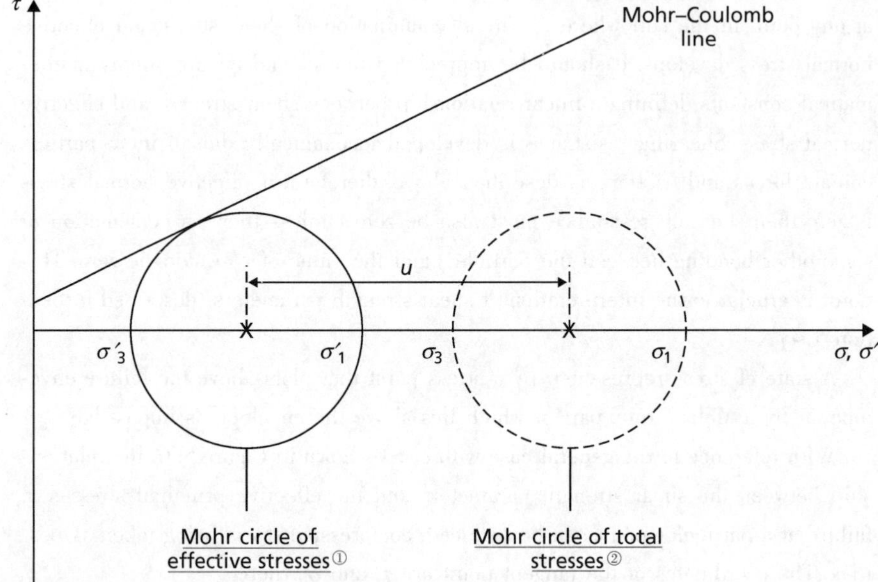

Figure 5.7 Mohr circles for total and effective stresses.[③]

③ 总应力与有效应力的莫尔圆。
④ 破坏包络线根据有效应力指标 ϕ' 与 c' 确定，所以它与土样是否处于排水条件没有关系。排水条件导致的差异体现在当施加附加应力时，在不排水加载中超孔隙水压力随之产生，从而改变土中的有效应力（在排水条件下，超孔隙水压力为零，类似固结已经完成）。
⑤ 不排水（抗剪）强度。
⑥ 抗剪强度参数。
⑦ 低渗透性的细粒土中（如黏土、粉土），短期加载（如几周或更短）是不排水的，而当长期加载时，则为排水条件。

drained conditions. The failure envelope is defined in terms of effective stresses by ϕ' and c', and so is the same irrespective of whether the soil is under drained or undrained conditions; the difference is that under a given set of applied total stresses, in undrained loading excess pore pressures are generated which change the effective stresses within the soil (under drained conditions excess pore pressures are zero as consolidation is complete).[④] Therefore, two identical samples of soil which are subjected to the same changes in total stress but under different drainage conditions will have different internal effective stresses and therefore different strengths according to the Mohr – Coulomb criterion. Rather than have to determine the pore pressures and effective stresses under undrained conditions, the **undrained strength**[⑤] can be expressed in terms of total stress. The failure envelope will still be linear, but will have a different gradient and inter cept; a Mohr – Coulomb model can therefore still be used, but the shear strength parameters[⑥] are different and denoted by c_u and $\phi_u(=0$, see Section 5.6), with the subscripts denoting undrained behaviour. The drained strength is expressed directly in terms of the effective stress parameters c' and ϕ' described previously.

When using these strength parameters to subsequently analyse geotechnical constructions in practice, the principal consideration is the rate at which the changes in total stress (due to construction operations) are applied in relation to the rate of dissipation of excess pore water pressure (consolidation), which in turn is related to the permeability of the soil as described in Chapter 4. In fine-grained soils of low permeability (e. g. clay, silt), loading in the short term (e. g. of the order of weeks or less) will be undrained, while in the long-term, conditions will be drained.[⑦] In coarse grained soils (e. g. sand, gravel) both short and long-term loading will result in drained conditions due to the higher permeability, allowing consolidation to take place rapidly. Under dynamic loading (e. g. earthquakes), loading may be fast enough to generate an undrained response in coarse-grained material. ' Short-term' is taken to be synonymous with ' during construction', while ' long-term' usually relates to the design life of the construction (usually many tens of years).

5.4 Laboratory shear tests①

① 室内剪切试验。

The shear stiffness (G) and strength parameters (c', ϕ', c_u) for a particular soil can be determined by means of laboratory tests on specimens taken from representative samples of the in-situ soil. Great care and judgement are required in the sampling operation and in the storage and handling of samples prior to testing, especially in the case of **undisturbed samples**② where the object is to preserve the in-situ structure and water content of the soil. In the case of clays, test specimens may be obtained from tube or block samples③, the latter normally being subjected to the least disturbance. Swelling of a clay specimen will occur due to the release of the in-situ total stresses.④ Sampling techniques are described in more detail in Chapter 6.

② 原状土（不扰动土）试样。

③ 管状或块状试样。

④ 由于原位总应力的释放，黏土试样发生回弹。

The direct shear test⑤

⑤ 直剪试验。

Test procedures are detailed in BS 1377, Parts 8 (UK), CEN ISO/TS 17892 – 10 (Europe) and ASTM D3080 (US). The specimen, of cross-sectional area A, is confined in a metal box (known as the **shear-box**⑥ or direct shear apparatus) of square or circular cross-section split horizontally at mid-height, a small clearance being maintained between the two halves of the box. At failure within an element of soil under principal stresses σ_1' and σ_3', a slip plane will form within the element at an angle θ as shown in Figure 5.6. The shear box is designed to represent the stress conditions along this slip plane. Porous plates are placed below and on top of the specimen if it is fully or partially saturated to allow free drainage: if the specimen is dry, solid metal plates may be used. The essential features of the apparatus are shown in Figure 5.8. In Figure 5.8 (a), a vertical force (N) is applied to the specimen through a loading plate, under which the sample is allowed to consolidate. Shear displacement is then gradually applied on a horizontal plane by causing the two halves of the box to move relative to each other, the shear force required (T) being measured together with the corresponding shear displacement (Δl). The induced shear stress within the sample on the slip plane is equal to that required to shear the two halves of the box. Suggested rates of shearing⑦ for fully drained conditions to be achieved are approximately 1 mm/min for sand, 0.01 mm/min for silt and 0.001 mm/min for clay (Bolton, 1991). Normally, the change in thickness (Δh) of the specimen is also measured. If the initial thickness of the specimen is h_0, then the shear strain (γ) can be approximated by $\Delta l/h_0$ and the volumetric strain (ε_v) by $\Delta h/h_0$.

⑥ 剪切盒。

⑦ 剪切速率。

① 破坏状态。
② 加载板。
③ 多孔透水(或固体)板。

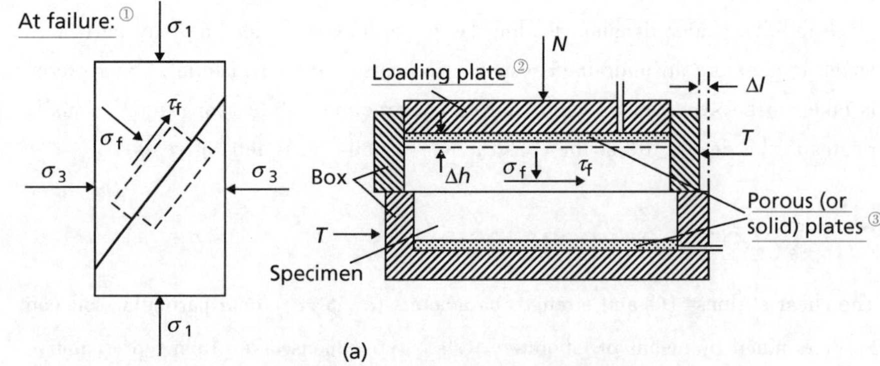

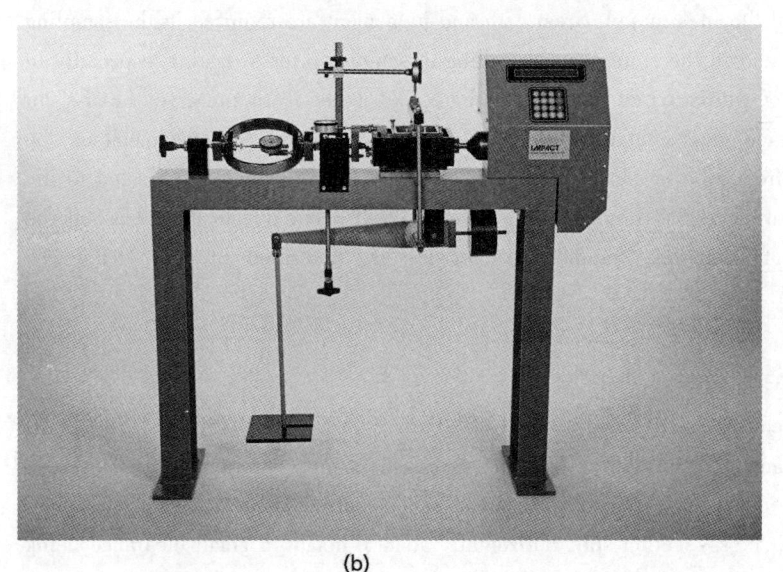

(b)

Figure 5.8 Direct shear apparatus: ④ (a) schematic ⑤, (b) standard direct shear apparatus ⑥ (image courtesy of Impact Test Equipment Ltd.).

④ 直剪仪。
⑤ 原理图。
⑥ 标准直剪仪。

Interpretation of direct shear test data ⑦

⑦ 直剪试验数据的分析。

A number of specimens of the soil are tested, each under a different vertical force, and the value of shear stress at failure ($\tau_f = T/A$) is plotted against the normal effective stress ($\sigma_f' = N/A$) for each test. The Mohr – Coulomb shear strength parameters c' and ϕ' are then obtained from the straight line which best fits the plotted points. The shear stress throughout the test may also be plotted against the shear strain; the gradient of the initial part of the curve before failure (peak shear stress) gives a crude approximation of the shear modulus G.

The test suffers from several disadvantages, the main one being that drainage conditions cannot be controlled. ⑧ As pore water pressure cannot be measured, only the total normal stress can be determined, although this is equal to the effective normal stress if the pore water pressure is zero (i. e. by shearing slowly enough to achieve drained conditions). Only an approximation to the state of pure shear defined in Figure 5.2 is produced in the specimen and shear stress on the failure

⑧ 此试验有许多不足之处，主要是排水条件无法控制。

plane is not uniform, failure occurring progressively from the edges towards the centre of the specimen. Furthermore, the cross-sectional area of the sample under the shear and vertical loads does not remain constant throughout the test. The advantages of the test are its simplicity and, in the case of coarse-grained soils, the ease of specimen preparation.[①]

① 试样制备。

The triaxial test

The **triaxial apparatus**[②] is the most widely used laboratory device for measuring soil behaviour in shear, and is suitable for all types of soil. The test has the advantages that drainage conditions can be controlled, enabling saturated soils of low permeability to be consolidated, if required, as part of the test procedure, and pore water pressure measurements can be made.[③] A cylindrical specimen, generally having a length/diameter ratio of 2, is used in the test; this sits within a chamber of pressurised water. The sample is stressed axially by a loading ram and radially by the confining fluid pressure under conditions of axial symmetry in the manner shown in Figure 5.9. The most common test, **triaxial compression**[④], involves applying shear to the soil by holding the confining pressure constant and applying compressive axial load through the loading ram.

② 三轴仪。

③ 此试验的优点是排水条件可控制，使低渗透性的饱和土可以固结。若有相关要求，在试验过程中可以测量孔隙水压力。

④ 三轴压缩。

The main features of the apparatus are also shown in Figure 5.9. The circular base has a central pedestal on which the specimen is placed, there being access through the pedestal for drainage and for the measurement of pore water pressure. A Perspex cylinder, sealed between a ring and the circular cell top, forms the body of the cell. The cell top has a central bush through which the loading ram passes. The cylinder and cell top clamp onto the base, a seal being made by means of an O-ring.

The specimen is placed on either a porous or a solid disc on the pedestal of the apparatus. Typical specimen diameters (in the UK) are 38 and 100mm. A loading cap is placed on top of the specimen, and the specimen is then sealed in a rubber membrane, O-rings under tension being used to seal the membrane to the pedestal and the loading cap to make these connections watertight. In the case of sands, the specimen must be prepared in a rubber membrane inside a rigid former[⑤] which fits around the pedestal. A small negative pressure[⑥] is applied to the pore water to maintain the stability of the specimen while the former is removed prior to the application of the all-round confining pressure. A connection may also be made through the loading cap to the top of the specimen, a flexible plastic tube leading from the loading cap to the base of the cell; this connection is normally used for the application of back pressure (as described later in this section). Both the top of the loading cap and the lower end of the loading ram have coned seatings, the load being transmitted through a steel ball. The specimen is subjected to an all-round fluid pressure in the cell, consolidation is allowed to take place, if appropriate, and then the axial stress is gradually increased by the application of compressive load through the ram until failure of the specimen takes place, usually on a diagonal

⑤ 刚性模具。

⑥ 小的负压（吸力）。

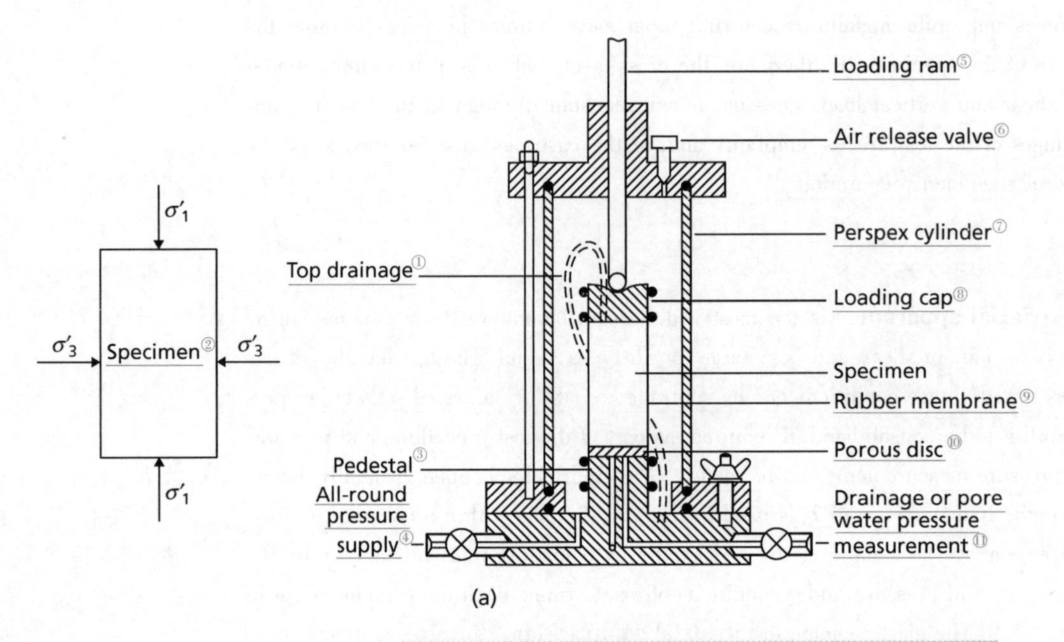

① 顶端排水。
② 试样。
③ 底座。
④ 围压供给装置。
⑤ 加载杆。
⑥ 排气阀。
⑦ 有机玻璃围压室。
⑧ 加载帽。
⑨ 橡胶膜。
⑩ 透水石。
⑪ 排水口或孔隙水压力测量仪。
⑫ 三轴仪。
⑬ 原理图。
⑭ 标准三轴室。

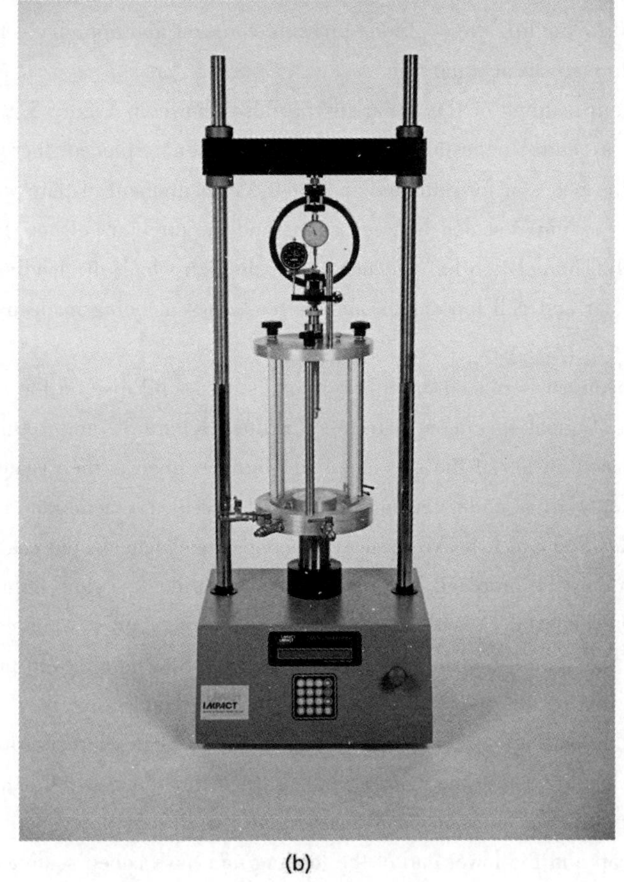

(b)

Figure 5.9 The triaxial apparatus:[12] (a) schematic,[13] (b) a standard triaxial cell[14] (image courtesy of Impact Test Equipment Ltd.)

plane through the sample (see Figure 5.6). The load is measured by means of a load ring[①] or by a load transducer[②] fitted to the loading ram either inside or outside the cell. The system for applying the all-round pressure must be capable of compensating for pressure changes due to cell leakage or specimen volume change.[③]

① 量力环。
② 荷载传感器。
③ 试样体积变化。

Prior to triaxial compression, sample consolidation may be permitted under equal increments of total stress normal to the end and circumferential surfaces of the specimen, i. e. by increasing the confining fluid pressure within the triaxial cell. Lateral strain in the specimen is not equal to zero during consolidation under these conditions (unlike in the oedometer test, as described in Section 4.2). This is known as **isotropic consolidation**[④]. Dissipation of excess pore water pressure takes place due to drainage through the porous disc at the bottom (or top, or both) of the specimen. The drainage connection leads to an external volume gauge, enabling the volume of water expelled from the specimen to be measured. Filter paper drains, in contact with the end porous disc, are sometimes placed around the circumference of the specimen; both vertical and radial drainage then take place and the rate of dissipation of excess pore water pressure is increased to reduce test time for this stage.

④ 等向固结。

The pore water pressure within a triaxial specimen can usually be measured, enabling the results to be expressed in terms of effective stresses within the sample, rather than just the known applied total stresses; conditions of no flow either out of or into the specimen must be maintained, otherwise the correct pressure will be modified. Pore water pressure is normally measured by means of an electronic pressure transducer.[⑤]

⑤ 孔隙水压力通常由电子压力传感器来测量。

If the specimen is partially saturated, a fine porous ceramic disc[⑥] must be sealed into the pedestal of the cell if the correct pore water pressure is to be measured. Depending on the pore size of the ceramic, only pore water can flow through the disc, provided the difference between the pore air and pore water pressures is below a certain value, known as the **air entry value**[⑦] of the disc. Under undrained conditions the ceramic disc will remain fully saturated with water, provided the air entry value is high enough, enabling the correct pore water pressure to be measured. The use of a coarse porous disc, as normally used for a fully saturated soil, would result in the measurement of the pore air pressure in a partially saturated soil.

⑥ 多孔陶瓷片。
⑦ 进气值。

Test limitations and corrections[⑧]

⑧ 试验的局限性与修正。

The average cross-sectional area (A) of the specimen does not remain constant throughout the test, and this must be taken into account when interpreting stress data from the axial ram load measurements. If the original cross-sectional area of the specimen is A_0, the original length is l_0 and the original volume is V_0, then, if the volume of the specimen decreases during the test,

$$A = A_0 \frac{1 - \varepsilon_v}{1 - \varepsilon_a} \tag{5.16}$$

① 体积应变。
② 轴向应变。
③ 径向应变。

where ε_v is the volumetric strain[①] ($\Delta V/V_0$) and ε_a is the axial strain[②] ($\Delta l/l_0$). If the volume of the specimen increases during a drained test, the sign of ΔV will change and the numerator in Equation 5.16 becomes $(1+\varepsilon_v)$. If required, the radial strain[③] (ε_r) could be obtained from the equation

$$\varepsilon_v = \varepsilon_a + 2\varepsilon_r \tag{5.17}$$

④ 试样端部的润滑措施。
⑤ 橡胶膜的顺变性。

In addition, the strain conditions in the specimen are not uniform due to frictional restraint produced by the loading cap and pedestal disc; this results in dead zones at each end of the specimen, which becomes barrel-shaped as the test proceeds. Non-uniform deformation of the specimen can be largely eliminated by lubrication of the end surfaces.[④] It has been shown, however, that non-uniform deformation has no significant effect on the measured strength of the soil, provided the length/diameter ratio of the specimen is not less than 2. The compliance of the rubber membrane[⑤] must also be accounted for.

⑥ 三轴试验数据的处理强度。

Interpretation of triaxial test data: strength[⑥]

⑦ 注：此处"－"号应为"＋"号。
⑧ 偏应力不变量。
⑨ 平均应力不变量。

Triaxial data may be presented in the form of Mohr circles at failure; it is more straightforward, however, to present it in terms of stress invariants, such that a given set of effective stress conditions can be represented by a single point instead of a circle. Under 2-D stress conditions, the state of stress represented in Figure 5.6 could also be defined by the radius and centre of the Mohr circle. The radius is usually denoted by $t = 1/2\ (\sigma_1' - \sigma_3')$, with the centre point denoted by $s' = 1/2\ (\sigma_1' - \sigma_3')$.[⑦] These quantities ($t$ and s') also represent the maximum shear stress within the element and the average principal effective stress, respectively. The stress state could also be expressed in terms of total stress. It should be noted that

$$\frac{1}{2}(\sigma_1' - \sigma_3') = \frac{1}{2}(\sigma_1 - \sigma_3)$$

i.e. parameter t, much like shear stress τ is independent of u. This parameter is known as the **deviatoric stress invariant**[⑧] for 2-D stress conditions, and is analogous to shear stress τ acting on a shear plane (alternatively, τ may be thought of as the deviatoric stress invariant in direct shear). Parameter s' is also known as the **mean stress invariant**[⑨], and is analogous to the normal effective stress acting on a shear plane (i.e. causing volumetric change but no shear). Substituting Terzaghi's Principle into the definition of s', it is apparent that

$$\frac{1}{2}(\sigma_1' + \sigma_3') = \frac{1}{2}(\sigma_1 + \sigma_3) - u$$

or, re-written, $s' = s - u$, i.e. Terzaghi's Principle rewritten in terms of the 2-D stress invariants.

The stress conditions and Mohr circle for a 3-D element of soil under a general distribution of stresses is shown in Figure 5.10. Unlike 2-D conditions when there are three unique stress components (σ_x', σ_z', τ_{zx}), in 3-D there are six stress components (σ_x', σ_y', σ_z', τ_{xy}, τ_{yz}, τ_{zx}). These can, however, be reduced to a set of three principal stresses, σ_1', σ_2' and σ_3'. As before, σ_1' and σ_3' are the major (largest) and minor (smallest) principal stresses; σ_2' is known as the **intermediate principal stress**[①]. For the general case where the three principal stress components are different ($\sigma_1' > \sigma_2' > \sigma_3'$, also described as **true triaxial** conditions[②]), a set of three Mohr circles can be drawn as shown in Figure 5.10.

① 中主应力。
② 真三轴条件。
③ 破坏包络线。

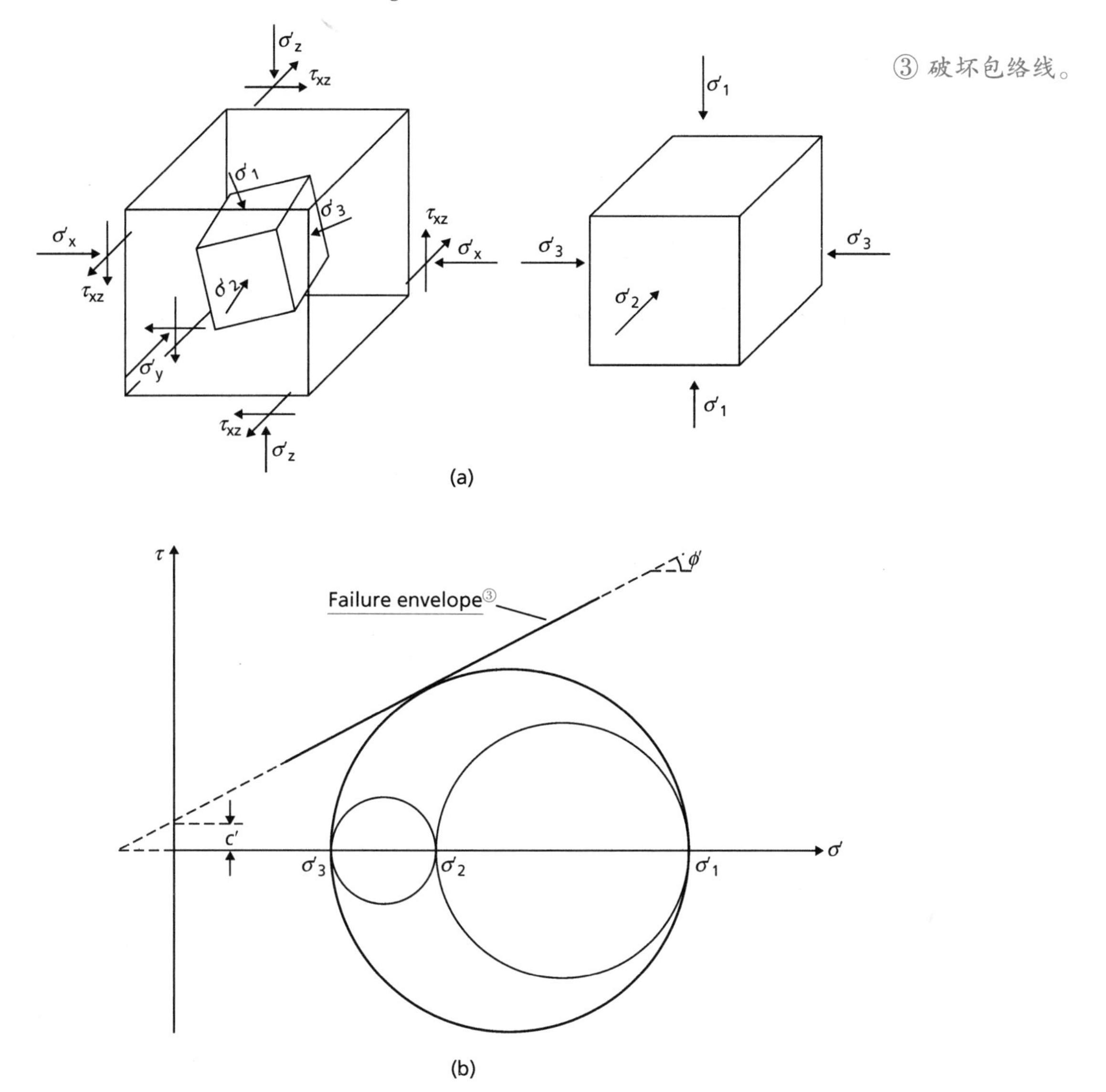

Figure 5.10 Mohr circles for triaxial stress conditions.[④]

④ 三轴应力条件下的莫尔圆。

As, in the case of 2-D stress conditions, it was possible to describe the stress state in terms of a mean and deviatoric invariant (s' and t respectively), so it is possible in 3-D conditions. To distinguish between 2-D and 3-D cases, the triaxial mean invariant is denoted p' (effective stress) or p (total stress), and the deviatoric invariant by q. As before, the mean stress invariant causes only volumetric change (does not induce shear), and is the average of the three principal stress components:[⑤]

$$p' = \frac{\sigma_1' + \sigma_2' + \sigma_3'}{3} \tag{5.18}$$

⑤ 如前，平均应力不变量只产生体积变化（不产生剪切），平均应力不变量为三个主应力的平均值。

Equation 5.18 may also be written in terms of total stresses. Similarly, q, as the deviatoric invariant, induces shearing within the sample and is independent of the pore fluid pressure, u:

$$q = \frac{1}{\sqrt{2}}[(\sigma_1 - \sigma_2)^2 + (\sigma_2 - \sigma_3)^2 + (\sigma_3 - \sigma_1)^2]^{\frac{1}{2}} \quad (5.19)$$

A full derivation of Equation 5.19 may be found in Atkinson and Bransby (1978). In the triaxial cell, the stress conditions are simpler than the general case shown in Figure 5.10. During a standard compression test, $\sigma_1 = \sigma_a$, and due to the axial symmetry[①], $\sigma_2 = \sigma_3 = \sigma_r$, so that Equations 5.18 and 5.19 reduce to:

① 轴对称条件。

$$p = \frac{\sigma_a + 2\sigma_r}{3} \qquad p' = \frac{\sigma_a' + 2\sigma_r'}{3} \quad (5.20)$$

$$q = \sigma_a' - \sigma_r' = \sigma_a - \sigma_r \quad (5.21)$$

The confining fluid pressure within the cell (σ_r) is the minor principal stress in the standard triaxial compression test. The sum of the confining pressure and the applied axial stress from the loading ram is the major principal stress (σ_a), on the basis that there are no shear stresses applied to the surfaces of the specimen. The applied axial stress component from the loading ram is thus equal to the deviatoric stress q, also referred to as the **principal stress difference**.[②] As the intermediate principal stress is equal to the minor principal stress, the stress conditions at failure can be represented by a single Mohr circle, as shown in Figure 5.6 with $\sigma_1' = \sigma_a'$ and $\sigma_3' = \sigma_r'$. If a number of specimens are tested, each under a different value of confining pressure, the failure envelope can be drawn and the shear strength parameters for the soil determined.

② 主应力差值。

Because of the axial symmetry in the triaxial test, both 2-D and 3-D invariants are in common usage, with stress points represented by s', t or p', q respectively. The parameters defining the strength of the soil (ϕ' and c') are not affected by the invariants used; however, the interpretation of the data to find these properties does vary with the set of invariants used. Figure 5.11 shows the Mohr – Coulomb failure envelope for an element of soil plotted in terms of direct shear (σ', τ), 2-D (s', t) and 3-D/triaxial (p', q) stress invariants. Under direct shear conditions, it has already been described in Section 5.3 that the gradient of the failure envelope τ/σ' is equal to the tangent of the angle of shearing resistance and the intercept is equal to c' (Figure 5.11 (a)). For 2-D conditions, Equation 5.15a gives:

$$(\sigma_1' - \sigma_3') = (\sigma_1' + \sigma_3')\sin\phi' + 2c'\cos\phi'$$
$$2t = 2s'\sin\phi' + 2c'\cos\phi'$$
$$t = s'\sin\phi' + c'\cos\phi' \quad (5.22)$$

Therefore, if the points of ultimate failure in triaxial tests are plotted in terms of s' and t, a straight-line failure envelope will be obtained – the gradient of this line is equal to the sine of the angle of shearing resistance and the intercept $= c'\cos\phi'$[③] (Figure 5.11 (b)). Under triaxial conditions with $\sigma_2 = \sigma_3$, from the definitions of p' and q (Equations 5.20 and 5.21) it is clear that:

③ 这条直线的斜率等于抗剪摩擦角的正弦值，截距为 $c'\cos\phi'$。

$$(\sigma_1' + \sigma_3') = (\sigma_a' + \sigma_r') = \frac{6p' + q}{3} \quad (5.23)$$

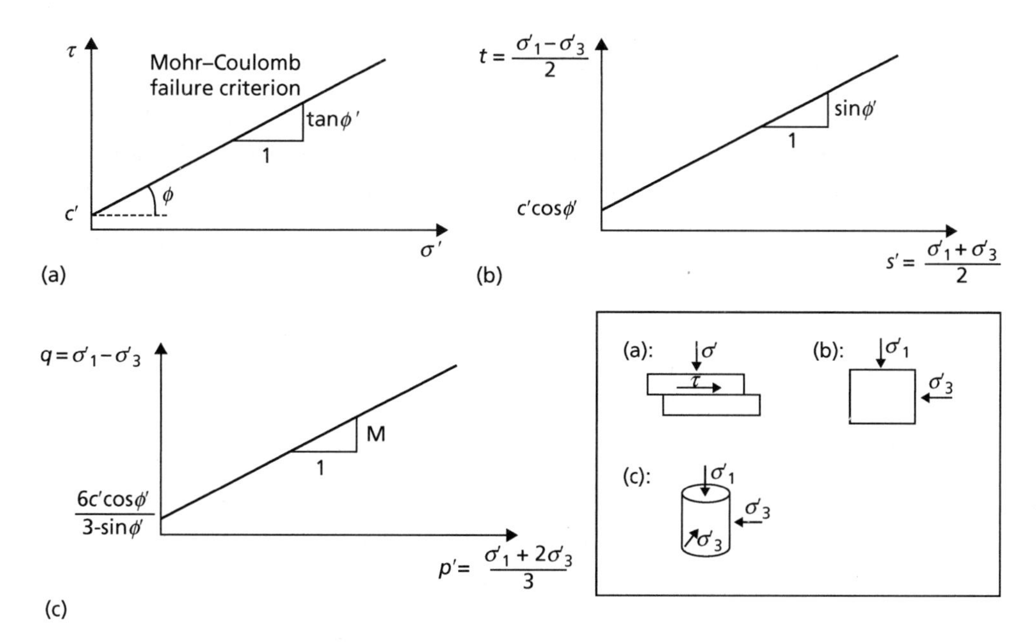

Figure 5.11 Interpretation of strength parameters c' and ϕ' using stress invariants.[1]

and

$$(\sigma_1' - \sigma_3') = (\sigma_a' - \sigma_r') = q \tag{5.24}$$

The gradient of the failure envelope in triaxial conditions is described by the parameter $M = q/p'$, (Figure 5.11 (c)) so that from Equation 5.15a:

$$(\sigma_1' - \sigma_3') = (\sigma_1' + \sigma_3')\sin\phi' + 2c'\cos\phi'$$

$$q = \left(\frac{6p' + q}{3}\right)\sin\phi' + 2c'\cos\phi' \tag{5.25}$$

$$q = \left(\frac{6\sin\phi'}{3 - \sin\phi'}\right)p' + \frac{6c'\cos\phi'}{3 - \sin\phi'}$$

Equation 5.25 represents a straight line when plotted in terms of p' and q with a gradient given by

$$M = \frac{q}{p'} = \frac{6\sin\phi'}{3 - \sin\phi'} \tag{5.26}$$

$$\therefore \sin\phi' = \frac{3M}{6 + M}$$

Equations 5.25 and 5.26 apply only to triaxial compression, where $\sigma_a > \sigma_r$. For samples under **triaxial extension**[2], $\sigma_r > \sigma_a$, the cell pressure[3] becomes the major principal stress. This does not affect Equation 5.23, but Equation 5.24 becomes

$$(\sigma_1' - \sigma_3') = (\sigma_r' - \sigma_a') = -q \tag{5.27}$$

giving:

$$\sin\phi' = \frac{3M}{6 - M} \tag{5.28}$$

As the friction angle of the soil must be the same whether it is measured in triaxial compression or extension, this implies that different values of M will be observed in compression and extension.[4]

① 利用应力不变量表达强度参数 c' 与 ϕ'。

② 三轴拉伸。

③ 围压。

④ 由于无论是用三轴压缩还是用三轴拉伸测量，土的内摩擦角应该是一致的，因此在三轴压缩或拉伸中 M 值不同。

① 三轴试验的数据处理刚度。

② 各向同性线弹性。

③ 三轴剪切应变。

④ 此关系式指出，在q-ε_s（即偏应力与偏应变）的关系曲线中，破坏前曲线的导数值等于剪切模量的3倍。

Interpretation of triaxial test data: stiffness[1]

Triaxial test data can also be used to determine the stiffness properties (principally the shear modulus, G) of a soil. Assuming that the soil is isotropically linear elastic[2], for 3-D stress conditions Equation 5.7 can be extended to give:

$$\begin{bmatrix}\varepsilon_x\\ \varepsilon_y\\ \varepsilon_z\\ \gamma_{xy}\\ \gamma_{yz}\\ \gamma_{zx}\end{bmatrix}=\frac{1}{2G(1+v)}\begin{bmatrix}1 & -v & -v & 0 & 0 & 0\\ -v & 1 & -v & 0 & 0 & 0\\ -v & -v & 1 & 0 & 0 & 0\\ 0 & 0 & 0 & 2(1+v) & 0 & 0\\ 0 & 0 & 0 & 0 & 2(1+v) & 0\\ 0 & 0 & 0 & 0 & 0 & 2(1+v)\end{bmatrix}\begin{bmatrix}\sigma_x'\\ \sigma_y'\\ \sigma_z'\\ \tau_{xy}\\ \tau_{yz}\\ \tau_{zx}\end{bmatrix} \tag{5.29}$$

This may alternatively be simplified in terms of principal stresses and strains to give:

$$\begin{bmatrix}\varepsilon_1\\ \varepsilon_2\\ \varepsilon_3\end{bmatrix}=\frac{1}{2G(1+v)}\begin{bmatrix}1 & -v & -v\\ -v & 1 & -v\\ -v & -v & 1\end{bmatrix}\begin{bmatrix}\sigma_1'\\ \sigma_2'\\ \sigma_3'\end{bmatrix} \tag{5.30}$$

The deviatoric shear strain ε_s within a triaxial cell (i. e. the strain induced by the application of the deviatoric stress, q, also known as the **triaxial shear strain**[3]) is given by:

$$\varepsilon_s=\frac{2}{3}(\varepsilon_1-\varepsilon_3) \tag{5.31}$$

Substituting for ε_1 and ε_3 from Equation 5.30 and using Equation 5.6, Equation 5.31 reduces to:

$$\varepsilon_s=\frac{1}{3G}q \tag{5.32}$$

This implies that on a plot of q versus ε_s (i. e. deviatoric stress versus deviatoric strain), the value of the gradient of the curve prior to failure is equal to three times the shear modulus.[4] Generally, in order to determine ε_s within a triaxial test, it is necessary to measure both the axial strain ε_a ($=\varepsilon_1$ for triaxial compression) and the radial strain ε_r ($=\varepsilon_3$). While the former is routinely measured, direct measurement of the latter parameter requires sophisticated sensors to be attached directly to the sample, though the volume change during drained shearing may also be used to infer ε_r using Equation 5.17. If, however, the test is conducted under undrained conditions, then there will be no volume change ($\varepsilon_v=0$) and hence from Equation 5.17:

$$\varepsilon_r=-\frac{1}{2}\varepsilon_a \tag{5.33}$$

From Equation 5.31 it is then clear that for undrained conditions, $\varepsilon_s=\varepsilon_a$. A plot of q versus ε_a for an undrained test will thus have a gradient equal to $3G$. Undrained triaxial testing is therefore extremely useful for determining shear modulus, using measurements which can be made on even the most basic triaxial cells.[①] As G is independent of the drainage conditions within the soil, the value obtained applies equally well for subsequent analysis of the soil under drained loadings. In addition to reducing instrumentation requirements, undrained testing is also much faster than drained testing, particularly for saturated clays of low permeability.

① 不排水三轴试验对于确定剪切模量非常有用，这在最基本的三轴室中都可进行。

If drained triaxial tests are conducted in which volume change is permitted, radial strain measurements should be made such that G can be determined from a plot of q versus ε_s. Under these test conditions the drained Poisson's ratio (ν')[②] can also be determined, being:

② 排水条件下泊松比。
注：即土体骨架泊松比。

$$\nu' = -\frac{\varepsilon_r}{\varepsilon_a} \tag{5.34}$$

Under undrained conditions it is not necessary to measure Poisson's ratio (ν_u)[③], as, comparing Equations 5.33 and 5.34, it is clear that $\nu_u = 0.5$ for there to be no volume change.

③ 不排水条件下泊松比。

Testing under back pressure[④]

④ 反压条件下试验。

Testing under back pressure involves raising the pore water pressure within the sample artificially by connecting a source of constant fluid pressure through a porous disc to one end of a triaxial specimen. In a drained test this connection remains open throughout the test, drainage taking place against the back pressure; the back pressure is then the datum for excess pore water pressure measurement. In a consolidated-undrained test[⑤] (described later) the connection to the back pressure source is closed at the end of the consolidation stage, before the application of the principal stress difference is commenced.

⑤ 固结不排水试验。

The object of applying a back pressure is to ensure full saturation of the specimen or to simulate insitu pore water pressure conditions.[⑥] During sampling, the degree of saturation of a fine-grained soil may fall below 100% owing to swelling on the release of in-situ stresses. Compacted specimens will also have a degree of saturation below 100%. In both cases, a back pressure is applied which is high enough to drive the pore air into solution in the pore water.

⑥ 施加反压的目的是保证试样的完全饱和或模拟原位孔隙水压力条件。

It is essential to ensure that the back pressure does not by itself change the effective stresses in the specimen. It is necessary, therefore, to raise the cell pressure simultaneously with the application of the back pressure and by an equal increment. Consider an element of soil, of volume V and porosity n, in equilibrium under total principal stresses σ_1, σ_2 and σ_3, as shown in Figure 5.12, the pore pressure being u_0. The element is subjected to equal increases in confining pressure $\Delta\sigma_3$ in each direction i.e. an isotropic increase in stress, accompanied by an increase Δu_3 in pore pressure.

The increase in effective stress in each direction $= \Delta\sigma_3 - \Delta u_3$

Reduction in volume of the soil skeleton $= C_s V(\Delta\sigma_3 - \Delta u_3)$

where C_s is the compressibility of the soil skeleton under an isotropic effective stress increment.

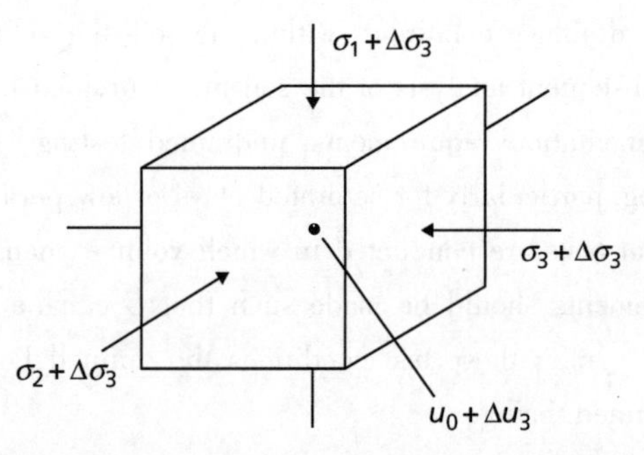

Figure 5.12 Soil element under isotropic stress increment. ①

① 各向等压增长下的土单元体。

Reduction in volume of the pore space = $C_v nV\Delta u_3$

where C_v is the compressibility of pore fluid under an isotropic pressure increment.

If the soil particles are assumed to be incompressible and if no drainage of pore fluid takes place, then the reduction in volume of the soil skeleton must equal the reduction in volume of the pore space, ② i. e.

② 如果假设土颗粒是不可压缩的，并且没有孔隙间流体的排水发生，那么土骨架体积的减少必等于孔隙空间体积的减少量。

$$C_s V(\Delta\sigma_3 - \Delta u_3) = C_v nV\Delta u_3$$

Therefore,

$$\Delta u_3 = \Delta\sigma_3 \left(\frac{1}{1 + n(C_v/C_s)} \right)$$

Writing $1/[1 + n(C_v/C_s)] = B$, defined as a **pore pressure coefficient** ③,

③ 孔隙压力系数（孔压系数）。

$$\Delta u_3 = B\Delta\sigma_3 \tag{5.35}$$

In fully saturated soils the compressibility of the pore fluid (water only) is considered negligible compared with that of the soil skeleton, and therefore $C_v/C_s \rightarrow 0$ and $B \rightarrow 1$. In partially saturated soils the compressibility of the pore fluid is high due to the presence of pore air, and therefore $C_v/C_s > 0$ and $B < 1$. The variation of B with degree of saturation for a particular soil is shown in Figure 5.13.

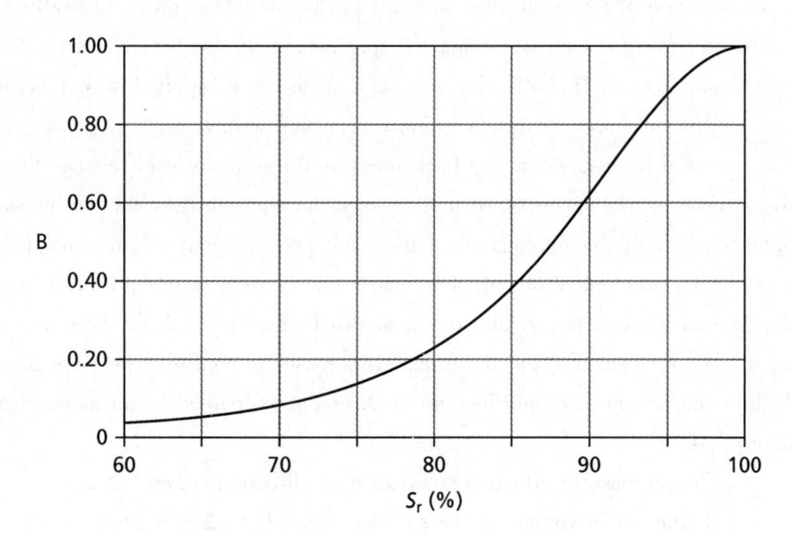

Figure 5.13 Typical relationship between B and degree of saturation. ④

④ B 值与饱和度之间的典型关系。

The value of B can be measured in the triaxial apparatus (Skempton, 1954). A specimen is set up under any value of all-round pressure and the pore water pressure measured. Under undrained conditions the all-round pressure is then increased (or reduced) by an amount $\Delta\sigma_3$ and the change in pore water pressure (Δu) from the initial value is measured, enabling the value of B to be calculated from Equation 5.35. A specimen is normally considered to be saturated if the pore pressure coefficient B has a value of at least 0.95.[①]

① 如果孔隙压力系数 B 达到 0.95，通常认为试样已饱和。

Types of triaxial test[②]

② 三轴试验的种类。

Many variations of test procedure are possible with the triaxial apparatus, but the three principal types of test are as follows:

1 *Unconsolidated – Undrained (UU)*[③]. The specimen is subjected to a specified confining pressure and then the principal stress difference is applied immediately, with no drainage/consolidation being permitted at any stage of the test. The test procedure is standardised in BS1377, Part 7 (UK), CEN ISO/TS 17892 – 8 (Europe) and ASTM D2850 (US).

③ 不固结不排水试验。

2 *Consolidated – Undrained (CU)*[④]. Drainage of the specimen is permitted under a specified confining pressure until consolidation is complete; the principal stress difference is then applied with no further drainage being permitted[⑤]. Pore water pressure measurements may be made during the undrained part of the test to determine strength parameters in terms of effective stresses. The consolidation phase is isotropic in most standard testing, denoted by CIU. Modern computer-controlled triaxial machines (also known as **stress path cells**[⑥]) use hydraulic pressure control units to control the cell (confining) pressure, back pressure and ram load (axial stress) independently (Figure 5.14). Such an apparatus can therefore apply a 'no-lateral strain'[⑦] condition where stresses are anisotropic, mimicking the one-dimensional compression that occurs in an oedometer test[⑧]. These tests are often denoted by CAU (the 'A' standing for anisotropic). The test procedure is standardised in BS1377, Part 8 (UK), CEN ISO/TS 17892 – 9 (Europe) and ASTM D4767 (US).

④ 固结不排水试验。

⑤ 允许试样先在围压下排水直至固结完成，然后在不排水条件下施加偏应力。

⑥ 应力路径室。

⑦ 无侧向变形。

⑧ 固结试验。

3 *Consolidated – Drained (CD)*[⑨]. Drainage of the specimen is permitted under a specified confining pressure until consolidation is complete; with drainage still being permitted, the principal stress difference is then applied at a rate slow enough to ensure that the excess pore water pressure is maintained at zero. The test procedure is standardised in BS1377, Part 8 (UK), CEN ISO/TS 17892 – 9 (Europe) and ASTM D7181 (US).

⑨ 固结排水试验。

The use of these test procedures for determining the strength and stiffness properties of both coarse and fine grained soils will be discussed in the following sections (5.5 and 5.6).

Other tests

Although direct shear and triaxial tests are the most commonly used laboratory tests

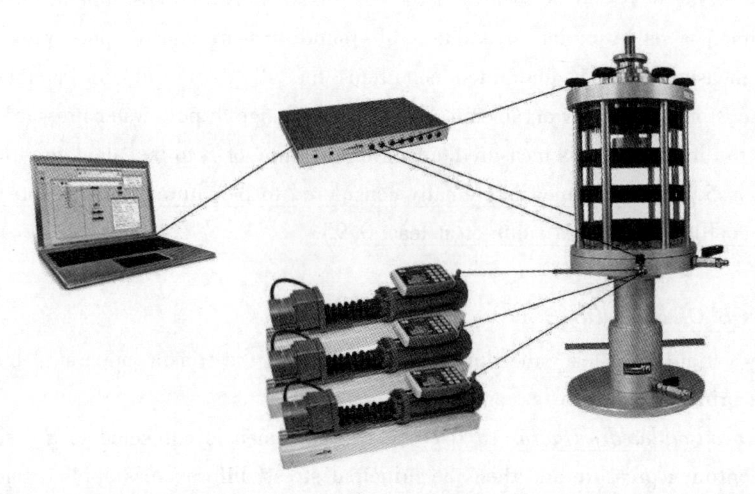

① 应力路径三轴室。 **Figure 5.14** Stress path triaxial cell[①] (image courtesy of GDS Instruments).

② 无侧限抗压试验。

③ 无侧限抗压强度。

for quantifying the constitutive behaviour of soil, there are other tests which are used routinely. An **unconfined compression test**[②] is essentially a triaxial test in which the confining pressure $\sigma_3 = 0$. The reported result from such a test is the **unconfined compressive strength** (UCS)[③], which is the major principal (axial) stress at failure (which, because $\sigma_3' = 0$, is also the deviatoric stress at failure). As with the triaxial test, a Mohr circle may be plotted for the test as shown in Figure 5.15; however, as only one test is conducted it is not possible to define the Mohr–Coulomb shear strength envelope without conducting further triaxial tests. The test is not suitable for cohesionless soils ($c' \approx 0$), which would fail immediately without the application of confining pressure. It is usually used with fine-grained soils, and is particularly popular for testing rock.

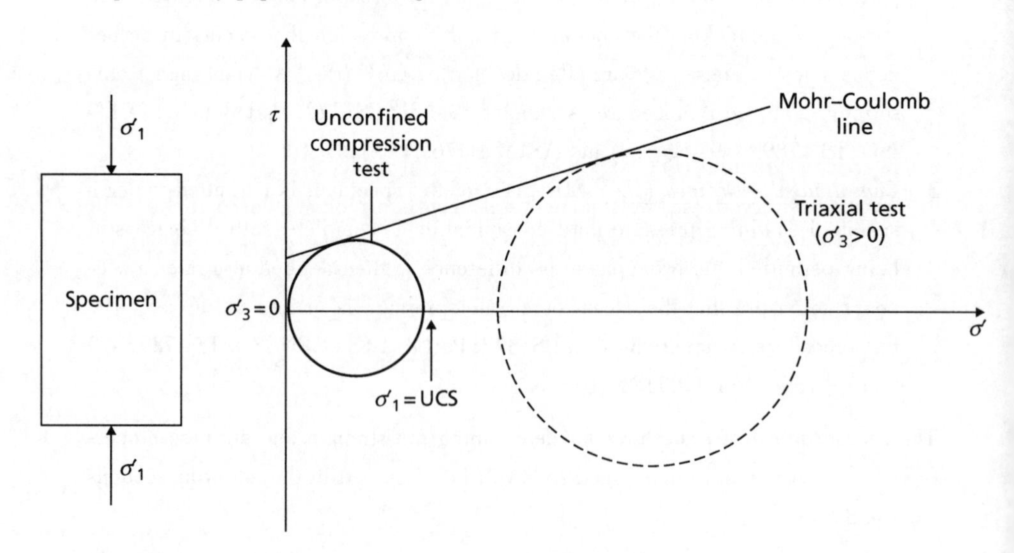

Figure 5.15 Unconfined compression test interpretation.

A **laboratory vane** is sometimes used in fine-grained soils for measuring the undrained shear strength parameter c_u[1]. It is much quicker and simpler for this purpose than conducting a UU triaxial test. It is not discussed further in this chapter, as its principle of operation is similar to the field vane, which is described in detail in Section 7.3.

The **simple shear apparatus** (SSA)[2] is an alternative to the direct shear apparatus (DSA)[3] described earlier in this section. Rather than using a split shear box, the side walls of the simple shear apparatus are usually formed of either plates which can rotate, or a series of laminations which can move relative to each other or are flexible, such arrangements allowing rotation of the sides of the sample which imposes a state of pure simple shear as defined in Figures 5.1 and 5.2[4]. Analysis of the test data is identical to the DSA, although the approximations and assumptions made for the conditions within the DSA are truer for the SSA. While the SSA is better for testing soil, it cannot conduct interface shear tests[5] and is therefore not as versatile or popular as the DSA.

① 室内十字板剪切试验常被用于测定细粒土的不排水抗剪强度参数 c_u。

② 单剪仪。

③ 直剪仪。

④ 不采用可滑动的剪切盒，单剪仪的侧壁由可转动的两个平板组成，或是一组可相对移动或有柔性的薄板。当施加图 5.1 和图 5.2 定义的纯单剪状态，该装置可使试样侧壁转动。

⑤ 界面剪切试验。

5.5 Shear strength of coarse-grained soils[6]

The shear strength characteristics of coarse-grained soils such as sands and gravels can be determined from the results of either direct shear tests or drained triaxial tests, only the drained strength of such soils normally being relevant in practice[7]. The characteristics of dry and saturated sands or gravels are the same, provided there is zero excess pore water pressure generated in the case of saturated soils, as strength and stiffness are dependent on effective stress. Typical curves relating shear stress and shear strain for initially dense and loose sand specimens in direct shear tests are shown in Figure 5.16 (a). Similar curves are obtained relating principal stress difference and axial strain in drained triaxial compression tests.

In dense deposits (high relative density,[8] I_D, see Chapter 1) there is a considerable degree of interlocking between particles. Before shear failure can take place, this interlocking must be overcome in addition to the frictional resistance at the points of inter-granular contact.[9] In general, the degree of interlocking is greatest in the case of very dense, well-graded soils consisting of angular particles. The characteristic stress – strain curve for initially dense sand shows a peak stress[10] at a relatively low strain, and thereafter, as interlocking is progressively overcome, the stress decreases with increasing strain. The reduction in the degree of interlocking produces an increase in the volume of the specimen during shearing as characterised by the relationship between volumetric strain and shear strain in the direct shear test, shown in Figure 5.16 (c)[11]. In the drained triaxial test, a similar relationship would be obtained between volumetric strain and axial strain. The change in volume is also shown in terms of void ratio (e) in Figure 5.16 (d). Eventually the specimen becomes loose enough to allow particles to move over and around their neighbours without any further net volume change, and the shear stress reduces to an ultimate value. However, in the triaxial test non-uniform deformation of the

⑥ 粗粒土的抗剪强度。

⑦ 粗粒土如砂和砾的抗剪强度可由直剪试验或排水三轴试验的结果得到，只有粗粒土的排水强度在实际工程中有用。

⑧ 相对密实度。

⑨ 在剪切破坏发生前，除了粒间接触的摩擦阻力外还需克服粒间的咬合力。

⑩ 峰值应力。

⑪ 粒间咬合程度的减弱致使试样在剪切过程中体积增大，这一特性体现在直剪试验的体应变与剪应变的关系曲线中，如图 5.16(c)所示。

specimen becomes excessive as strain is progressively increased, and it is unlikely that the ultimate value of principal stress difference can be reached.

① 剪胀性。

② 剪胀角。

The term **dilatancy**① is used to describe the increase in volume of a dense coarse-grained soil during shearing, and the rate of dilation can be represented by the gradient $d\varepsilon_v/d\gamma$, the maximum rate corresponding to the peak stress (Figure 5.16 (c)). The **angle of dilation**② (ψ) is defined as $\tan^{-1}(d\varepsilon_v/d\gamma)$. The concept of dilatancy can be illustrated in the context of the direct shear test by considering the shearing of dense and loosely packed spheres (idealised soil particles) as shown in Figure 5.17. During shearing of a dense soil (Figure 5.17 (a)), the macroscopic shear plane is horizontal but sliding between individual particles takes place on numerous microscopic planes inclined at various angles above the horizontal, as the particles move up and over their neighbours. The angle of dilation represents an average value of these angles for the specimen as a whole. The loading plate of the apparatus is thus forced upwards, work being done against the normal stress on the shear plane. For a dense soil, the maximum (or peak) angle of shearing resistance (ϕ'_{max}) determined from peak stresses (Figure 5.16 (b)) is significantly greater than the true angle of friction (ϕ_μ) between the surfaces of individual particles, the difference representing the work required to overcome interlocking and rearrange the particles.

③ 峰值破坏包络线。

④ 临界状态线。

⑤ 体积增加。

⑥ 体积减少。

⑦ 粗粒土的抗剪强度特性。

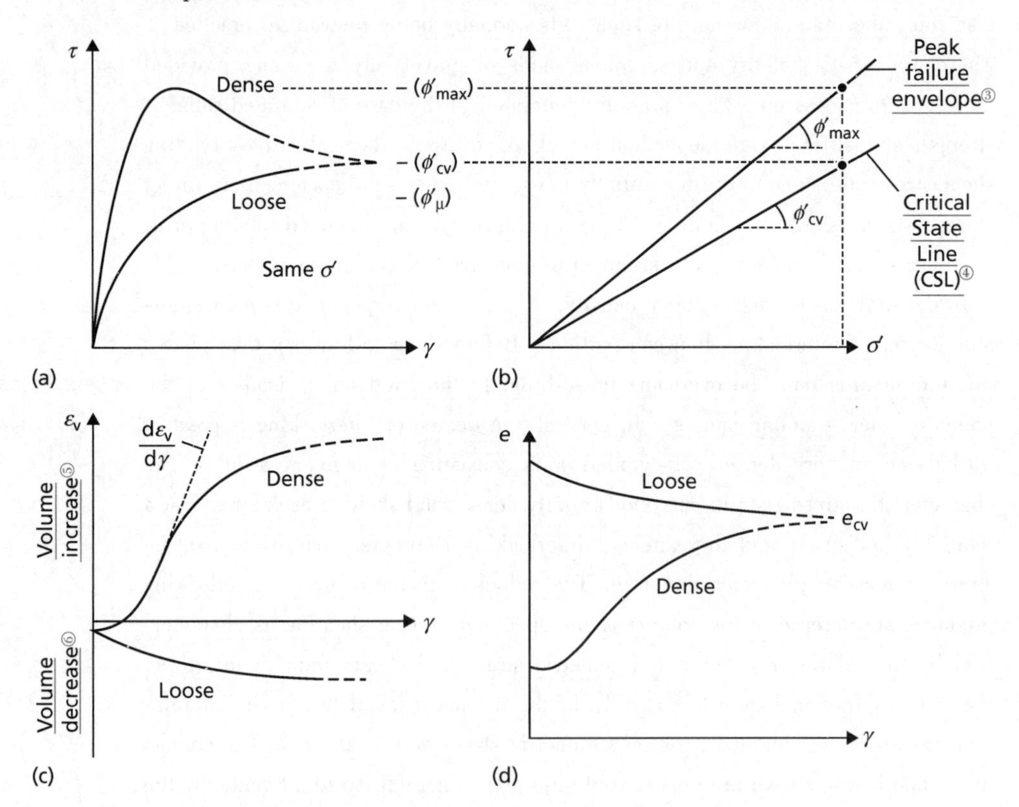

Figure 5.16 Shear strength characteristics of coarse-grained soils⑦.

In the case of initially loose soil (Figure 5.17 (b)), there is no significant particle interlocking to be overcome and the shear stress increases gradually to an ultimate value without a prior peak, accompanied by a decrease in volume. The ultimate values of shear stress and void ratio for dense and loose specimens of the same soil under the same values of normal stress in the direct shear test are essentially equal, as indicated in Figures 5.16 (a) and (d). The ultimate resistance occurs when there is no further change in volume or shear stress (Figures 5.16 (a) and (c)), which is known as the **critical state**①. Stresses at the critical state define a straight line (Mohr – Coulomb) failure envelope intersecting the origin, known as the **critical state line (CSL)**, the slope of which is tan ϕ'_{cv} (Figure 5.16 (b)). The corresponding angle of shearing resistance at critical state (also called the **critical state angle of shearing resistance**)② is usually denoted ϕ'_{cv} or ϕ'_{crit}. The difference between ϕ'_{μ} and ϕ'_{cv} represents the work required to rearrange the particles③. The friction angles ϕ'_{cv} and ϕ'_{max} are related to ψ after the relationship given by Bolton (1986):

$$\phi'_{max} = \phi'_{cv} + 0.8\psi \tag{5.36}$$

Equation 5.36 applies for conditions of plane strain within soil, such as those induced within the DSA or SSA. Under triaxial conditions, the final term becomes approximately 0.5ψ.

① 临界状态。

② 临界状态剪切角（内摩擦角）。

③ ϕ'_{μ}和ϕ'_{cv}的差代表了颗粒重排列需要的能量大小。

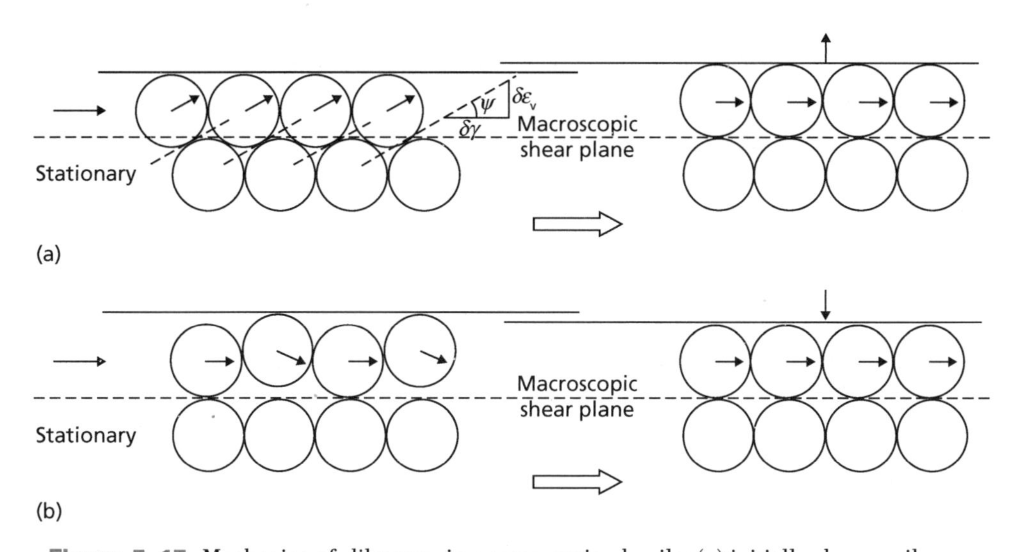

Figure 5.17 Mechanics of dilatancy in coarse-grained soils: (a) initially dense soil, exhibiting dilation, (b) initially loose soil, showing contraction.④

④ 粗粒土剪胀的力学机理：(a) 初始密实土，表现为剪胀，(b) 初始松散土，表现为剪缩。

It can be difficult to determine the value of the parameter ϕ'_{cv} from laboratory tests because of the relatively high strain required to reach the critical state. In general, the critical state is identified by extrapolation of the stress – strain curve to the point of constant stress, which should also correspond to the point of zero rate of dilation ($d\varepsilon_v/d\gamma = 0$) on the volumetric strain – shear strain curve.

An alternative method of representing the results from laboratory shear tests is to plot the **stress ratio**⑤ (τ/σ' in direct shear) against shear strain. Plots of stress ratio against shear strain representing tests on three specimens of sand in a direct

⑤ 应力比。

shear test, each having the same initial void ratio, are shown in Figure 5. 18 (a), the values of effective normal stress (σ') being different in each test. The plots are labelled A, B and C, the effective normal stress being lowest in test A and highest in test C. Corresponding plots of void ratio against shear strain are also shown in Figure 5. 18 (b). Such results indicate that both the maximum stress ratio and the ultimate (or critical) void ratio decrease with increasing effective normal stress[①]. Thus, dilation is suppressed by increasing mean stress (normal stress σ' in direct shear). This is descried in greater detail by Bolton (1986). The ultimate values of stress ratio ($= \tan \phi'_{cv}$), however, are the same. From Figure 5. 18 (a) it is apparent that the difference between peak and ultimate stress decreases with increasing effective normal stress; therefore, if the maximum shear stress is plotted against effective normal stress for each individual test, the plotted points will lie on an envelope which is slightly curved, as shown in Figure 5. 18 (c). Figure 5. 18 (c) also shows the **stress paths**[②] for each of the three specimens leading up to failure. For any type of shear test, two stress paths may be plotted: the **total stress path (TSP)**[③] plots the variation of σ and τ through the test; the **effective stress path (ESP)**[④] plots the variation of σ' and τ. If both stress paths are plotted on the same axis, the horizontal distance between the two paths at a given value of τ (i. e. $\sigma - \sigma'$) represents the pore water pressure in the sample from Terzaghi's Principle (Equation 3. 1). In direct shear tests, the pore water pressure is approximately zero such that

① 结果表明，最大应力比与最终（临界）孔隙比都随着有效正应力的增加而下降。

② 应力路径。

③ 总应力路径。

④ 有效应力路径。

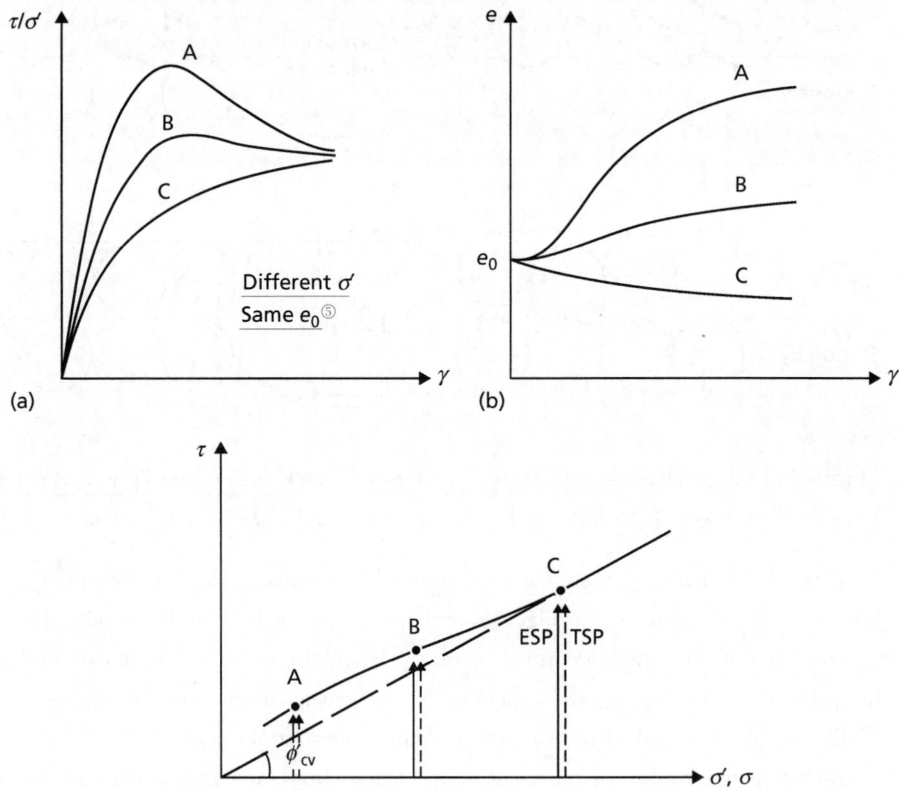

Figure 5. 18 Determination of peak strengths from direct shear test data[⑥].

⑤ σ'不同，e_0相同。

⑥ 由直剪试验数据确定峰值强度。

the TSP and ESP lie on the same line, as shown in Figure 5.18 (c). Remembering that it is the effective (not total) stresses that govern soil shear strength (Equation 5.11), failure occurs when the ESP reaches the failure envelope.

The value of ϕ'_{max} for each test can then be represented by a secant parameter: in the shear box test, $\phi'_{max} = \tan^{-1}(\tau_{max}/\sigma')$. The value of ϕ'_{max} decreases with increasing effective normal stress until it becomes equal to ϕ'_{cv}. The reduction in the difference between peak and ultimate shear stress with increasing normal stress is mainly due to the corresponding decrease in ultimate void ratio①. The lower the ultimate void ratio, the less scope there is for dilation. In addition, at high stress levels some fracturing or crushing of particles② may occur, with the consequence that there will be less particle interlocking to be overcome. Crushing thus causes the suppression of dilatancy and contributes to the reduced value of ϕ'_{max}.

① 最终的孔隙比。

② 颗粒破碎。

In the absence of any cementation or bonding between particles, the curved peak failure envelopes for coarse-grained soils would show zero shear strength at zero normal effective stress. Mathematical representations of the curved envelopes may be expressed in terms of power laws, i.e. of the form $\tau_f = A\gamma^B$. These are not compatible with many standard analyses for geotechnical structures which require soil strength to be defined in terms of a straight line (Mohr – Coulomb model). It is common, therefore, in practice to fit a straight line to the peak failure points to define the peak strength in terms of an angle of shearing resistance ϕ' and a cohesion intercept c'. It should be noted that the parameter c' is only a mathematical line-fitting constant used to model the peak states, and should not be used to imply that the soil has shear strength at zero normal effective stress③. This parameter is therefore also commonly referred to as the **apparent cohesion**④ of the soil. In soils which do have natural cementation/bonding⑤, the cohesion intercept will represent the combined effects of any apparent cohesion and the **true cohesion**⑥ due to the interparticle bonding.

③ 需注意到参数 c' 仅是用于模拟破坏状态的数学线性拟合常数，并不意味在有效应力为零时土具有抗剪强度。

④ 假黏聚力。

⑤ 天然的黏结或胶结。

⑥ 真黏聚力。

Once soil has been sheared to the critical state (ultimate conditions), the effects of any true or apparent cohesion are destroyed. This is important when selecting strength properties for use in design, particularly where soil has been tested under its in-situ condition (where line-fitting may suggest $c' > 0$), then sheared during excavation⑦ and subsequently placed to support a foundation or used to backfill behind a retaining structure. In such circumstances the excavation/placement imposes large shear strains within the soil such that critical state conditions (with $c' = 0$) should be assumed in design⑧.

⑦ 开挖。

⑧ 在这样的情况下，应在设计中考虑到开挖或布置中土因极大应变而达到临界状态条件（$c'=0$）。

Figure 5.19 shows the behaviour of soils A, B and C as would be observed in a drained triaxial test. The main differences compared to the behaviour in direct shear (Figure 5.18) lie in the stress paths and failure envelope shown in Figure 5.19 (c). In standard triaxial compression, radial stress is held constant ($\Delta\sigma_r = 0$) while axial stress is increased (by $\Delta\sigma_a$). From Equations 5.20 and 5.21, this gives $\Delta p = \Delta\sigma_a/3$ and $\Delta q = \Delta\sigma_a$. The gradient of the TSP is therefore $\Delta q/\Delta p = 3$. In a drained test there is no change in pore water pressure, so the ESP is parallel to the

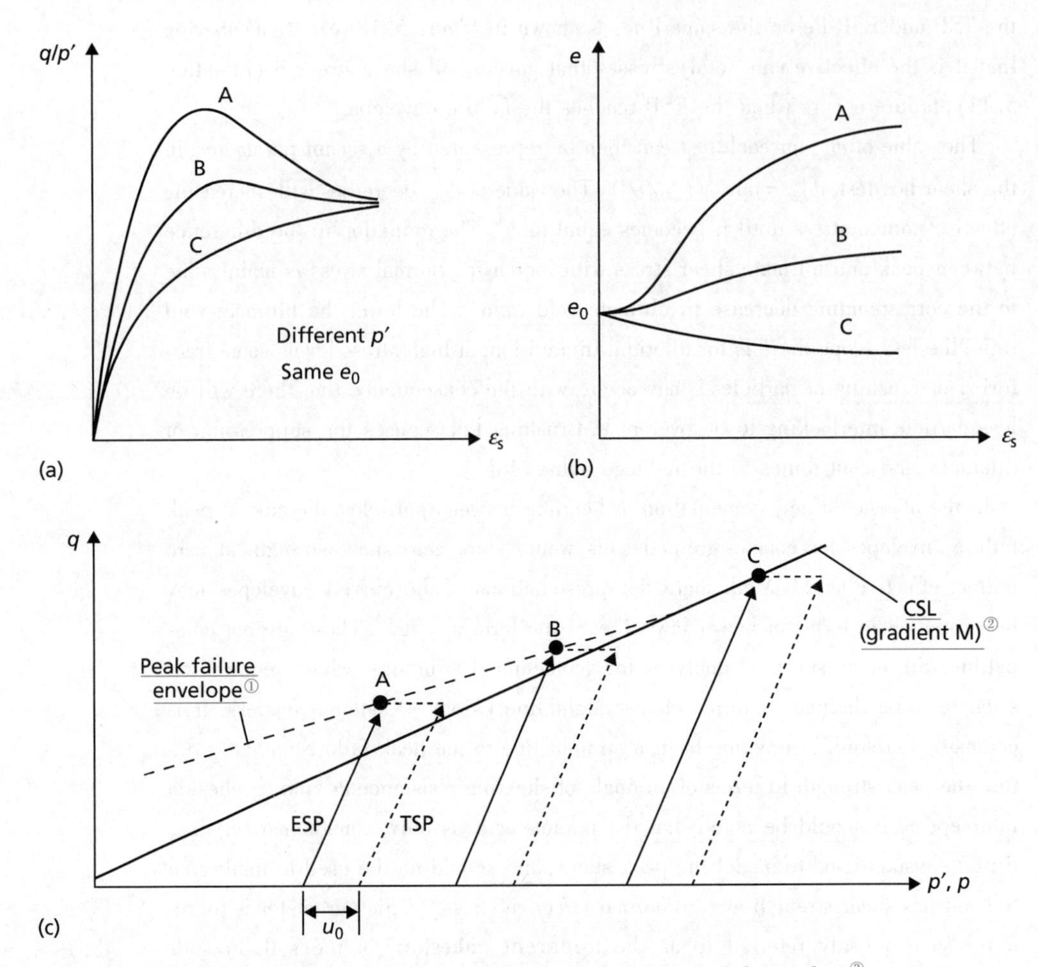

Figure 5.19 Determination of peak strengths from drained triaxial test data[3].

① 峰值破坏包络线。
② 临界状态线(斜率为 M)。
③ 排水三轴试验数据确定峰值强度。
④ 反压。

TSP. If the sample is dry, the TSP and ESP lie along the same line; if the sample is saturated and a back pressure[4] of u_0 applied, the TSP and ESP are parallel, maintaining a constant horizontal separation of u_0 throughout the test, as shown in Figure 5.19(c). As before, failure occurs when the ESP meets the failure envelope. The value of ϕ'_{max} for each test is determined by finding M $(= q/p')$ at failure and using Equation 5.26, with the resulting value of $\phi' = \phi'_{max}$.

⑤ 非扰动样。

In practice, the routine laboratory testing of sands is difficult because of the problem of obtaining undisturbed specimens[5] and setting them up, still undisturbed, in the test apparatus. If required, tests can be undertaken on specimens reconstituted in the apparatus at appropriate densities, but the in-situ structure is then unlikely to be reproduced.

Example 5.1

The results shown in Figure 5.20 were obtained from direct shear tests on reconstituted specimens of sand taken from loose and dense deposits and compacted to the in – situ density in each case. The raw data from the tests and the use of a spreadsheet to process the test data may be found on the

Companion Website. Plot the failure envelopes of each sand for both peak and ultimate states, and hence determine the critical state friction angle ϕ'_{cv}.

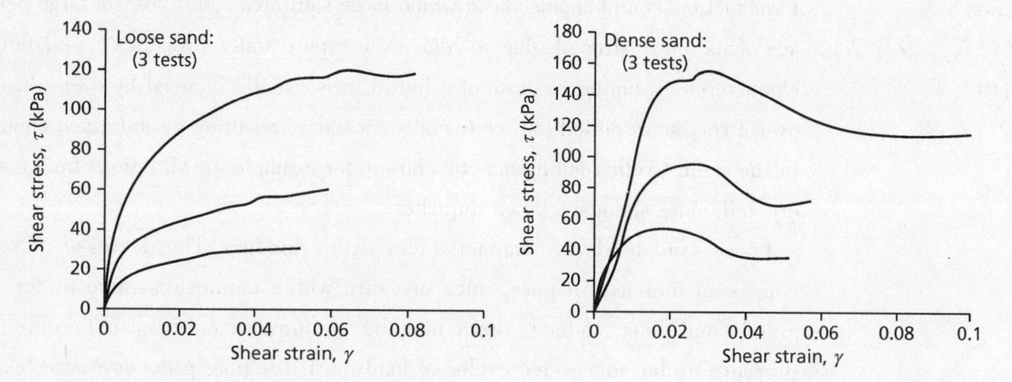

Figure 5.20 Example 5.1.

Solution

The values of shear stress at peak and ultimate states are read from the curves in Figure 5.20 and are plotted against the corresponding values of normal stress, as shown in Figure 5.21. The failure envelope is the line having the best fit to the plotted points; for ultimate conditions a straight line through the origin is appropriate (CSL). From the gradients of the failure envelopes, $\phi'_{cv} = 33.4°$ for the dense sand and 32.6° for the loose sand. These are within 1° of each other, and confirm that the crit-ical state friction angle is an intrinsic soil property which is independent of state (i.e. density). The loose sand does not exhibit peak behaviour, while the peak failure envelope for the dense sand may be characterised by $c' = 15.4$kPa and $\phi' = 38.0°$ (tangent value) or by secant values as given in Table 5.1.

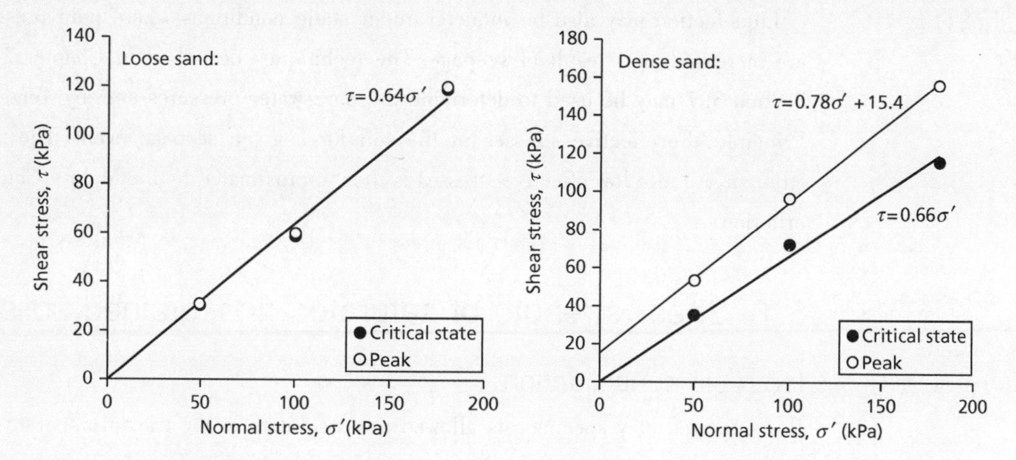

Figure 5.21 Example 5.1: Failure envelopes for (a) loose, and (b) dense sand samples.

Table 5.1 Example 5.1

Normal stress (kPa)	*Secant friction angle (°)*
50	46.8
100	43.7
181	40.8

① 液化。
② 饱和松砂。
③ 超静孔隙水压力。
④ 循环荷载。

Liquefaction[①]

Liquefaction is a phenomenon in which loose saturated sand[②] loses a large percentage of its shear strength due to high excess pore water pressures[③], and develops characteristics similar to those of a liquid. It is usually induced by cyclic loading[④] over a very short period of time (usually seconds), resulting in undrained conditions in the sand. Cyclic loading may be caused, for example, by vibrations from machinery and, more seriously, by earthquakes.

Loose sand tends to compact under cyclic loading. The decrease in volume causes an increase in pore water pressure which cannot dissipate under undrained conditions. Indeed, there may be a cumulative increase in pore water pressure under successive cycles of loading. If the pore water pressure becomes equal to the maximum total stress component, normally the overburden pressure[⑤], σ_v, the value of effective stress will be zero by Terzaghi's Principle, as described in Section 3.7 – i.e. interparticle forces will be zero, and the sand will exist in a liquid state with negligible shear strength. Even if the effective stress does not fall to zero, the reduction in shear strength may be sufficient to cause failure.

⑤ 上覆应力。

Liquefaction may develop at any depth in a sand deposit where a critical combination of in-situ density and cyclic deformation occurs. The higher the void ratio of the sand and the lower the confining pressure, the more readily liquefaction will occur.[⑥] The larger the strains produced by the cyclic loading, the lower the number of cycles required for liquefaction.

⑥ 砂土孔隙比越大，围压越低，越易发生液化。

Liquefaction may also be induced under static conditions where pore pressures are increased as a result of seepage. The techniques described in Chapter 2 and Section 3.7 may be used to determine the pore water pressures and, by Terzaghi's Principle, the effective stresses in the soil for a given seepage event. The shear strength at these low effective stresses is then approximated by the Mohr – Coulomb criterion.

⑦ 饱和细粒土的抗剪强度。
⑧ 等向固结。

5.6 Shear strength of saturated fine-grained soils[⑦]

Isotropic consolidation[⑧]

If a saturated clay specimen is allowed to consolidate in the triaxial apparatus under a sequence of equal confining cell pressures (σ_3), sufficient time being allowed between successive increments to ensure that consolidation is complete, the relationship between void ratio and effective stress can be obtained. This is similar to an oedometer test, though triaxial data are conventionally expressed in terms of specific volume (v) rather than void ratio (e). Consolidation in the triaxial apparatus under equal cell pressure is referred to as isotropic consolidation[⑨]. Under these conditions, $\sigma_a = \sigma_r = \sigma_3$, such that $p = p' = \sigma_3$, from Equation 5.20 and $q = 0$. As

⑨ 三轴仪中，在等量围压下的固结被称为等向固结。

the deviatoric stress is zero, there is no shear induced in the specimen ($\varepsilon_s = 0$), though volumetric strain (ε_v) does occur under the increasing p'[①]. Unlike one-dimensional (anisotropic) consolidation (discussed in Chapter 4), the soil element will strain both axially and radially by equal amounts (Equation 5.31).

① 当偏应力为零时，试样中将不会产生剪切（$\varepsilon_s = 0$），但 p' 的增长会产生体积应变（ε_v）。

The relationship between void ratio and effective stress during isotropic consolidation depends on the stress history[②] of the clay, defined by the overconsolidation ratio[③] (Equation 4.6), as described in Section 4.2. Overconsolidation is usually the result of geological factors, as described in Chapter 4; overconsolidation can also be due to higher stresses previously applied to a specimen in the triaxial apparatus.

② 应力历史。

③ 超固结比。

One-dimensional and isotropic consolidation characteristics are compared in Figure 5.22. The key difference between the two relationships is the use of the triaxial mean stress invariant p' for isotropic conditions and the one-dimensional normal stress σ' for 1-D consolidation. This has an effect on the gradients of the virgin compression line (here defined as the **isotropic compression line, ICL**[④]) and the unload – reload lines, denoted by λ and κ respectively, which are different to the values of C_c and C_e. It should be realised that a state represented by a point to the right of the ICL is impossible[⑤].

④ 等向压缩线。

⑤ 需要注意到，土的状态不可能处于等向压缩线右侧。

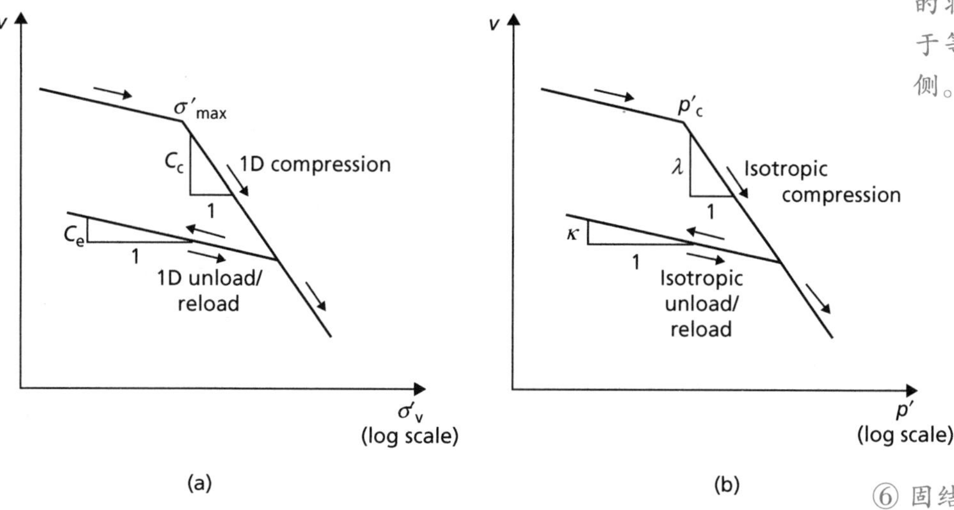

Figure 5.22 Consolidation characteristics: (a) one-dimensional, (b) isotropic[⑥].

⑥ 固结特性：(a)一维，(b)等向固结。

As a result of the similarity between the two processes, it is possible to use values of λ and κ determined from the consolidation stage of a triaxial test to directly estimate the 1-D compression relationship as $C_c \approx 2.3\lambda$ and $C_e \approx 2.3\kappa$.

Strength in terms of effective stress[⑦]

⑦ 基于有效应力的强度。

The strength of a fine-grained soil in terms of effective stress, i.e. for drained or long-term loading, can be determined by either the consolidated – undrained triaxial test with pore water pressure measurement, or the drained triaxial test[⑧]. In a drained test, the stress path concept can be used to determine the failure envelope as described earlier for sands, by determining where the ESP reaches its peak and

⑧ 例如对于排水或长期加载条件，有效应力下的细粒土强度可由能测量孔隙水压力的固结不排水三轴试验得到，也可由固结排水试验得到。

ultimate (critical state) values, and using these to determine ϕ'_{max} and ϕ'_{cv} respectively. However, because of the low permeability of most fine- grained soils, drained tests are rarely used for such materials because consolidated undrained (CU) tests can provide the same information in less time as consolidation does not need to occur during the shearing phase. The undrained shearing stage of the CU test must, however, be run at a rate of strain slow enough to allow equalisation of pore water pressure① throughout the specimen, this rate being a function of the permeability of the clay.

① 孔隙水压力的均匀化（重分布）。

The key principle in using undrained tests (CU) to determine drained properties is that the ultimate state will always occur when the ESP reaches the critical state line. Unlike in the drained test, significant excess pore pressures will develop in an undrained test, resulting in divergence of the TSP and ESP. Therefore, in order to determine the ESP for undrained conditions from the known TSP applied to the sample, pore water pressure in the specimen must be measured.

② 固结不排水和排水三轴试验的典型结果。

Typical test results for specimens of normally consolidated and overconsolidated clays are shown in Figure 5. 23. In CU tests, axial stress and pore water pressure

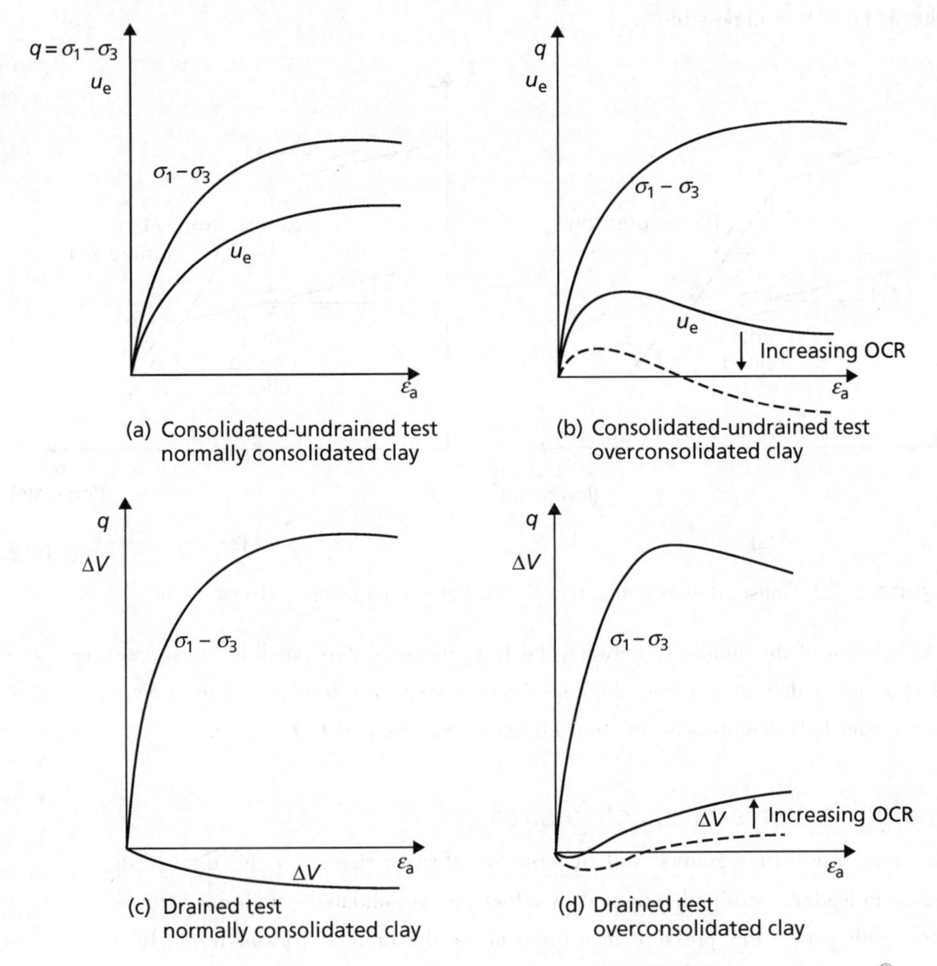

Figure 5. 23 Typical results from consolidated – undrained and drained triaxial tests②.

are plotted against axial strain. For normally consolidated clays, axial stress reaches an ultimate value at relatively large strain, accompanied by an increase in pore water pressure to a steady value[①]. For overconsolidated clays, axial stress increases to a peak value and then decreases with subsequent increase in strain. However, it is not usually possible to reach the ultimate stress due to excessive specimen deformation. Pore water pressure increases initially and then decreases; the higher the overconsolidation ratio, the greater the decrease[②]. Pore water pressure may become negative in the case of heavily overconsolidated clays, as shown by the dotted line in Figure 5.23 (b).

In drained tests, axial stress and volume change are plotted against axial strain. For normally consolidated clays, an ultimate value of stress is again reached at relatively high strain. A decrease in volume takes place during shearing, and the clay hardens[③]. For overconsolidated clays, a peak value of axial stress is reached at relatively low strain. Subsequently, axial stress decreases with increasing strain but, again, it is not usually possible to reach the ultimate stress in the triaxial apparatus. After an initial decrease, the volume of an overconsolidated clay increases prior to and after peak stress and the clay softens.[④] For overconsolidated clays, the decrease from peak stress towards the ultimate value becomes less pronounced as the overconsolidation ratio decreases.

Failure envelopes for normally consolidated and overconsolidated clays are of the forms shown in Figure 5.24. This figure also shows typical stress paths for samples of soil which are initially consolidated isotropically to the same initial mean stress p'_c; the overconsolidated sample is then partially unloaded isotropically prior to shearing to induce overconsolidation. For a normally consolidated or lightly overconsolidated clay with negligible cementation (Figure 5.24 (a)), the failure envelope should pass through the origin (i. e. $c' \approx 0$)[⑤]. The envelope for a heavily overconsolidated clay is likely to exhibit curvature over the stress range up to

① 对于正常固结黏土，当应变达到某相对较大值时轴向应力可达极限值，伴随着孔隙水压力增加并保持在一稳定值。

② 对于超固结黏土，孔隙水压力先增长而后下降，超固结比越高，下降程度越大。

③ 在剪切时，体积缩小，黏土硬化。

④ 超固结土的体积在经历初始缩小后，而在应力峰值前后都保持增大，同时黏土软化。

⑤ 对于正常固结或轻微超固结并可忽略胶结的黏土，其破坏包络线应过原点（如$c' \approx 0$）。

⑥ 三轴试验中的破坏包络线与应力路径：(a) 正常固结黏土，(b) 超固结黏土。

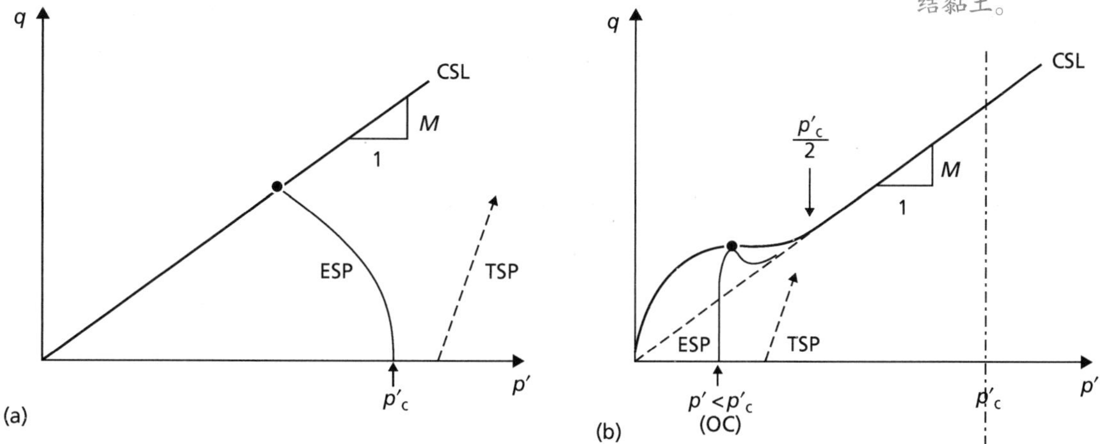

Figure 5.24 Failure envelopes and stress paths in triaxial tests for: (a) normally consolidated (NC) clays, (b) overconsolidated (OC) clays[⑥].

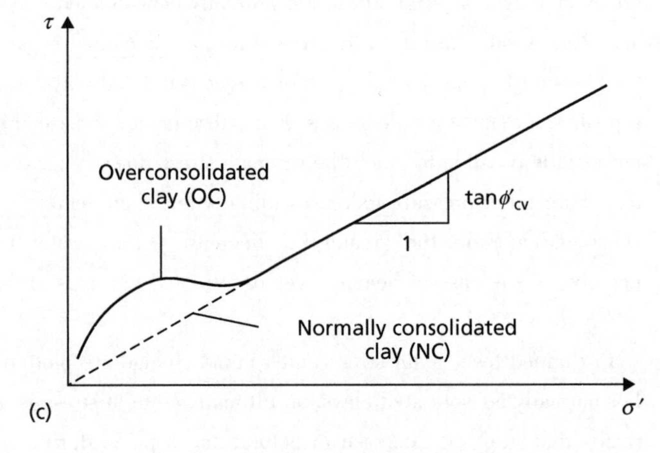

Figure 5.24 Failure envelopes and stress paths in triaxial tests for: (c) corresponding Mohr – Coulomb failure envelope.[①]

① 三轴试验中的破坏包络线与应力路径：（c）相应的莫尔-库仑破坏包络线。

② 重超固结黏土的破坏包络线在应力水平接近 $p_c'/2$ 之前表现为曲线。

③ 先期固结压力。

④ 由泥浆制备的重塑试样。

approximately $p_c'/2$[②] (Figure 5.24 (b)). The corresponding Mohr – Coulomb failure envelopes in terms of σ' and τ for use in subsequent geotechnical analyses are shown in Figure 5.24 (c). The gradient of the straight part of the failure envelope is approximately tan ϕ_{cv}'. The value of ϕ_{cv}' can be found from the gradient of the critical state line M using Equation 5.26. If the critical state value of ϕ_{cv}' is required for a heavily overconsolidated clay, then, if possible, tests should be performed at stress levels that are high enough to define the critical state envelope, i.e. specimens should be consolidated at all-round pressures in excess of the preconsolidation value[③]. Alternatively, an estimated value of ϕ_{cv}' can be obtained from tests on normally consolidated specimens reconsolidated from a slurry[④].

Example 5.2

The results shown in Table 5.2 were obtained for peak failure in a series of consolidated – undrained triaxial tests, with pore water pressure measurement, on specimens of a saturated clay. Determine the values of the effective stress strength parameters defining the peak failure envelope.

Table 5.2 Example 5.2

All – round pressure (kPa)	*Principal stress difference (kPa)*	*Pore water pressure (kPa)*
150	192	80
300	341	154
450	472	222

Solution

Values of effective principal stresses σ_3' and σ_1' at failure are calculated by subtracting pore water pressure at failure from the total principal stresses as shown in Table 5.3 (all stresses in kPa). The Mohr circles in terms of effective stress are drawn in Figure 5. 25. In this case, the failure envelope is slightly curved and a different value of the secant parameter ϕ' applies to each circle. For circle (a) the value of ϕ' is the slope of the line OA, i. e. 35°. For circles (b) and (c) the values are 33° (OB) and 31° (OC), respectively.

Table 5.3 Example 5.2 (contd.)

σ_3 (kPa)	σ_1 (kPa)	σ_3' (kPa)	σ_1' (kPa)
150	342	70	262
300	641	146	487
450	922	228	700

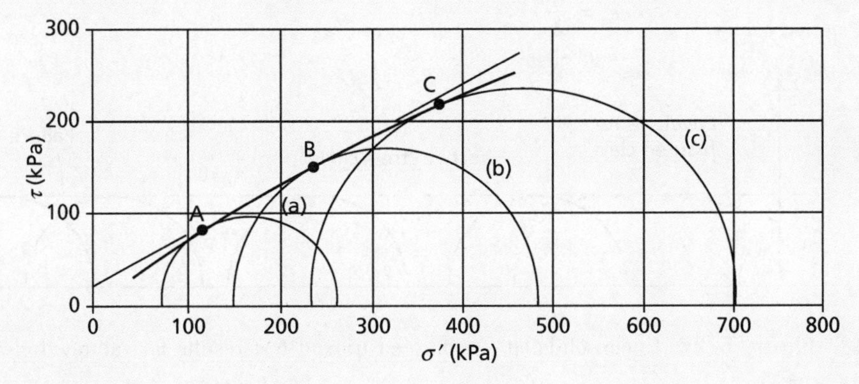

Figure 5.25 Example 5.2.

Tangent parameters can be obtained by approximating the curved envelope to a straight line over the stress range relevant to the problem. In Figure 5. 25 a linear approximation has been drawn for the range of effective normal stress 200 – 300kPa, giving parameters $c' = 20$kPa and $\phi' = 29°$.

Undrained strength[①]

In principle, the unconsolidated – undrained (UU) triaxial test enables the undrained strength of a finegrained soil in its in-situ condition to be determined, the void ratio of the specimen at the start of the test being unchanged from the in-situ value at the depth of sampling. In practice, however, the effects of sampling and preparation result in a small increase in void ratio, particularly due to swelling when the in-situ stresses are removed. Experimental evidence (e. g. Duncan and Seed, 1966) has shown that the in-situ undrained strength of saturated clays is significantly anisotropic, the strength depending on the direction of the major principal stress relative to the in-situ orientation of the specimen. Thus, undrained strength is not a unique parameter, unlike the critical state angle of shearing resistance[②].

When a specimen of saturated fine-grained soil is placed on the pedestal of the triaxial cell the initial pore water pressure is negative due to capillary tension, total stresses being zero and effective stresses positive. After the application of confining pressure the effective stresses in the specimen remain unchanged because, for a

① 不排水强度。

② 试验证据（Duncan和Seed,1996）表明，饱和黏土的原位不排水强度有较强的各向异性，其强度取决于加载最大主应力方向与试样原位沉降方向的相对关系，因此与临界状态摩擦角不同，不排水强度不是一个唯一的参数。

fully saturated soil under undrained conditions, any increase in all-round pressure results in an equal increase in pore water pressure (see Figure 4.11). Assuming all specimens to have the same void ratio and composition, a number of UU tests, each at a different value of confining pressure, should result, therefore, in equal values of principal stress difference at failure①. The results are expressed in terms of total stress as shown in Figure 5.26, the failure envelope being horizontal, i. e. $\phi_u = 0$, and the shear strength is given by $\tau_f = c_u$, where c_u is the undrained shear strength②. Undrained strength may also be determined without the use of Mohr circles; the principal stress difference at failure (q_f) is the diameter of the Mohr circle, while τ_f is its radius, therefore, in an undrained triaxial test:

① 假设所有试样有相同的孔隙比与结构，在不同围压下进行的不固结不排水试验应得出相同的破坏主应力差值。

② 不排水抗剪强度。

$$c_u = \frac{q_f}{2} \tag{5.37}$$

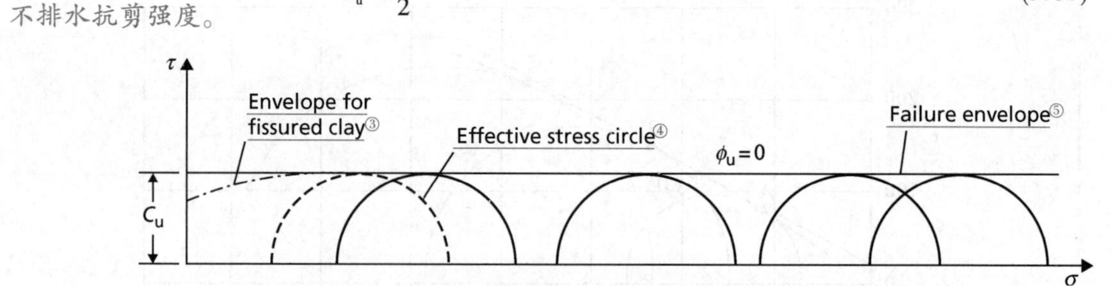

Figure 5.26 Unconsolidated – undrained triaxial test results for saturated clay⑥.

③ 含裂隙黏土包络线。

④ 有效应力圆。

⑤ 破坏包络线。

⑥ 饱和黏土的不固结不排水三轴试验结果。

It should be noted that if the values of pore water pressure at failure were measured in the series of tests, then in principle only one effective stress circle, shown dotted in Figure 5.26, would be obtained. The circle representing an unconfined compression test (i. e. with a cell pressure of zero) would lie to the left of the effective stress circle in Figure 5.26 because of negative pore water pressure (suction) in the specimen. The unconfined strength of a soil is due to a combination of friction and pore water suction.

If the best common tangent to the Mohr circles obtained from a series of UU tests is not horizontal, then the inference is that there has been a reduction in void ratio during each test due to the presence of air in the voids – i. e. the specimen had not been fully saturated at the outset. It should never be inferred that $\phi_u > 0$. It could also be that an initially saturated specimen has partially dried prior to testing, or has been repaired. Another reason could be the entrapment of air between the specimen and the membrane.

In the case of fissured clays⑦ the failure envelope at low values of confining pressure is curved, as shown in Figure 5.26. This is because the fissures open to some extent on sampling, resulting in a lower strength, and only when the confining pressure becomes high enough to close the fissures again does the strength become constant. Therefore, the unconfined compression test is not appropriate in the case of fissured clays. The size of a fissured clay specimen must also be large enough to represent the mass structure, i. e. to contain fissures representative of those in-situ,

⑦ 含裂隙黏土。

otherwise the measured strength will be greater than the in-situ strength. Large specimens are also required for clays exhibiting other features of macro-fabric. Curvature of the undrained failure envelope at low values of confining pressure may also be exhibited in heavily overconsolidated clays due to relatively high negative pore water pressure at failure causing cavitation[①] (pore air coming out of solution in the pore water).

The results of unconsolidated – undrained tests are usually presented as a plot of c_u against the corresponding depth from which the specimen originated. Considerable scatter can be expected on such a plot as the result of sampling disturbance and macro-fabric features[②] (if present). For normally consolidated fine-grained soils, the undrained strength will generally increase linearly with increasing effective vertical stress σ_v' (i. e. with depth if the water table is at the surface). If the water table is below the surface of the clay the undrained strength between the surface and the water table will be significantly higher than that immediately below the water table, due to drying.

The consolidated – undrained (CU) triaxial test can be used to determine the undrained strength of the clay after the void ratio has been changed from the initial value by consolidation[③]. If this is undertaken, it should be realised that clays in-situ have generally been consolidated under conditions of zero lateral strain, the effective vertical and horizontal stresses being unequal – i. e. the clay has been consolidated one-dimensionally (see Chapter 4). A stress release then occurs on sampling. In the standard CU triaxial test the specimen is consolidated again, though usually under isotropic conditions, to the value of the effective vertical stress in-situ. Isotropic consolidation in the triaxial test under a pressure equal to the in-situ effective vertical stress results in a void ratio lower than the in-situ (one-dimensional) value, and therefore an undrained strength higher than the (actual) in-situ value.

The unconsolidated – undrained test and the undrained part of the consolidated – undrained test can be carried out rapidly (provided no pore water pressure measurements are to be made), failure normally being produced within a period of 10 – 15 minutes. However, a slight decrease in strength can be expected if the time to failure is significantly increased, and there is evidence that this decrease is more pronounced the greater the plasticity index of the clay. Each test should be continued until the maximum value of principal stress difference has been passed or until an axial strain of 20% has been attained.

Sensitivity

Some fine-grained soils are very sensitive to remoulding, suffering considerable loss of strength due to their natural structure being damaged or destroyed[④]. The **sensitivity**[⑤] of a soil is defined as the ratio of the undrained strength in the undisturbed state to the undrained strength, at the same water content, in the remoulded state, and is denoted by S_t. Remoulding for test purposes is normally brought about by

① 由于破坏时负孔隙水压力相对较高，造成了气穴现象（气泡从水中析出），强超固结土的破坏包络线在低围压时也会表现出曲线的形态。

② 宏观土体结构。

③ 固结不排水的三轴试验可以通过固结步骤测定孔隙比初始值已改变的黏土不排水强度。

④ 细粒土对于重塑非常敏感，由于其天然结构已被破坏或损伤，其强度会有很大削弱。

⑤ 灵敏度。

① 灵敏的。
② 高灵敏的。
③ 快黏土。
④ 重塑。

the process of kneading. The sensitivity of most clays is between 1 and 4. Clays with sensitivities between 4 and 8 are referred to as **sensitive**[①], and those with sensitivities between 8 and 16 as **extrasensitive**[②]. **Quick clays**[③] are those having sensitivities greater than 16; the sensitivities of some quick clays may be of the order of 100. Typical values of S_t are given in Section 5.9 (Figure 5.37).

Sensitivity can have important implications for geotechnical structures and soil masses. In 1978, a landslide occurred in a deposit of quick clay at Rissa in Norway. Spoil from excavation works was deposited on the gentle slope forming the shore of Lake Botnen. This soil already had a significant in-situ shear stress applied due to the ground slope, and the additional load caused the undisturbed undrained shear strength to be exceeded, resulting in a small slide. As the soil strained, it remoulded[④] itself, breaking down its natural structure and reducing the amount of shear stress which it could carry. The surplus was transferred to the adjacent undisturbed soil, which then exceeded its undisturbed undrained shear strength. This process continued progressively until, after a period of 45 minutes, 5 – 6 million cubic metres of soil had flowed out into the lake, at some points reaching a flow velocity of 30 – 40km/h, destroying seven farms and five homes (Figure 5.27). It is fortunate that the slide occurred beneath a sparsely populated rural area. Further information regarding the Rissa landslide can be found on the Companion Website.

Figure 5.27 Damage observed following the Rissa quick clay flowslide (photo: Norwegian Geotechnical Institute – NGI).

Example 5.3

The results shown in Figure 5.28 were obtained at failure in a series of UU triaxial tests on specimens taken from the same approximate depth within a layer of soft saturated clay. The raw data from the tests and the use of a spreadsheet to interpret the test data may be found on the Companion Website. Determine the undrained shear strength at this depth within the soil.

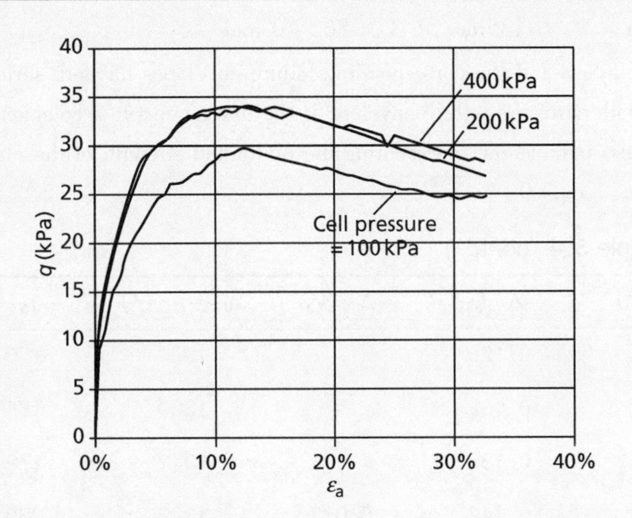

Figure 5.28 Example 5.3.

Solution

The maximum value of q $(= q_f)$ is read from Figure 5.28 for each test, and Equation 5.37 used to determine c_u for each sample. As the samples are all from the same depth but tested at different confining pressures, c_u should theoretically be the same for all samples, so an average is taken as c_u = 16.3kPa.

Example 5.4

The results shown in Table 5.4 were obtained at failure in a series of triaxial tests on specimens of a saturated clay initially 38mm in diameter by 76mm long. Determine the values of the shear strength parameters with respect to (a) total stress and (b) effective stress.

Table 5.4 Example 5.4

Type of test	*Confining pressure (kPa)*	*Axial load (N)*	*Axial deformation (mm)*	*Volume change (ml)*
(a) Undrained (UU)	200	222	9.83	–
	400	215	10.06	–
	600	226	10.28	–
(b) Drained (D)	200	403	10.81	6.6
	400	848	12.26	8.2
	600	1265	14.17	9.5

Solution

The principal stress difference at failure in each test is obtained by dividing the axial load by the cross – sectional area of the specimen at failure (Table 5.5). The corrected cross – sectional area is calculated from Equation 5.16. There is, of course, no volume change during an undrained test on a saturated clay. The initial values of length, area and volume for each specimen are:

$l_0 = 76\text{mm}, \quad A_0 = 1135\text{mm}^2, \quad V_0 = 86 \times 10^3\text{mm}^3$

The Mohr circles at failure and the corresponding failure envelopes for both series of tests are shown in Figure 5.29. In both cases the failure envelope is the line nearest to a common tangent to the Mohr circles. The total stress parameters, representing the undrained strength of the clay, are

$c_u = 85\text{kPa}, \quad \phi_u = 0$

Table 5.5 Example 5.4 (contd.)

	σ_3 (kPa)	$\Delta l/l_0$	$\Delta V/V_0$	Area (mm^2)	$\sigma_1 - \sigma_3$ (kPa)	σ_1 (kPa)
(a)	200	0.129	–	1304	170	370
	400	0.132	–	1309	164	564
	600	0.135	–	1312	172	772
(b)	200	0.142	0.077	1222	330	530
	400	0.161	0.095	1225	691	1091
	600	0.186	0.110	1240	1020	1620

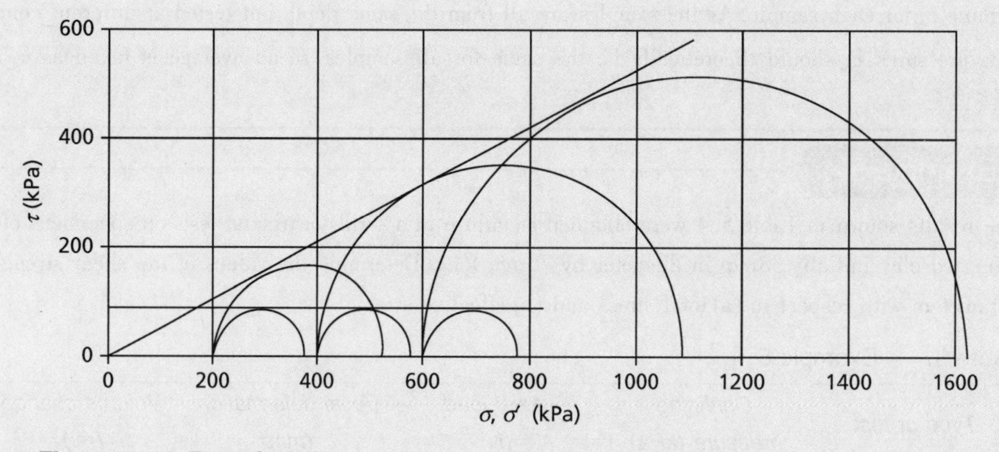

Figure 5.29 Example 5.4.

The effective stress parameters, representing the drained strength of the clay, are

$c' = 0, \quad \phi' = 27°$

① 临界状态理论体系。

5.7 The critical state framework①

The critical state concept, originally presented by Roscoe *et al*. (1958), represents an idealisation of the observed patterns of behaviour of saturated clays under applied principal stresses. However, the critical state concept applies to all soils, both coarse- (as described in Section 5.5) and fine-grained. The concept relates the effective stresses and the corresponding specific volume ($v = 1 + e$) of a clay during shearing under drained or undrained conditions, thus unifying the characteristics of shear strength and deformation (volumetric change). It was demonstrated that a characteristic surface exists which limits all possible states of the clay, and that all effective stress paths reach or approach a line on that surface when yielding occurs at constant volume under constant effective stress[①]. This line represents the critical states of the soil, and links p', q and v at these states. The model was originally derived based on observations of behaviour from triaxial tests, and in this section the critical state concept will be 'rediscovered' in a similar way by considering the triaxial behaviour that has previously been discussed in this chapter.

① 当屈服发生于恒定有效应力和恒定体积时，所有有效应力路径抵达或接近上述界面的一条线。

Volumetric behaviour during undrained shear[②]

② 不排水剪切中的体积变化特性。

The behaviour of a saturated fine-grained soil during undrained shearing in the triaxial cell is first considered, as presented in Figure 5. 23. As the pore pressure within the sample change during undrained shearing, the ESP cannot be determined only from the TSP as for the drained test (Figure 5. 19 (c)). However, during undrained shearing it is known that there must be no change in volume ($\Delta v = 0$), so use can be made of the volumetric behaviour shown in Figure 5. 22 (b). As q is independent of u, any excess pore water pressure generated during undrained shearing must result from a change in mean stress ($\Delta p'$), i. e. $u_e = \Delta p'$. The normally consolidated and overconsolidated clays from Figure 5. 23 are plotted in Figure 5. 30 (a), assuming that the samples are all consolidated under the same cell pressure prior to undrained shearing. The normally consolidated (NC) clay has an initial state on the ICL and shows a large positive ultimate value of u_e at critical state (ultimate strength), corresponding to a reduction in p'[③]. The lightly overconsolidated (LOC) clay has a higher initial volume, having swelled a little, with its initial state lying on the unload – reload line. At critical state, the excess pore water pressure is lower than the NC clay, so the change in p' is less. The heavily overconsolidated (HOC) clay starts at a higher initial volume, having swelled more than the LOC sample. The sample exhibits negative excess pore water pressure at critical state, i. e. an increase in p' due to shearing. It will be seen from Figure 5. 30 (a) that the points representing the critical states of the three samples all lie on a line parallel to the ICL and slightly below it. This is the critical state line (CSL) that has been described earlier, but represented in $v - p'$ space, and represents the specific volume of the soil at critical state, i. e. when q is at its ultimate value. This line is a projection of the set of critical states of a soil in the $v - p'$ plane, just as the critical state line on the $q - p'$ plane ($q = Mp'$, Figure 5. 11 (c)) is a projection of the same set of critical states in terms of the stress parameters q and p'.

③ 正常固结黏土初始位于等向压缩线，在临界状态时（最终强度），正孔压为较大值，相应地p'下降。

Volumetric behaviour during drained shear[④]

④ 排水剪切中的体积变化特性。

Considering the behaviour of the coarse-grained soil under triaxial compression described in Figure 5. 19, the volumetric behaviour is plotted in Figure 5. 30. Samples A and B both exhibit dilation (increase in v) due to shearing; as the confining

① 试样 A 和 B 由于剪切表现出剪胀，当围压增加，剪胀量下降。

② 临界状态是土体的内在特性。

stress increases, the amount of dilation reduces[①]. As the confining stress is increased still further (sample C), compression occurs in place of dilation. Again, the final points at critical state lie on a straight line (the CSL) parallel to the ICL; however, the CSL for this coarsegrained soil will have different numerical values of the intercept and gradient compared to the finegrained soil considered in Figure 5.30 (a), the CSL being an intrinsic soil property[②].

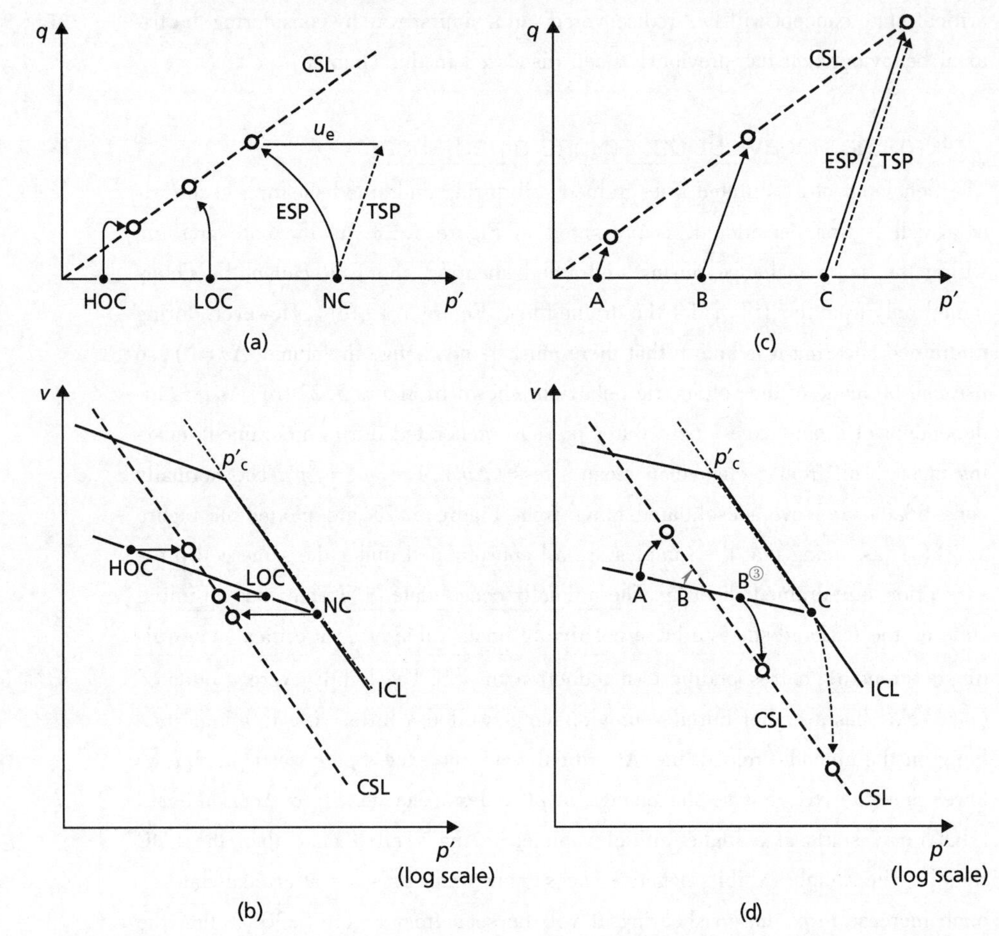

Figure 5.30 Volumetric behaviour of soils during (a) undrained tests, (b) drained tests.[④]

③ 注：原书中 B 的标注位置有误，应为图中蓝色标注。

④ 土的体积变化特性。
(a) 不排水试验，
(b) 排水试验。

⑤ 临界状态线定义。

⑥ 在 q-p' 面的投影。

⑦ 对数坐标轴。

Defining the critical state line[⑤]

Figure 5.31 shows the CSL plotted in three dimensions, along with the projections in the $q-p'$ and $v-p'$ planes. The projection in the $q-p'$ plane[⑥] is a straight line with gradient M. As described previously in both this and the previous chapters, it is common to plot $v-p'$ data with a logarithmic axis[⑦] for the mean stress p'. The CSL in this plane is defined by

$$v_{cv} = \Gamma - \lambda \ln p' \qquad (5.38)$$

where Γ is the value of v on the critical state line at $p' = 1$kPa and is an intrinsic soil property; λ is the gradient of the ICL as defined previously. In order to use the critical state framework, it is also necessary to know the initial state of the soil,

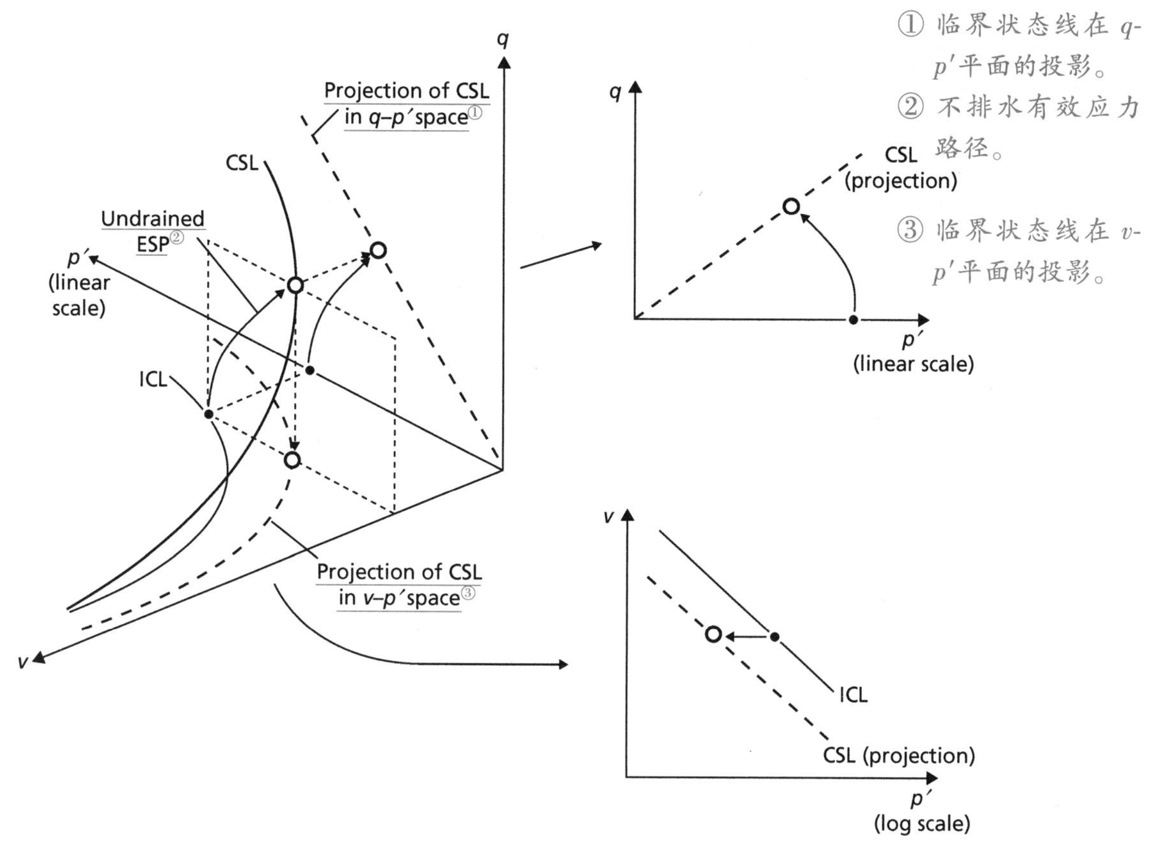

① 临界状态线在 q-p'平面的投影。
② 不排水有效应力路径。
③ 临界状态线在 v-p'平面的投影。

Figure 5.31 Position of the Critical State Line (CSL) in $p' - q - v$ space. The effective stress path in an undrained triaxial test is also shown④.

④ $p' - q - v$ 空间中临界状态线位置，以及不排水三轴试验的有效应力路径。

which depends on its stress history; in a triaxial test this is usually governed by previous consolidation and swelling under isotropic conditions, prior to shearing⑤. The equation of the normal consolidation line (ICL) is

$$v_{\text{icl}} = N - \lambda \ln p' \tag{5.39}$$

where N is the value of v at $p' = 1\,\text{kPa}$. The swelling and recompression (unload – reload) relationships can be approximated to a single straight line of slope $-\kappa$, represented by the equation

$$v = v_{\text{k}} - \kappa \ln p' \tag{5.40}$$

where v_{k} is the value of v at $p' = 1\,\text{kPa}$, and will depend on the preconsolidation pressure (i. e. it is not a material constant). The initial volume prior to shearing (v_0) can alternatively be defined by a single equation:

$$v_0 = N - \lambda \ln p_c' + \kappa \ln\left(\frac{p_c'}{p'}\right) \tag{5.41}$$

In order to use the critical state framework to predict the strength of soil, it is therefore necessary to define five constants: N, λ, κ, Γ, M. These will allow for the determination of the initial state and the CSL in both the $q - p'$ and $v - \ln(p')$ planes. These parameters may theoretically all be determined from a single CU test, though, typically, multiple tests will be carried out at different values of p_c'.

⑤ 为了使用临界状态准则，有必要知道土的初始状态取决于土的应力历史，在三轴试验中，初始状态取决于剪切前在各向等压条件下的预固结与回弹。

Example 5.5

The data shown in Figure 5.32 were obtained from a series of CU triaxial compression tests on a soft saturated clay in a modern computer-controlled stres-path cell. The samples were isotropically consolidated to confining pressures of 250, 500 and 750kPa prior to undrained shearing. Determine the critical state parameters N, λ, Γ and M. The raw data from these tests and their interpretation using a spreadsheet is provided on the Companion Website.

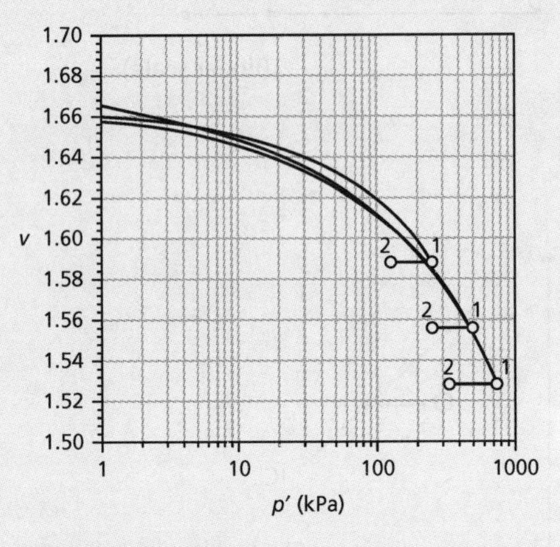

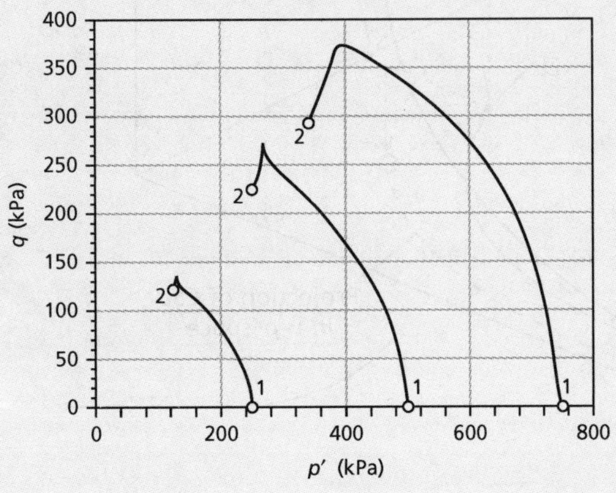

Figure 5.32 Example 5.5.

Solution

The values of p', q and v at the end of both the consolidation stage (1) and the undrained shearing stage (2) are read from the figures and summarised in Table 5.6. Assuming that all of the consoli-dation pressures are high enough to exceed any pre-existing preconsolidation stress, at the end of consolidation the samples should all lie on the ICL. By plotting v_1 versus p'_1, N and λ may be found by fitting a straight line as shown in Figure 5.33, the intercept being $N = 1.886$ and the gradient $\lambda = 0.054$. If the points at the end of shearing (p'_2, v_2) are plotted in a similar way, the straight line fit lies almost exactly parallel to the ICL – this is the CSL (assuming that the strain induced in the shearing stage has been sufficient to reach the critical state in each case). The intercept is $\Gamma = 1.867$ and the gradient $\lambda = 0.057$. To find M, the points (p'_2, q_2) at the end of shearing (critical state) are plotted as shown in Figure 5.33. The best-fit straight line goes through the origin with gradient $M = 0.88$.

Table 5.6 Example 5.5

Confining pressure, σ_3 (kPa)	*After consolidation:*			*After shearing:*		
	p'_1	q_1	v_1	p'_2	q_2	v_2
250	250	0	1.588	125	121	1.588
500	500	0	1.556	250	225	1.556
750	750	0	1.528	340	292	1.528

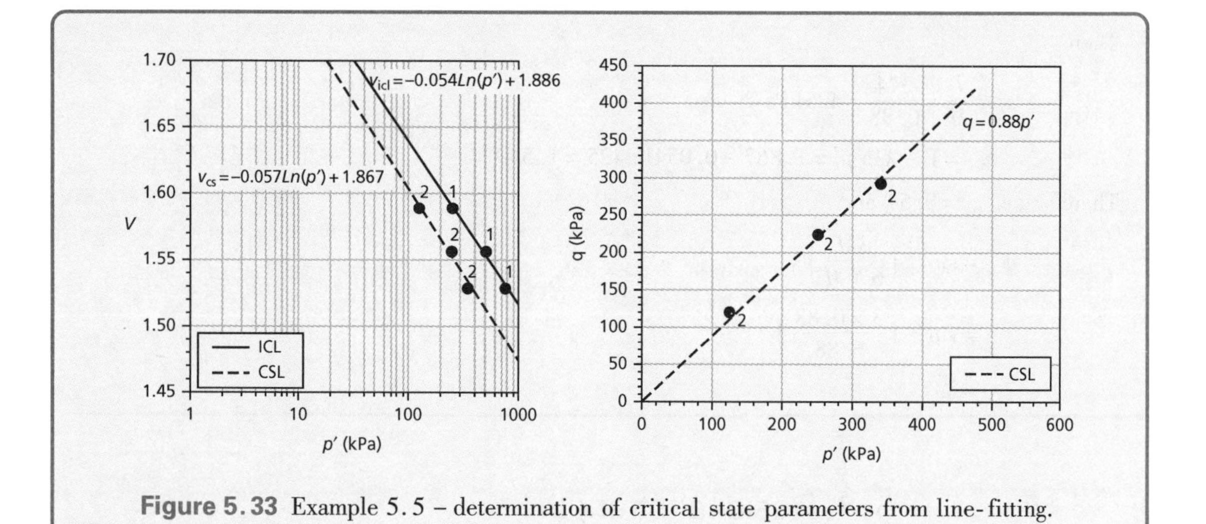

Figure 5.33 Example 5.5 – determination of critical state parameters from line-fitting.

Example 5.6

Estimate the values of principal stress difference and void ratio at failure in undrained and drained triaxial tests on specimens of the clay described in Example 5.5, isotropically consolidated under a confining pressure of 300kPa. What would be the expected value of ϕ'_{cv}?

Solution

After normal consolidation to $p'_c = 300$kPa the sample will be on the ICL with specific volume (v_0) given by

$$v_0 = N - \lambda \ln p'_c = 1.886 - 0.054\ln 300 = 1.578$$

In an undrained test the volume change is zero, and therefore the specific volume at critical state (v_{cs}) will also be 1.58, i.e. the corresponding void ratio will be $e_{cs(U)} = 0.58$.

Assuming failure to take place on the critical state line,

$$q'_f = Mp'_f$$

and the value of p'_f can be obtained from Equation 5.38. Therefore

$$q'_{f(U)} = M \exp\left(\frac{\Gamma - v_0}{\lambda}\right)$$

$$= 0.88\exp\left(\frac{1.867 - 1.578}{0.054}\right)$$

$$= 186\text{kPa}$$

For a drained test the slope of the stress path on a $q - p'$ plot is 3, i.e.

$$q'_f = 3\,(p'_f - p'_c) = 3\left(\frac{q'_f}{M} - p'_c\right)$$

Therefore,

$$q'_{f(D)} = \frac{3Mp'_c}{3 - M} = \frac{3 \times 0.88 \times 300}{3 - 0.88}$$

$$= 374\text{kPa}$$

Then

$$p_f' = \frac{q_f}{M} = \frac{374}{0.88} = 425\text{kPa}$$

$$\therefore v_{cs} = \Gamma - \lambda \ln p_f' = 1.867 - 0.054 \ln 425 = 1.540$$

Therefore, $e_{cs(D)} = 0.54$ and

$$\phi_{cv}' = \sin^{-1}\left(\frac{3M}{6+M}\right)$$

$$= \sin^{-1}\left(\frac{3 \times 0.88}{6.88}\right)$$

$$= 23°$$

① 残余强度。

5.8 Residual strength[①]

In the drained triaxial test, most clays would eventually show a decrease in shear strength with increasing strain after the peak strength has been reached. However, in the triaxial test there is a limit to the strain which can be applied to the specimen. The most satisfactory method of investigating the shear strength of clays at large strains is by means of the ring shear apparatus[②] (Bishop *et al.*, 1971; Bromhead, 1979), an annular direct shear apparatus. The annular specimen[③] (Figure 5.34(a)) is sheared, under a given normal stress, on a horizontal plane by the rotation of one half of the apparatus relative to the other; there is no restriction to the magnitude of shear displacement between the two halves of the specimen. The rate of rotation must be slow enough to ensure that the specimen remains in a drained condition. Shear stress, which is calculated from the applied torque[④], is plotted against shear displacement as shown in Figure 5.34(b).

② 环剪仪。
③ 环形试样。
④ 施加的扭矩。
⑤ 峰值。
⑥ 临界状态。

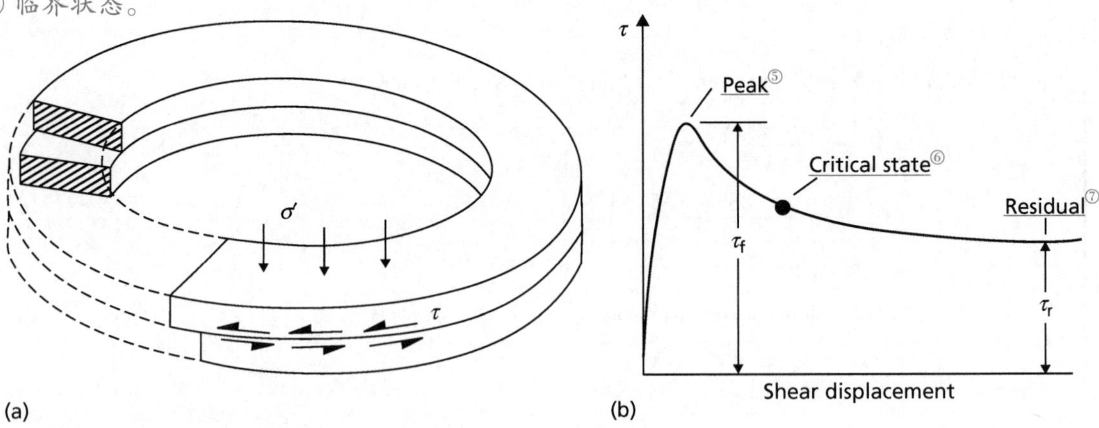

Figure 5.34 (a) Ring shear test[⑧], and (b) residual strength[⑨].

⑦ 残余强度。
⑧ 环剪试验。
⑨ 残余强度。

The shear strength falls below the peak value, and the clay in a narrow zone adjacent to the failure plane will soften and reach the critical state. However, because of non-uniform strain in the specimen, the exact point on the curve corresponding to the critical state is uncertain. With continuing shear displacement the shear strength continues to decrease, below the critical state value, and eventually reaches a residual value at a relatively large displacement. If the soil contains a relatively high proportion of plate-like particles①, a reorientation of these particles② parallel to the failure plane will occur (in the narrow zone adjacent to the failure plane) as the strength decreases towards the residual value. However, reorientation may not occur if the plate-like particles exhibit high interparticle friction. In this case, and in the case of soils containing a relatively high proportion of bulky particles, rolling and translation of particles takes place as the residual strength is approached. It should be appreciated that the critical state concept envisages continuous deformation of the specimen as a whole, whereas in the residual condition there is preferred orientation or translation of particles in a narrow shear zone. The original soil structure③ in this narrow shear zone is completely destroyed as a result of particle reorientation. A remoulded specimen④ can therefore be used in the ring shear apparatus if only the residual strength (and not the peak strength) is required.

① 偏平状颗粒。
② 颗粒方向重排列。
③ 初始的土体结构。
④ 重塑土样。

The results from a series of tests, under a range of values of normal stress, enable the failure envelope for both peak and residual strength to be obtained, the residual strength parameters in terms of effective stress being denoted c_r' and ϕ_r'. Residual strength data from ring shear testing for a large range of soils have been published (e.g. Lupini *et al.*, 1981; Mesri and Cepeda-Diaz, 1986; Tiwari and Marui, 2005), which indicate that the value of c_r' can be taken to be zero. Thus, the residual strength can be expressed as

$$\tau_r = \sigma_f' \tan\phi_r' \tag{5.42}$$

Typical values of ϕ_r' are given in Section 5.9 (Figure 5.39).

5.9 Estimating strength parameters from index tests⑤

⑤ 基于土性指标的强度参数估算。

In order to obtain reliable values of soil strength parameters from triaxial and shear box testing, undisturbed samples of soil are required. Methods of sampling are discussed in Section 6.3. No sample will be fully undisturbed, and obtaining high quality samples is almost always difficult and often expensive. As a result, the principle strength properties of soils (ϕ_{cv}', c_u, S_t, ϕ_r') are here correlated to the basic index properties described in Chapter 1 (I_P and I_L for fine-grained soils and e_{max} and e_{min} for coarse-grained soils). These simple tests can be undertaken using disturbed samples⑥. The spoil generated from drilling a borehole⑦ provides essentially a continuous disturbed sample for testing so a large amount of information describing the strength of the ground can be gathered without performing detailed laboratory or field testing⑧. The use of correlations can therefore be very useful during the preliminary stages of a Ground Investigation (GI)⑨. Ground investigation can also be cheaper/more efficient using mainly disturbed samples, supplemented by fewer undisturbed samples to verify the properties.

⑥ 使用扰动土样。
⑦ 钻孔。
⑧ 现场试验。
⑨ 地基勘察。

It should be noted that the use of any of the approximate correlations presented here should be regarded as estimates only; they should be used to support, and never replace, laboratory tests on undisturbed samples, particularly when a given strength property is critical to the design or analysis of a given geotechnical construction. They are most useful for checking the results of laboratory tests, for increasing the amount of data from which the determination of strength properties is made and for estimating parameters before the results of laboratory tests are known (e. g. for feasibility studies[1] and when planning a ground investigation).

① 可行性研究。

Critical state angle of shearing resistance (ϕ'_{cv})[2]

② 临界状态摩擦角。

Figure 5. 35 (a) shows data of ϕ'_{cv} for 65 coarse- grained soils determined from shear box and triaxial testing collected by Bolton (1986), Miura *et al.* (1998) and Hanna (2001). The data are correlated against $e_{max} - e_{min}$. This correlation is appropriate as both e_{max} and e_{min} are independent of the soil state/density (defined by void ratio e), just as ϕ'_{cv} is an intrinsic property and independent of e[3]. A correlation line, representing a best fit to the data is also shown in Figure 5. 35(a), which suggests that ϕ'_{cv} increases as the potential for volumetric change ($e_{max} - e_{min}$) increases. The large scatter in the data may be attributed to the influence of particle shape[4] (angularity), which is only partially captured by $e_{max} - e_{min}$, and grain roughness (which is a function of the parent material (s) from which the soil was produced).

③ 当 e_{max} 和 e_{min} 与土体状态/密度（由孔隙比 e 定义）无关，ϕ'_{cv}是固有特性且与 e 无关时，这个相关关系是恰当的。

④ 颗粒形状。

Figure 5. 35(b) shows data of ϕ'_{cv} for 32 undisturbed and 32 remoulded fine-grained soils from triaxial testing collected by Kenney (1959), Parry (1960) and Zhu and Yin (2000). The data are correlated against plasticity index (I_P) which is also independent of the current soil state (defined by the current water content w). A correlation line, representing a best fit to the undisturbed data is also shown in Figure 5.35(b), which suggests that ϕ'_{cv} reduces as the plasticity index increases.

⑤ 土体 ϕ'_{cv}与其指标特性的相关关系。(a) 粗粒土，(b) 细粒土。

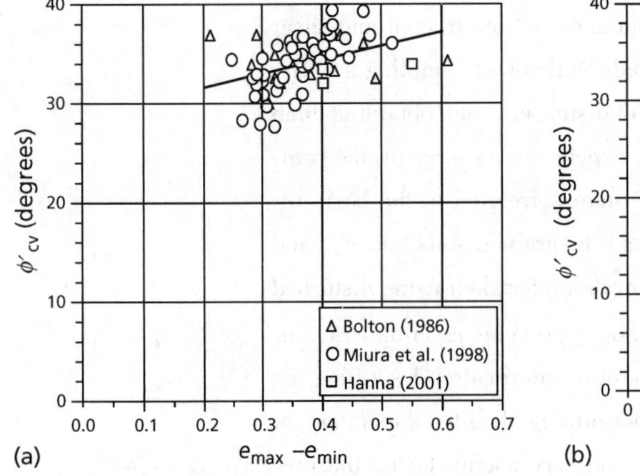

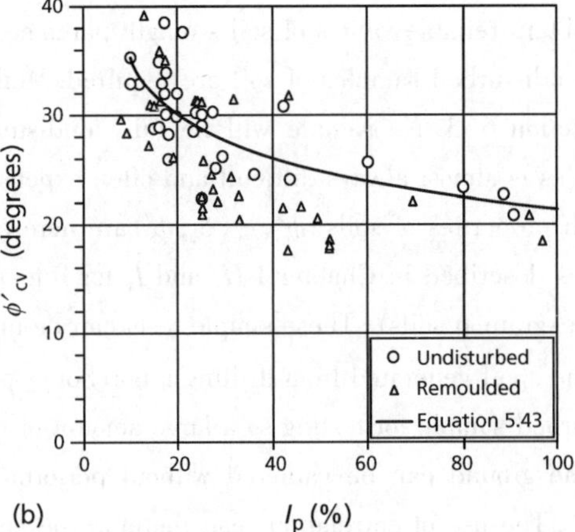

Figure 5. 35 Correlation of ϕ'_{cv} with index properties for (a) coarse- grained, and (b) fine- grained soils.[5]

This may be attributed to the increase in clay fraction (i. e. fine platy particles) as I_P increases, with these particles tending to have lower friction than the larger particles within the soil matrix The scatter in the data may be attributed to the minerality of these clay-sized particles – as outlined in Chapter 1, the main clay minerals (kaolinite, smectite, illite, montmorillonite) have very different particle shapes, specific surface and frictional properties. The equation of the line is given by:

$$\phi'_{cv} = 57(I_P)^{-0.21} \tag{5.43}$$

Undrained shear strength and sensitivity[①] (c_u, S_t)

① 不排水抗剪强度与灵敏度。

As described in Section 5.6, the undrained strength of a fine-grained soil is not an intrinsic material property, but is dependent on the state of the soil (as defined by w) and the stress level[②]. The undrained shear strength should therefore be correlated to a state-dependent index parameter, namely liquidity index I_L[③]. Figure 5.36 shows data of c_u plotted against I_L for 62 remoulded fine-grained soils collected by Skempton and Northey (1953), Parry (1960), Leroueil *et al.* (1983) and Jardine *et al.* (1984). At the liquid limit, $c_{ur} \approx 1.7$kPa. The undrained shear strength at the plastic limit is defined as 100 times that at the liquid limit, which suggests that the relationship between c_u (in kPa) and I_L should be of the form

② 如 5.6 节所述，细粒土的不排水强度不是材料的固有特性，而与土的状态（由 w 定义）和应力水平有关。

③ 液性指数。

$$c_{ur} \approx 1.7 \times 100^{(1-I_L)} \tag{5.44}$$

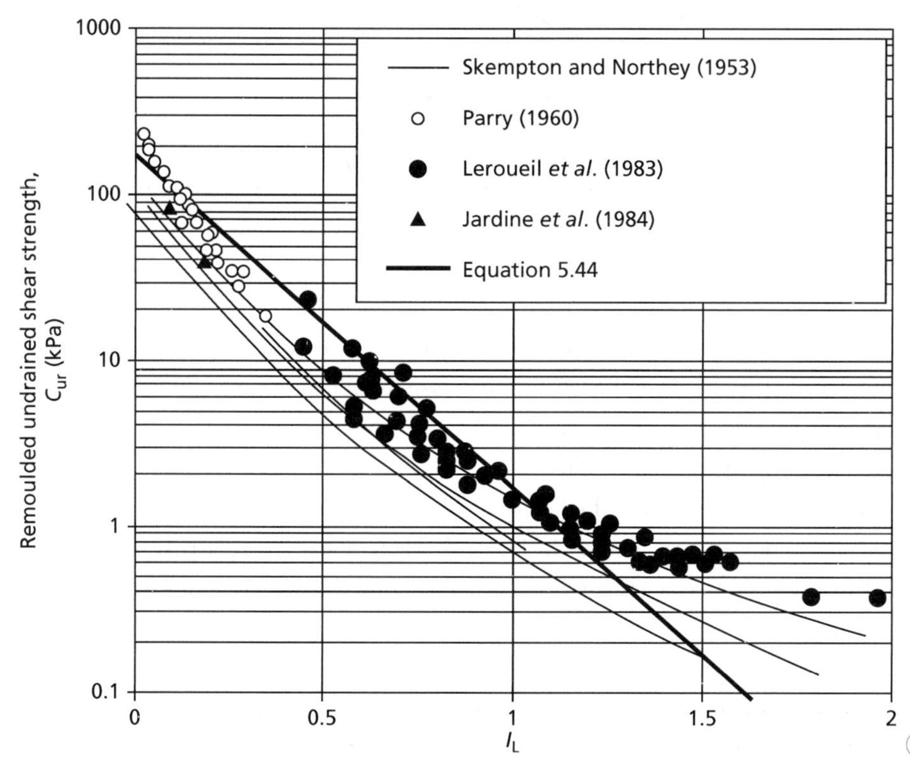

Figure 5.36 Correlation of remoulded undrained shear strength c_{ur} with index properties[④].

④ 重塑后的土体不排水，抗剪强度 c_{ur} 与液性指数的关系。

It should be noted that the undrained strength in Equation 5.44 is that appropriate to the soil in a remoulded (fully disturbed) state. Sensitive clays may therefore exhibit much higher apparent undrained strengths in their undisturbed condition than would be predicted by Equation 5.44.

Figure 5.37 shows data for 49 sensitive clays collated from Skempton and Northey (1953), Bjerrum (1954) and Bjerrum and Simons (1960), in which sensitivity is also correlated to I_L. The data suggest a linear correlation described by:

$$S_t \approx 100^{(0.43 I_L)} \tag{5.45}$$

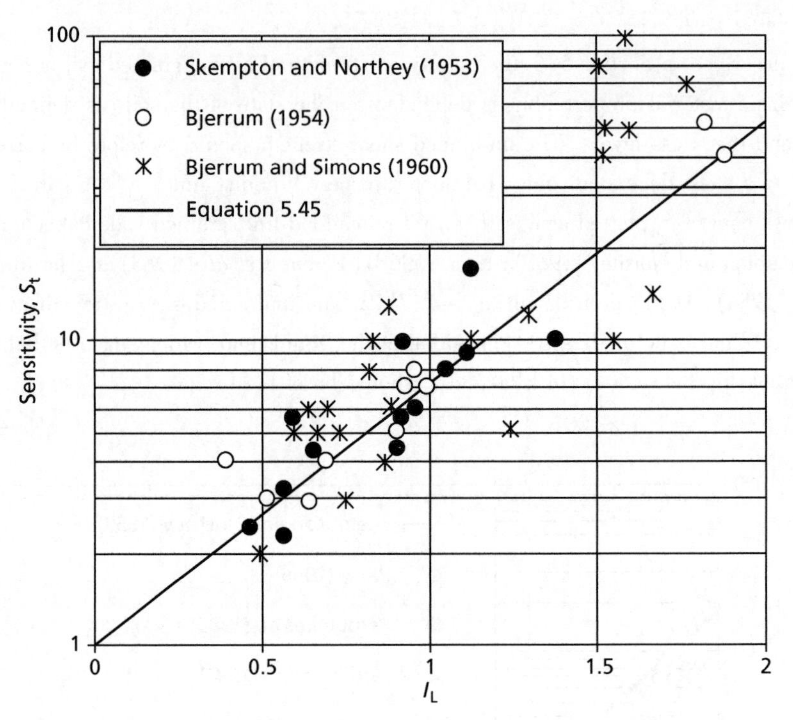

Figure 5.37 Correlation of sensitivity S_t with index properties[①].

① 土体灵敏度 S_t 与液性指数的相互关系。

A correlation for the undisturbed (in-situ) undrained shear strength may also be estimated from the definition of sensitivity as the ratio of undisturbed to remoulded strength (i.e. by combining Equations 5.44 and 5.45):

$$c_u = S_t c_{ur} \approx 1.7 \times 100^{(1-0.57 I_L)} \tag{5.46}$$

To demonstrate how useful Equations 5.44 – 5.46 can be, Figure 5.38 shows examples of two real soil profiles. In Figure 5.38(a), the soil is a heavily overconsolidated insensitive clay (Gault clay, near Cambridge). Applying Equation 5.45, it can be seen that $S_t \approx 1.0$ everywhere, such that Equations 5.44 and 5.46 give almost identical values (solid markers). This is compared with the results of UU triaxial tests which measure the undisturbed undrained shear strength. It will be seen that although there is significant scatter (in both datasets), both the magnitude and variation with depth is captured well using the approximate correlations[②]. Figure 5.38(b) shows data for a normally consolidated clay with $S_t \approx 5$ (Bothkennar clay, near Edinburgh). For sensitive clay such as this, the undisturbed strength is likely

② 可以看到，虽然数据间有很大的离散性，但是其大小及随深度的变化可由该关系很好地描述。

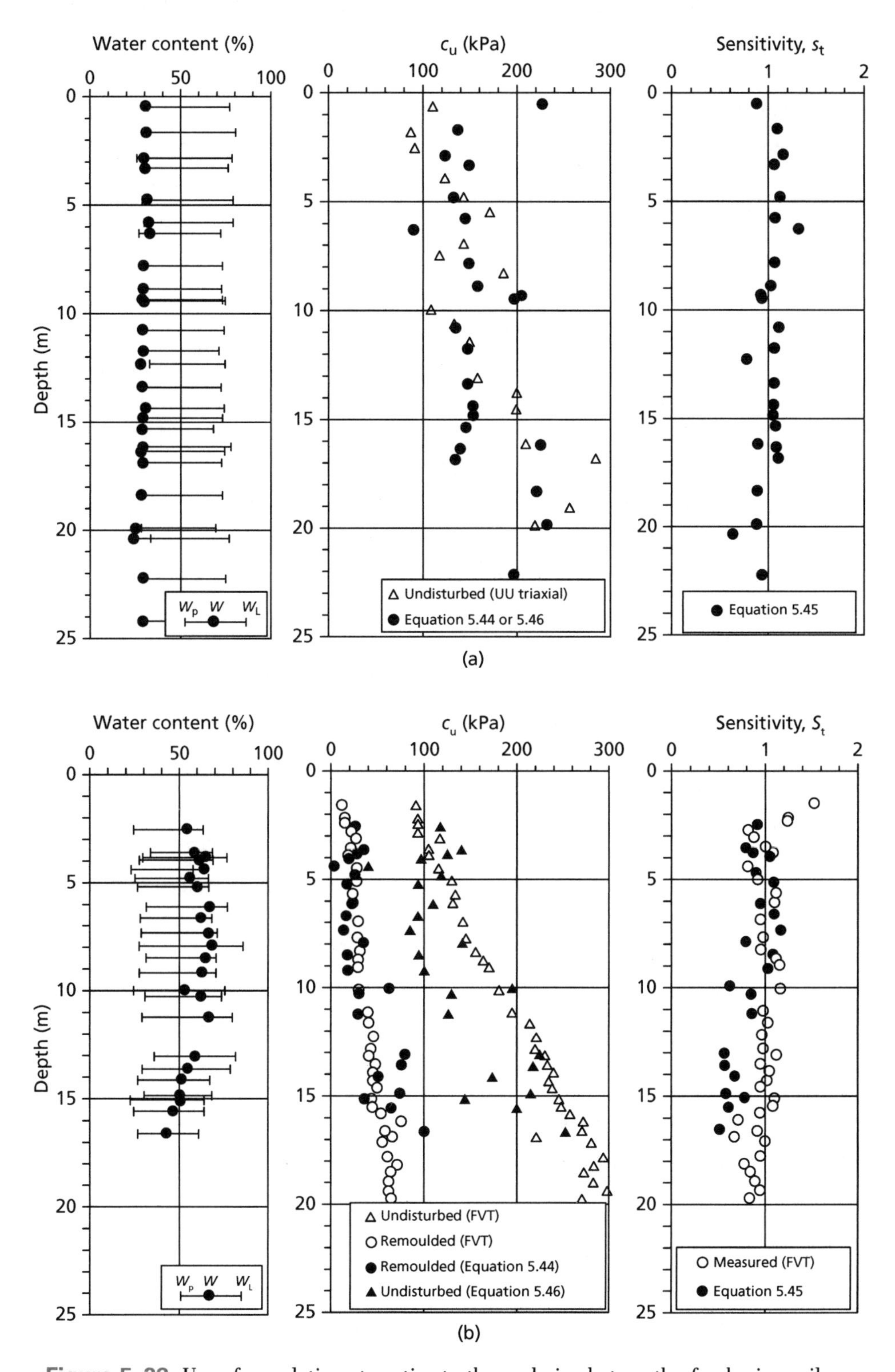

Figure 5.38 Use of correlations to estimate the undrained strength of cohesive soils: (a) Gault clay, (b) Bothkennar clay.

to be much higher than the remoulded strength, and it will be seen that the predictions of Equations 5.44 and 5.46 are different. To validate these predictions, Field Vane Testing (FVT)[①], which is further described in Section 7.3, was used as this can obtain both the undisturbed and remoulded strengths (and therefore also the sensitivity) in soft fine-grained soils. It can be seen from Figure 5.38(b) that the correlations presented here do reasonably predict the values measured by the more reliable in-situ testing.

① 现场十字板剪切试验。

Residual friction angle (ϕ'_r)[②]

② 残余摩擦角。

As noted in Section 5.8, the residual strength of a soil reduces with an increasing proportion of platy particles (e.g. clays) and is therefore correlated with clay fraction and plasticity index. Figure 5.39 shows data of ϕ'_r determined from ring shear tests plotted against I_P for 89 clays and tills and 23 shales, collected by Lupini *et al.* (1981), Mesri and Cepeda-Diaz (1986) and Tiwari and Marui (2005). As observed for ϕ'_{cv} (Figure 5.35(b)), there is considerable scatter[③] around the correlation line which may be attributed to the different characteristics of the clay minerals present within these soils. The data appears to follow a power law where:

③ 离散性。

$$\phi'_r = 93(I_P)^{-0.56} \tag{5.47}$$

with ϕ'_r given in degrees.

In most conventional geotechnical applications, the strains are normally small

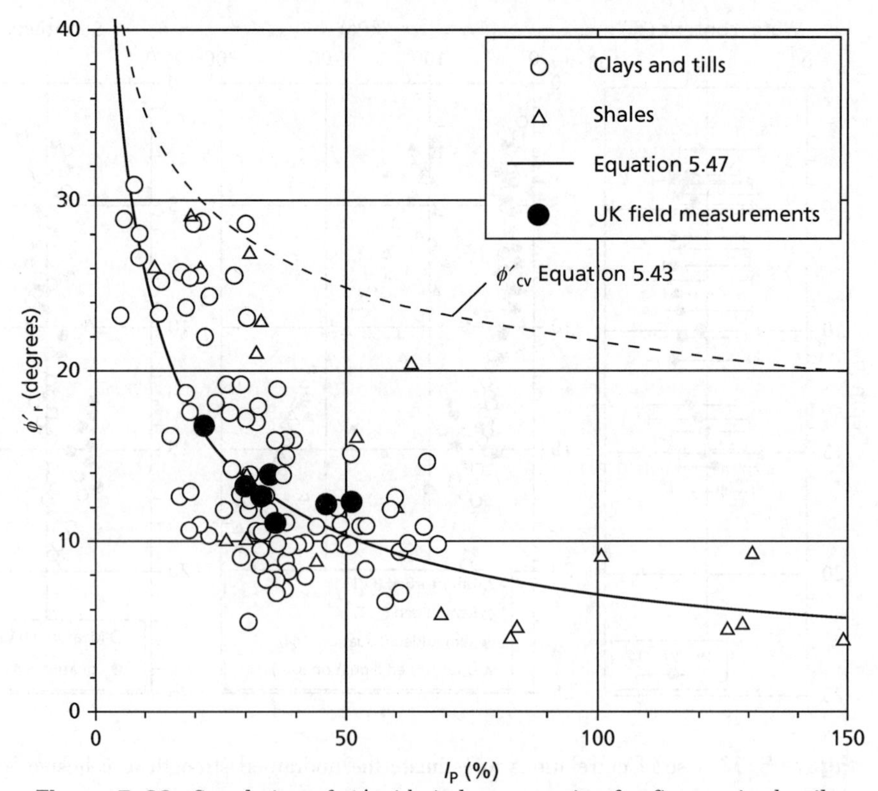

Figure 5.39 Correlation of ϕ'_r with index properties for fine-grained soils, showing application to UK slope case studies.

enough that residual conditions will never be reached. Residual strength is particularly important in the study of slope stability[①], however, where historical slip[②] may have been of sufficient magnitude to align the particles along the slip plane[③] (slope stability is discussed further in Chapter 12). Under these conditions, a slip which is reactivated (e. g. due to reduction of effective stress on the shear plane due to an increase in the water table (rain) or seepage) will be at residual conditions and ϕ'_r is the most appropriate measure of strength to use in analysis. As an example of the use of this correlation, Figure 5.39 also shows a series of data points for a range of slopes around the UK with historical slips, as reported by Skempton (1985). It will be seen that using I_P alone, the predicted values of ϕ'_r are within 2° of the values measured by large direct shear tests on the slip surfaces in the field.

① 边坡稳定。
② 历史上发生的滑动。
③ 滑动面。

Summary

1 Knowledge of the strength and stiffness of soil (its constitutive behaviour) is fundamental for assessing the stability and performance of geotechnical constructions. It relates ground stresses (which are in equilibrium with the applied loads) to ground strains (giving compatible deformations). The constitutive behaviour of soil is highly non-linear and dependent on the level of confining stress, but for most practical problems, it can be modelled/idealised using isotropic linear elasticity coupled with Mohr-Coulomb (stress dependent) plasticity[④].

2 The strength and stiffness of soil may be directly measured in the laboratory by direct shear tests, triaxial tests or residual (ring) shear tests (amongst others). The method of operation, set-up and strengths/weaknesses of each test have been described. These tests may be used to derive strength properties for the Mohr–Coulomb model (c and ϕ)[⑤] and the shear stiffness of the soil (G)[⑥]. Digital examplars provided on the Companion Website have demonstrated how digital data from modern computerised test apparatus may be efficiently processed.

3 For most common rates of loading or geotechnical processes, coarse-grained soils will behave in a drained way. The peak shear strength of such soils is governed by dilatancy (volumetric change) which is density (state) dependent. This behaviour may be modeled by secant peak friction angles or using a linearised Mohr-Coulomb model ($\tau = c'\tau + \sigma' \tan \phi'$)[⑦]. If such soils are sheared to large enough strains, volume change will cease and the soil will reach a critical state (ultimate strength). Soil properties may typically be obtained from drained direct shear or drained triaxial tests. Fine-grained soils will behave in a similar way if they

④ 土体的特性表现为高度非线性和应力相关特性，但对于大多数实际工程问题，土体特性可用等向线弹性并结合莫尔-库仑塑性模型描述或理想化。
⑤ 莫尔-库仑模型（c 和 ϕ 两强度参数）。
⑥ 土剪切模量 G。
⑦ 注：此式原书有误应为 $\tau = c' + \sigma' \tan\phi'$。

are allowed to drain (i. e. for slow processes or long-term conditions). If loaded rapidly, they will respond in an undrained way, and the strength is defined in terms of total, rather than effective stresses using the Mohr-Coulomb model ($\tau = c_u$). Drained and undrained strength properties and the shear modulus G (which is independent of drainage conditions) are typically quantified using triaxial tests (CD, CU, UU) in such soils. At very large strains, the strength of finegrained soils may reduce below the critical state value to a residual value defined by a friction angle ϕ'_r[①], which may be measured using a ring shear apparatus.

① 土体应变非常大时，细粒土的强度会达到低于临界状态值的残余值（用ϕ'_r表示）。

4 With increasing strain with any drainage conditions, all saturated soils will move towards a critical state where they achieve their ultimate shear strength. The Critical State Line defines the critical states for any initial state of the soil, and is therefore an intrinsic soil property. By linking volume change to shear strength, the critical state concept shows that drained and undrained strength both represent the effective stress path arriving at the critical state line (for the two extreme amounts of drainage).

5 Simple index tests (described in Chapter 1) may be used with empirical correlations to estimate values of a range of strength properties (ϕ'_{cv}, c_u and ϕ'_r). Such data can be useful when high quality laboratory test data are unavailable and for providing additional data to support the results of such tests. While useful, empirical correlations should never be used to replace a comprehensive programme of laboratory testing.

Problems

5.1 What is the shear strength in terms of effective stress on a plane within a saturated soil mass at a point where the total normal stress is 295kPa and the pore water pressure 120kPa? The strength parameters of the soil for the appropriate stress range are $\phi' = 30°$ and $c' = 12$kPa.

5.2 Three separate direct shear tests are carried out on identical samples of dry sand. The shear box is 60 × 60mm in plan. The hanger loads, peak forces and ultimate forces measured during the tests are summarised below:

TABLE K

Test	*Hanger load (N)*	*Peak shear force (N)*	*Ultimate shear force (N)*
1	180	162	108
2	360	297	216
3	540	423	324

Determine the Mohr coulomb parameters (ϕ', c') for modeling the peak strength of the soil, the critical state angle of shearing resistance, and the dilation angles in the three tests.

5.3 A series of drained triaxial tests with zero back pressure were carried out on specimens of a sand prepared at the same porosity, and the following results were obtained at failure.

TABLE L

Confining pressure (kPa)	100	200	400	800
Principal stress difference (kPa)	452	908	1810	3624

Determine the value of the angle of shearing resistance ϕ'.

5.4 In a series of unconsolidated-undrained (UU) triaxial tests on specimens of a fully saturated clay, the following results were obtained at failure. Determine the values of the shear strength parameters c_u and ϕ_u.

TABLE M

Confining pressure (kPa)	200	400	600
Principal stress difference (kPa)	222	218	220

5.5 The results below were obtained at failure in a series of consolidated-undrained (CU) triaxial tests, with pore water pressure measurement, on specimens of a fully saturated clay. Determine the values of the shear strength parameters c' and ϕ'. If a specimen of the same soil were consolidated under an all-round pressure of 250kPa and the principal stress difference applied with the all-round pressure changed to 350kPa, what would be the expected value of principal stress difference at failure?

TABLE N

σ_3 (kPa)	150	300	450	600
$\sigma_1 - \sigma_3$ (kPa)	103	202	305	410
u (kPa)	82	169	252	331

5.6 A consolidated-undrained (CU) triaxial test on a specimen of saturated clay was carried out under an all-round pressure of 600kPa. Consolidation took place against a back pressure of 200kPa. The following results were recorded during the test:

TABLE O

$\sigma_1 - \sigma_3$ (kPa)	0	80	158	214	279	319
u (kPa)	200	229	277	318	388	433

Draw the stress paths (total and effective). If the clay has reached its critical state at the end of the test, estimate the critical state friction angle.

5.7 The following results were obtained at failure in a series of consolidated-drained (CD) triaxial tests on fully saturated clay specimens originally 38mm in diameter by 76mm long, with a back pressure of zero. Determine the secant value of ϕ' for each test and the values of tangent parameters c' and ϕ' for the stress range 300 – 500kPa.

TABLE P

All-round pressure (kPa)	200	400	600
Axial compression (mm)	7.22	8.36	9.41
Axial load (N)	565	1015	1321
Volume change (ml)	5.25	7.40	9.30

5.8 In a triaxial test, a soil specimen is allowed to consolidate fully under an all-round pressure of 200kPa. Under undrained conditions the all-round pressure is increased to 350kPa, the pore water pressure then being measured as 144kPa. Axial load is then applied under undrained conditions until failure takes place, the following results being obtained.

TABLE Q

Axial strain (%)	0	2	4	6	8	10
Principal stress difference (kPa)	0	201	252	275	282	283
Pore water pressure (kPa)	144	211	228	222	212	209

Determine the value of the pore pressure coefficient B and determine whether the test can be considered as saturated. Plot the stress – strain ($q - \varepsilon_s$) curve for the test, and hence determine the shear modulus (stiffness) of the soil at each stage of loading. Draw also the stress paths (total and effective) for the test and estimate the critical state friction angle.

References

ASTM D2850 (2007) *Standard Test Method for Unconsolidated-Undrained Triaxial Compression Test on Cohesive Soils*, American Society for Testing and Materials, West Conshohocken, PA.

ASTM D3080 (2004) *Standard Test Method for Direct Shear Test of Soils Under Consolidated Drained Conditions*, American Society for Testing and Materials, West Conshohocken, PA.

ASTM D4767 (2011) *Standard Test Method for Consolidated Undrained Triaxial Compression Test for Cohesive Soils*, American Society for Testing and Materials, West Conshohocken, PA.

ASTM D7181 (2011) *New Test Method for Consolidated Drained Triaxial Compression Test for Soils*, American Society for Testing and Materials, West Conshohocken, PA.

Atkinson, J. H. (2000) Non-linear soil stiffness in routine design, *Géotechnique*, **50** (5), 487 – 508.

Atkinson, J. H. and Bransby, P. L. (1978) *The Mechanics of Soils: An Introduction to Critical State Soil Mechanics*, McGraw-Hill Book Company (UK) Ltd, Maidenhead, Berkshire.

Bishop, A. W., Green, G. E., Garga, V. K., Andresen, A. and Brown, J. D. (1971) A new ring shear apparatus and its application to the measurement of residual strength, *Géotechnique*, **21** (4), 273 – 328.

Bjerrum, L. (1954) Geotechnical properties of Norwegian marine clays, *Géotechnique*, **4** (2), 49 – 69.

Bjerrum, L. and Simons, N. E. (1960) Comparison of shear strength characteristics of normally consolidated clays, in *Proceedings of Research Conference on Shear Strength of Cohesive Soils, Boulder, Colorado*, pp. 711 – 726.

Bolton, M. D. (1986) The strength and dilatancy of sands, *Géotechnique*, **36** (1), 65 – 78.

Bolton, M. D. (1991) *A Guide to Soil Mechanics*, Macmillan Press, London.

British Standard 1377 (1990) *Methods of Test for Soils for Civil Engineering Purposes*, British Standards Institution, London.

Bromhead, E. N. (1979) A simple ring shear apparatus, *Ground Engineering*, **12** (5), 40 – 44.

Calladine, C. R. (2000) *Plasticity for Engineers: Theory and Applications*, Horwood Publishing, Chichester, W. Sussex.

CEN ISO/TS 17892 (2004) *Geotechnical Investigation and Testing – Laboratory Testing of Soil*, International Organisation for Standardisation, Geneva.

Duncan, J. M. and Seed, H. B. (1966) Strength variation along failure surfaces in clay, *Journal of the Soil Mechanics & Foundations Division, ASCE*, **92**, 81 – 104.

Hanna, A. (2001) Determination of plane-strain shear strength of sand from the results of triaxial tests, *Canadian Geotechnical Journal*, **38**, 1231 – 1240.

Hill, R. (1950) *Mathematical Theory of Plasticity*, Oxford University Press, New York, NY.

Jardine, R. J., Symes, M. J. and Burland J. B. (1984) The measurement of soil stiffness in the triaxial apparatus, *Géotechnique* **34** (3), 323 – 340.

Kenney, T. C. (1959) Discussion of the geotechnical properties of glacial clays, *Journal of the Soil Mechanics and Foundations Division, ASCE*, **85**, 67 – 79.

Leroueil, S. Tavenas, F. and Le Bihan, J. P. (1983) Propriétés caractéristiques des argiles de l'est du Canada, *Canadian Geotechnical Journal*, **20**, 681 – 705 (in French).

Lupini, J. F., Skinner, A. E. and Vaughan, P. R. (1981) The drained residual strength of cohesive soils, *Géotechnique*, **31** (2), 181 – 213.

Mesri, G., and Cepeda-Diaz, A. F. (1986) Residual shear strength of clays and shales, *Géotechnique*, **36** (2), 269 – 274.

Miura, K., Maeda, K., Furukawa, M. and Toki, S. (1998) Mechanical characteristics of sands with different primary properties, *Soils and Foundations*, **38** (4), 159 – 172.

Parry, R. H. G. (1960) Triaxial compression and extension tests on remoulded saturated clay, *Géotechnique*, **10** (4), 166 – 180.

Roscoe, K. H., Schofield, A. N. and Wroth, C. P. (1958) On the yielding of soils, *Géotechnique*, **8** (1), 22 – 53.

Skempton, A. W. (1954) The pore pressure coefficients A and B, *Géotechnique*, **4** (4), 143 – 147.

Skempton, A. W. (1985) Residual strength of clays in landslides, folded strata and the laboratory, *Géotechnique*, **35** (1), 1 – 18.

Skempton, A. W. and Northey, R. D. (1953) The sensitivity of clays, *Géotechnique*, **3** (1), 30 – 53.

Tiwari, B., and Marui, H. (2005) A new method for the correlation of residual shear strength of the soil with mineralogical composition, *Journal of Geotechnical and Geoenvironmental Engineering*, ASCE, **131** (9), 1139 – 1150.

Zhu, J. and Yin, J. (2000) Strain-rate-dependent stress – strain behaviour of overconsolidated Hong Kong marine clay, *Canadian Geotechnical Journal*, **37** (6), 1272 – 1282.

Further reading

Atkinson, J. H. (2000) Non-linear soil stiffness in routine design, *Géotechnique*, **50** (5), 487 – 508.

This paper provides a state-of-the-art review of soil stiffness (including non-linear behaviour) and the selection of an appropriate shear modulus in geotechnical design.

Head, K. H. (1986) *Manual of Soil Laboratory Testing*, three volumes, Pentech, London.

This book contains comprehensive descriptions of the set-up of the major laboratory tests for soil testing and practical guidance on test procedures, to accompany the various test standards.

Muir Wood, D. (1991) *Soil Behaviour and Critical State Soil Mechanics*, Cambridge University Press, Cambridge.

This book covers the behaviour of soil entirely within the critical state framework, building substantially on the material presented herein and also forming a useful reference volume.

Chapter 6

Ground investigation[①]

① 地基勘察。

Learning outcomes

After working through the material in this chapter, you should be able to:

1 Specify a basic site investigation[②] strategy to identify soil deposits and determine the depth, thickness and areal extent of such deposits within the ground;

2 Understand the applications and limitations of a wide range of methods available for profiling the ground, and interpret their findings (Sections 6.2, 6.5-6.7);

3 Appreciate the effects of sampling[③] on the quality of soil samples taken for laboratory testing, and the implications of these effects for the interpretation of such test data (Section 6.3).

② 现场勘察。

③ 取样。

6.1 Introduction

An adequate ground investigation is an essential preliminary to the execution of a civil engineering project. Sufficient information must be obtained to enable a safe and economic design to be made and to avoid any difficulties during construction. The principal objects of the investigation are: (1) to determine the sequence, thicknesses and lateral extent of the soil strata and, where appropriate, the level of bedrock; (2) to obtain representative samples of the soils (and rock) for identification and classification, and, if necessary, for use in laboratory tests to determine relevant soil parameters; (3) to identify the groundwater conditions.[④] The investigation may also include the performance of **in-situ** tests[⑤] to assess appropriate soil characteristics. In situ testing will be discussed in Chapter 7. Additional considerations arise if it is suspected that the ground may be contaminated. The results of a ground investigation should provide adequate information, for example, to enable the most suitable type of foundation for a proposed structure to be selected, and to indicate if special problems are likely to arise during construction.

④ 勘察的主要目的如下：(1) 确定土层次序、厚度与水平向延展，有需要时确定基岩深度；(2) 取得土（和岩石）的代表样本，用于鉴定与分类，若有必要，需进行室内试验以确定相关土性参数；(3) 确定地下水文条件。

⑤ 原位测试。

① 资料研究。
② 航空照片。
③ 地形图。
④ 卫星图像。
⑤ 钻孔。

Before any ground investigation work is started on site, a **desk study**[1] should be conducted. This involves collating available relevant information about the site to assist in planning the subsequent fieldwork. A study of geological maps and memoirs, if available, should give an indication of the probable soil conditions of the site in question. If the site is extensive and if no existing information is available, the use of aerial photographs[2], topographical maps[3] or satellite imagery[4] can be useful in identifying existing features of geological significance. Existing borehole[5] or other site investigation data may have been collected for previous uses of the site; in the UK, for example, the National Geological Records Centre may be a useful source of such information. Links to online sources of desk study materials are provided on the Companion Website. Particular care must be taken for sites that have been used previously where additional ground hazards may exist, including buried foundations, services, mine workings etc. Such previous uses may be obtained by examining historical mapping data.

⑥ 现场踏勘。

Before the start of fieldwork an inspection of the site[6] and the surrounding area should be made on foot. River banks, existing excavations, quarries and road or railway cuttings, for example, can yield valuable information regarding the nature of the strata and groundwater conditions; existing structures should be examined for signs of settlement damage. Previous experience of conditions in the area may have been obtained by adjacent owners or local authorities. Consideration of all of the information obtained in the desk study enables the most suitable type of investigation to be selected, and allows the fieldwork to be targeted to best characterise the site. This will ultimately result in a more effective site investigation.

⑦ 钻孔。
⑧ 探坑。
⑨ 静力触探试验。
⑩ 勘察点。
⑪ 地层变异程度。

The actual investigation procedure depends on the nature of the strata and the type of project, but will normally involve the excavation of boreholes or trial pits. The number and location of **boreholes**[7], **trial pits**[8] and **CPT soundings**[9] (Section 6.5) should be planned to enable the basic geological structure of the site to be determined and significant irregularities in the subsurface conditions to be detected. Approximate guidance on the spacing of these **investigation points**[10] is given in Table 6.1. The greater the degree of variability of the ground conditions[11], the greater the number of boreholes or pits required. The locations should be offset from areas on which it is known that foundations are to be sited. A preliminary investigation on a modest scale may be carried out to obtain the general characteristics of the strata, followed by a more extensive and carefully planned investigation including sampling and in situ testing.

⑫ 深基础。

It is essential that the investigation is taken to an adequate depth. This depth depends on the type and size of the project, but must include all strata liable to be significantly affected by the structure and its construction. The investigation must extend below all strata which might have inadequate shear strength for the support of foundations, or which would give rise to significant settlement. If the use of deep foundations[12] (Chapter 9) is anticipated the investigation will thus have to extend to a considerable depth below the surface. If rock is encountered, it should be penetrated by at least 3m in more than one location to confirm that bedrock (and not a large boulder) has been reached, unless geological knowledge indicates otherwise. The investigation may have to be taken to depths greater than normal in areas of old mine workings or other underground cavities. Boreholes and trial pits should be backfilled after use. Backfilling with compacted soil may be adequate in many cases, but if the groundwater conditions are altered by a borehole and the resultant flow could produce adverse effects then it is necessary to use a cement based grout to seal the hole.①

① 回填压实土在很多情况下已能满足要求，但若钻孔使地下水情况改变，并且钻孔存在导致的渗流有可能产生负面影响，则有必要使用水泥灌浆封孔。

Table 6.1 Guidance on spacing of ground investigation points (Eurocode 7, Part 2:2007)

Type of construction	*Spacing of investigation points*
High-rise and industrial structures	Grid pattern, at spacing of 15-40m
Large-area structures	Grid pattern, spacing≤60m
Linear structures② (e. g. roads, railways, retaining walls etc.)	Along route, at spacing of 20-200m
Special structures (e. g. bridges, chimneys/stacks, machine foundations)	2-6 investigation points per foundation
Dams and weirs③	25- to 75-m spacing, along relevant sections

② 线状结构。

③ 坝与围堰。

The cost of an investigation depends on the location and extent of the site, the nature of the strata and the type of project under consideration. In general, the larger the project, and the less critical the ground conditions are to the design and construction of the project, the lower the cost of the ground investigation as a percentage of the total cost. The cost of a ground investigation is generally within the range 0.1-2% of the project cost. To reduce the scope of an investigation for financial reasons alone is never justified. Chapman (2008) provides an interesting example of this, considering the development of a six-storey office building in central London with a construction cost of £30 million. It is demonstrated that in halving the cost of site investigation (from £45000 to £22500), poorer foundation design due to less available ground information carries a cost of around £210000; If project completion is delayed by more than one month due to unforeseen ground conditions (this happens in approximately 20% of projects), the associated cost could exceed £800000 in lost building rents and re-design. Clients' costs are therefore very vulnerable to the unexpected. The aim of a good ground investigation should be to ensure that unforeseen ground conditions do not increase this vulnerability.④

④ 好的地基勘探应当保证不能预见的地质情况不会增加风险性。

6.2 Methods of intrusive investigation

Trial pits

The excavation of trial pits is a simple and reliable method of investigation, but is limited to a maximum depth of 4-5m. The soil is generally removed by means of the back-shovel of a mechanical excavator. Before any person enters the pit, the sides must always be supported unless they are sloped at a safe angle or are stepped; the excavated soil should be placed at least 1m from the edge of the pit (see Section 12.2 for a discussion of stability of trial pits and trenches). If the pit is to extend below the water table, some form of dewatering[①] is necessary in the more permeable soils, resulting in increased costs. The use of trial pits enables the in-situ soil conditions to be examined visually, and thus the boundaries between strata and the nature of any macro-fabric can be accurately determined.[②] It is relatively easy to obtain disturbed or undisturbed soil samples: in fine grained soils block samples[③] can be cut by hand from the sides or bottom of the pit, and tube samples[④] can be obtained below the bottom of the pit. Trial pits are suitable for investigations in all types of soil, including those containing cobbles or boulders.

① 降水。

② 探坑可以使我们从视觉上判断原位土体情况，可准确判别土层分界与一些土体宏观结构。

③ 块状试样。

④ 管状试样。

Shafts and headings

Deep pits or shafts are usually advanced by hand excavation, the sides being supported by timbering. **Headings**[⑤] or adits are excavated laterally from the bottom of shafts or from the surface into hillsides, both the sides and roof being supported. It is unlikely that shafts or headings would be excavated below the water table.[⑥] Shafts and headings are very costly, and their use would be justified only in investigations for very large structures, such as dams, if the ground conditions could not be ascertained adequately by other means.

⑤ 导坑。

⑥ 竖井与导坑几乎不可能挖到地下水位以下。

Percussion boring

The boring rig (Figure 6.1) consists of a derrick, a power unit, and a winch carrying a light steel cable which passes through a pulley on top of the derrick. Most rigs are fitted with road wheels, and when folded down can be towed behind a vehicle. Various boring tools can be attached to the cable. The borehole is advanced by the percussive action of the tool which is alternately raised and dropped (usually over a distance of 1-2m) by means of the winch unit. The two most widely used tools are the **shell** (Figure 6.1 (b)) and the **clay cutter**[⑦] (Figure 6.1 (c)). If necessary, a heavy steel element called a sinker bar can be fitted immediately above the tool to increase the impact energy.

⑦ 黏土切削器。

The shell, which is used in sands and other coarse grained soils, is a heavy steel tube fitted with a flap or clack valve at the lower end. Below the water table, the percussive action of the shell loosens the soil and produces a slurry[①] in the borehole. Above the water table, a slurry is produced by introducing water into the borehole. The slurry passes through the clack valve during the downward movement of the shell and is retained by the valve during the upward movement. When full, the shell is raised to the surface to be emptied. In cohesionless soils (e. g. sands and gravels), the borehole must be cased to prevent collapse. The casing, which consists of lengths of steel tubing screwed together, is lowered into the borehole and will normally slide down under its own weight; however, if necessary, installation of the casing can be aided by driving. On completion of the investigation the casing is recovered by means of the winch or by the use of jacks: excessive driving during installation may make recovery of the casing difficult.

① 泥浆。

The clay cutter, which is used in fine-grained soils (e. g. clays, silts and tills), is an open steel tube with a cutting shoe and a retaining ring at the lower end; the tool is used in a dry borehole. The percussive action of the tool cuts a plug of soil which eventually fractures near its base due to the presence of the retaining ring. The ring also ensures that the soil is retained inside the cutter when it is raised to the surface to be emptied.

Small boulders, cobbles and hard strata can be broken up by means of a **chisel**[②], aided by the additional weight of a sinker bar if necessary.

② 凿刀。

Borehole diameters can range from 150 to 300mm. The maximum borehole depth is generally between 50 and 60m. Percussion boring can be employed in most types of soil, including those containing cobbles and boulders. However, there is generally some disturbance of the soil below the bottom of the borehole, from which sam-

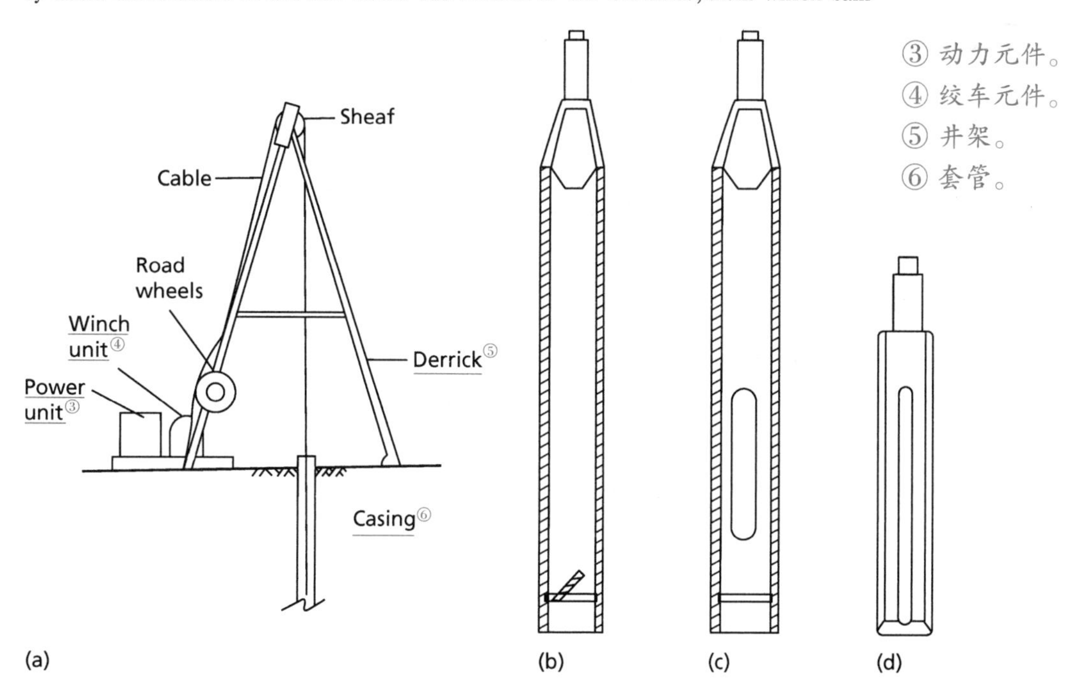

③ 动力元件。
④ 绞车元件。
⑤ 井架。
⑥ 套管。

Figure 6.1 (a) Percussion boring rig[⑦], (b) shell, (c) clay cutter, and (d) chisel.

⑦ 贯入式钻孔机。

① 用途广泛的。

② 链动螺旋钻。

③ 超固结黏土。

④ 膨润土泥浆。

⑤ 螺旋。

ples are taken, and it is extremely difficult to detect thin soil layers and minor geological features with this method. The rig is extremely versatile[1], and can normally be fitted with a hydraulic power unit and attachments for mechanical augering, rotary core drilling and in situ testing (Chapter 7).

Mechanical augers

Power-operated augers are generally mounted on vehicles or in the form of attachments to the derrick used for percussion boring. The power required to rotate the auger depends on the type and size of the auger itself, and the type of soil to be penetrated. Downward pressure on the auger can be applied hydraulically, mechanically or by dead weight. The types of tool generally used are the **flight auger**[2] and the **bucket auger**. The diameter of a flight auger is usually between 75 and 300mm, although diameters as large as 1m are available; the diameter of a bucket auger can range from 300mm to 2m. However, the larger sizes are used principally for excavating shafts for bored piles. Augers are used mainly in soils in which the borehole requires no support and remains dry, i. e. mainly in stiffer, overconsolidated clays[3]. The use of casing would be inconvenient because of the necessity of removing the auger before driving the casing; however, it is possible to use bentonite slurry[4] to support the sides of unstable holes (Section 12.2). The presence of cobbles or boulders creates difficulties with the smaller sized augers.

Short-flight augers (Figure 6.2 (a)) consist of a helix of limited length, with cutters below the helix[5]. The auger is attached to a steel shaft, known as a

(a) (b) (c) (d)

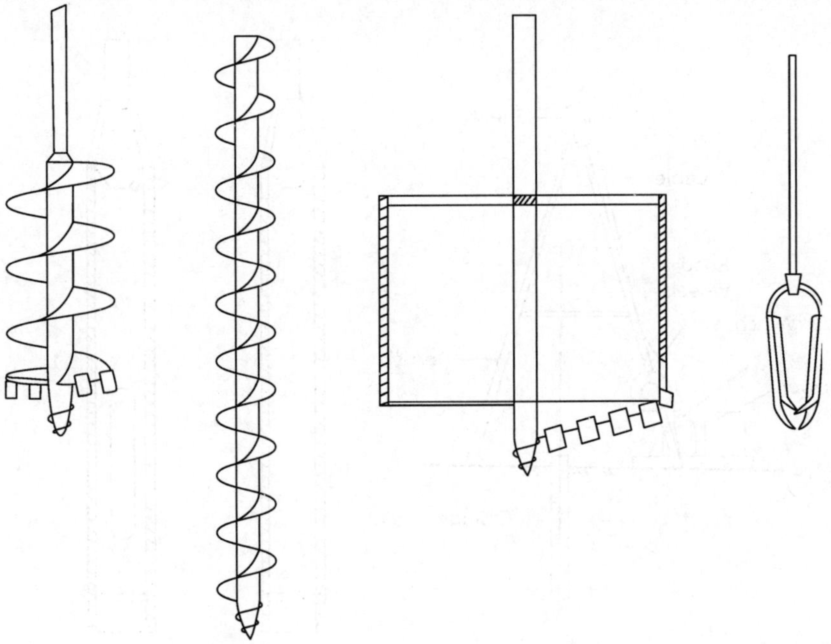

Figure 6.2 (a) Short-flight auger, (b) continuous-flight auger, (c) bucket auger, and (d) Iwan (hand) auger.

kelly bar[1], which passes through the rotary head of the rig. The auger is advanced until it is full of soil, then it is raised to the surface where the soil is ejected by rotating the auger in the reverse direction. Clearly, the shorter the helix, the more often the auger must be raised and lowered for a given borehole depth. The depth of the hole is limited by the length of the kelly bar.

① 钻杆。

Continuous-flight augers (Figure 6. 2 (b)) consist of rods with a helix covering the entire length. The soil rises to the surface along the helix, obviating the necessity for withdrawal; additional lengths of auger are added as the hole is advanced. Borehole depths up to 50m are possible with continuous-flight augers, but there is a possibility that different soil types may become mixed as they rise to the surface, and it may be difficult to determine the depths at which changes of strata occur.

Continuous-fight augers[2] with hollow stems are also used. When boring is in progress, the hollow stem is closed at the lower end by a plug fitted to a rod running inside the stem. Additional lengths of auger (and internal rod[3]) are again added as the hole is advanced. At any depth the rod and plug may be withdrawn from the hollow stem to allow undisturbed samples to be taken, a sample tube mounted on rods being lowered down the stem and driven into the soil below the auger. If bedrock is reached, drilling can also take place through the hollow stem. The internal diameter of the stem can range from 75 to 150mm. As the auger performs the function of a casing it can be used in permeable soils (e. g. sands) below the water table, although difficulty may be experienced with soil being forced upwards into the stem by hydrostatic pressure; this can be avoided by filling the stem with water up to water table level.

② 长螺旋连续钻。

③ 内杆。

Bucket augers (Figure 6. 2 (c)) consist of a steel cylinder, open at the top but fitted with a base plate on which cutters arc mounted, adjacent to slots in the plate: the auger is attached to a kelly bar. When the auger is rotated and pressed downwards, the soil removed by the cutters passes through the slots into the bucket. When the bucket is full, it is raised to the surface to be emptied by releasing the hinged base plate.

Augered holes of 1-m diameter and larger can be used for the examination of the soil strata in-situ, by lowering a remote CCTV camera into the borehole.

Hand and portable augers

Hand augers[4] can be used to excavate boreholes to depths of around 5m using a set of extension rods. The auger is rotated and pressed down into the soil by means of a T-handle on the upper rod. The two common types are the Iwan or post-hole auger (Figure 6. 2 (d)), with diameters up to 200mm, and the small helical auger, with diameters of about 50mm. Hand augers are generally used only if the sides of the hole require no support and if particles of coarse gravel size and above are absent. The auger must be withdrawn at frequent intervals for the removal of soil. Undisturbed samples[5] can be obtained by driving small-diameter tubes below the bottom of the borehole.

④ 手工钻。

⑤ 原状土样。

Small portable power augers, generally transported and operated by two persons, are suitable for boring to depths of 10-15m; the hole diameter may range from 75 to 300mm. The borehole may be cased if necessary, and therefore the auger can be used in most soil types provided the larger particle sizes are absent.

① 冲洗钻探。

Wash boring[①]

In this method, water is pumped through a string of hollow boring rods and is released under pressure through narrow holes in a chisel attached to the lower end of the rods (Figure 6.3). The soil is loosened and broken up by the water jets and the up-and-down movement of the chisel. There is also provision for the manual rotation of the chisel by means of a tiller attached to the boring rods above the surface. The soil particles are washed to the surface between the rods and the side of the borehole, and are allowed to settle out in a sump. The rig consists of a derrick with a power unit, a winch and a water pump[②]. The winch carries a light steel cable which passes through the sheaf of the derrick and is attached to the top of the boring rods. The string of rods is raised and dropped by means of the winch unit, producing the chopping action of the chisel. The borehole is generally cased, but the method can be used in uncased holes. Drilling fluid may be used as an alternative to water in the method, eliminating the need for casing.

② 水泵。

Wash boring can be used in most types of soil, but progress becomes slow if particles of coarse gravel size and larger are present. The accurate identification of soil types is difficult, due to particles being broken up by the chisel and to mixing as the material is washed to the surface; in addition, segregation of particles takes place as they settle out in the sump. However, a change in the feel of the boring tool can sometimes be detected, and there may be a change in the colour of the water rising to the surface, when the boundaries between different strata are reached. The method is unacceptable as a means of obtaining soil samples. It is used only as a means of advancing a borehole to enable tube samples to be taken or in-situ tests to be carried out below the bottom of the hole. An advantage of the method is that the soil immediately below the hole remains relatively undisturbed.

Rotary drilling

Although primarily intended for investigations in rock, the method is also used in soils. The drilling tool, which is attached to the lower end of a string of hollow drilling rods (Figure 6.4), may be either a cutting bit or a coring bit; the coring bit is fixed to the lower end of a core barrel[③], which in turn is carried by the drilling rods. Water or drilling fluid is pumped down the hollow rods and passes under pressure through narrow holes in the bit or barrel; this is the same principle as used in wash boring. The drilling fluid cools and lubricates the drilling tool, and carries the loose debris to the surface between the rods and the side of the hole. The fluid also provides some support to the sides of the hole if no casing is used.

③ 芯筒。

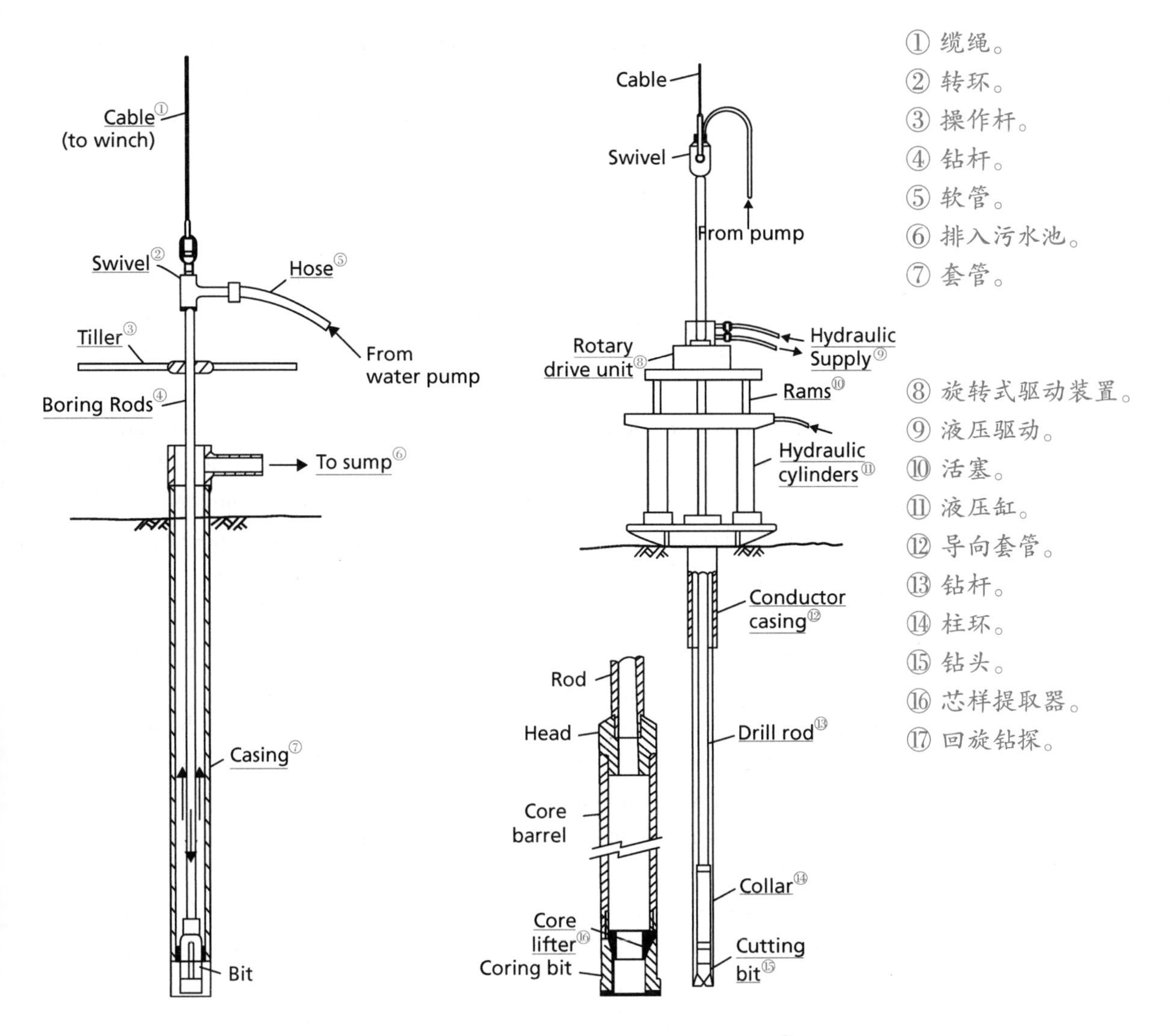

Figure 6.3 Wash boring. **Figure 6.4** Rotary drilling.⑰

The rig consists of a derrick, power unit, winch, pump and a drill head to apply high-speed rotary drive and downward thrust to the drilling rods. A rotary head attachment can be supplied as an accessory to a percussion boring rig.

There are two forms of rotary drilling: open-hole drilling⑱ and core drilling⑲. Open-hole drilling, which is generally used in soils and weak rock, uses a cutting bit to break down all the material within the diameter of the hole. Open-hole drilling can thus be used only as a means of advancing the hole; the drilling rods can then be removed to allow tube samples to be taken or in-situ tests to be carried out. In core drilling, which is used in rocks and hard clays, the diamond or tungsten carbide bit cuts an annular hole in the material and an intact core enters the barrel, to be removed as a sample. However, the natural water content of the material is liable to be increased due to contact with the drilling fluid. Typical core diameters are 41, 54 and 76mm, but can range up to 165mm.

⑱ 敞口钻进。
⑲ 岩芯钻进。

The advantage of rotary drilling in soils is that progress is much faster than with other investigation methods and disturbance of the soil below the borehole is slight. The method is not suitable if the soil contains a high percentage of gravel-sized (or larger) particles, as they tend to rotate beneath the bit and are not broken up.

① 地下水观测。

Groundwater observations①

② 地下水位及特定土层承压水压力的确定，是所有地基勘察的重要内容之一。

An important part of any ground investigation is the determination of water table level and of any artesian pressure in particular strata.② The variation of level or pore water pressure over a given period of time may also require determination. Groundwater observations are of particular importance if deep excavations are to be carried out.

③ 响应时间。

④ 上层滞水位。

Water table level can be determined by measuring the depth to the water surface in a borehole. Water levels in boreholes may take a considerable time to stabilise; this time, known as the **response time**③, depends on the permeability of the soil. Measurements, therefore, should be taken at regular intervals until the water level becomes constant. It is preferable that the level should be determined as soon as the borehole has reached water table level. If the borehole is further advanced it may penetrate a stratum under artesian pressure (Section 2. 1), resulting in the water level in the hole being above water table level. It is important that a stratum of low permeability below a perched water table (Section 2. 1) should not be penetrated before the water level has been established. If a perched water table④ exists, the borehole must be cased in order that the main water table level is correctly determined; if the perched aquifer is not sealed, the water level in the borehole will be above the main water table level.

⑤ 孔压计。

⑥ 高进气值陶瓷。

When it is required to determine the pore water pressure in a particular stratum, a **piezometer**⑤ should be used. A piezometer consists of an element filled with de-aired water and incorporating a porous tip which provides continuity between the pore water in the soil and the water within the element. The element is connected to a pressure-measuring system. A high air-entry ceramic⑥ tip is essential for the measurement of pore water pressure in partially saturated soils (e. g. compacted fills), the air-entry value being the pressure difference at which air would bubble through a saturated filter. Therefore the air-entry value must exceed the difference between the pore air and pore water pressures, otherwise pore air pressure will be recorded. A coarse porous tip can only be used if it is known that the soil is fully saturated. If the pore water pressure is different from the pressure of the water in the measuring system, a flow of water into or out of the element will take place. This, in turn, results in a change in pressure adjacent to the tip, and consequent seepage of pore water towards or away from the tip. Measurement involves balancing the pressure in the measuring system with the pore water pressure in the vicinity of the tip. However the response time taken for the pressures to equalise depends on the permeability of the soil and the flexibility of the measuring system. The response time of a piezometer should be as short as possible. Factors governing flexibility are the volume change required to actuate the measuring device, the expansion of the connections, and the presence of entrapped air. A de-airing① unit forms an essential part of the equipment: efficient de-airing during installation is essential if errors in pressure measurement are to be avoided. To achieve a rapid response time in soils of low permeability the measuring system must be as stiff as possible, requiring the use of a closed hydraulic system in which virtually no flow of water is required to operate the measuring device.

① 排气。

The simplest type is the open standpipe piezometer② (Figure 6.5), which is used in a cased borehole and is suitable if the soil is fully saturated and the permeability is relatively high. The water level is normally determined by means of an electrical dipper③, a probe with two conductors on the end of a measuring tape; the battery-operated circuit closes, triggering an indicator, when the conductors come into contact with the water. The standpipe is normally a plastic tube of 50 mm diameter or smaller, the lower end of which is either perforated or fitted with a porous element. A relatively large volume of water must pass through the porous element to change the standpipe level, therefore a short response time will only be obtained in soils of relatively high permeability. Sand or fine gravel is packed around the lower end,

② 敞口侧压管。

③ 水位蜂鸣器。

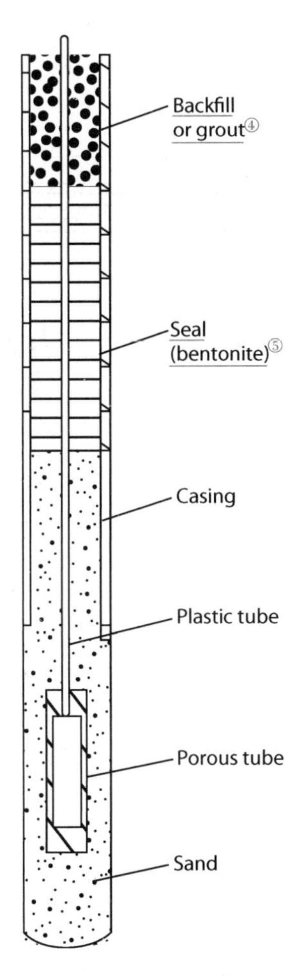

Figure 6.5 Open standpipe piezometer.

④ 回填或灌浆。

⑤ 密封（膨润土）。

and the standpipe is sealed in the borehole with clay (generally by the use of bentonite pellets) immediately above the level at which pore pressure is to be measured. The remainder of the borehole is backfilled with sand except near the surface, where a second seal is placed to prevent the inflow of surface water. The top of the standpipe is fitted with a cap, again to prevent ingress of water. Open standpipe piezometers which can be pushed or driven into the ground have also been developed. Piezometers may also be used to collect water samples for further chemical analysis① in the laboratory on sites which may be contaminated②(Section 6.8).

① 水化分析。
② 被污染的。
③ 低渗透性。

The open standpipe piezometer has a long response time in soils of low permeability③, and in such cases it is preferable to install a hydraulic piezometer having a relatively short response time. To achieve a rapid response time in soils of low permeability the measuring system must be as stiff as possible, requiring the use of a closed hydraulic system in which no flow of water is required to operate the measuring device. Three types of piezometer for use with a closed hydraulic system are illustrated in Figure 6.6. The piezometers consist of a brass or plastic body into which a porous tip of ceramic④, bronze⑤ or stone is sealed. Two tubes lead from the device, one of which is connected to a transducer⑥, allowing results to be recorded automatically. The tubes are of nylon⑦ coated with polythene⑧, nylon being impermeable to air and polythene to water.

④ 陶瓷。
⑤ 铜。
⑥ 传感器。
⑦ 尼龙。
⑧ 聚乙烯。

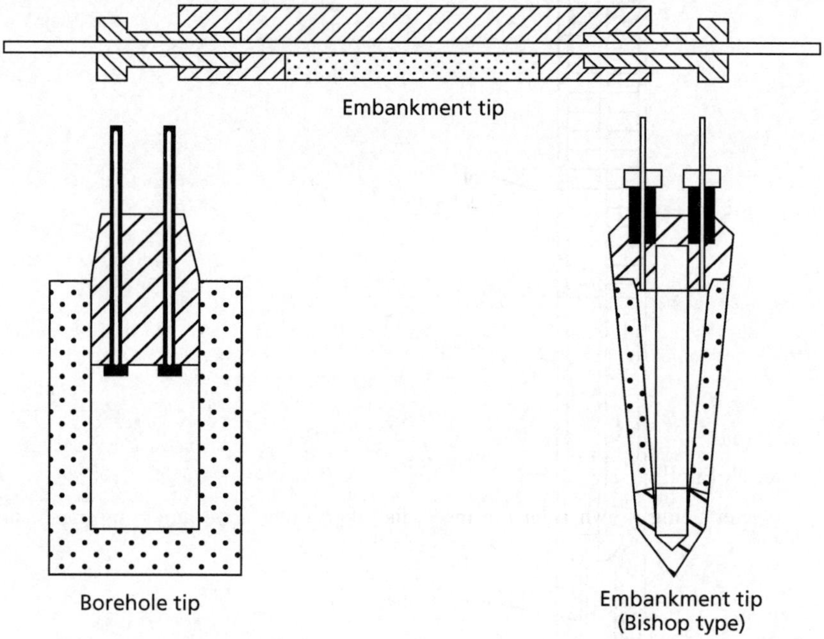

Embankment tip

Borehole tip

Embankment tip (Bishop type)

Figure 6.6 Piezometer tips

The two tubes enable the system to be kept air-free by the periodic circulation of de-aired water. Allowance must be made for the difference in level between the tip and the measuring instrument, which should be sited below the tip whenever possible.

6.3 Sampling[1]

Once an opening (trial pit or borehole) has been made in the ground, it is often desirable to recover samples of the soil to the surface for laboratory testing. Soil samples are divided into two main categories, **undisturbed**[2] and **disturbed**[3]. Undisturbed samples, which are required mainly for shear strength[4] and consolidation tests[5] (Chapters 4 and 5), are obtained by techniques which aim at preserving the insitu structure and water content of the soil as far as is practically possible. In boreholes, undisturbed samples can be obtained by withdrawing the boring tools (except when hollow-stem continuous-flight augers are used) and driving or pushing a sample tube into the soil at the bottom of the hole. The sampler is normally attached to a length of boring rod which can be lowered and raised by the cable of the percussion rig. When the tube is brought to the surface, some soil is removed from each end and molten wax is applied, in thin layers, to form a water-tight seal[6] approximately 25mm thick; the ends of the tube are then covered by protective caps. Undisturbed block samples can be cut by hand from the bottom or sides of a trial pit. During cutting, the samples must be protected from water, wind and sun to avoid any change in water content; the samples should be covered with molten wax[7] immediately they have been brought to the surface. It is impossible to obtain a sample that is completely undisturbed, no matter how elaborate or careful the ground investigation and sampling technique might be. In the case of clays, for example, swelling will take place adjacent to the bottom of a borehole due to the reduction in total stresses when soil is removed, and structural disturbance may be caused by the action of the boring tools; subsequently, when a sample is removed from the ground the total stresses are reduced to zero.

Soft clays are extremely sensitive to sampling disturbance, the effects being more pronounced in clays of low plasticity than in those of high plasticity.[8] The central core of a soft clay sample will be relatively less disturbed than the outer zone adjacent to the sampling tube. Immediately after sampling, the pore water pressure in the relatively undisturbed core will be negative due to the release of the in-situ total stresses. Swelling of the relatively undisturbed core will gradually take place due to water being drawn from the more disturbed outer zone and resulting in the dissipation of the negative excess pore water pressure[9]; the outer zone of soil will consolidate due to the redistribution of water within the sample. The dissipation of the negative excess pore water pressure is accompanied by a corresponding reduction in effective stresses. The soil structure of the sample will thus offer less resistance to shear, and will be less stiff than the in-situ soil.

A disturbed sample is one having the same particle size distribution as the in-situ soil but in which the soil structure has been significantly damaged or completely destroyed; in addition, the water content may be different from that of the in-situ soil. Disturbed samples, which are used mainly for soil classification tests (Chapter

① 取样。

② 非扰动的。

③ 非扰动的。

④ 抗剪强度。

⑤ 固结试验。

⑥ 不透水密封。

⑦ 融蜡。

⑧ 软黏土对于试样扰动非常敏感，低塑性黏土中这一现象比高塑性黏土更明显。

⑨ 超负孔隙水压力。

1), visual classification and compaction tests, can be excavated from trial pits or obtained from the tools used to advance boreholes (e. g. from augers and the clay cutter). The soil recovered from the shell in percussion boring will be deficient in fines and will be unsuitable for use as a disturbed sample. Samples in which the natural water content has been preserved should be placed in airtight[①], non-corrosive containers; all containers should be completely filled so that there is negligible air space above the sample.

① 气密的。

Samples should be taken at changes of stratum (as observed from the soil recovered by augering/drilling) and at a specified spacing within strata of not more than 3m. All samples should be clearly labelled to show the project name, date, location, borehole number, depth and method of sampling; in addition, each sample should be given a unique serial number[②]. Special care is required in the handling, transportation and storage[③] of samples (particularly undisturbed samples) prior to testing.

② 唯一的土样编号。
③ 运输和存储。

The sampling method used should be related to the quality of sample required. Quality can be classified as shown in Table 6. 2, with Class 1 being most useful and of the highest quality, and Class 5 being useful only for basic visual identification of soil type. For Classes 1 and 2, the sample must be undisturbed. Samples of Classes 3, 4 and 5 may be disturbed. The principal types of tube samplers are described below.

Table 6.2 Sample quality related to end use (after EC7-2:2007)

Soil property	*Class 1*	*Class 2*	*Class 3*	*Class 4*	*Class 5*
Sequence of layers[④]	•	•	•	•	•
Strata boundaries[⑤]	•	•	•	•	
Particle size distribution[⑥]	•	•	•	•	
Atterberg limits, organic content[⑦]	•	•	•	•	
Water content[⑧]	•	•	•		
(Relative) Density, porosity[⑨]	•	•			
Permeability[⑩]	•	•			
Compressibility, shear strength[⑪]	•				

④ 土层顺序。
⑤ 土层边界。
⑥ 粒径分布。
⑦ 界限含水量，有机物含量。
⑧ 含水量。
⑨ (相对) 密度，孔隙率。
⑩ 渗透性。
⑪ 压实性，抗剪强度。

Open drive sampler

An open drive sampler[⑫](Figure 6.7 (a)) consists of a long steel tube with a screw thread at each end. A cutting shoe is attached to one end of the tube. The other end of the tube screws into a sampler head, to which, in turn, the boring rods are connected. The sampler head also incorporates a nonreturn valve to allow air and water to escape as the soil fills the tube, and to help retain the sample as the tube is withdrawn. The inside of the tube should have a smooth surface, and must be maintained in a clean condition.

⑫ 敞口贯入取土器。

The internal diameter of the cutting edge (d_c) should be approximately 1% smaller than that of the tube to reduce frictional resistance between the tube and the sample. This size difference also allows for slight elastic expansion of the sample on entering the tube, and assists in sample retention. The external diameter of the cutting shoe (d_w) should be slightly greater than that of the tube to reduce the force required to withdraw the tube. The volume of soil displaced by the sampler as a proportion of the sample volume is represented by the area ratio[①] (C_a) of the sampler, where

① 面积比。

$$C_a = \frac{d_w^2 - d_c^2}{d_c^2} \tag{6.1}$$

② 钻孔杆。

The area ratio is generally expressed as a percentage. Other factors being equal, the lower the area ratio, the lower is the degree of sample disturbance.

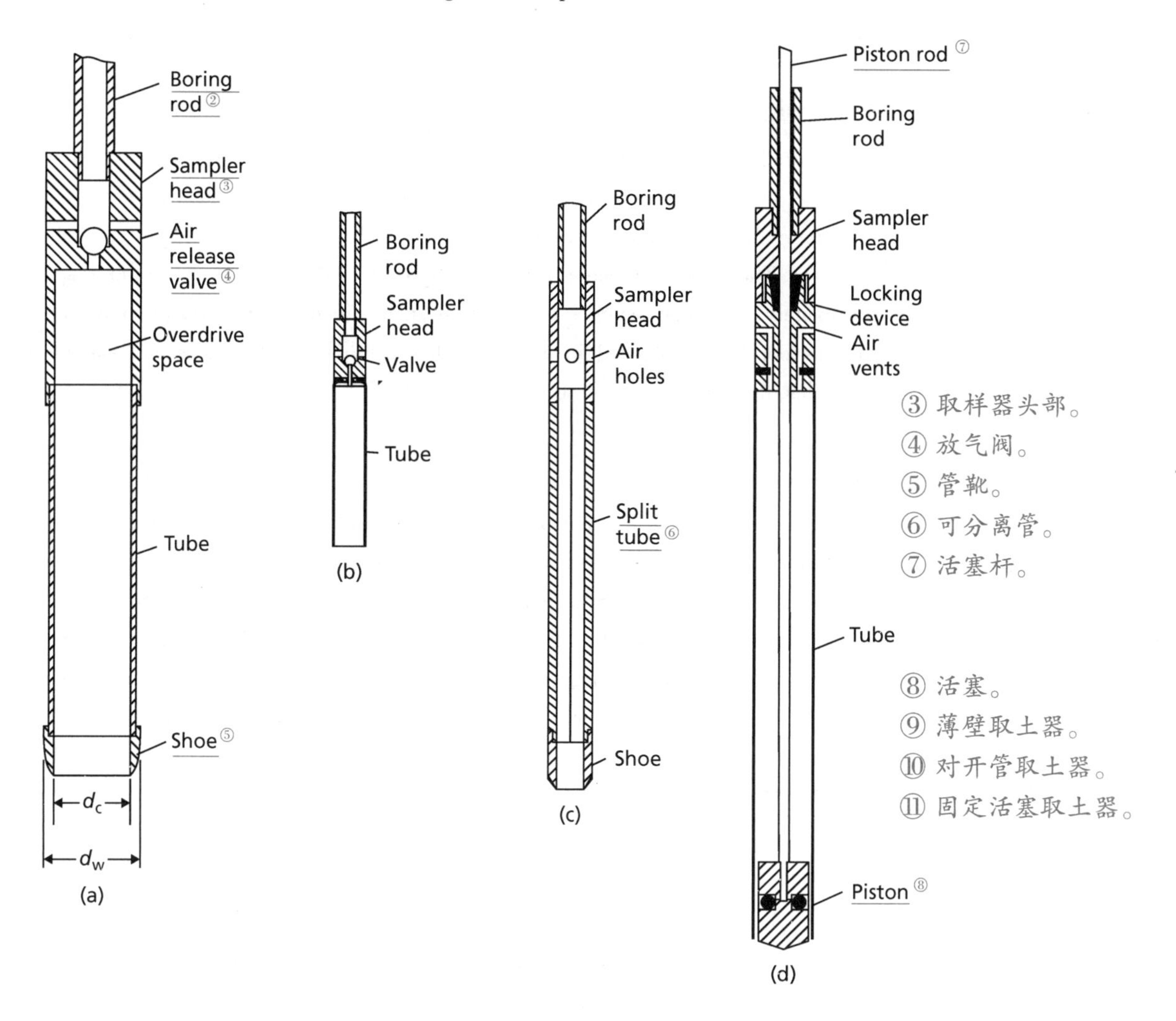

③ 取样器头部。
④ 放气阀。
⑤ 管靴。
⑥ 可分离管。
⑦ 活塞杆。
⑧ 活塞。
⑨ 薄壁取土器。
⑩ 对开管取土器。
⑪ 固定活塞取土器。

Figure 6.7 Types of sampling tools: (a) open drive sampler, (b) thin-walled sampler[⑨], (c) split-barrel sampler[⑩], and (d) stationary piston sampler[⑪].

The sampler can be driven dynamically by means of a drop weight or sliding hammer, or statically by hydraulic or mechanical jacking. Prior to sampling, all loose soil should be removed from the bottom of the borehole. Care should be taken to ensure that the sampler is not driven beyond its capacity, otherwise the sample will be compressed against the sampler head. Some types of sampler head have an overdrive space below the valve to reduce the risk of sample damage. After withdrawal, the cutting shoe and sampler head are detached and the ends of the sample are sealed.

The most widely used sample tube has an internal diameter of 100mm and a length of 450mm; the area ratio is approximately 30%. This sampler is suitable for all clay soils. When used to obtain samples of sand, a **core-catcher**[①], a short length of tube with spring-loaded flaps, should be fitted between the tube and cutting shoe to prevent loss of soil. The class of sample[②] obtained depends on soil type.

① 取芯器。

② 土样等级。

Thin-walled sampler

Thin-walled samplers[③] (Figure 6.7 (b)) are used in soils which are sensitive to disturbance, such as soft to firm clays and plastic silts. The sampler does not employ a separate cutting shoe, the lower end of the tube itself being machined to form a cutting edge. The internal diameter may range from 35 to 100mm. The area ratio is approximately 10%, and samples of first-class quality can be obtained provided the soil has not been disturbed in advancing the borehole. In trial pits and shallow boreholes, the tube can often be driven manually.

③ 薄壁取土器。

Split-barrel sampler

Split-barrel samplers[④] (Figure 6.7 (c)) consist of a tube which is split longitudinally into two halves; a shoe and a sampler head incorporating air-release holes are screwed onto the ends. The two halves of the tube can be separated when the shoe and head are detached to allow the sample to be removed. The internal and external diameters are 35 and 50mm, respectively, the area ratio being approximately 100%, with the result that there is considerable disturbance of the sample (Class 3 or 4). This sampler is used mainly in sands, being the tool specified in the **Standard Penetration Test**[⑤] (SPT, see Chapter 7).

④ 对开管取土器。

⑤ 标准贯入试验。

Stationary piston sampler

Stationary piston samplers[⑥] (Figure 6.7 (d)) consist of a thin-walled tube fitted with a piston. The piston is attached to a long rod which passes through the sampler head and runs inside the hollow boring rods. The sampler is lowered into the borehole with the piston located at the lower end of the tube, the tube and piston being locked together by means of a clamping device at the top of the rods. The piston prevents water or loose soil from entering the tube. In soft soils the sampler can be pushed below the bottom of the borehole, bypassing any disturbed soil. The piston

⑥ 固定活塞取土器。

is held against the soil (generally by clamping the piston rod to the casing) and the tube is pushed past the piston (until the sampler head meets the top of the piston) to obtain the sample. The sampler is then withdrawn, a locking device in the sampler head holding the piston at the top of the tube as this takes place. The vacuum[①] between the piston and the sample helps to retain the soil in the tube; the piston thus serves as a non-return valve.

① 真空。

Piston samplers should always be pushed down by hydraulic or mechanical jacking; they should never be driven. The diameter of the sampler is usually between 35 and 100mm, but can be as large as 250mm. The samplers are generally used for soft clays, and can produce samples of first-class quality up to 1m in length.

Continuous sampler

The continuous sampler[②] is a highly specialised type of sampler which is capable of obtaining undisturbed samples up to 25m in length; the sampler is used mainly in soft clays. Details of the soil fabric can be determined more easily if a continuous sample is available. An essential requirement of continuous samplers is the elimination of frictional resistance between the sample and the inside of the sampler tube. In one type of sampler, developed by Kjellman *et al.* (1950), this is achieved by superimposing thin strips of metal foil between the sample and the tube. The lower end of the sampler (Figure 6.8 (a)) has a sharp cutting edge above which the external diameter is enlarged to enable 16 rolls of foil to be housed in recesses within the wall of the sampler. The ends of the foil are attached to a piston, which fits loosely inside the sampler; the piston is supported on a cable which is fixed at the surface. Lengths of sample tube (68mm in diameter) are attached as required to the upper end of the sampler.

② 连续取土器。

As the sampler is pushed into the soil the foil unrolls and encases the sample, the piston being held at a constant level by means of the cable. As the sampler is withdrawn the lengths of tube are uncoupled and a cut is made, between adjacent tubes, through the foil and sample. Sample quality is generally Class 1 or 2.

Another type is the Delft continuous sampler, of either 29 or 66mm in diameter. The sample feeds into an impervious nylon stockinette sleeve[③]. The sleeved sample, in turn, is fed into a fluid-supported thin-walled plastic tube.

③ 不透水尼龙织物套筒。

Compressed air sampler

The compressed air sampler[④] (Figure 6.8 (b)) is used to obtain undisturbed samples of sand (generally Class 2) below the water table. The sample tube, usually 60mm in diameter, is attached to a sampler head having a relief valve which can be closed by a rubber diaphragm[⑤]. Attached to the sampler head is a hollow guide rod surmounted by a guide head. An outer tube, or bell, surrounds the sample tube, the bell being attached to a weight which slides on the guide rod. The boring rods fit loosely into a plain socket in the top of the guide head, the weight of the bell and

④ 压缩空气取土器。

⑤ 橡胶膜。

sampler being supported by means of a shackle which hooks over a peg in the lower length of boring rod; a light cable, leading to the surface, is fixed to the shackle. Compressed air, produced by a foot pump, is supplied through a tube leading to the guide head, the air passing down the hollow guide rod to the bell.

① 缆绳。
② 隔膜扣。
③ 活塞。
④ 取样管。
⑤ 隔膜。
⑥ 取样头。
⑦ 隔膜卷。
⑧ 气线。
⑨ 弹簧。
⑩ 钻孔杆。
⑪ 钉。
⑫ 导头。
⑬ 导杆。
⑭ 间距控制块。
⑮ 配重。
⑯ 钟形罩。
⑰ 连续取土器。
⑱ 压缩空气取土器。

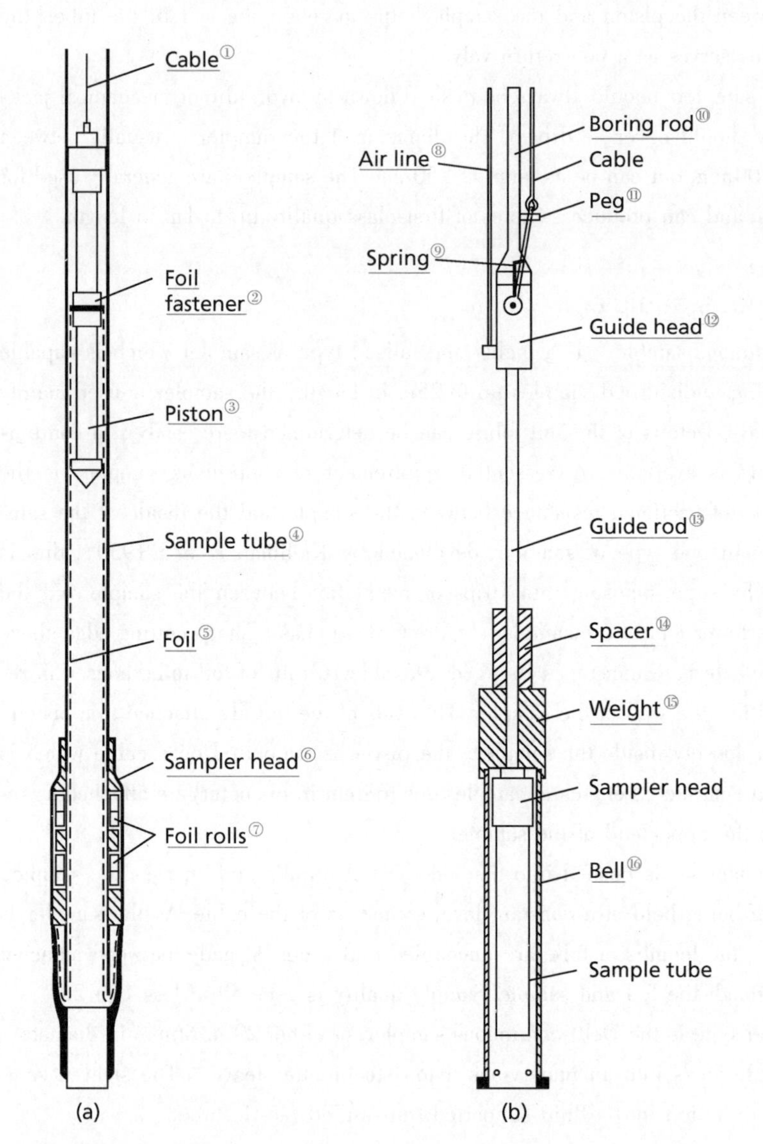

Figure 6.8 (a) Continuous sampler[17], (b) compressed air sampler[18].

The sampler is lowered on the boring rods to the bottom of the borehole, which will contain water below the level of the water table. When the sampler comes to rest at the bottom of the borehole the shackle springs off the peg, removing the connection between the sampler and the boring rods. The tube is pushed into the soil by means of the boring rods, a stop on the guide rod preventing overdriving; the boring rods are then withdrawn. Compressed air is now introduced to expel the water from the bell and to close the valve in the sampler head by pressing the diaphragm downwards. The tube is withdrawn into the bell by means of the cable, and then the tube and bell together are raised to the surface. The sand sample remains in the tube by virtue of arching and the slight negative pore water pressure in the soil.[①] A plug is placed at the bottom of the tube before the suction is released and the tube is removed from the sampler head.

① 砂样由于土拱效应和土中轻微的负孔隙水压力而保持在管中。

Window sampler

This sampler, which is most suited to dry fine grained soils, employs a series of tubes, usually 1m in length and of different diameters (typically 80, 60, 50 and 36mm). Tubes of the same diameter can be coupled together. A cutting shoe is attached to the end of the bottom tube. The tubes are driven into the soil by percussion using either a manual or rig-supported device, and are extracted either manually or by means of the rig. The tube of largest diameter is the first to be driven and extracted with its sample inside. A tube of lesser diameter is then driven below the bottom of the open hole left by extraction of the larger tube. The operation is repeated using tubes of successively lower diameter, and depths of up to 8m can be reached. There are longitudinal slots or 'windows' in the walls at one side of the tubes to allow the soil to be examined and enable disturbed samples of Class 3 or 4 to be taken.

6.4 Selection of laboratory test method (s)

The ultimate aim of careful sampling is to obtain the mechanical characteristics of soil for use in subsequent geotechnical analyses and design.[②] These basic characteristics and the laboratory tests for their determination have been described in detail in Chapters 1-5. Table 6.3 summarises the mechanical characteristics which can be obtained from each type of laboratory test discussed previously using undisturbed samples (i. e. Class 1/2). The table demonstrates why the triaxial test is so popular in soil mechanics, with modern computer-controlled machines being able to conduct various different test stages on a single sample of soil, thereby making best use of the material recovered from site.

② 认真取样的最终目的是获得土的力学特性，并用于岩土工程分析与设计。

Disturbed samples may be effectively used to support the tests described in Table 6.3 by determining the index properties (w, w_L, w_P, I_P and I_L for fine-grained soils; e_{max}, e_{min} and e for coarse-grained soils) and using the empirical correlations[③] presented in Section 5.9 (i. e. Equations 5.43-5.47 inclusive).

③ 经验公式。

6.5 Borehole logs

After an investigation has been completed and the results of any laboratory tests are available, the ground conditions discovered in each borehole (or trial pit) are summarised in the form of a borehole (or trial pit) log. An example of such a log appears in Table 6.4, but details of the layout can vary. The last few columns are originally left without headings to allow for variations in the data presented. The method of investigation and details of the equipment used should be stated on each log. The location, ground level and diameter of the hole should be specified, together with details of any casing used. The names of the client and the project should be stated.

The log should enable a rapid appraisal of the soil profile[1] to be made. The log is prepared with reference to a vertical depth scale. A detailed description of each stratum is given, and the levels of strata boundaries clearly shown; the level at which boring was terminated should also be indicated. The different soil (and rock) types are represented by means of a legend using standard symbols. The depths (or ranges of depth) at which samples were taken or at which in-situ tests were performed are recorded; the type of sample is also specified. The results of certain laboratory or in-situ tests may be given in the log-Table 6.4 shows N values which are the result of Standard Penetration tests (SPT), which are described in Chapter 7. The depths at which groundwater was encountered and subsequent changes in levels, with times, should be detailed.

① 土层信息。
② 固结仪。
③ 剪切盒。
④ 三轴仪。
⑤ 渗透仪。
⑥ 粒径分布。
⑦ 压缩特性。
⑧ 刚度特性。
⑨ 排水强度。
⑩ 不排水强度。
⑪ 渗透系数。

Table 6.3 Derivation of key soil properties from undisturbed samples tested in the laboratory

Parameter	*Oedometer*[2] *(Chapter 4)*	*Shear box*[3] *(Chapter 5)*	*Triaxial cell*[4] *(Chapter 5)*	*Permeameter*[5] *(Chapter 2)*	*Particle size distribution*[6] *(PSD) (Chapter 1)*
Consolidation characteristics[7]: m_v, C_c	YES	C_c from λ			
Stiffness properties[8]: G, G_0		YES			
Drained strength properties[9]: ϕ', c'		YES	CD/CU tests		
Undrained strength properties[10]: c_u (in-situ)			UU tests		
Permeability[11]: k			FH*-sands & gravels CH*-finer soils	FH-sands & gravels CH-finer soils	sands & gravels (Eq. 2.4)

Notes: * FH = Falling head test; CH = Constant head test. These tests can be conducted in a modern stress-path cell by controlling the back pressure and maintaining a zero lateral strain condition.

Table 6.4 Sample borehole log[1]

BOREHOLE LOG

Location: Barnhill | Borehole No. 1
Client: RFG Consultants | Sheet 1 of 1
Boring method: Shell and auger to 14.4m | Ground level: 36.30
Rotary core drilling to 17.8m | Date: 30.7.77
Diameter: 150mm | Scale: 1:100
NX
Casing: 150mm to 5m

Description of strata	Level	Legend	Depth	Samples	N	C_u (kN/m^2)	
TOPSOIL[2]	35.6		0.7				
Loose, light brown SAND[3]				D	6		
	33.7		2.6				
Medium dense, brown gravelly SAND[4] ▽	32.5			D	15		
	31.9		4.4				
Firm, yellowish-brown, closely fissured CLAY[5]				U		80	
				U		86	
				U		97	
				U		105	
	24.1		12.2				
Very dense, red, silty SAND with decomposed SANDSTONE[6]				D	50 for 210mm		
	21.9		14.4				
Red, medium-grained, granular, fresh SANDSTONE,[7] moderately weak, thickly bedded							
	18.5		17.8				

U: Undisturbed sample
D: Disturbed sample
B: Bulk disturbed sample
W: Water sample
▽: Water table

REMARKS

Water level (0930h)	
29.7.77	32.2m
30.7.77	32.5m
31.7.77	32.5m

① 钻孔记录。
② 表层土。
③ 松散，淡棕色，砂。
④ 中密，棕色，砾砂。
⑤ 密实，黄棕色，多裂缝黏土。
⑥ 极密的，含有风化砂岩的红色粉砂。
⑦ 新鲜砂岩。

The soil description should be based on particle size distribution and plasticity, generally using the rapid procedure in which these characteristics are assessed by means of visual inspection and feel; disturbed samples are generally used for this purpose. The description should include details of soil colour, particle shape and composition; if possible, the geological formation and type of deposit should be given. The structural characteristics of the soil mass should also be described, but this requires an examination of undisturbed samples or of the soil in-situ (e. g. in a trial pit). Details should be given of the presence and spacing of bedding features, fissures and other relevant characteristics. The density index of sands and the consistency of clays (Table 1.3) should be indicated.

6.6 Cone penetration testing (CPT)

① 圆锥静力触探仪。

② 深基础。

③ 压入机。

④ 顶杆。

⑤ 锥尖阻力。

⑥ 侧摩阻力。

The **Cone Penetrometer**① is one of the most versatile tools available for soil exploration (Lunne *et al.*, 1997). In this section, the use of the CPT to identify stratigraphy and the materials which are present in the ground will be presented. However, the technique can also be used to determine a wide range of standard geotechnical parameters for these materials instead of or in addition to the laboratory tests summarised in Section 6.4. This will be further discussed in Chapter 7. Furthermore, because of the close analogy between a CPT and a pile under vertical loading, CPT data may also be used directly in the design of deep foundations② (Chapter 9).

The penetrometer consists of a short cylindrical element, at the end of which is a cone-shaped tip. The cone has an apex angle of 60° and a cross-sectional area of $1000mm^2$. This is pushed vertically into the ground using a **thrust machine**③ at a constant rate of penetration of 20mm/s (ISO, 2006). For onshore applications, the thrust machine is commonly a CPT truck which provides a reaction mass due to its self-weight (typically 15-20 tonnes). A 20-tonne (200-kN) thrust will normally permit penetration to around 30m in dense sands or stiff clays (Lunne *et al.*, 1997). For deeper investigations where higher resistances to penetration will be encountered, this may be further ballasted or temporarily anchored to the ground (Figure 6.9). As the instrument penetrates, additional **push rods**④ of the same diameter as the instrument are attached to extend the string. Cables passing up through the centre of the push rods carry data from instruments within the penetrometer to the surface. In a standard electrical cone (CPT), a load cell between the cone and the body of the instrument continuously records the resistance to penetration of the cone (cone tip resistance⑤ q_c), and a friction sleeve is used to measure the interface shearing resistance⑥ (f_s) along the cylindrical body of the instrument. Different types of soil will exhibit different proportions of sleeve friction to end resistance: for example, gravels generally have low f_s and high q_c, while clays have high f_s and low q_c. By examining an extensive database of CPT test data, Robertson (1990) proposed a chart which may be used for identifying soil types based on normalised versions of these parameters (Q_t and F_r), which is shown in Figure 6.10.

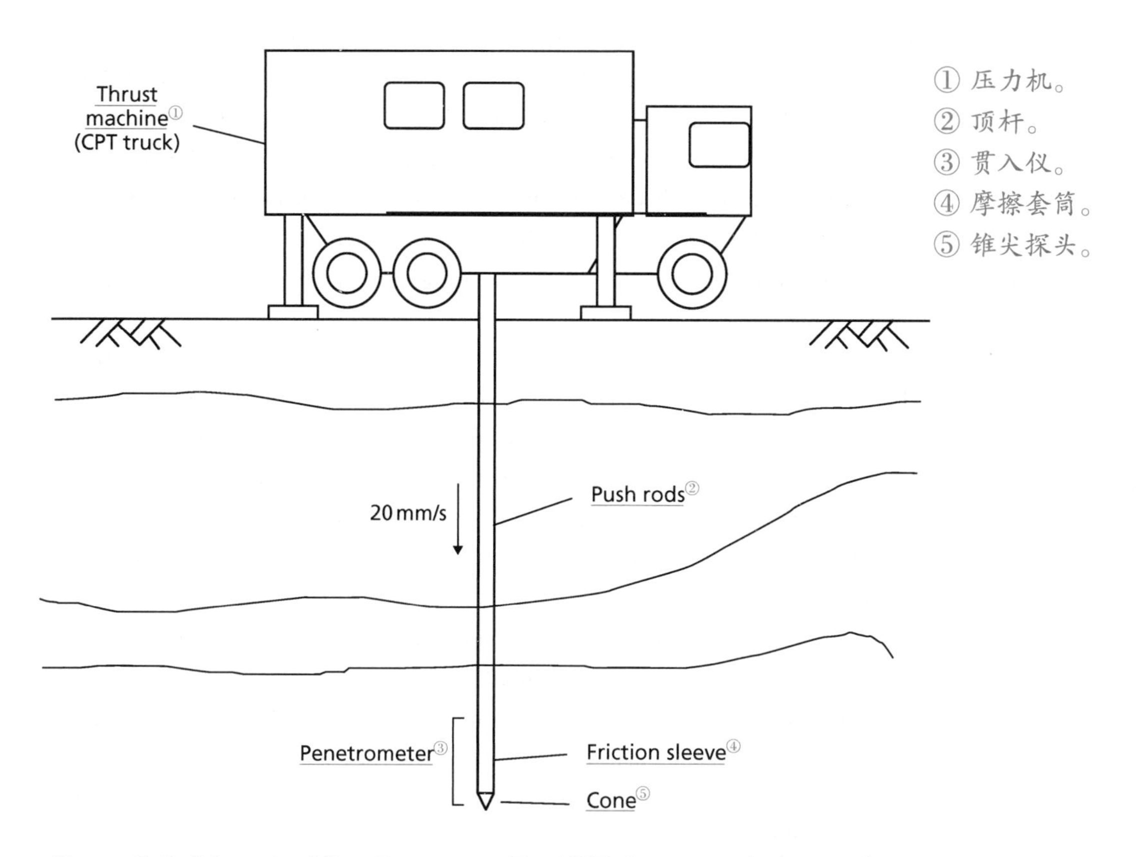

① 压力机。
② 顶杆。
③ 贯入仪。
④ 摩擦套筒。
⑤ 锥尖探头。

Figure 6.9 Schematic of Cone Penetrometer Test (CPT) showing standard terminology.

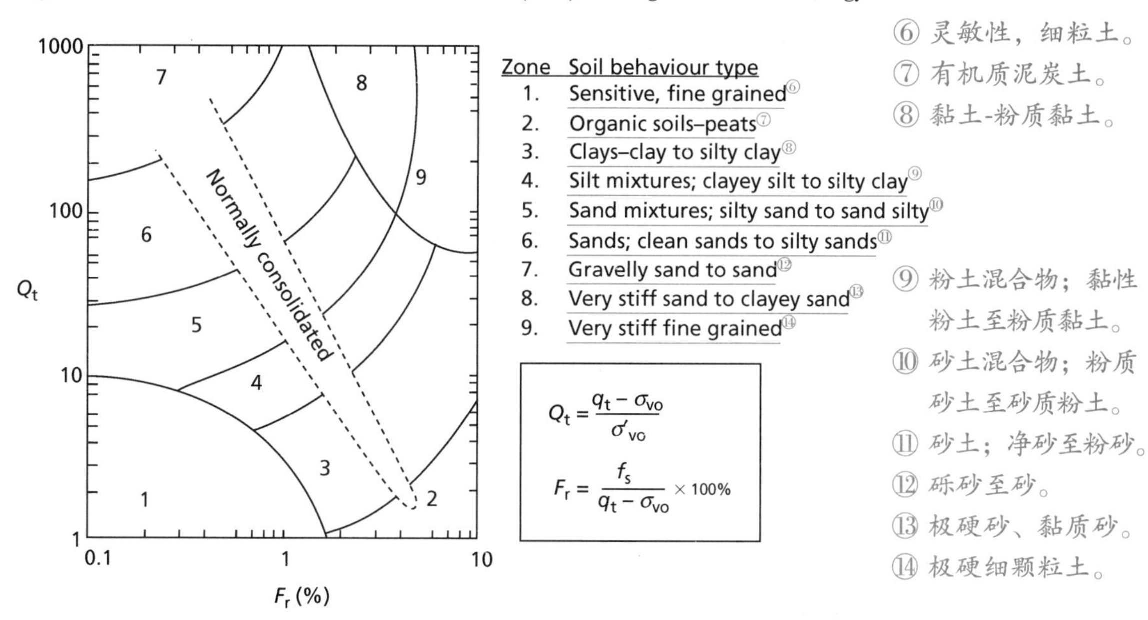

⑥ 灵敏性，细粒土。
⑦ 有机质泥炭土。
⑧ 黏土-粉质黏土。
⑨ 粉土混合物；黏性粉土至粉质黏土。
⑩ 砂土混合物；粉质砂土至砂质粉土。
⑪ 砂土；净砂至粉砂。
⑫ 砾砂至砂。
⑬ 极硬砂、黏质砂。
⑭ 极硬细颗粒土。

Figure 6.10 Soil behaviour type classification chart based on normalised CPT data (reproduced after Robertson, 1990).

For a standard CPT cone, q_t is approximated by q_c. During penetration, however, excess pore water pressures around the cone will increase, particularly in fine-grained soils, which artificially reduce q_c. More sophisticated **piezocones**① (CPTU) include localised measurement of the excess pore water pressures around the cone which are induced by penetration. These are most commonly measured immediately behind the cone (u_2), though measurements may also be made on the cone itself (u_1) and/or at the other end of the friction sleeve (u_3), as shown in Figure 6.11. When such measurements are made, the corrected cone resistance② q_t is determined using

① 孔压静力触探仪。

② 修正锥头阻力。

$$q_t = q_c + u_2 (1 - a) \tag{6.2}$$

The parameter a is an area correction factor depending on the penetrometer, and typically varies between 0.5-0.9. As before, different soils will experience different changes in pore water pressure during penetration: coarse-grained soils (sands and gravels) will exhibit little excess pore water pressure generation due to their high permeability, while fine-grained soils of lower permeability typically exhibit larger values of u_2.

If the soil is heavily overconsolidated, u_2 may be negative. The pore water pressure measurement u_2 therefore provides a third continuous parameter which may be used to identify soil types, using the chart shown in Figure 6.12 (Robertson, 1990). If CPTU data are available, both Figures 6.10 and 6.12 should be used to determine soil type. In some instances, the two charts may give different interpretations of the ground conditions. In these cases, judgement is required to correctly identify the ground conditions. CPT soundings may also struggle to identify inter-bedded soil layers③ (Lunne *et al.*, 1997).

③ 夹层土。

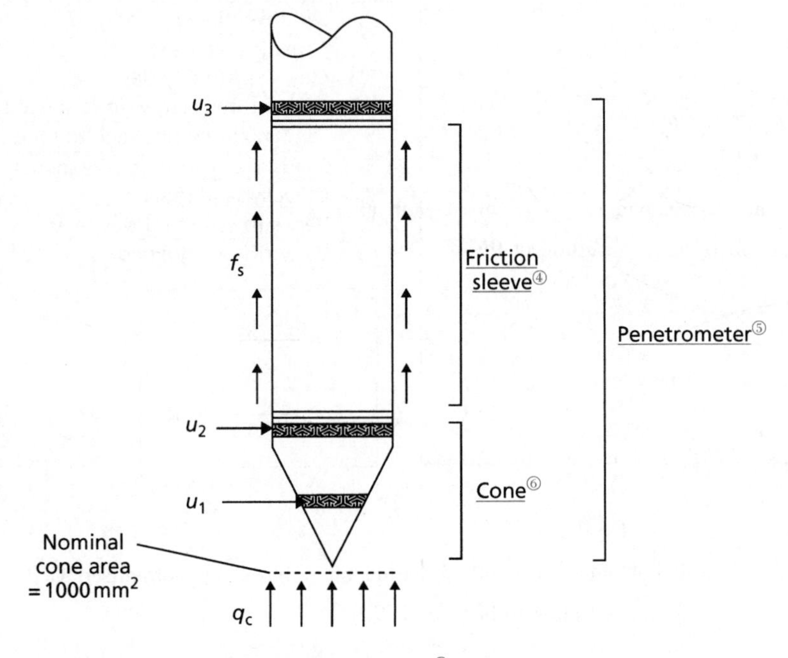

④ 摩擦套管。
⑤ 贯入仪。
⑥ 触探圆锥。
⑦ 带孔压静力触探。

Figure 6.11 Schematic of piezocone⑦ (CPTU).

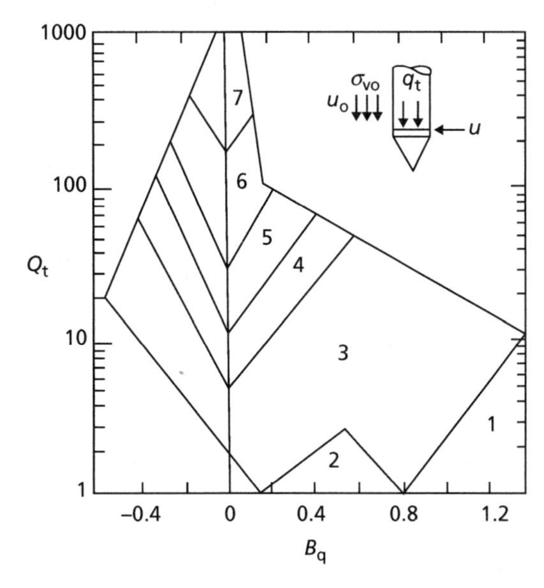

Figure 6.12 Soil behaviour type classification chart based on normalised CPTU data.

As data are recorded continuously with increasing depth during a CPT sounding, the technique can be used to produce a ground profile showing soil stratigraphy and classification, similar to a borehole log. An example of the use of the foregoing identification charts is shown in Figure 6.13. The CPT test is quick, relatively cheap, and has the advantage of not leaving a large void in the ground as in the case of a borehole or trial pit.[①] However, due to the difficulties in interpretation which have previously been discussed, CPT data are most effective at 'filling in the gaps' between widely spaced boreholes. Under these conditions, use of the CPT (U) soil identification charts can be informed by the observations from the boreholes. The CPT (U) data will then provide useful information as to how the levels of different soil strata vary across a site, and may identify localised hard inclusions[②] or voids which may have been missed by the boreholes.

① 圆锥触探试验具有快速且相对便宜的优点，并且可以避免如钻孔或探坑法在场地上留下较大的孔洞。

② 局部硬夹层。

Given the large amount of data that is generated from a continuous CPT sounding, and that soil behaviour types (zones 1-9 in Figures 6.10 and 6.12) fall within ranges defined by the magnitude of the measured parameters, interpretation of soil stratigraphy from CPT data benefits greatly from automation[③]/computerisation[④]. It may be seen from Figure 6.10 that the boundaries between zones 2-7 are close to being circular arcs with an origin around the top left corner of the plot. Robertson and Wride (1998) quantified the radius of these arcs by a parameter I_c:

③ 自动化。

④ 计算机化。

$$I_c = \sqrt{(3.47 - \log Q_t)^2 + (\log F_r + 1.22)^2} \tag{6.3}$$

where Q_t and F_r are the normalised tip resistance[⑤] and friction ratio[⑥] as defined in Figure 6.10. Figure 6.14 shows curves plotted using Equation 6.3 for values of I_c which most closely represent the soil boundaries in Figure 6.10. Using a single equation to define the soil behaviour type makes the processing of CPT data amenable to automated analysis using a spreadsheet, by applying Equation 6.3 to each

⑤ 归一化锥尖阻力。

⑥ 摩阻比。

① 孔压。
② 侧摩阻力。
③ 锥尖阻力。
④ 土层剖面。

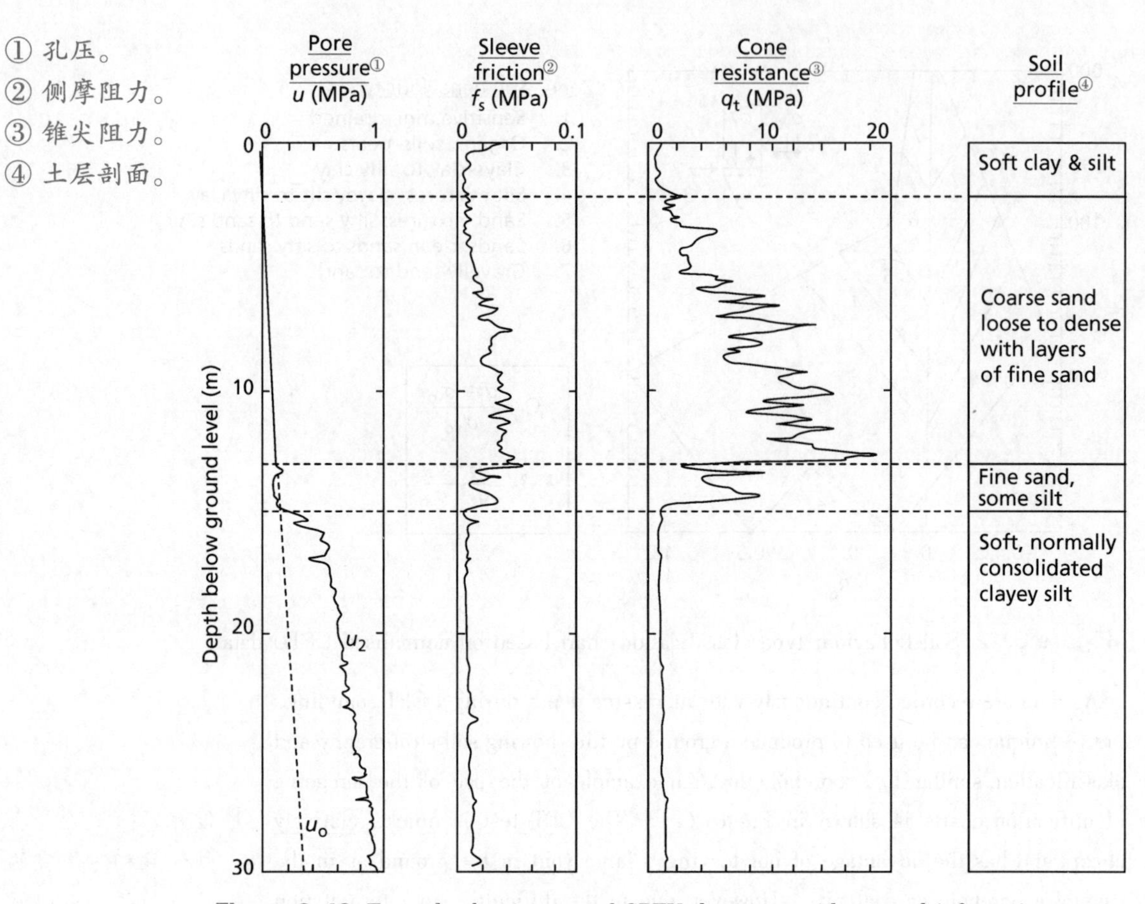

Figure 6.13 Example showing use of CPTU data to provide ground information.

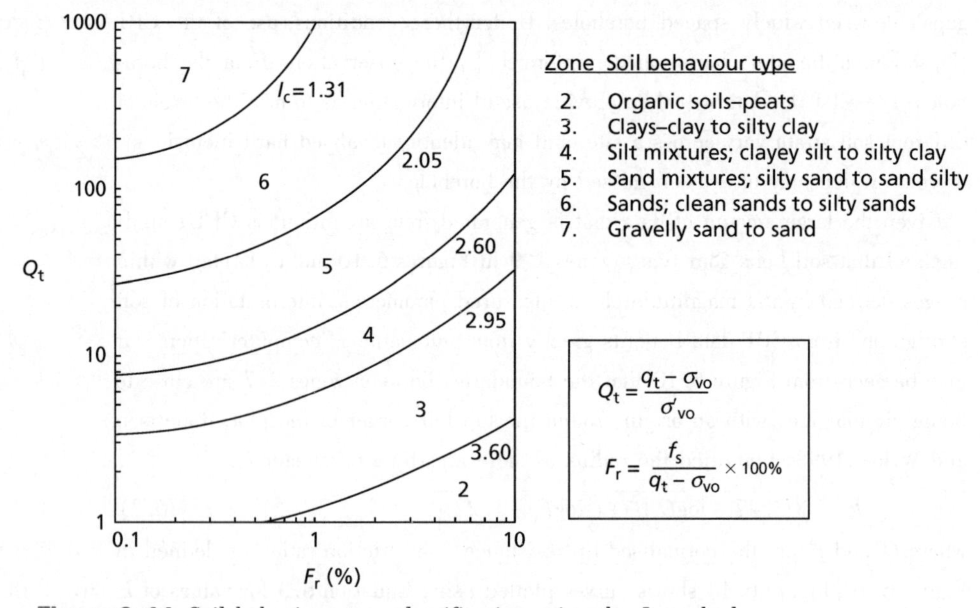

Figure 6.14 Soil behaviour type classification using the I_c method.

test datapoint. A spreadsheet tool, CPTic_CSM8. xls, which implements the I_c method may be found on the Companion Website. Care must be taken when using the method, as it will not correctly identify soil types 1, 8 and 9 (Figure 6.10); however, for most routine use, the I_c method provides a valuable tool for the interpretation of stratigraphic information from CPT soundings.

6.7 Geophysical methods

Under certain conditions geophysical methods may be useful in ground investigation, especially at the reconnaissance stage[①]. However, the methods are not suitable for all ground conditions and there are limitations to the information that can be obtained; thus they must be considered mainly as supplementary methods. It is possible to locate strata boundaries only if the physical properties of the adjacent materials are significantly different. It is always necessary to check the results against data obtained by direct methods such as boring or CPT soundings. Geophysical methods[②] can produce rapid and economic results, making them useful for the filling in of detail between widely spaced boreholes or to indicate where additional boreholes may be required. The methods can also be useful in estimating the depth to bedrock or to the water table, or for locating buried metallic objects (e.g. unexploded ordnance) and voids. They can be particularly useful in investigating sensitive contaminated sites, as the methods are non-intrusive[③] - unlike boring or CPT. There are several geophysical techniques, based on different physical principles. Three of these techniques are described below.

① 初勘阶段。

② 地球物理方法。

③ 非侵入式。

Seismic refraction

The seismic refraction[④] method depends on the fact that seismic waves have different velocities in different types of soil (or rock), as shown in Table 6.5; in addition, the waves are refracted when they cross the boundary between different types of soil. The method enables the general soil types and the approximate depths to strata boundaries, or to bedrock, to be determined.

④ 地震波折射法。

Table 6.5 Shear wave velocities[⑤] of common geotechnical materials (after Borcherdt, 1994)

Soil/rock type	*Shear wave velocity, V_s (m/s)*
Hard rocks (e.g. metamorphic)[⑥]	1400 +
Firm to hard rocks (e.g. igneous, conglomerates, competent sedimentary)[⑦]	700-1400
Gravelly soils and soft rocks (e.g. sandstone, shale, soils with >20% gravel)[⑧]	375-700
Stiff clays and sandy soils[⑨]	200-375
Soft soils (e.g. loose submerged fills and soft clays)[⑩]	100-200
Very soft soils (e.g. marshland, reclaimed soil)[⑪]	50-100

⑤ 剪切波速。

⑥ 硬质岩（如变质岩）。

⑦ 较硬岩至硬质岩。

⑧ 碎石土与软质岩（如砂岩、页岩、含20%以上碎石的土）。

⑨ 硬黏土与砂土。

⑩ 软土（如松散水下填土和软黏土）。

⑪ 极软土（如沼泽地、围垦土）。

Waves are generated either by the detonation of explosives or by striking a metal plate with a large hammer. The equipment consists of one or more sensitive vibration transducers, called geophones, and an extremely accurate time-measuring device called a seismograph[①]. A circuit between the detonator[②] or hammer and the seismograph starts the timing mechanism at the instant of detonation or impact. The geophone is also connected electrically to the seismograph: when the first wave reaches the geophone[③] the timing mechanism stops and the time interval is recorded in milliseconds.

① 地震仪。
② 雷管。
③ 地震检波器。

When detonation or impact takes place, waves are emitted in every direction. One particular wave, called the direct (or surface) wave, will travel parallel to the surface in the direction of the geophone. Other waves travel in a downward direction, at various angles to the horizontal, and will be refracted if they pass into a stratum of different **shear wave** (or seismic) **velocity** (V_s). [④] If the shear wave velocity of the lower stratum is higher than that of the upper stratum, one particular wave will travel along the top of the lower stratum, parallel to the boundary, as shown in Figure 6.15 (a): this wave continually 'leaks' energy back to the surface. Energy from this refracted wave can be detected by geophones at the surface.

④ 剪切波速。

The test procedure consists of installing either a single geophone in turn at a number of points in a straight line, at increasing distances from the source of wave generation, or, alternatively, using a single geophone position and producing a series of vibration sources at increasing distances from the geophone (as shown in Figure 6.15 (a)). The length of the line of points should be three to five times the required depth of investigation, and the spacing between measurement/shot points is approximately 3m. For each shot (detonation or impact), the arrival time of the first wave at the geophone position is recorded. When the distance between the source and the geophone is short, the arrival time will be that of the direct wave. When the distance between the source and the geophone exceeds a certain value (depending on the thickness of the upper stratum), the refracted wave[⑤] will be the first to be detected by the geophone. This is because the path of the refracted wave, although longer than that of the direct wave, is partly through a stratum of higher shear wave velocity. The use of explosives is generally necessary if the source-geophone distance exceeds 30-50m or if the upper soil stratum is loose.

⑤ 折射波。

Arrival time[⑥] is plotted against the distance between the source and the geophone, a typical plot being shown in Figure 6.15 (b). If the source-geophone spacing is less than d_1, the direct wave reaches the geophone in advance of the refracted wave and the time-distance relationship is represented by a straight line through the origin. On the other hand, if the source-geophone distance is greater than d_1 but less than d_2, the refracted wave arrives in advance of the direct wave and the time-distance relationship is represented by a straight line at a different slope. For still larger spacings, a third straight line may be observed representing the third layer, and so on. The slopes of the lines which are read from the graph are the reciprocals[⑦] of the shear wave velocities (V_{s1}, V_{s2} and V_{s3}) of the upper, middle

⑥ 到达时间。
⑦ 倒数。

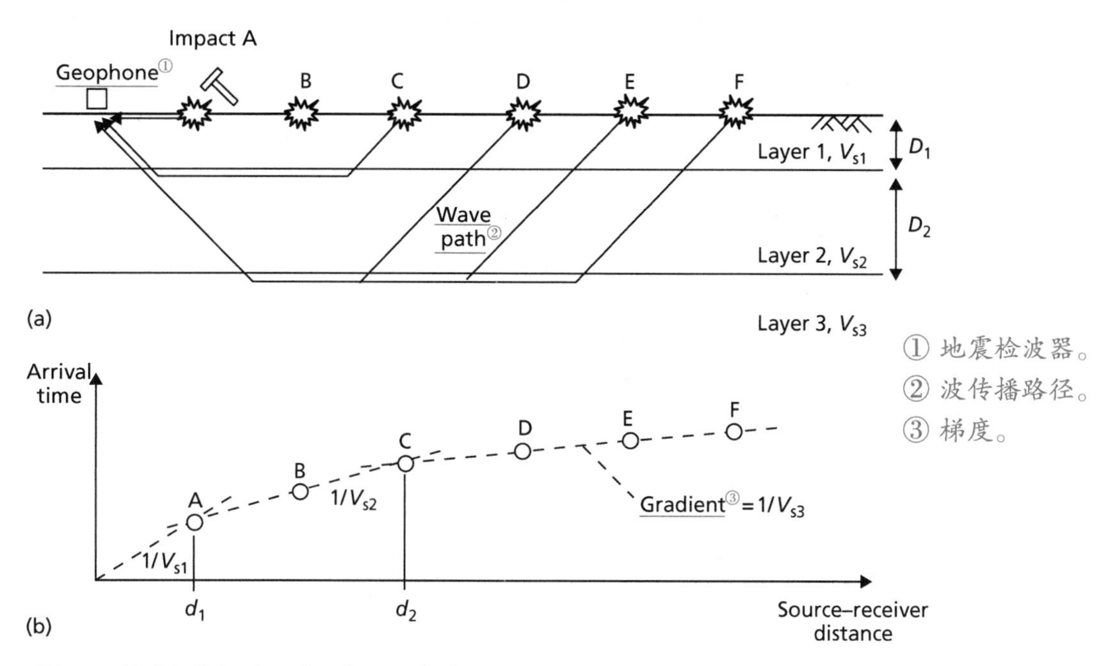

Figure 6.15 Seismic refraction method.

① 地震检波器。
② 波传播路径。
③ 梯度。

and lower strata, respectively. The general types of soil or rock can be determined from a knowledge of these velocities, for example by using Table 6.5. The depths (D_1 and D_2) of the boundaries between the soil strata (provided the thickness of the strata is constant) can be estimated from the formulae

$$D_1 = \frac{d_1}{2}\sqrt{\left(\frac{V_{s2} - V_{s1}}{V_{s2} + V_{s1}}\right)} \tag{6.4}$$

$$D_2 = 0.8D_1 + \frac{d_2}{2}\sqrt{\left(\frac{V_{s3} - V_{s2}}{V_{s3} + V_{s2}}\right)} \tag{6.5}$$

The method can also be used where there are more than three strata and procedures exist for the identification of inclined strata boundaries and vertical discontinuities.

The formulae used to estimate the depths of strata boundaries are based on the assumptions that each stratum is homogeneous④ and isotropic⑤, the boundaries are plane, each stratum is thick enough to produce a change in slope on the time-distance plot, and the shear wave velocity increases in each successive stratum from the surface downwards. Softer, lower-velocity material (e.g. soft clay) can therefore be masked by overlying stronger, higher-velocity material (e.g. gravel). Other difficulties arise if the velocity ranges of adjacent strata overlap, making it difficult to distinguish between them, and if the velocity increases with depth in a particular stratum. It is important that the results are correlated with data from borings.

④ 均质的。
⑤ 各向同性的。

Knowledge of V_s can also be used to directly determine the stiffness of the soil.⑥ As shear waves induce only very small strains in the materials as they pass, the material behaviour is elastic. In an elastic medium, the small strain shear modulus⑦ (G_0) is related to shear wave velocity by:

$$G_0 = \rho V_s^2 \tag{6.6}$$

⑥ 剪切波速 V_s 可直接用于确定土体刚度。
⑦ 小应变剪切模量。

Once G_0 is known, the full non linear G-γ relationship may be inferred using normalised relationships such as Equation 5.10 (Section 5.2).

① 表面波频谱分析法。

Spectral analysis of surface waves (SASW)[①]

This is a relatively new method which in a practical sense is very similar to seismic refraction; indeed, the test procedure is essentially identical. However, instead of measuring arrival times, a detailed analysis of the frequency content of the signal received at the geophone is conducted, and only the lower frequency surface wave is considered. The analysis is automated using a spectrum analyser to produce a dispersion curve, and a computer is used to develop a ground model (a series of layers of different shear wave velocity) which would lead to the measured signal, a process known as **inversion**[②]. In the SASW procedure only a single geophone is used, which can make it unreliable in urban areas where there is substantial low-frequency background noise which is difficult to distinguish from the signal of interest. A modified technique which overcomes these difficulties uses a set of geophones in a linear array. This is known as **multi-channel analysis of surface waves**[③](MASW), and has the additional advantage that analysis can be reliably automated.

② 反演。

③ 多道表面波分析法。

Both techniques are quick to conduct, and by conducting tests in different locations a two-dimensional map of the ground layering (based on shear wave velocities) can be determined- a process known as tomography[④]. The main advantages of MASW over the refraction method is that the test can infer layers of low shear wave velocity beneath material of high shear wave velocity, which would be masked in the refraction method. As with the previous method, the MASW data should always be correlated with data from borings. It may similarly be used to determine G_0 from V_s using Equation 6.6.

④ 断层成像法。

⑤ 电阻率法。

Electrical resistivity[⑤]

This method depends on differences in the electrical resistance of different soil (and rock) types. The flow of current through a soil is mainly due to electrolytic action, and therefore depends on the concentration of dissolved salts in the pore water: the mineral particles of a soil are poor conductors of current[⑥]. The resistivity of a soil therefore decreases as both the water content and the concentration of salts[⑦] increase. A dense, clean sand above the water table, for example, would exhibit a high resistivity due to its low degree of saturation and the virtual absence of dissolved salts. A saturated clay of high void ratio, on the other hand, would exhibit a low resistivity due to the relative abundance of pore water and the free ions in that water.

⑥ 电流的不良导体。

⑦ 盐浓度。

In its usual form (Figure 6.16 (a)), the method involves driving four electrodes[⑧] into the ground at equal distances (L) apart in a straight line. Current (I), from a battery, flows through the soil between the two outer electrodes, producing an electric field within the soil. The potential drop[⑨](E) is then measured between the

⑧ 电极。

⑨ 势能差。

two inner electrodes. The apparent resistivity[①] (R_Ω) is given by the equation

① 表观电阻。

$$R_\Omega = \frac{2\pi LE}{I} \tag{6.7}$$

The apparent resistivity represents a weighted average of true resistivity in a large volume of soil, the soil close to the surface being more heavily weighted than the soil at depth. The presence of a stratum of soil of high resistivity lying below a stratum of low resistivity forces the current to flow closer to the surface, resulting in a higher voltage drop[②] and hence a higher value of apparent resistivity. The opposite is true if a stratum of low resistivity lies below a stratum of high resistivity.

② 电压降。

When the variation of resistivity with depth is required, the following method can be used to make rough estimates of the types and depths of strata. A series of readings are taken, the (equal) spacing of the electrodes being increased for each successive reading; however, the centre of the four electrodes remains at a fixed point. As the spacing is increased, the apparent resistivity is influenced by a greater depth of soil. If the resistivity increases with increasing electrode spacing, it can be concluded that an underlying stratum of higher resistivity is beginning to influence the readings. If increased separation produces decreasing resistivity, on the other hand, a stratum of lower resistivity is beginning to influence the readings. The greater the thickness of a layer, the greater the electrode spacing over which its influence will be observed, and vice versa.

Apparent resistivity is plotted against electrode spacing, preferably on log- log paper[③]. Characteristic curves for a two-layer structure are illustrated in Figure 6.16(b). For curve A, the resistivity of layer 1 is lower than that of layer 2; for curve B, layer 1 has a higher resistivity than layer 2. The curves become asymptotic to lines representing the true resistivities R_1 and R_2 of the respective layers. Approximate layer thicknesses can be obtained by comparing the observed curve of resistivity versus electrode spacing with a set of standard curves. Other methods of interpretation have also been developed for two-layer and three-layer systems.

③ 双对数坐标纸。

The procedure known as profiling is used in the investigation of lateral variation of soil types. A series of readings is taken, the four electrodes being moved laterally as a unit for each successive reading; the electrode spacing remains constant for each reading. Apparent resistivity is plotted against the centre position of the four electrodes, to natural scales; such plots can be used to locate the positions of soil of high or low resistivity. Contours of resistivity[④] can be plotted over a given area.

④ 电阻率等值线图。

The apparent resistivity for a particular soil or rock type can vary over a wide range of values, as shown in Table 6.6; in addition, overlap occurs between the ranges for different types. This makes the identification of soil or rock type and the location of strata boundaries extremely uncertain. The presence of irregular features near the surface and of stray potentials can also cause difficulties in interpretation. It is essential, therefore, that the results obtained are correlated with borehole data. The method is not considered to be as reliable as the seismic methods described previously.

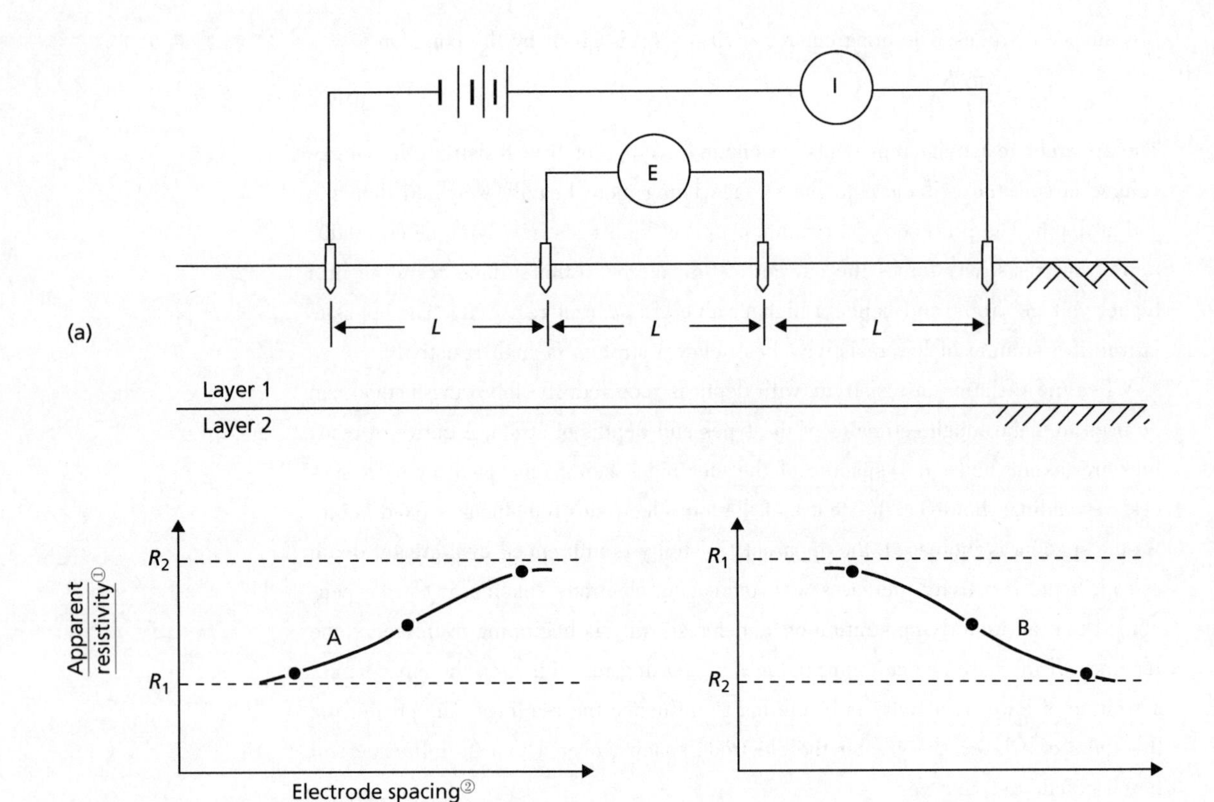

Figure 6.16 (a) Electrical resistivity method[3], (b) identification of soil layers by sounding[4].

① 表观电阻。
② 电极间隙。
③ 电阻率法。
④ 声波法。
⑤ 基岩（大体积）。
⑥ 基岩（破碎），砂和碎石干沉积物。
⑦ 泥岩。
⑧ 砂土（饱和）。
⑨ 黏土/粉土（饱和）。
⑩ 冰碛土。
⑪ 垃圾渗滤液。
⑫ 净水。

Table 6.6 Typical resistivities of common geotechnical materials (collated after Campanella and Weemes, 1990; McDowell *et al.*, 2002; Hunt, 2005)

Material	*Resistivity, R_Ω (Ωm)*
Bedrock (massive)[5]	>2400
Bedrock (fractured), dry deposits of sand and gravel[6]	300-2400
Mudstone[7]	20-60
Sandy soils (saturated)[8]	15-300
Clayey/silty soils (saturated)[9]	1.5-15
Glacial till[10]	20-30
Landfill leachate[11]	0.5-10
Freshwater[12]	20-60
Sea water	0.18-0.24

⑬ 受污染场地。

6.8 Contaminated ground[13]

The scope of an investigation must be extended if it is known or suspected that the ground in question has been contaminated. In such cases the soil and groundwater may contain potentially harmful substances, such as organic or inorganic chemicals, fibrous materials such as asbestos, toxic or explosive gases, biological agents, and/or radioactive elements. The contaminant[①] may be in solid, liquid or gaseous form. Chemical contaminants may be adsorbed on the surfaces of fine soil particles. The presence of contamination influences all other aspects of ground investigation, and may have consequences for foundation design and the general suitability of the site for the project under consideration. Adequate precautions must be taken to ensure the safety from health hazards of all personnel working on the site and when dealing with samples. During the investigation, precautions must be taken to prevent the spread of contaminants by personnel, by surface or groundwater flow and by wind.

① 污染物。

At the outset, possible contamination may be predicted from information on previous uses of the site or adjacent areas, such as by certain types of industry, mining operations, or reported leakage of hazardous liquids on the surface or from underground pipelines. This information should be collected during the initial desk study. The visual presence of contaminants and the presence of odours give direct evidence of potential problems. Remote sensing[②] and geophysical techniques (e. g. infra-red photography[③] and electrical resistivity testing, respectively) can be useful in assessing possible contamination.

② 遥感。

③ 红外线照相法。

Soil and groundwater samples are normally obtained from shallow trial pits or boreholes, as contamination from incidents of pollution is often highest towards the ground surface. The depths at which samples are taken depend on the probable source of contamination, and details of the types and structures of the strata. Experience and judgement are thus required in formulating the sampling programme. Solid samples, which would normally be taken at depth intervals of 100-150mm, are obtained by means of stainless steel tools, which are easily cleaned and are not contaminated, or in driven steel tubes. Samples should be sealed in water-tight containers made of material that will not react with the sample. Care must be taken to avoid the loss of volatile contaminants[④] to the atmosphere. Groundwater samples may be required for chemical analysis to determine if they contain agents that would chemically attack steel and concrete which may be used in the subsurface works (e. g. sulphates, chlorides, magnesium and ammonium[⑤]), or substances which may be harmful to potential users of the site and the environment. It is important to ensure that samples are not contaminated or diluted. Samples can be taken directly from trial pits or by means of specially designed sampling probes; however, it is preferable to obtain samples from standpipe piezometers if these have been installed to measure the pore water pressure regime. In cases where groundwater contamination is present, care must be taken to ensure that boring does not lead to preferential flow paths being created within the ground which allow the contaminants to spread. A sample should be taken immediately after the water-bearing stratum is reached in boring, and subsequent samples should be taken over a suitable period to determine if properties are constant or variable. This may be used to im-

④ 挥发性污染物。

⑤ 硫酸盐，氯化物，镁和铵。

ply whether pollution or migration of the contaminants is ongoing. Gas samples can be obtained from tubes suspended in a perforated standpipe in a borehole, or from special probes. There are several types of receptacle suitable for collecting gases. Details of the sampling process should be based on the advice of specialist analysts who will undertake the testing programme and report on the results.

When designing a site investigation in contaminated ground, the locations of the investigation points should be related to the position of the contaminants. If the position of a polluting source is known, or can be inferred from a desk study (e. g. the position of a factory waste outfall or spoil heap), **targeted sampling**① should be undertaken to determine the spread of contaminant from the source. This typically involves radial lines of sampling out from the contaminant source. An initially wide spacing can be used initially to determine the extent of the contaminant, with subsequent investigation 'filling in the gaps' as necessary. Some **non-targeted sampling**② should always be additionally conducted, where the site is split into smaller areas within each of which a single investigation point is made. The aim of this type of sampling is to identify potentially unexpected contaminants within the ground. The spacing of ground investigation points for non-targeted sampling should typically be between 18 and 24m (BSI, 2001).

① 靶向抽样。

② 非靶向抽样。

Summary

1 Ground investigation using intrusive methods (trial pits, boreholes, CPT) or nonintrusive geophysical methods can be used to determine the location, extent and identity of soil deposits within the ground.

2 Boreholes and trial pits are essential for visually confirming the geotechnical materials within the ground, and for obtaining samples for laboratory testing (using the methods outlined in Chapters 2, 4 and 5). To be most effective, these measurements should ideally be used in combination with CPT and geophysical methods to minimise the chances of encountering unforeseen ground conditions once construction has started, which may significantly increase project duration and cost.

3 The quality (and often also the cost) of the sampling that will be required will depend on the geotechnical characteristics/properties that are required from laboratory tests on such samples.

References

Borcherdt, R. (1994) Estimates of site-dependent response spectra for design (methodology and justification), *Earthquake Spectra*, **10**, 617 – 653.

BSI (2001) *Investigation of Potentially Contaminated Sites – Code of Practice BS 10175: 2001*, British Standards Institution, London.

Campanella, R. G. and Weemes, I. A. (1990) Development and use of an electrical resistivity cone for groundwater contamination studies, in *Geotechnical Aspects of Contaminated Sites, 5th Annual Symposium of the Canadian Geotechnical Society, Vancouver*.

Chapman, T. J. P. (2008) The relevance of developer costs in geotechnical risk management, in *Proceedings of the 2nd BGA International Conference on Foundations, Dundee*.

EC7 – 2 (2007) *Eurocode 7: Geotechnical Design – Part 2: Ground Investigation and Testing, BS EN 1997 – 2: 2007*, British Standards Institution, London.

Hunt, R. E. (2005) *Geotechnical Engineering Investigation Handbook* (2nd edn), Taylor & Francis Group, Boca Raton, FL.

ISO (2006) *Geotechnical Investigation and Testing – Field Testing – Part 1: Electrical Cone and Piezocone Penetration Tests, ISO TC 182/SC 1*, International Standards Organisation, Geneva.

Kjellman, W., Kallstenius, T. and Wager, O. (1950) Soil sampler with metal foils, in *Proceedings of the Royal Swedish Geotechnical Institute*, No. 1.

Lunne, T., Robertson, P. K. and Powell, J. J. M. (1997) *Cone Penetration Testing in Geotechnical Practice*, E & FN Spon, London.

McDowell, P. W., Barker, R. D., Butcher, A. P., Culshaw, M. G., Jackson, P. D., McCann, D. M., Skipp, B. O., Matthews, S. L. and Arthur, J. C. R. (2002) *Geophysics in Engineering Investigations*, CIRIA Publication C562, CIRIA, London.

Robertson, P. K. (1990) Soil classification using the cone penetration test, *Canadian Geotechnical Journal*, **27** (1), 151 – 158.

Robertson, P. K. and Wride, C. E. (1998). Evaluating cyclic liquefaction potential using the cone penetration test, *Canadian Geotechnical Journal*, **35** (3): 442 – 459.

Further reading

BSI (2002) *Geotechnical Investigation and Testing – Identification and Classification of Soil. Identification and Description (AMD Corrigendum 14181& 16930) BS EN ISO 14688 – 1: 2002*, British Standards Institution, London.

Standard (in UK and Europe) by which soil is identified and described when in the field and for producing borehole logs/ground investigation reports. ASTM D2488 and D5434 are the equivalent standards used in the US and elsewhere.

BSI (2006) *Geotechnical Investigation and Testing – Sampling Methods and Groundwater Measurements. Technical Principles for Execution BS EN ISO 22475 – 1: 2006*, British Standards Institution, London.

Standard (in UK and Europe) describing the detailed procedures and practical requirements for collecting soil samples of high quality for laboratory testing. In the US and elsewhere, a range of standards cover a similar area, including ASTM D5730, D1452, D6151, D1587 and D6519.

Clayton, C. R. I., Matthews, M. C. and Simons, N. E. (1995) *Site Investigation* (2nd edn), Blackwell, London.

A comprehensive book on site investigation which contains much useful practical advice and greater detail than can be provided in this single chapter.

Lerner, D. N. and Walter, R. G. (eds) (1998) Contaminated land and groundwater: future directions, *Geological Society, London, Engineering Geology Special Publication*, **14**, 37 – 43.

This article provides a more comprehensive introduction to the additional ground investigation issues associated with contaminated land.

Rowe, P. W. (1972) The relevance of soil fabric to site investigation practice, *Géotechnique*, **22** (2), 195 – 300.

Presents 35 case studies demonstrating how the depositional/geological history of the soil deposits influences the selection of laboratory tests and the associated requirements of sampling techniques.

Chapter 7

In-situ testing[①]

① 原位测试。

Learning outcomes

After working through the material in this chapter, you should be able to:

1. Understand the rationale behind testing soils in-situ to obtain their constitutive properties, and appreciate the part it plays with laboratory testing and the use of empirical correlations in establishing a reliable ground model (Section 7.1);
2. Understand the principle of operation of four common in-situ testing devices, their applicability and the constitutive properties that can be reliably obtained from them (Sections 7.2-7.5);
3. Process the test data from these methods with the help of a computer and use this to derive key strength and stiffness properties (Sections 7.2-7.5).

7.1 Introduction

In Chapter 5, laboratory tests for determining the constitutive behaviour[②] of soil (strength and stiffness properties) were described. While such tests are invaluable in quantifying the mechanical behaviour of an element of soil, there remain a number of disadvantages. First, to obtain high quality data through triaxial testing, undisturbed samples must be obtained, which can be difficult and expensive in some deposits (e.g. sands and sensitive clays, see Chapter 6). Second, in deposits where there are significant features within the macro-fabric[③] (e.g. fissuring in stiff clays) the response of a small element of soil may not represent the behaviour of the complete soil mass, if the sample happens to be taken such that it does not contain any of these features. As a result of such limitations, in-situ testing methods have been developed which can overcome these limitations and provide a rapid assessment of key parameters which can be conducted during the ground investigation phase.

② 本构行为。

③ 宏观结构。

In this chapter the four principal in-situ testing techniques will be considered, namely:

① 标准贯入试验。
② 现场十字板试验。
③ 旁压试验。
④ 圆锥静力触探试验。

- the Standard Penetration Test (SPT); ①
- the Field Vane Test (FVT); ②
- the Pressuremeter Test (PMT); ③
- the Cone Penetration Test (CPT). ④

In each case the testing methodology will be briefly described, but the focus will be on the parameters which can be measured/estimated from each test and the theoretical/empirical models for achieving this, the range of application to different soils, interpretation of constitutive properties (e. g. ϕ', c_u, G), stress history (OCR) and stress state (K_0) from the test data, and limitations of the data collected. The worked examples in the main text and problems at the end of the chapter are all based on actual test data from real sites which have been collated from the literature.

⑤ 膨胀计试验。
⑥ 平板载荷试验。
⑦ 浅基础。
⑧ 最终施工方案。
⑨ 表面波/多道表面波。
⑩ 地震波折射。

The four techniques listed above are not the only in-situ testing techniques; a Dilatometer Test (DMT) ⑤, for example, is similar in principle of operation and in the properties which can be measured to the PMT, involving expanding a cavity within the soil to determine mechanical properties. This common test will not be discussed herein, but references are provided at the end of the chapter for further reading on this topic. Plate Loading Tests (PLT) ⑥ are also in common usage-these involve performing a load test on a small plate which is essentially a model shallow foundation⑦, and are most commonly used to derive soil data for foundation works due to the close similarity of the test procedure to the ultimate construction⑧. Soil parameters are then back-calculated using standard techniques for analysing shallow foundations, which are described in detail in Chapter 8. It should also be noted that geophysical methods for profiling which use seismic methods (e. g. SASW/MASW⑨, seismic refraction⑩) measure the shear wave velocity (V_s) in-situ from which G_0 can be determined and are therefore in-situ tests in their own right (these methods were previously described in Chapter 6).

⑪ 原位试验中获得的数据，应被视作室内试验与取样的补充，而不能替代它们。

The data collected from in-situ tests should always be considered as complementing rather than replacing sampling and laboratory testing. ⑪ Indeed, three of the tests that will be discussed (SPT, FVT, PMT) require the prior drilling of a borehole, so a single borehole may be used very efficiently to gain visual identification of materials from spoil (Section 6.4), disturbed samples for index testing and use of subsequent empirical correlations (Section 5.9), undisturbed samples for laboratory testing (Chapters 5 and 6) and in-situ measurements of soil properties (this chapter). These independent observations should be used to support each other in identifying and characterising the deposits of soil in the ground and producing a detailed and accurate ground model for subsequent geotechnical analyses (Part 2 of this book).

7.2 Standard Penetration Test (SPT)

The SPT is one of the oldest and most widely used in-situ tests worldwide. The technical standards governing its use are EN ISO 22476, Part 3 (UK and Europe) and ASTM D1586 (US). Its popularity is largely due to its low cost and simplicity, and the fact that testing may be conducted rapidly as a borehole is drilled. A borehole is first drilled (using casing where appropriate) to just above the test depth. A **split-barrel sampler**[①] (Figure 7.1) with a smaller diameter than the borehole is then attached to a string of rods and driven into the soil at the base of the borehole by a **drop hammer**[②] (a known mass falling under gravity from a known height). An initial seating drive to 150mm penetration is first performed to embed the sampler into the soil. This is followed by the test itself, in which the sampler is driven further into the soil by 300mm (this is usually marked off on the rod string at the surface). The number of blows of the hammer to achieve this penetration is recorded; this is the (uncorrected) **SPT blowcount, N**[③].

① 对开筒取土器。

② 落锤。

③ 标准贯入击数。

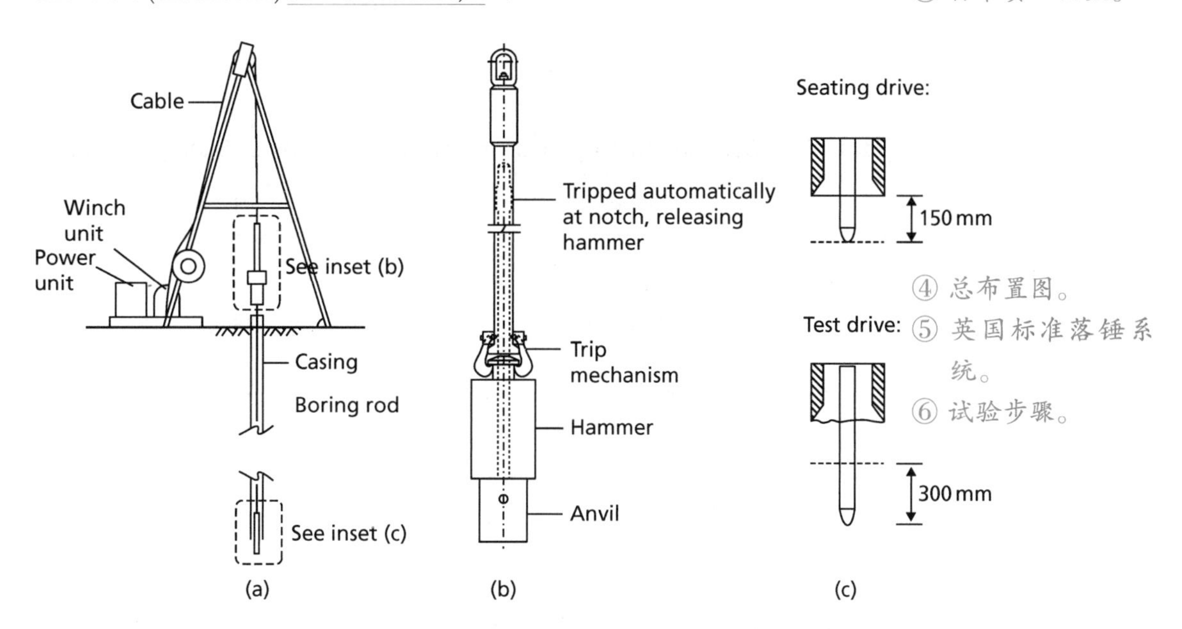

Figure 7.1 The SPT test: (a) general arrangement[④], (b) UK standard hammer system[⑤], (c) test procedure[⑥].

④ 总布置图。

⑤ 英国标准落锤系统。

⑥ 试验步骤。

A wide range of equipment is used worldwide to undertake testing which influences the amount of energy transferred to the sampler with each blow of the drop hammer. The constitutive properties of a given soil deposit should not vary with the equipment used, and so N is conventionally corrected to a value N_{60}, representing a standardised energy ratio of 60%. The blowcount also needs to be corrected for the size of the borehole and for tests done at shallow depths (< 10m). These corrections are achieved using

$$N_{60} = N\zeta\left(\frac{ER}{60}\right) \tag{7.1}$$

① 杆长修正系数。
② 能量比。

where ζ is the correction factor for rod length[①] (i. e. test depth) and borehole size from Table 7.1, and ER is the Energy Ratio[②] of the equipment used. BS EN ISO 22476 (2005) describes how ER may be measured for any SPT apparatus; for most purposes, however, it is sufficient to use the values given in Table 7.2 (after Skempton, 1986).

Table 7.1 SPT correction factor ζ (after Skempton, 1986)

Rod length/ depth (m)	*Borehole diameter (mm)*		
	65-115	*150*	*200*
3-4	0.75	0.79	0.86
4-6	0.85	0.89	0.98
6-10	0.95	1.00	1.09
>10	1.00	1.05	1.15

Table 7.2 Common Energy ratios in use worldwide (after Skempton, 1986)

Country	*ER (%)*
UK	60
USA	45-55
China	55-60
Japan	65-78

Interpretation of SPT data in coarse-grained soils (I_D, ϕ'_{max})

③ 上覆应力。
④ 归一化的标准贯入击数。
⑤ 上覆应力修正系数。

The SPT is most suited to the investigation of coarse-grained soils. In sands and gravels, the corrected blowcounts are further normalised to account for the overburden pressure[③] (σ'_{v0}) at the test depth, as the penetration resistance will naturally increase with stress level and this may mask smaller changes in constitutive properties. The **normalised blowcount**[④] $(N_1)_{60}$ is obtained from

$$(N_1)_{60} = C_N N_{60} \tag{7.2}$$

where C_N is the overburden correction factor[⑤] and is given by:

$$C_N = \frac{A}{B + \sigma'_{v0}} \tag{7.3}$$

⑥ 正常固结。

In Equation 7.3, σ'_{v0} should be entered in kPa, and A and B vary with density, coarseness and OCR. For normally consolidated[⑥] (NC) fine sands (D_{50} < 0.5mm) of medium relative density ($I_D \approx 40-60\%$), $A=200$ and $B=100$; for overconsolidated[①] (OC) fine sands, $A=170$ and $B=70$; for dense coarse sands ($D_{50}>0.5$mm, $I_D \approx 60\%-80\%$) $A=300$ and $B=200$ (Skempton, 1986). The differences between the three classes of soil are most pronounced[②] at lower values of σ'_{v0}(i. e. at shallower depths), as shown in Figure 7.2.

① 超固结。
② 显著的。

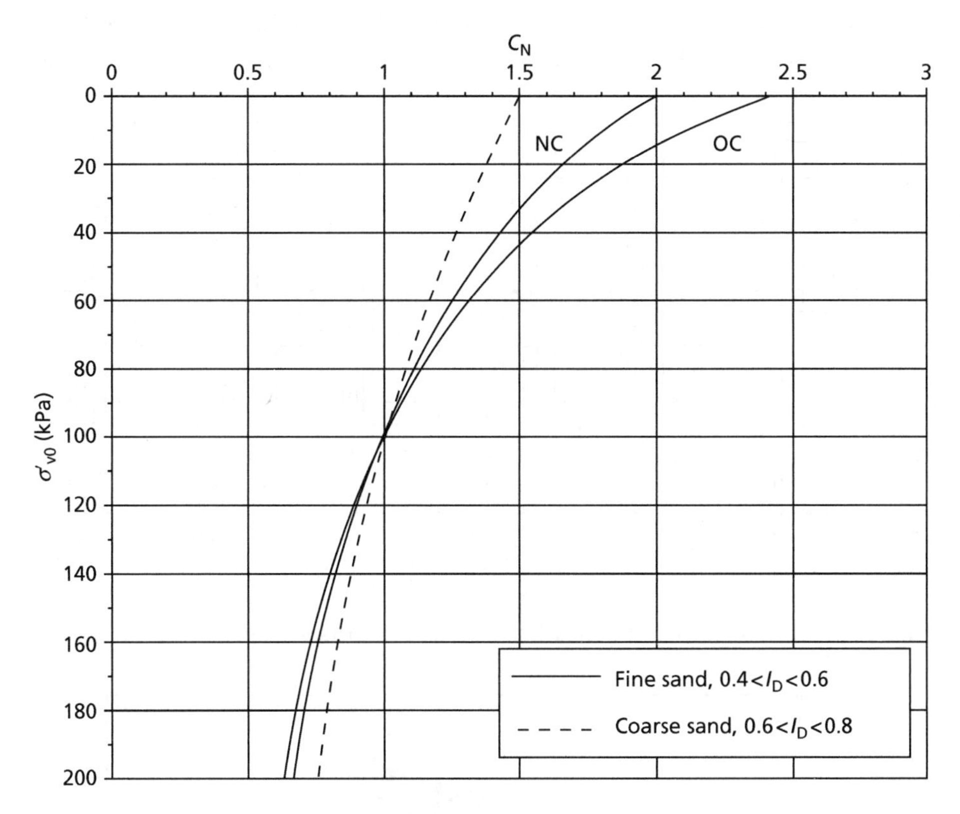

Figure 7.2 Overburden correction factors for coarse-grained soils (after Skempton, 1986).

The density of a coarse-grained deposit may then be found from $(N_1)_{60}$. Skempton (1986) proposed that for most natural deposits $(N_1)_{60}/I_D^2 \approx 60$ would provide a reasonable estimate of I_D[③] if nothing else is known about the deposit. As a cautionary note however, Skempton's own case studies showed $35<(N_1)_{60}/I_D^2<85$, so the SPT can only provide an estimate at best of the relative density. However as determination of a more precise density in coarse-grained soils requires estimation of the in-situ void ratio e and the use of Equation 1.23 (requiring high quality, usually frozen or resin-injected samples, which is expensive), SPT data are extremely valuable. Skempton further showed that for more recent depositions, $(N_1)_{60}/I_D^2$ reduces as shown in Figure 7.3. This is important when analysing test data from hydraulic fills, such as those used to build artificial islands[④] (e. g. The World and Palm developments in Dubai; Chek Lap Kok airport, Hong Kong), or when testing in river sediments.

③ 相对密实度。
④ 人工岛。

In Chapter 5 it was demonstrated that density (i. e. I_D) is one of the key parameters governing the peak strength of coarse- grained soil. As the normalised SPT blowcount[①] can be correlated against density, it follows that it may further be correlated to the peak angle of shearing resistance, ϕ'_{max}. This is also mechanically reasonable, as the SPT involves penetration of the soil (i. e. continuously exceeding its capacity) such that the soil resistance is expected to be governed by the peak strength. Figure 7.4 shows correlations for silica sands and gravels from Stroud

① 归一化的标准贯入击数。

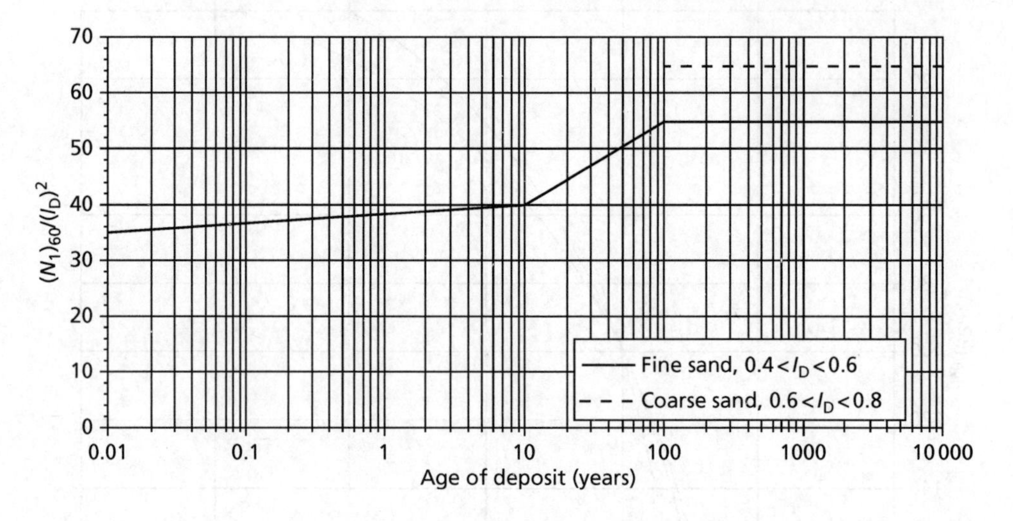

Figure 7.3 Effect of age on SPT data interpretation in coarse- grained soils.

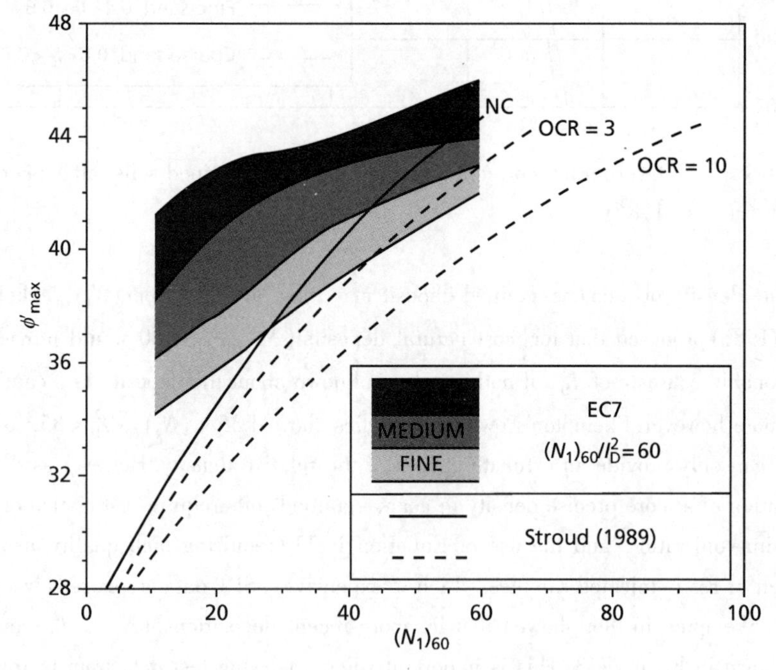

Figure 7.4 Determination of ϕ'_{max} from SPT data in coarse-grained soils.

(1989) and based on the suggestions from Eurocode 7: Part 2 (2007), assuming $(N_1)_{60}/I_D^2 = 60$. The former data shows the effect of overconsolidation on the interpretation of ϕ'_{max}, where increasing OCR is shown to reduce the peak friction angle①, as expected from Chapter 5. The ranges for the latter represent bounds on the uniformity of the soil from uniform (lower bound) to well-graded② (upper bound).

① 峰值摩擦角。
② 级配良好。

Interpretation of SPT data in fine-grained soils (c_u)

The SPT may, in principle, be used in fine-grained soils to determine an estimate of the in-situ undrained shear strength③ (the test is rapid so undrained conditions can be assumed). Correlations between c_u and blowcount depend on a number of factors in such soils, including OCR and any resultant fissuring, soil plasticity (I_P) and sensitivity (S_t). This high level of dependency means that SPT data should normally by used qualitatively in such soils, to support other in-situ and laboratory test data. However, if enough experience can be gained in a certain type of soil, reliable soil-specific correlations can be developed to provide quantitative data (EC7-2, 2007).

③ 不排水抗剪强度。

As an example, Stroud (1989) demonstrated that for overconsolidated UK clays, $c_u/N_{60} \approx 5$ for $I_P > 30\%$. For clays at lower plasticity index, this value increases to approximately $c_u/N_{60} = 7$ at $I_P = 15\%$. This grouping of clays has a number of similarities, generally being overconsolidated, fissured and insensitive. The scatter in the correlation is consequently low and the SPT can be used in such soils with some confidence, which explains its popularity in UK practice. It should be noted that the non-normalised blow-count④ (N_{60}) is used as c_u is a total stress parameter, and is therefore independent of the effective stress in the ground. Clayton (1995) showed, by using London Clay (which was part of Stroud's database), that if the fissuring is removed by remoulding such soils, $c_u/N_{60} \approx 11$ - i. e. in unfissured soils, a higher value of c_u/N_{60} should be used when interpreting SPT data in fine-grained soils. This is consistent with US practice, where $c_u/N_{60} = 10$ is routinely used (Terzaghi and Peck, 1967). In sensitive soils, Schmertman (1979) suggested that the sides of the SPT sampler, which contribute approximately 70% of the penetration resistance⑤ in clays, are generally governed by the remoulded strength⑥, while the base is influenced by the undisturbed undrained shear strength in the soil beneath the sampler. This suggests that in sensitive soils,

④ 未归一化的贯入击数。
⑤ 贯入阻力。
⑥ 重塑强度。

$$\frac{c_u}{N_{60}} = \frac{CS_t}{0.7 + 0.3S_t} \tag{7.4}$$

where C is the value of c_u/N_{60} for an insensitive clay ($S_t = 1$). To use Equation 7.4, the sensitivity may be estimated using Equation 5.45. Combining Equation 7.4 and the other recommendations outlined in this section, a tentative interpretation for SPT data in fine-grained soils is shown in Figure 7.5.

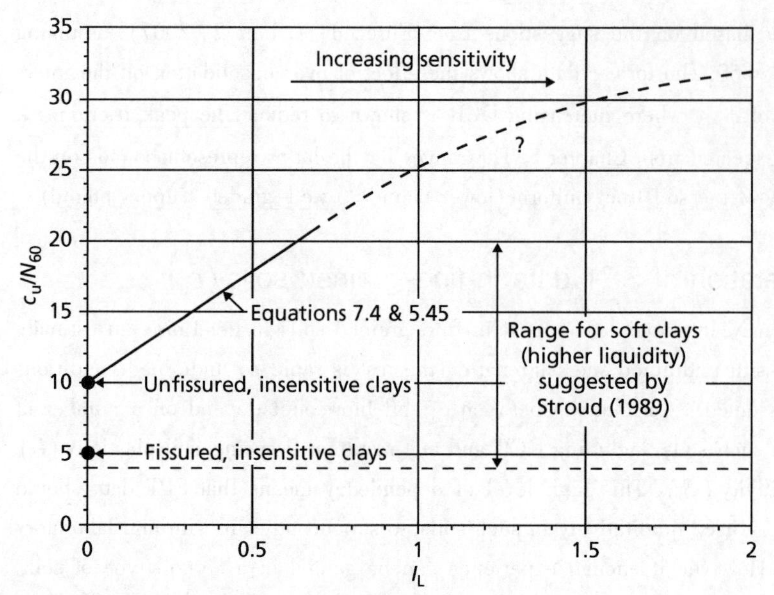

Figure 7.5 Estimation of c_u from SPT data in fine-grained soils.

7.3 Field Vane Test (FVT)[①]

① 现场十字板剪切试验。

② 不同于主要针对粗颗粒土的标准贯入试验，现场十字板剪切试验主要用于原位测试未受损饱和黏土的不排水强度特性。

③ 不锈钢十字板。

④ 高强度钢杆。

In contrast to the SPT test, which is predominantly used for coarse-grained soils, the FVT test is principally used for the in-situ determination of the undrained strength characteristics of intact, fully saturated clays.[②] Silts and glacial tills may also be characterised using this method, though the reliability of such data is more questionable and should be supported with other test data where possible. The test is not suitable for coarse-grained soils. In particular, the FVT is very suitable for soft clays, the shear strength of which, if measured in the laboratory, may be significantly altered by the sampling process and subsequent handling. Generally, this test is only used in clays with $c_u < 100$kPa. The test may not give reliable results if the clay contains sand or silt laminations.

The technical standards governing its use are EN ISO 22476, Part 9 (UK and Europe) and ASTM D2573 (US). The equipment consists of a stainless steel **vane**[③] (Figure 7.6) of four thin rectangular blades at 90° to each other, carried on the end of a high-tensile steel rod[④]; the rod is enclosed by a sleeve packed with grease. The length of the vane is equal to twice its overall width, typical dimensions being 150mm by 75mm and 100mm by 50mm. Preferably, the diameter of the rod should not exceed 12.5mm.

The vane and rod are pushed into the soil below the bottom of a borehole to a depth of at least three times the borehole diameter; if care is taken this can be done without appreciable disturbance of the clay. Steady bearings are used to keep the rod and sleeve central in the borehole casing. In soft clays, tests may be conducted without a borehole by direct penetration of the vane from ground level; in this case a shoe is required to protect the vane during penetration. Small, hand-operated vanes are also available for use in exposed clay strata.

Torque is applied gradually to the upper end of the rod until the clay fails in shear due to rotation of the vane.[①] Shear failure takes place over the surface and ends of a cylinder having a diameter equal to the overall width of the vane. The rate of rotation[②] of the vane should be within the range of 6-12° per minute. The shear strength is calculated from the expression

① 在杆件的上端逐渐施加扭矩，直至黏土由于十字板旋转而发生剪切破坏。

② 旋转速率。

$$T = \pi c_{uFV}\left(\frac{d^2 h}{2} + \frac{d^3}{6}\right) \tag{7.5}$$

where T is the torque at failure, d the overall vane width and h the vane length (see Figure 7.6). However, the shear strength over the cylindrical vertical surface may be different from that over the two horizontal end surfaces, as a result of anisotropy[③]. The shear strength is normally determined at intervals over the depth of interest. If, after the initial test, the vane is rotated rapidly through several revolutions, the soil will become remoulded; the shear strength in this condition can then be determined if required. The ratio of the in-situ and fully remoulded[④] values of c_u determined in this way gives the sensitivity of the soil.

③ 各向异性。

④ 完全重塑。

⑤ 旋转驱动装置和旋转测量。

⑥ 十字板。

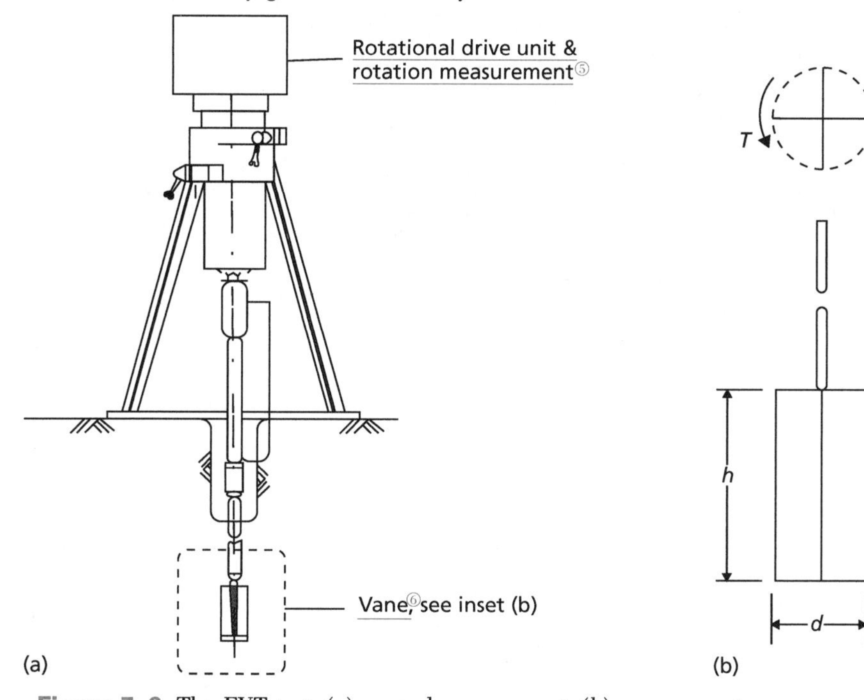

Figure 7.6 The FVT test: (a) general arrangement, (b) vane geometry.

The undrained strength as measured by the vane test is generally greater than the average strength mobilised along a failure surface in a field situation (Bjerrum, 1973). The discrepancy was found to be greater the higher the plasticity index of the clay, and is attributed primarily to differences in the rate of loading[⑦] between the two cases. In the vane test shear failure occurs within a few minutes, whereas in a field situation the stresses are usually applied over a period of a few weeks or months. A secondary factor may be anisotropy. Bjerrum and, later, Azzouz *et al*. (1983) presented correction factors[⑧] (μ), correlated empirically with I_P, as shown in Figure 7.7. The probable field strength[⑨] (c_u) is then determined from the measured FVT strength (c_{uFV})[⑩] using

⑦ 加载速率。

⑧ 修正系数。

⑨ 可靠的现场强度。

⑩ 现场十字板剪切试验的测试强度。

$$c_u = \mu \cdot c_{uFV} \tag{7.6}$$

The FVT can further be used to estimate the overconsolidation ratio (OCR) of the soil, as demonstrated by Mayne and Mitchell (1988). This is achieved using a second empirical factor, α_{FV}, where:

$$OCR = \alpha_{FV} \cdot \left(\frac{c_{uFV}}{\sigma'_{v0}}\right) \tag{7.7}$$

By considering a large database of test results from 96 different sites, it has been shown that

$$\alpha_{FV} \approx 22\,(I_P)^{-0.48} \tag{7.8}$$

where I_P is entered in percent. The relationship between α_{FV} and I_P in Equation 7.8 is similar in shape to that between μ and I_P (Figure 7.7), such that $\alpha_{FV} \approx 4\mu$. Mayne and Mitchell (1988) further demonstrated good agreement between this method and the results of conventional oedometer tests for determining OCR (as previously described in Chapter 4) over a range of I_P = 8-100%.

① 由现场十字板剪切试验确定的土体不排水抗剪强度的修正系数μ。

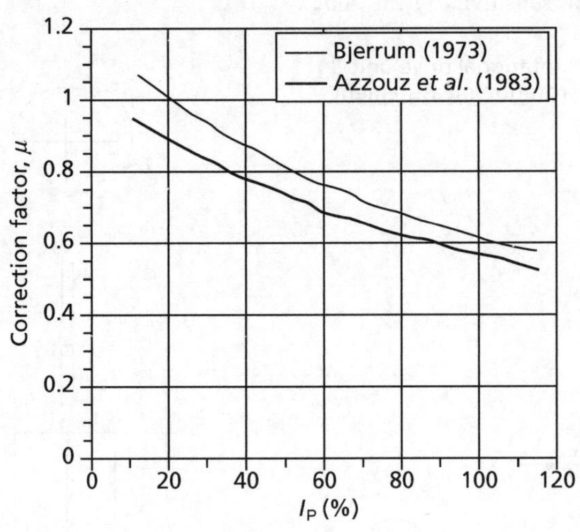

Figure 7.7 Correction factor μ for undrained strength as measured by the FVT. ①

Example 7.1

The FVT test data for the Bothkennar clay of Figure 5.38 (b) has been provided in electronic form on the Companion Website (both peak and remoulded strengths are given). The index test data (w, w_P and w_L) from Figure 5.38 (b) are also provided in electronic form. The water table is 0.8m below ground level (BGL); the soil above the water table has a bulk unit weight of γ = 18.7kN/m^3, while the soil below the water table has γ = 16kN/m^3. Using these data, estimate the variation of S_t and OCR with depth. The latter should be compared with oedometer test data on the same soil shown in Figure 7.8 (a).

Solution

The detailed calculations were conducted using the spreadsheet provided on the Companion Website. The sensitivity can be directly found for each test by dividing the peak strength by the remoulded value. The results of these calculations are shown in Figure 5.38 (b), where they agree well with values estimated using the empirical correlations of Section 5.9. To determine OCR from the FVT data,

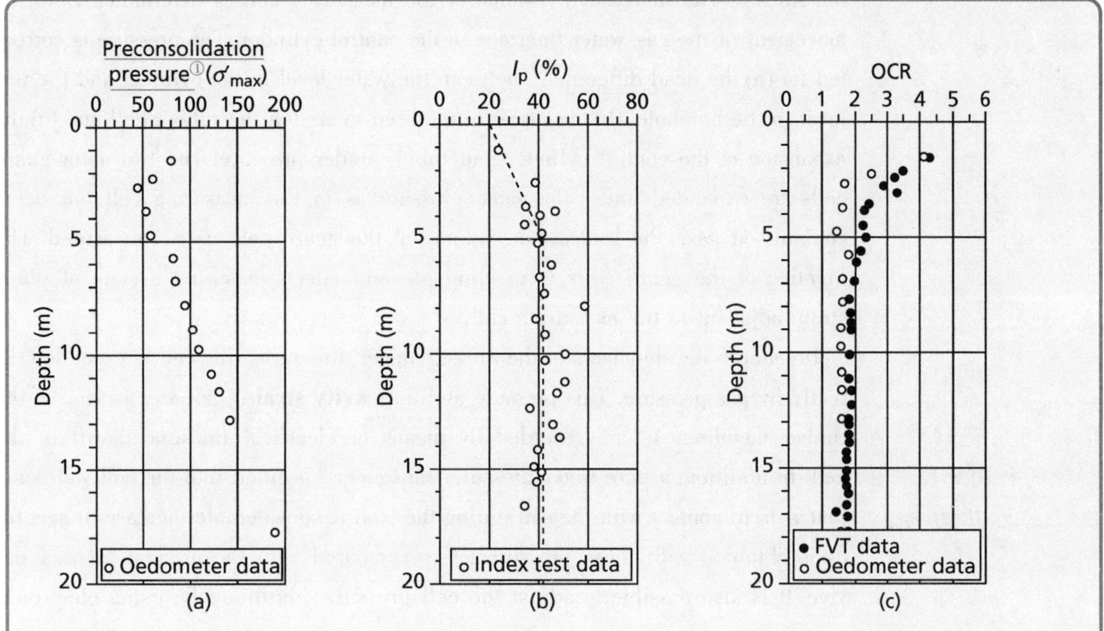

Figure 7.8 Example 7.1 (a) Oedometer test data, (b) I_p calculated from index test data, (c) OCR from FVT and oedometer data.

the in-situ stress conditions② (σ_{v0}, u_0 and σ'_{v0}) are found at each test depth. As the index test data were not conducted at the same depths as the FVT tests (this is common in practice), I_P is determined at each depth and an average trend determined as shown in Figure 7.8 (b). This is defined by $I_P = 4.4z + 20\%$ for $0 \leqslant z \leqslant 5$m and $I_P = 42\%$ for $z \geqslant 5$m. Using these values of I_P, α_{FV} is calculated for each FVT depth from Equation 7.8. The OCR at each depth is then found using Equation 7.7. The resulting data are compared to the oedometer test data in Figure 7.8 (c), where the two sets of data show similar trends, though the FVT data slightly overpredicts OCR compared to the oedometer test data.

① 先期固结压力。
② 原位应力条件。

7.4 Pressuremeter Test (PMT)

The **pressuremeter**③ was developed in the 1950s by Ménard to provide a high quality in-situ test which could be used to derive both strength and stiffness parameters for soil as an alternative to triaxial testing. As the test is in-situ, it overcame the problem of sampling disturbance associated with the latter; as the pressuremeter influences a much larger volume of soil than normal laboratory tests, it also ensures that the macro-fabric of the soil is adequately represented. Ménard's original design, illustrated in Figure 7.9 (a), consists of three cylindrical rubber cells④ of equal diameter arranged coaxially. The device is lowered into a (slightly oversize) borehole to the required depth and the central measuring cell is expanded against the borehole wall by means of water pressure, measurements of the applied pressure and the corresponding increase in cell volume being recorded. Pressure is applied to the water by compressed gas (usually nitrogen⑤) in a control cylinder at

③ 旁压仪。
④ 圆柱形橡胶室。
⑤ 氮气。

the surface. The increase in volume of the measuring cell is determined from the movement of the gas-water interface in the control cylinder. The pressure is corrected for (a) the head difference[①] between the water level in the cylinder and the test level in the borehole, (b) the pressure required to stretch the rubber cell and (c) the expansion of the control cylinder and tubing under pressure. The two outer guard cells are expanded under the same pressure as in the measuring cell but using compressed gas; the increase in volume of the guard cells is not measured. The function of the guard cells is to eliminate end effects, ensuring a state of plane strain adjacent to the measuring cell.

In modern developments of the pressuremeter, the measuring cell is expanded directly by gas pressure. This pressure and the **cavity strain** (radial expansion of the rubber membrane)[②] are recorded by means of electrical transducers within the cell. In addition, a pore water pressure transducer[③] is fitted into the cell wall such that it is in contact with the soil during the test. A considerable increase in accuracy is obtained with these pressuremeters compared with the original Ménard device. It is also possible to adjust the cell pressure continuously, using electronic

① 水头差。
② 孔洞应变（橡胶膜的径向膨胀）。
③ 孔隙水压力传感器。
④ 体积指示器。
⑤ 保护腔。
⑥ 测试腔。
⑦ 空心轴。
⑧ 橡胶膜。
⑨ 孔压传感器。
⑩ 刀片。
⑪ 梅纳德旁压仪。
⑫ 自钻式旁压仪。

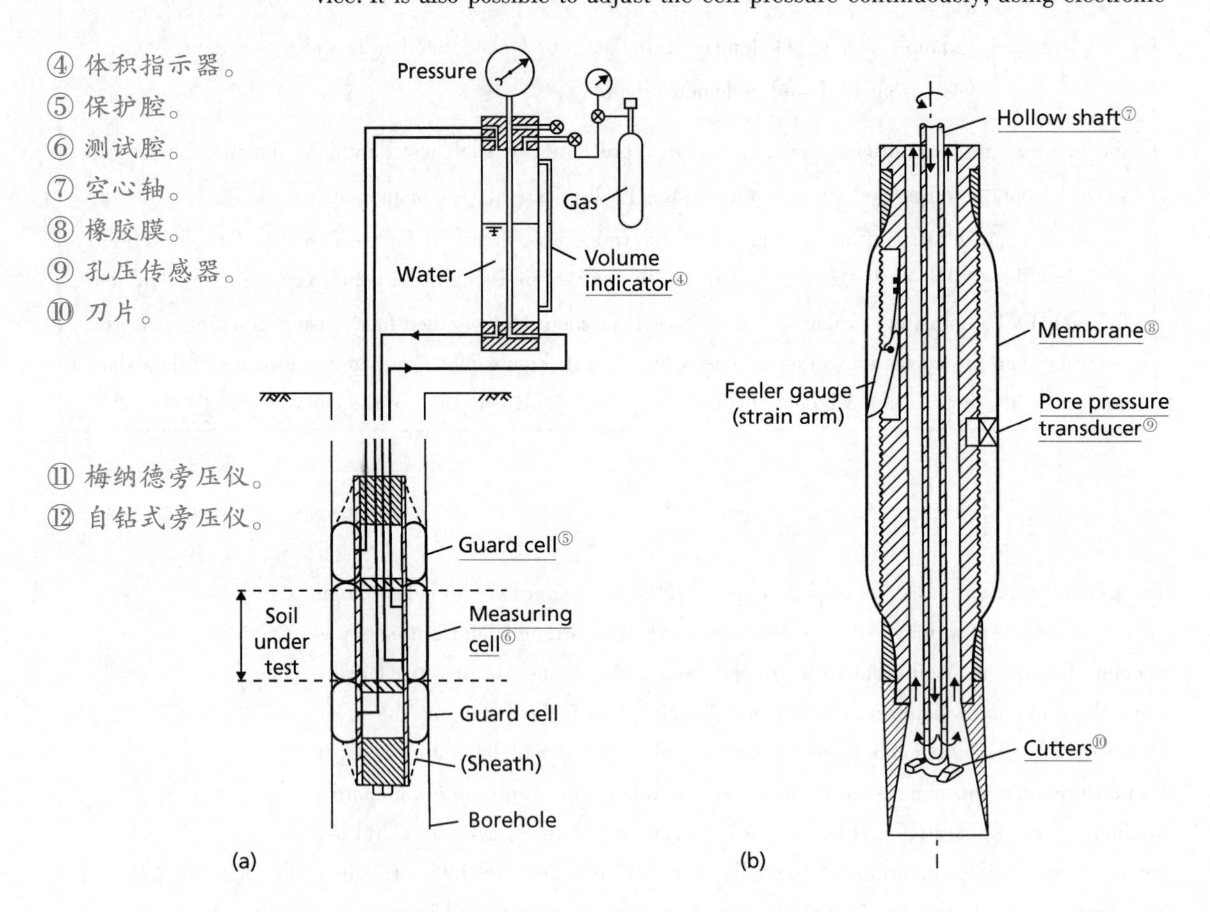

Figure 7.9 Basic features of (a) Ménard pressuremeter[⑪], and (b) self-boring pressuremeter.[⑫]

control equipment, to achieve a constant rate of increase in circumferential strain (i. e. a strain-controlled test), rather than applying the pressure in increments (a stress-controlled test). The technical standards governing the use of pressuremeters in pre-bored holes are EN ISO 22476, Part 5 (UK and Europe) and ASTM D4719 (US). The Ménard device is still popular in some parts of Europe; this is governed by EN ISO 22476, Part 4.

Some soil disturbance adjacent to a borehole is inevitable due to the boring process, and the results of pressuremeter tests in pre-formed holes can be sensitive to the method of boring. The **self-boring pressuremeter** (SBPM)[①] was developed to overcome this problem, and is suitable for use in most types of soil; however, special insertion techniques are required in the case of sands. This device, illustrated in Figure 7.9 (b), is jacked slowly into the ground and the soil is broken up by a rotating cutter fitted inside a cutting head at the lower end, the optimum position of the cutter being a function of the shear strength of the soil. Water or drilling fluid is pumped down the hollow shaft to which the cutter is attached, and the resulting slurry is carried to the surface through the annular space adjacent to the shaft; the device is thus inserted with minimal disturbance of the soil. The only correction required is for the pressure required to stretch the membrane. If a self-boring pressuremeter is used, EN ISO 22476, Part 6 is the relevant standard (an ASTM standard has not yet been released).

① 自钻式旁压仪。

The membrane of a pressuremeter may be protected against possible damage (particularly in coarse soils) by a thin stainless steel sheath with longitudinal cuts[②], designed to cause only negligible resistance to the expansion of the cell.

② 纵向切割。

Like the FVT described in Section 7.3, soil parameters are derived from the cell pressure and cell volume change (or cavity strain) using a theoretical model[③] (see Equation 7.5) rather than empirical correlations. These analyses will be described in the following sections.

③ 理论模型。

Interpretation of PMT data in fine-grained soils (G, c_u)

In fine-grained soils, the following analysis is derived from Gibson and Anderson (1961). During the pressuremeter test, the cavity (borehole) is expanded radially from its initial radius of r_c[④] to a new radius $r_c + y_c$ by an amount y_c (the displacement at the cavity wall[⑤]). There is no displacement or strain in the vertical direction (along the axis of the borehole) - these conditions are known as **plane strain**,[⑥] as the soil may only strain in a single plane (in this case, horizontal). During this expansion, the volume of the cavity increases by an amount dV. The soil at any radius r away from the centre of the cavity similarly expands from its initial radius r to a new radius $r + y$ by an amount y. In an undrained test there must be no overall change in the volume of the soil, so the change in volume of the soil expanding from r to $r + y$ must be equal to dV (i. e. the two shaded annuli in Figure 7.10 (a) must have equal areas), giving

④ 初始半径。
⑤ 孔壁位移。
⑥ 平面应变。

$$2\pi r y dz = dV$$

$$\gamma = \frac{\mathrm{d}V}{2\pi r \mathrm{d}z} \tag{7.9}$$

The soil may strain radially by an amount ε_r and circumferentially by an amount ε_θ.[①] The strain in the axial direction $\varepsilon_a = 0$ (plane strain)[②]. If the soil is isotropic, ε_r, ε_θ and ε_a are principal strains. The shear strain in the soil is then given by

$$\gamma = \varepsilon_r - \varepsilon_\theta \tag{7.10}$$

And the volumetric strain ε_v by

$$\varepsilon_v = \varepsilon_r + \varepsilon_\theta + \varepsilon_a \tag{7.11}$$

In an undrained test $\varepsilon_v = 0$, so from Equation 7.11, $\varepsilon_r = -\varepsilon_\phi$ and from Equation 7.10

$$\gamma = -2\varepsilon_\theta \tag{7.12}$$

The circumference of the soil annulus[③] is initially $2\pi r$ (Figure 7.10 (a)) and increases to $2\pi(r+y)$ giving an extension of $2\pi y$. Choosing compression of the soil as positive, the circumferential strain[④] is then:

$$\varepsilon_\theta = -\frac{2\pi y}{2\pi r} = -\frac{y}{r} \tag{7.13}$$

Substituting Equations 7.13 and 7.9 into Equation 7.12 gives the equation of compatibility for the pressuremeter test (cylindrical cavity expansion[⑤]):

$$\gamma = \frac{2y}{r} = \frac{\mathrm{d}V}{\pi r^2 \mathrm{d}z} \tag{7.14}$$

Equation 7.14 was derived by considering compatibility of the soil displacements from the cavity wall outwards. As described in Section 5.1, equilibrium must also be satisfied within the soil mass. The stresses acting on a segment of the soil along the annulus defined by r is shown in Figure 7.10 (b). For there to be radial equilibrium[⑥]:

$$(\sigma_r + \mathrm{d}\sigma_r)(r + \mathrm{d}r)\,\mathrm{d}\theta = \sigma_r r \mathrm{d}\theta + \sigma_\theta \mathrm{d}r \mathrm{d}\theta$$

$$\therefore r\frac{\mathrm{d}\sigma_r}{\mathrm{d}r} + (\sigma_r - \sigma_\theta) = 0 \tag{7.15}$$

The stresses σ_r and σ_θ are principal stresses[⑦], being associated with the principal strains ε_r and ε_ϕ[⑧] respectively. Considering the Mohr Circle represented by these stresses, the associated maximum shear stress is $\tau = (\sigma_r - \sigma_\theta)/2$, which when substituted into Equation 7.15 gives the equation of equilibrium for the pressuremeter test (cylindrical cavity expansion):

$$r\frac{\mathrm{d}\sigma_r}{\mathrm{d}r} + 2\tau = 0 \tag{7.16}$$

Having considered both compatibility and equilibrium, it only remains to consider the constitutive model to link the shear stresses (in Equation 7.16) with the shear strains (Equation 7.14).

Linear elastic soil behaviour

While the soil is behaving elastically (Figure 7.11 (a)), the constitutive relationship is given by

① 土体会产生径向应变 ε_r 和环向应变 ε_θ。
② 平面应变中轴向应变 $\varepsilon_a = 0$。
③ 土环。
④ 环向应变。
⑤ 圆柱体孔扩张。
⑥ 径向力平衡。
⑦ 主应力。
⑧ 注：原书中错误，此处应为 ε_θ。

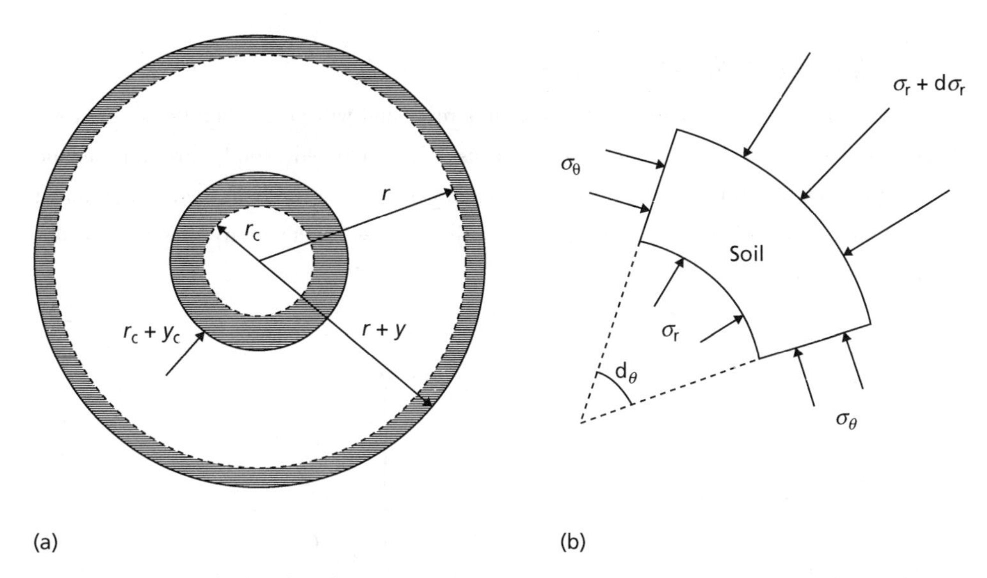

Figure 7.10 Idealised soil response during cavity expansion: (a) compatible displacement field, (b) equilibrium stress field.[①]

① 圆孔扩张时理想土体响应：(a) 相容位移场，(b) 平衡应力场。

$$\tau = G\gamma \tag{7.17}$$

Substituting Equations 7.14 and 7.17 into Equation 7.16 gives

$$r\frac{d\sigma_r}{dr} + 2G\frac{dV}{\pi r^2 dz} = 0 \tag{7.18}$$

This is the governing differential equation governing the response of all of the soil around the pressuremeter.

At the cavity wall, $r = r_c$ and $\sigma_r = p$ (where p is the pressure within the pressuremeter); far from the cavity the soil is unaffected by the pressuremeter expansion so that at $r = \infty$, $\sigma_r = \sigma_{h0}$ where σ_{h0} is the in-situ horizontal total stress[②] in the ground (also termed the **lift-off pressure**[③]). Equation 7.18 may then be integrated using these limits:

② 原位水平总应力。

③ 起升压力。

$$\int_{\sigma_{h0}}^{p} d\sigma_r = -\int_{\infty}^{r_c} \frac{2G dV}{\pi r^3 dz} dr$$

$$[\sigma_r]_{\sigma_{h0}}^{p} = -2G\frac{dV}{\pi}\left[-\frac{1}{2r^2}\right]_{\infty}^{r_c}$$

$$p - \sigma_{h0} = G\frac{dV}{\pi r_c^2}$$

Recognising that πr_c^2 is the volume of the cavity[④] (V), the following relationship is obtained:

④ 腔室体积。

$$p = G\frac{dV}{V} + \sigma_{h0} \tag{7.19}$$

Equation 7.19 suggests that, for elastic soil behaviour, if the cavity pressure p is plotted against the volumetric strain in the cavity dV/V, a straight line will be given, the gradient of which gives the soil stiffness G and the intercept gives the initial total horizontal stress at the test depth, σ_{h0}. This is shown in Figure 7.11 (b).

Elasto-plastic soil behaviour

In reality, the soil cannot remain elastic forever and will yield when the shear stress reaches τ_{max}.[①] Pressuremeter tests are usually conducted rapidly compared to the consolidation time required in most finegrained soils so the behaviour is undrained and $\tau_{max} = c_u$ (Figure 7.12 (a)). Yield will occur when the cavity pressure reaches p_y defined by

$$p_y = \sigma_{h0} + c_u \tag{7.20}$$

① 在实际中，土体不可能一直保持弹性，且会在剪应力达到 τ_{max} 时发生屈服。
② 本构模型（线弹性）。
③ 由测量的 p 和 dV/V 确定 G 和 σ_{h0}

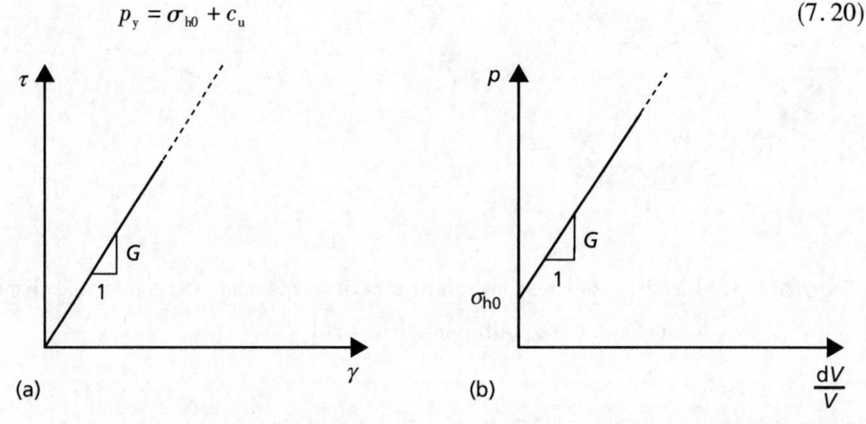

Figure 7.11 Pressuremeter interpretation during elastic soil behaviour: (a) constitutive model (linear elasticity)[②], (b) derivation of G and σ_{h0} from measured p and dV/V[③].

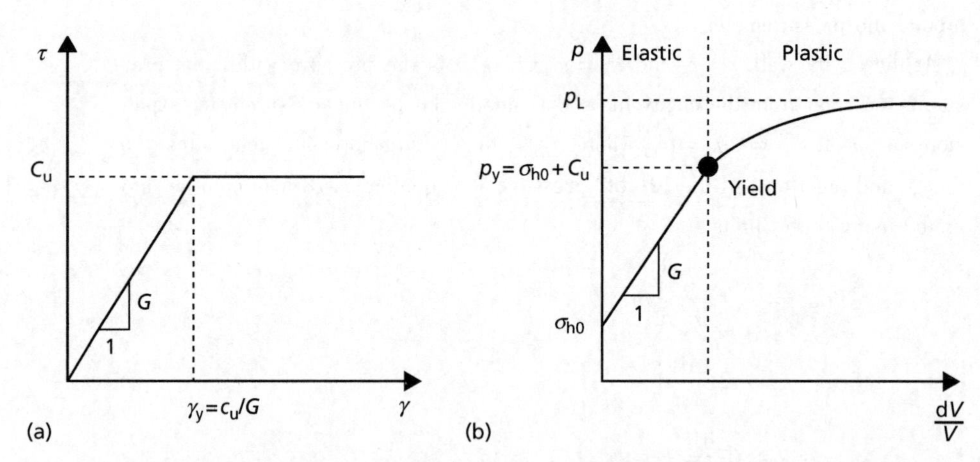

Figure 7.12 Pressuremeter interpretation in elasto-plastic soil: (a) constitutive model (linear elasticity, Mohr-Coulomb plasticity), (b) non-linear characteristics of measured p and dV/V.

This is shown in Figure 7.12 (b). Yielding will not occur in all of the soil (to $r = \infty$) simultaneously. There will be a zone of yielding soil immediately around the cavity out to a radius $r = r_y$ where the stresses are highest-in this zone the radial stresses will be $\sigma_r = p_y$ everywhere. Outside of this zone (i. e. from $r_y < r < \infty$), the soil will be elastic. In the plastic zone[④], $\tau = c_u$ everywhere. Substituting this into Equation

④ 塑性区。

7.16 gives

$$r\frac{\mathrm{d}\sigma_{\mathrm{r}}}{\mathrm{d}r}=2c_{\mathrm{u}}=0 \tag{7.21}$$

Equation 7.21 is only valid within the plastic zone, so that the equation should be integrated between the limits of $\sigma_{\mathrm{r}}=p_{\mathrm{y}}$ at $r=r_{\mathrm{y}}$ to $\sigma_{\mathrm{r}}=p$ at $r=r_{\mathrm{c}}$ (as before):

$$\int_{p_{\mathrm{y}}}^{p}\mathrm{d}\sigma_{\mathrm{r}}=-\int_{r_{\mathrm{y}}}^{r_{\mathrm{c}}}\frac{2c_{\mathrm{u}}}{r}\mathrm{d}r$$

$$p-p_{\mathrm{y}}=2c_{\mathrm{u}}\ln\left(\frac{r_{\mathrm{y}}}{r_{\mathrm{c}}}\right) \tag{7.22}$$

In order to make use of Equation 7.22, the parameter r_{y} needs to be expressed in terms of familiar parameters. Irrespective of whether the soil is elastic or plastic, there must be no overall change in volume in an undrained test. Therefore, from Equation 7.14

$$\gamma_{\mathrm{y}}=\frac{2y_{\mathrm{y}}}{r_{\mathrm{y}}} \tag{7.23}$$

and from Equation 7.9

$$y_{\mathrm{y}}r_{\mathrm{y}}=y_{\mathrm{c}}r_{\mathrm{c}} \tag{7.24}$$

Equations 7.23 and 7.24 are combined to eliminate the unknown y_{y}, giving

$$\gamma_{\mathrm{y}}=\frac{2y_{\mathrm{y}}}{r_{\mathrm{y}}}$$

$$\left(\frac{r_{\mathrm{y}}}{r_{\mathrm{c}}}\right)^{2}=\frac{1}{\gamma_{\mathrm{y}}}\left(\frac{2y_{\mathrm{c}}}{r_{\mathrm{c}}}\right) \tag{7.25}$$

From Figure 7.12 (a) $\gamma_{\mathrm{y}}=c_{\mathrm{u}}/G$, and from Equation 7.14 $2y_{\mathrm{c}}/r_{\mathrm{c}}=\mathrm{d}V/V$. Substituting these relationships into Equation 7.25 gives

$$\left(\frac{r_{\mathrm{y}}}{r_{\mathrm{c}}}\right)^{2}=\frac{G}{c_{\mathrm{u}}}\left(\frac{\mathrm{d}V}{V}\right) \tag{7.26}$$

Substituting Equations 7.20 and 7.26 into Equation 7.22 gives

$$p=c_{\mathrm{u}}\ln\left(\frac{\mathrm{d}V}{V}\right)+\sigma_{\mathrm{h0}}+c_{\mathrm{u}}+c_{\mathrm{u}}\ln\left(\frac{G}{c_{\mathrm{u}}}\right) \tag{7.27}$$

Equation 7.27 suggests that, for linearly elastic-perfectly plastic[1] soil behaviour, if the cavity pressure p is plotted against the logarithm of the volumetric strain[2] in the cavity ln ($\mathrm{d}V/V$), the data will approach a straight line asymptote as $\mathrm{d}V/V$ gets large (i. e. as ln ($\mathrm{d}V/V$) tends towards zero). The gradient of this asymptote will be the undrained shear strength c_{u}, as shown in Figure 7.13. Furthermore, ln ($\mathrm{d}V/V$) = 0 represents the limiting case of infinite expansion in the soil. This corresponds to an ultimate or **limit pressure**[3] p_{L}, which is shown in Figures 7.12 (b) and 7.13. It should be noted that this pressure is impossible to achieve in practice.

① 理想弹塑性。

② 对数体积应变。

③ 极限压力。

Practical derivation of soil parameters

In practice, pressuremeter test data are usually plotted as a graph of cavity pressure versus cavity strain ε_{c}[4] (Figure 7.14). By considering the strains at the cavity wall

④ 在实际中，旁压试验的数据通常以腔内压力与腔体应变为坐标轴来绘图。

due to a change in cavity volume from V to $V+\mathrm{d}V$, it can be shown that

$$\frac{\mathrm{d}V}{V}=1-\left(\frac{1}{1+\varepsilon_{c}}\right)^{2} \tag{7.28}$$

At most stages of a pressuremeter test, the strains are small enough that Equation 7.28 can be approximated by

$$\frac{\mathrm{d}V}{V}\approx 2\varepsilon_{c} \tag{7.29}$$

① 体积变化。

From Equation 7.29, the volumetric change① is directly proportional to the cavity strain at small strains. Therefore the graph of p versus ε_c will have the same shape as Figure 7.12 (b), so that σ_{h0} may be directly read from the graph, and G determined from the gradient of the curve at any point. It should be noted that unlike in Figure 7.11 (b) where the gradient was G, on a plot of p versus $\mathrm{d}V/V$, the gradient will be $2G$ from Equation 7.29, i. e.

$$G=\frac{1}{2}\left(\frac{\mathrm{d}p}{\mathrm{d}\varepsilon_{c}}\right) \tag{7.30}$$

② 斜率。

③ 由旁压试验数据确定不排水抗剪强度。

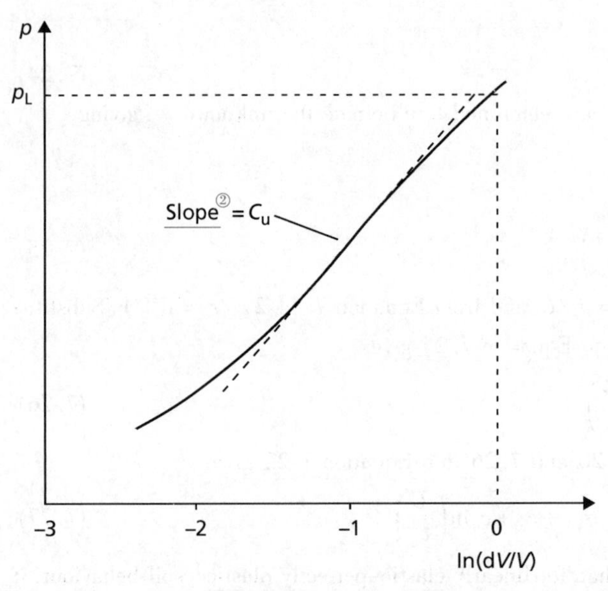

Figure 7.13 Determination of undrained shear strength from pressuremeter test data.③

after Palmer (1972). Rather than find G from the slope of the initial elastic portion of the p-ε_c curve as suggested by Figure 7.11 (b), in practice the modulus is obtained from the slope of an unloading-reloading cycle as shown in Figure 7.14, ensuring that the soil remains in the 'elastic' state during unloading. Wroth (1984) has shown that, in the case of a clay, this requirement will be satisfied if the reduction in pressure during the unloading stage is less than $2c_u$, i. e.

$$\Delta p<2c_{u} \tag{7.31}$$

Most modern pressuremeters have multiple strain arms arranged in diametrically opposite pairs around the cylindrical body of the pressuremeter. The lift-off pressure (σ_{h0}) is normally an average value inferred from all of the gauges. However the

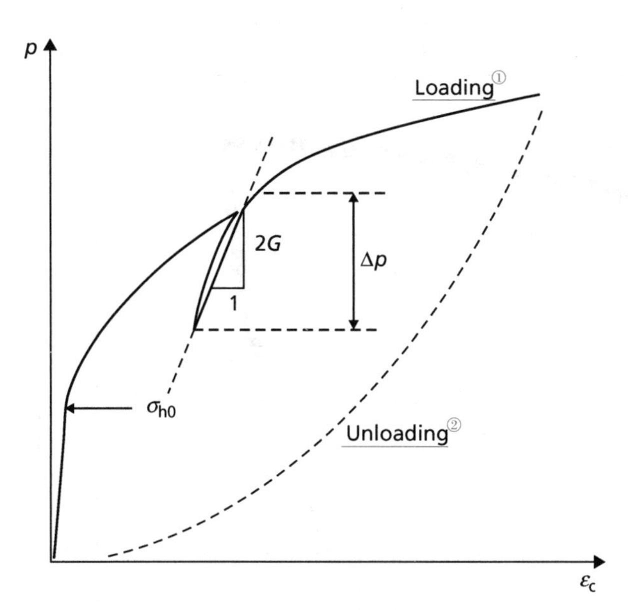

Figure 7.14 Direct determination of G and σ_{h0} in fine-grained soils from pressuremeter test data.③

① 加载。
② 卸载。
③ 由旁压试验数据直接确定细粒土的 G 与 σ_{h0}。
④ 原始数据。

data from individual pairs of gauges can be used to infer differences in in-situ σ_{h0} within the ground, i.e. stress anisotropy (Dalton and Hawkins, 1982).

⑤ 需注意的是，需要改变趋势线的截距使拟合直线沿滞回圈的主轴方向。

Example 7.2

Figure 7.15 shows data from a self-boring pressuremter test undertaken at a depth of 8m below ground level in the Gault clay shown in Figure 5.38 (a). The raw data④ for this test is provided in electronic form on the Companion Website. Determine the following parameters: σ_{h0}, c_u, G (for both unload-reload loops conducted during the test).

Solution

The in-situ horizontal total stress can be directly estimated from inspection of the p-ε_c curve in Figure 7.15, where the lift-off pressure $\sigma_{h0} \approx 395$kPa. The values of G for each of the unload-reload loops is found by plotting the p-ε_c data over the range of the loop only and then fitting a straight line to the data. The spreadsheet on the Companion Website shows this procedure using the trendline fit in MS Excel. Note that it was necessary to vary the intercept of the trendline to force the fitting line along the major axis of the loop.⑤ From Equation 7.30 the gradients of a trendline = $2G$, giving values of $G = 32.3$ and 27.0MPa for the first and second loops, respectively. To obtain the undrained shear strength, the cavity strains are converted into volumetric strain (dV/V) using Equation 7.28 and the data are replotted in the form of Figure 7.13. This is also demonstrated within the spreadsheet on the Companion Website. A straight line is fitted to the data, the gradient of which gives $c_u = 111$kPa. This compares favourably with c_u measured from triaxial tests at this depth from Figure 5.38 (a).

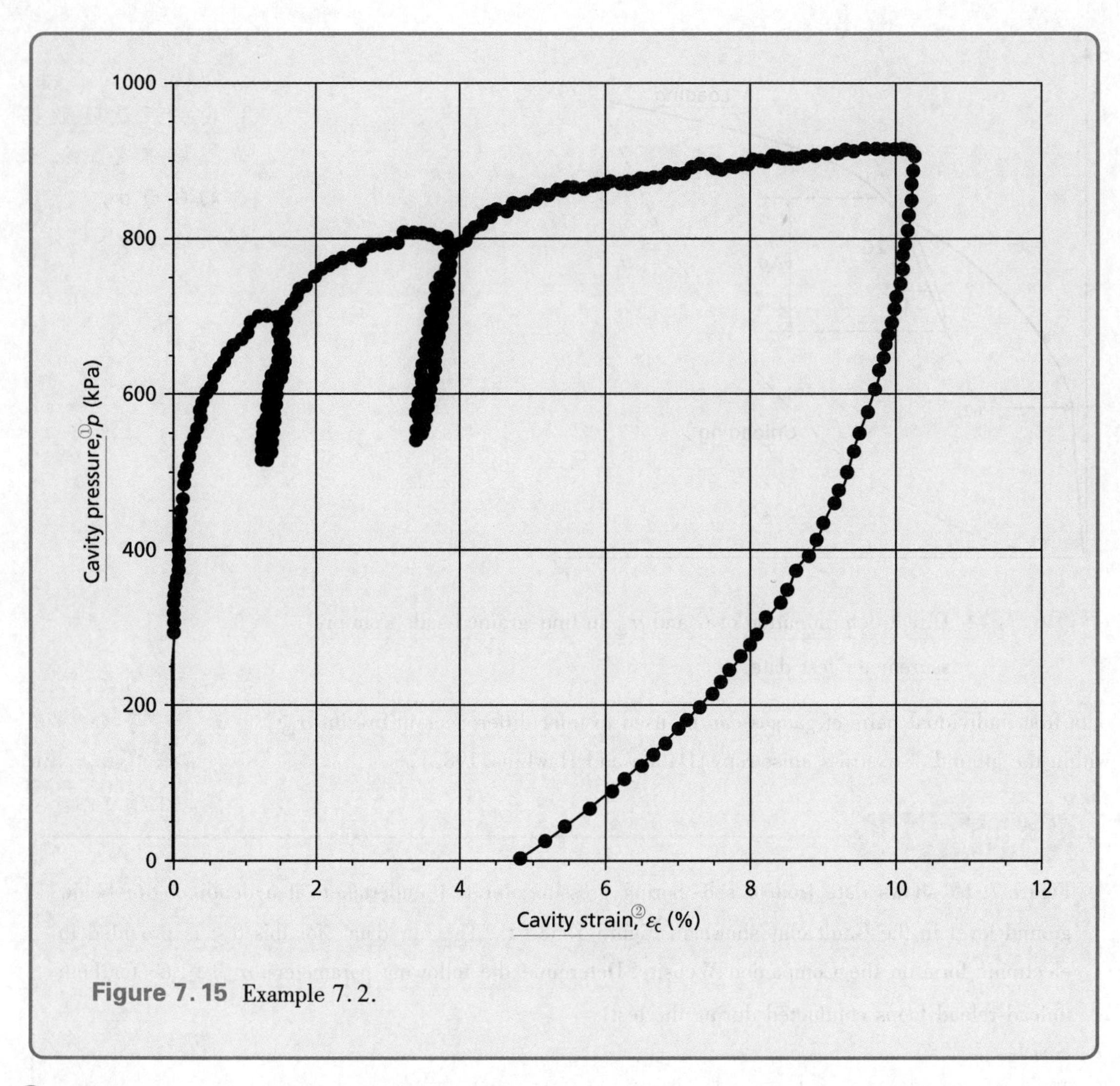

Figure 7.15 Example 7.2.

① 腔内压力。
② 腔体应变。
③ 剪胀。

Interpretation of PMT data in coarse-grained soils (G, ϕ', ψ)

The analysis of the pressuremeter test in a drained soil is similar to the analysis described in the previous section for undrained soil, involving the formation of equations of compatibility, equilibrium and constitutive behaviour, but with stresses now expressed in terms of effective rather than total components. However, use can no longer be made of the 'no volume change' criterion such that the dilatancy[3] of the soil must be considered. This makes the constitutive law more complex. A complete analysis is given by Hughes *et al.* (1977). The analysis enables values for the angle of shearing resistance (ϕ') and the angle of dilation (ψ) to be determined, and the derivation of these parameters from pressuremeter test data is described below.

It is typical, as for fine-grained soils, to plot cavity pressure versus cavity strain. A typical test curve is shown in Figure 7.16. If the total cavity pressure is plotted (Figure 7.16 (a)), the lift-off pressure defines σ_{h0} as before. If following cavity expansion the pressuremeter is completely unloaded, the cavity pressure at which ε_c returns to zero represents the initial pore pressure within the ground (u_0). This pore pressure is constant throughout the test, as no excess pore water pressures are generated during drained shearing of soil[①] (see Chapter 5). Unload-reload cycles[②] are usually conducted to determine G. The gradient of these loops is $2G$ as before (Equation 7.30). The soil will remain completely elastic during these stages as long as the reduction in pressure during the unloading stage satisfies

① 试验的孔压一直保持不变，所以土体排水剪切时没有超孔隙水压力产生。

② 加卸载循环。

$$\Delta p < \frac{2\sin\phi'}{1+\sin\phi'}(p-u_0) \tag{7.32}$$

It is common to subsequently correct all of the cavity pressures by subtracting the value of u_0 to give the effective cavity pressure $p-u_0$ (Figure 7.16 (b)). The lift-off pressure identified from this graph then represents the in-situ effective horizontal stress σ'_{h0}.

In order to determine the strength parameters (ϕ' and ψ), the data are replotted on different axes (see Figure 7.13). In the case of drained analysis, the data are replotted as log ($p-u_0$) versus log (ε_c); alternatively, the corrected data may be replotted on log-log axes. The data should then lie approximately on a straight line, the gradient of which is defined as s (Figure 7.17). Once the value of s has been determined, ϕ' and ψ can be estimated using

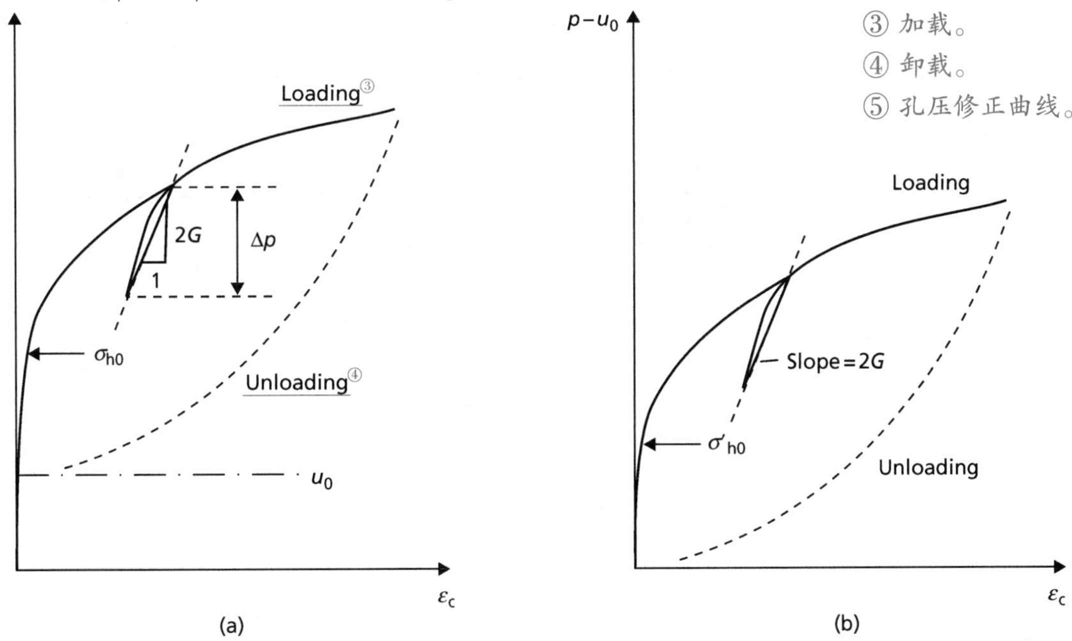

③ 加载。

④ 卸载。

⑤ 孔压修正曲线。

Figure 7.16 Direct determination of G, σ_{h0} and u_0 in coarse-grained soils from pressuremeter test data: (a) uncorrected curve, (b) corrected for pore pressure u_0.[⑤]

$$\sin\phi' = \frac{s}{1+(s-1)\sin\phi'_{cv}} \tag{7.33}$$

$$\sin\psi = s + (s-1)\sin\phi'_{cv} \tag{7.34}$$

The angle of shearing resistance ϕ' from Equation 7.33 represents the peak value (ϕ'_{max})[⑥]. Equations 7.33 and 7.34 have been plotted in Figure 7.18 for graphical

⑥ 峰值摩擦角。

solution. Interpretation of the strength data relies on knowing the critical state angle of shearing resistance for the soil (ϕ'_{cv}).[①] It is recommended that this is found from drained triaxial tests on loose samples of the soil such that the peak and critical state strengths are coincident (ϕ'_{cv} is independent of density, Chapter 5). If these data are not available, a value may be estimated from index test data using Figure 5.35.

① 土体临界状态剪切摩擦角。

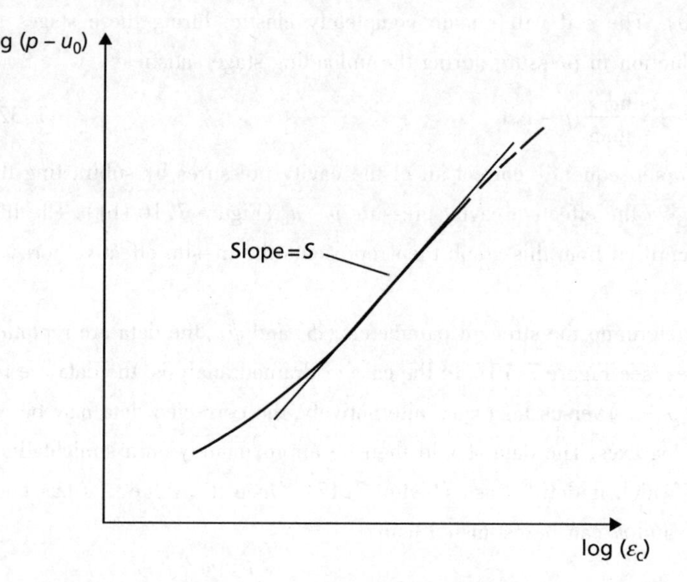

Figure 7.17 Determination of parameter s from pressuremeter test data.[②]

② 由旁压试验数据确定参数 s。

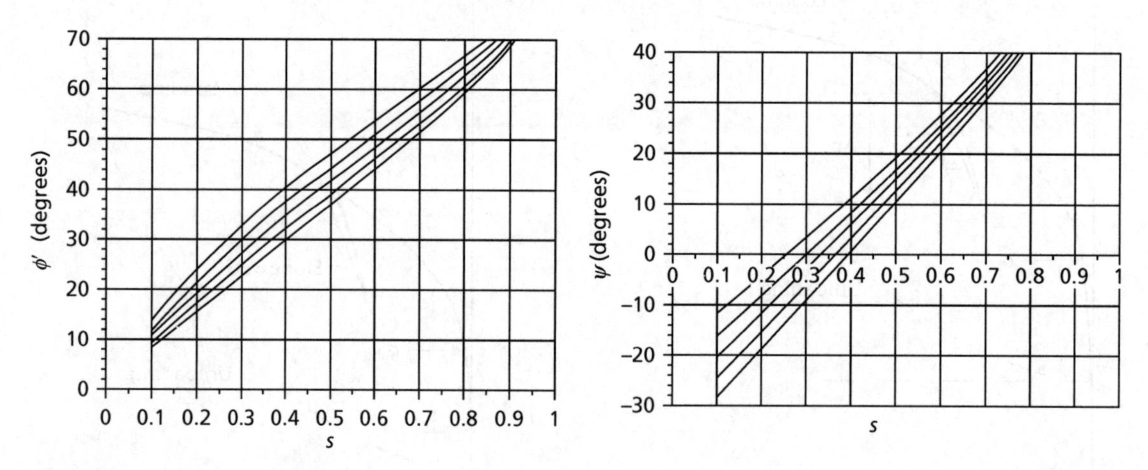

Figure 7.18 Determination of ϕ' and ψ from parameter s.[③]

③ 由 ϕ' 与 ψ 确定参数 s。

Example 7.3

Figure 7.19 shows data from a self-boring pressuremter test undertaken in a deposit of sand. The raw data for this test are provided in electronic form on the Companion Website. Determine the following parameters: σ_{h0}, u_0, σ'_{h0}, ϕ'_{max} and ψ. State also whether the unload-reload loops have been conducted over an appropriate (fully elastic)[④] stress range.

④ 完全弹性。

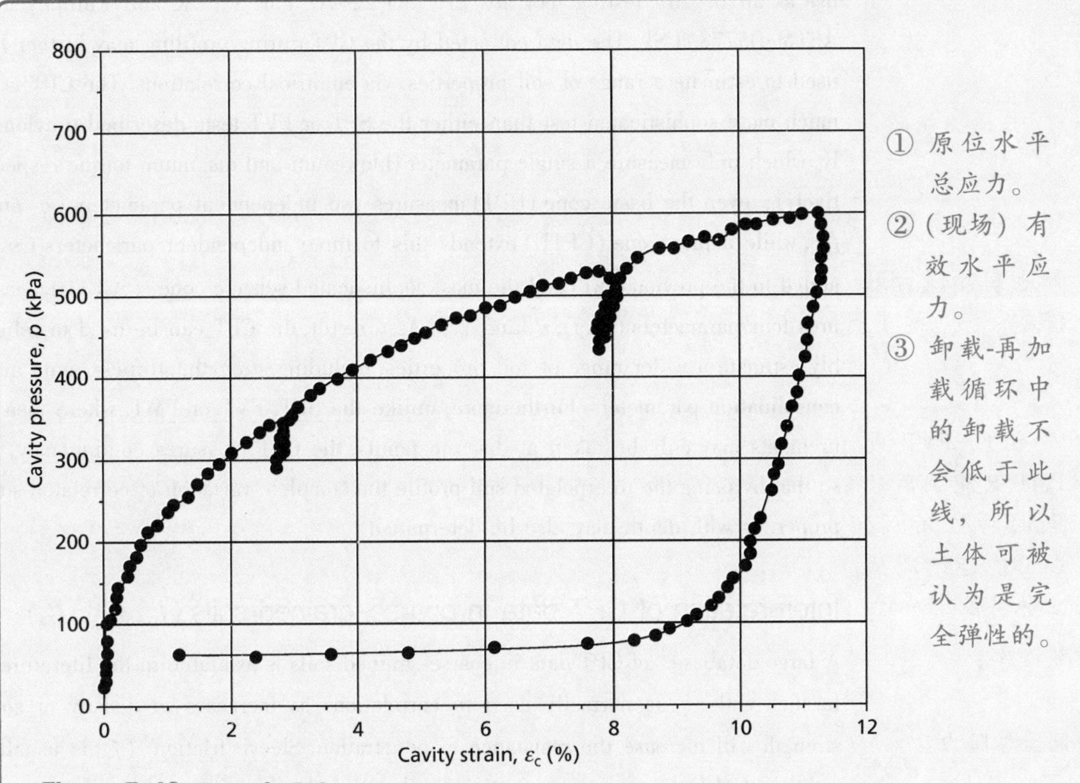

Figure 7.19 Example 7.3.

① 原位水平总应力。

② (现场) 有效水平应力。

③ 卸载-再加载循环中的卸载不会低于此线，所以土体可被认为是完全弹性的。

Solution

The in-situ horizontal total stress[①] can be directly estimated from inspection of the p-ε_c curve in Figure 7.19, where the lift-off pressure $\sigma_{h0} \approx 107$kPa. The in-situ pore pressure is given by the intercept of the unloading portion of the curve at $\varepsilon_c = 0$ (i.e. cavity pressure at the end of the test), giving $u_0 = 63$kPa. The effective horizontal stress[②] is then found using Terzaghi's Principle: $\sigma'_{h0} = 107 - 63 = 44$kPa. To obtain the strength parameters, the cavity pressures are corrected for u_0 and the data replotted in the form of Figure 7.17. This is shown within the spreadsheet on the Companion Website. A straight line is fitted to the data, the gradient of which gives $s = 0.50$. This is then used either in Equations 7.33 and 7.34, or in Figure 7.18, to give $\phi' = \phi'_{max} = 41.2°$ and $\psi = 14.5°$. Finally, Δp is calculated using Equation 7.32 for each data point using the corrected cavity pressure ($p' = p - u_0$) and the value of ϕ' found in the previous step. This is subtracted from the values of p (ignoring the unload-reload loops themselves) to give a curved locus, offset from the test data by an amount Δp (shown in the worksheet on the Companion Website). The unload-reload loops are not unloaded below this line, so the behaviour is expected to be fully elastic.[③]

7.5 Cone Penetration Test (CPT)[④]

④ 圆锥静力触探试验。

The CPT test was described in Chapter 6, where its use in identifying and profiling the different strata within the ground was demonstrated. The standards governing its use as an in-situ testing tool are EN ISO 22476, Part 1 (UK and Europe) and ASTM D5778 (US). The data collected by the CPT during profiling may further be used to estimate a range of soil properties, via empirical correlations. The CPT is a much more sophisticated test than either the SPT or FVT tests described previously, which only measure a single parameter (blowcount and maximum torque respectively); even the basic cone (CPT) measures two independent parameters (q_c and f_s), while a piezocone (CPTU) extends this to three independent parameters (u_2, added to the previous two) and the most sophisticated seismic cone[①] (SCPTU) measures four parameters (q_c, f_s, u_2 and V_s). As a result, the CPT can be used to reliably estimate a wider range of soil properties, including strength, stiffness, state and consolidation parameters. Furthermore, unlike the SPT, FVT or PMT, where measurements may only be taken at discrete points, the CPT measures continuously,[②] so that by using the interpolated soil profile the complete variation of correlated soil properties with depth may also be determined.

① 带地震波静力触探。

② 与 SPT、FVT 或 PMT 只能测离散点的特性不同，CPT 能连续测试土体特性。

Interpretation of CPT data in coarse-grained soils (I_D, ϕ'_{max}, G_0)

A large database[③] of CPT data in coarse-grained soils is available in the literature. In such soils, q_c is normally used in correlations as increases in density or soil strength will increase the resistance to penetration. Sleeve friction[④] (f_s) is usually small and of little use in interpretation, other than for identifying the soil in question as coarse-grained. As the permeability of coarse-grained deposits is usually high (Table 2.1) it is not necessary to correct q_c for pore pressure effects (Equation 6.2) such that $q_t \approx q_c$, and a basic cone is suitable for most testing. In most correlations it is usual to correct the cone resistance for over-burden stresses[⑤] by using the parameter $q_c/(\sigma'_{v0})0.5$.

③ 数据库。

④ 侧壁摩阻力。

⑤ 锥尖阻力的上覆应力修正。

Figure 7.20 shows correlations between I_D and $q_c/(\sigma'_{v0})^{0.5}$ for a database of nearly 300 tests in a range of normally consolidated (NC) silica and carbonate sands[⑥], collated by Jamiolkowski *et al.* (2001) and Mayne (2007). There is a considerable amount of scatter shown, which is predominantly a function of the compressibility of the soil. The best-fit lines[⑦] for use in the interpretation of new data are given by

$$I_D = D + E\log\left(\frac{q_c}{\sigma_{v0}'^{0.5}}\right) \tag{7.35}$$

where q_c and σ'_{v0} are in kPa. For silica sands of average compressibility, $D = -1.21$ and $E = 0.584$ (best-fit line shown in Figure 7.20). For highly compressible silica sands D may be as high as -1.06 (upper bound envelope to test data), while for very low compressibility soils D may be as low as -1.36 (lower bound envelope to test data); the value of E (the gradient of the line) is insensitive to soil compressibility. Carbonate sands are more highly compressible than silica sands due to their highly crushable grains[⑧], and the data for such soils therefore lie above the silica data (i.e. on the highly compressible side of the graph). The rela-

⑥ 石英砂与钙质砂。

⑦ 拟合曲线。

⑧ 易破碎颗粒。

tionship between I_D and q_c for these soils may still be characterised using Equation 7.35, but with $D = -1.97$ and $E = 0.907$ (line also shown in Figure 7.20).

As with the SPT test, CPT data in coarse-grained soils may further be correlated against ϕ'_{max} as shown in Figure 7.21. The data used in determining this correlation are from Mayne (2007), and show very low scatter for soils with low fines content[1]. The best-fit lines for use in the interpretation of new data are given by

$$\phi'_{max} = 6.6 + 11\log\left(\frac{q_c}{\sigma'^{0.5}_{v0}}\right) \tag{7.36}$$

where ϕ'_{max} is in degrees and q_c and σ'_{v0} are in kPa as before.

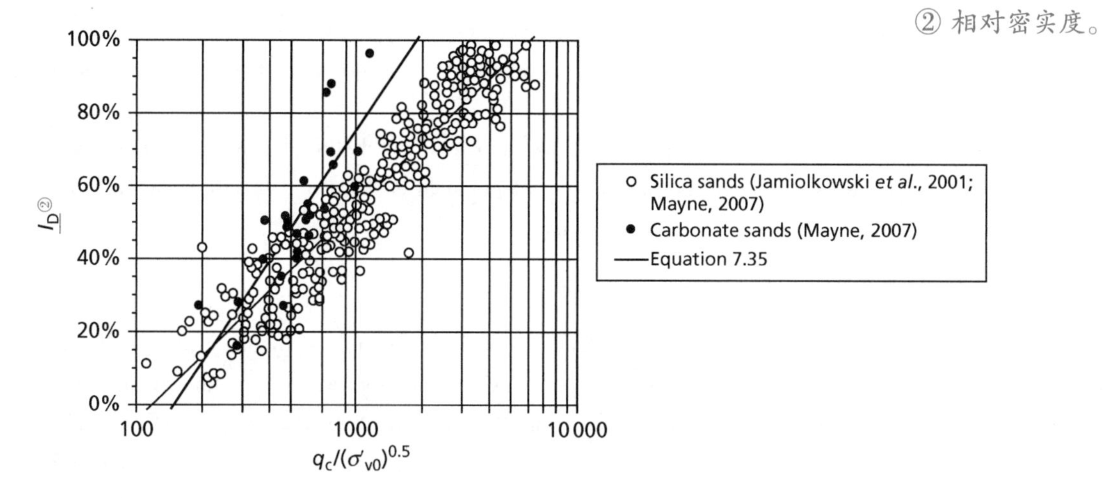

Figure 7.20 Determination of I_D from CPT/CPTU data.

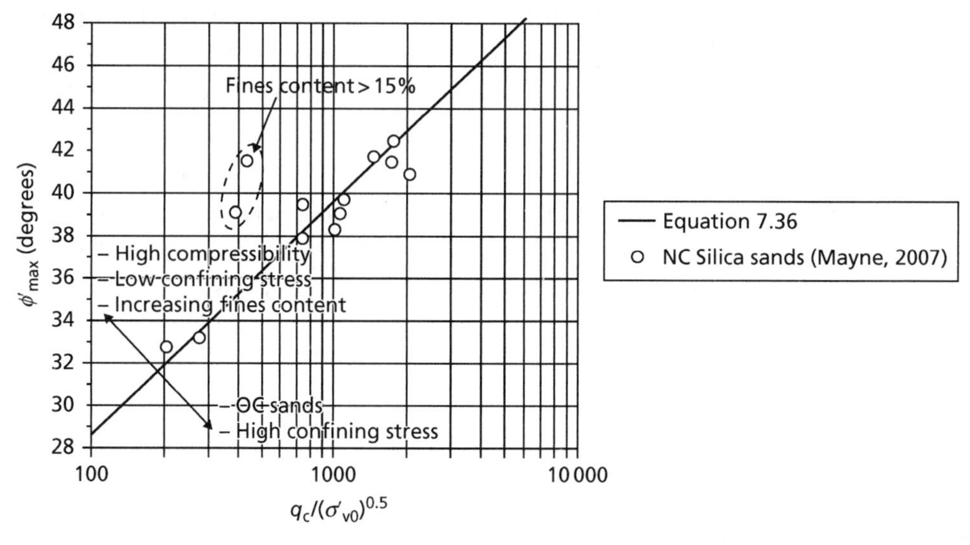

Figure 7.21 Determination of ϕ'_{max} from CPT/CPTU data.

The use of a seismic cone (SCPTU) allows for discrete seismic soundings to be made during a CPT test such that the shear wave velocity[3] is additionally determined. The small strain shear modulus (G_0)[4] can then be determined as for the ge-

① 细颗粒含量。

② 相对密实度。

③ 剪切波速。

④ 小应变剪切模量。

ophysical methods described in Section 6.7 using Equation 6.6:

$$G_0 = \rho V_s^2$$

Interpretation of CPT data in fine-grained soils (c_u, OCR, K_0, ϕ'_{max}, G_0)

In fine-grained soils, CPT data are most commonly used to assess the in-situ undrained shear strength of the soil.[①] As mentioned previously, the CPT provides these data continuously over the full depth of such a layer, unlike triaxial tests on undisturbed samples which can only give a limited number of discrete values. CPT data should always be calibrated[②] against another form of testing (e. g. UU triaxial compression tests or FVT data) in a given material, but once this has been done the CPT can then be used directly to determine c_u at other locations within the same geological unit.

① 土体原位不排水抗剪强度。

② 标定。

The calibration process described above varies slightly depending on the type of cone used, though the principle is identical in each case.[③] If only basic CPT data are available, c_u is determined using:

$$c_u = \frac{q_c - \sigma_{v0}}{N_k} \tag{7.37}$$

where N_k is the 'calibration factor'.[④] This is determined by using the results of a series of laboratory tests (e. g. UU triaxial test), from which c_u is known, and interpolating the value of q_c and σ_{v0} from the CPT log at the depths for which the laboratory tests were sampled (see Example 7.4). Once an appropriate average value of N_k has been determined for a given unit of soil, Equation 7.37 is then applied to the full CPT log to determine the variation of c_u with depth. Figure 7.22 (a) shows reported values of N_k as a function of plasticity index for different fine-grained soils for general guidance and checking of N_k. It will be seen that $N_k = 15$ is normally a good first approximation, though in fissured clays (those shown are from the UK) the value can be significantly higher (i. e. using a value of $N_k = 15$ would overestimate[⑤] c_u in a fissured clay).

③ 虽然每个实例的标定原理是一样的，但标定步骤通常会由于所用锥体的不同而有微小改变。

④ 标定系数。

⑤ 高估。

If CPTU data are available, the process is the same; however q_t replaces q_c so that excess pore pressures generated during penetration are corrected for. With this modification, calibration factors will not be the same as described above and it is conventional to then modify Equation 7.37 to read

$$c_u = \frac{q_t - \sigma_{v0}}{N_{kt}} \tag{7.38}$$

where N_{kt} is the calibration factor for CPTU data[⑥]. Figure 7.22 (b) shows reported values of N_{kt} which can be seen to be a function of the pore pressure parameter B_q (defined in Figure 6.12). The scatter in the data here is much lower as, between them, q_t and B_q implicitly include the effects of overconsolidation (see following discussion). The best fit line to the data is given by

$$N_{kt} = 7.2\,(B_q)^{-0.77} \tag{7.39}$$

⑥ 考虑孔压后的标定系数。

It should be noted that in Figure 7.22 (a), the reference value of c_u is that from UU triaxial compression tests; for fissured clays these were conducted on large (100-

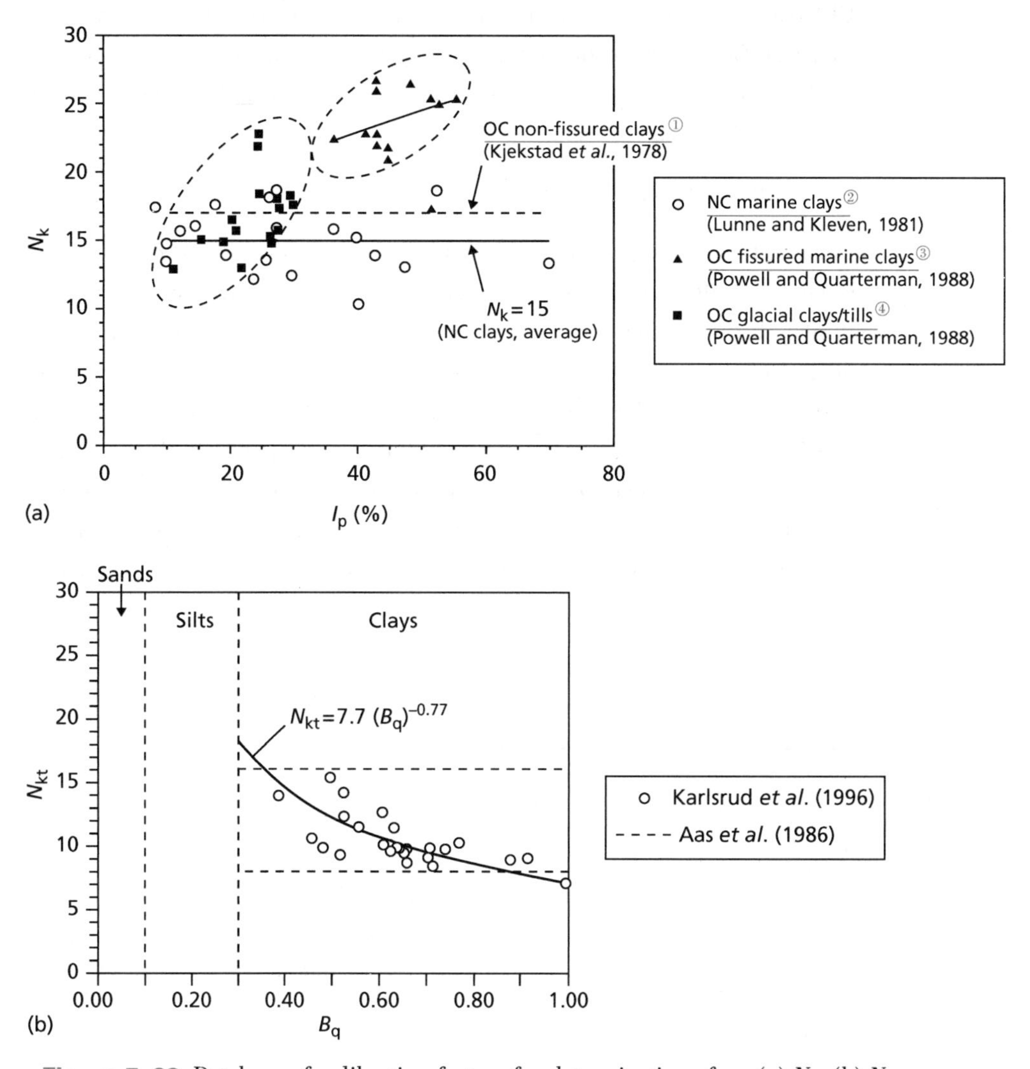

Figure 7.22 Database of calibration factors for determination of c_u: (a) N_k, (b) N_{kt}.

① 超固结无裂隙黏土。
② 正常固结海洋黏土。
③ 超固结有裂隙海洋黏土。
④ 超固结冰碛黏土。

mm diameter) samples to account for the effects of the fissures. In non-fissured clays, there should be little difference in the values of N_k or N_{kt} for different sizes of triaxial sample. If there are different units of clay indicated in a single log (e. g. soft marine deposited clay overlying fissured clay), it may be necessary to use different values of N_k or N_{kt} in the different strata.

In addition to determining the undrained strength properties of fine-grained soils, CPTU data can also be used to estimate the effective stress strength parameter ϕ'_{max}. Mayne and Campanella (2005) suggested that this parameter can be correlated to the normalised cone resistance⑤ $Q_t = (q_t - \sigma_{v0})/\sigma'_{v0}$ and the pore pressure parameter B_q by

⑤ 归一化的锥尖阻力。

$$\phi'_{max} \approx 29.5\,(B_q)^{0.121}\left[0.256 + 0.336B_q + \log\left(\frac{q_t - \sigma_{v0}}{\sigma'_{v0}}\right)\right] \tag{7.40}$$

Equation 7.40 is applicable for $0.1 < B_q < 1.0$. For soils with $B_q < 0.1$ (i. e. sands), Equation 7.36 should be used instead.

CPT data can also be reliably used in most fine-grained soils for detailed determination of OCR with depth, and thereby quantifying the stress history of the soil[1]. Based on a large database of test data for non-fissured clays, Mayne (2007) suggests that OCR may be estimated using

① 土的应力历史。

$$\mathrm{OCR} = 0.33\left(\frac{q_t - \sigma_{v0}}{\sigma'_{v0}}\right) \tag{7.41}$$

Equation 7.41 is compared in Figure 7.23 with data for marine deposited clays[2] from Lunne *et al.* (1989), and it can be seen that the method is reliable in non-fissured soils. Also shown in Figure 7.23 is a zone for fissured clays based on an additional smaller database from Mayne (2007), where the coefficient in Equation 7.41 should be increased to between 0.66 and 1.65. Given the wide extent of this zone, the CPT should be considered less reliable for determining OCR in fissured soils, and should always be supported by data from other tests (e.g. oedometer test data[3]).

② 海洋沉积黏土。

③ 固结试验数据。

The CPT may also be used to estimate the in-situ horizontal stresses[4] in the ground. The ratio of in-situ horizontal effective stress (σ'_{h0}) to in-situ vertical effective stress (σ'_{v0}) is expressed by

④ 原位水平向应力。

$$K_0 \approx \frac{\sigma'_{h0}}{\sigma'_{v0}} \tag{7.42}$$

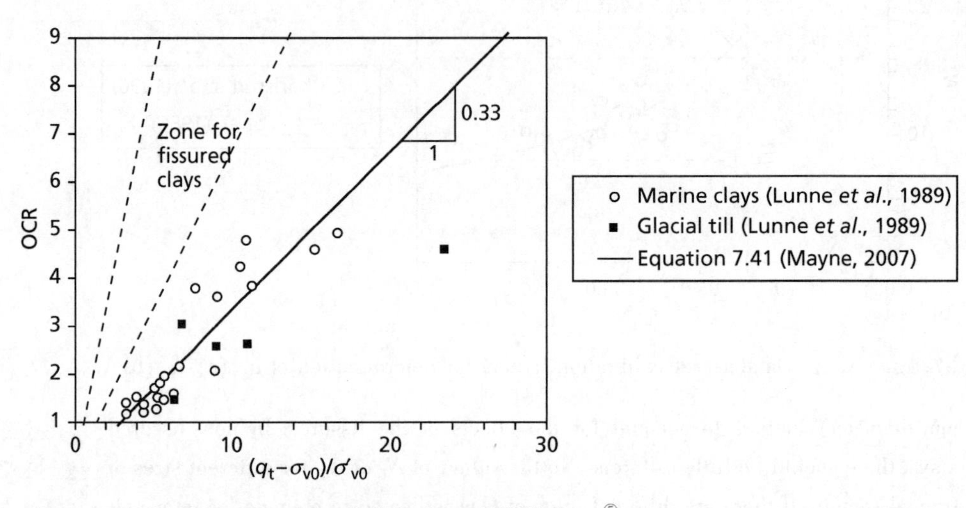

Figure 7.23 Determination of OCR from CPTU data.[5]

where K_0 is the **coefficient of lateral earth pressure (at rest)**[6]. Kulhawy and Mayne (1990) presented an empirical correlation for K_0 from CPTU data, meaning that the CPT can also provide information on the in-situ stress state within the ground:

⑤ 由 CPTU 试验数据确定超固结比。

⑥ 侧向土压力系数（静止）。

$$K_0 \approx 0.1\left(\frac{q_t - \sigma_{v0}}{\sigma'_{v0}}\right) \tag{7.43}$$

Once σ'_{h0} has been determined using Equation 7.42 and 7.43, the total horizontal stress may then be found by adding the in-situ pore pressure (Terzaghi's Principle). The correlation represented by Equation 7.43 is shown in Figure 7.24. There is considerable scatter in the data[7], such that the CPT should only be used to interpret K_0 if no other data are available. If a more accurate value is required,

⑦ 数据离散性较大。

the PMT[①] should be used to directly measure the in-situ horizontal stresses from which K_0 may be determined using Equation 7.42.

① 旁压试验。

As in coarse-grained soils, the use of a seismic cone (SCPTU) allows for measurements of G_0 from shear wave velocity to be made, using Equation 6.6.

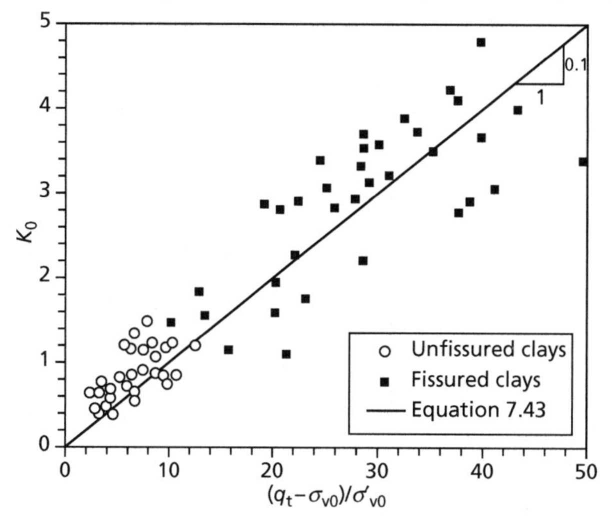

Figure 7.24 Estimation of K_0 from CPTU data.

Example 7.4

CPTU data are shown in Figure 7.25 for the Bothkennar clay of Example 7.1. Figure 7.26 shows laboratory test data consisting of UU triaxial test data and oedometer test data for the same soil. Both sets of data are provided electronically on the Companion Website. Using the two sets of data: (a) determine the value of N_{kt} appropriate for the CPT in this geological unit[②]; (b) determine the variation of OCR with depth from the CPTU data and compare this to the oedometer test data.

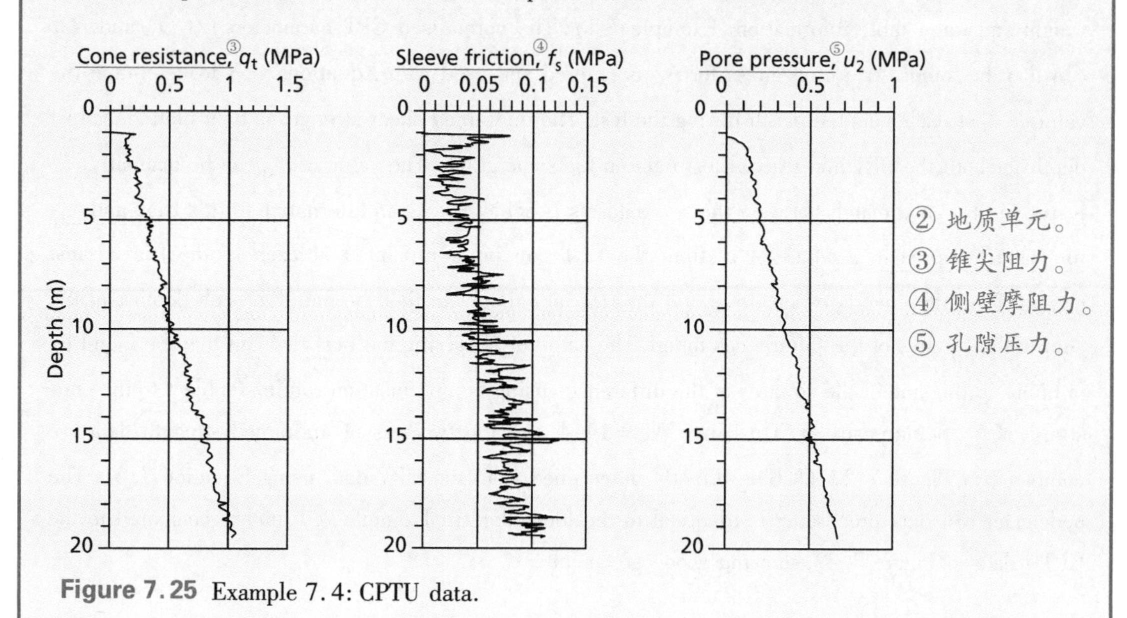

Figure 7.25 Example 7.4: CPTU data.

② 地质单元。
③ 锥尖阻力。
④ 侧壁摩阻力。
⑤ 孔隙压力。

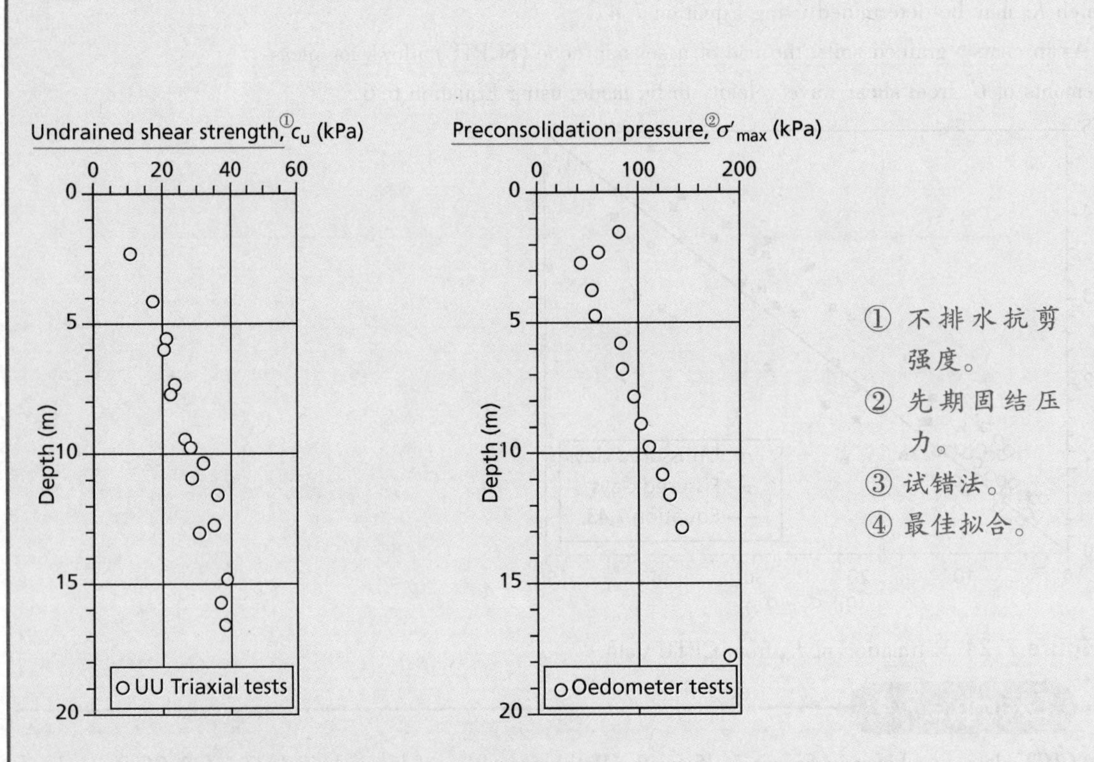

Figure 7.26 Example 7.4: Laboratory test data.

Solution

Values of σ_{v0}, u_0 and σ'_{v0} are first found at each depth sampled with the CPTU using the soil unit weight and water table information (Example 7.1). The normalised CPT parameters (Q_t, F_r and B_q) can then be found. An initial guess for N_{kt} is entered and used with Equation 7.38 to determine the value of c_u at each sampled depth during the test. The undrained shear strength is then plotted against depth for both the CPT and triaxial test data on the same graph. The value of N_{kt} can be manually adjusted until a good match between the two datasets is achieved. As an alternative to this trial-and-error manual approach[3], values of c_u from the CPT can be interpolated at each of the triaxial test depths. The difference between these and the triaxial values can then be found at each depth and the sum of the squares of the differences found. The value of N_{kt} giving the best fit[4] can then be found by minimising the sum of the squares of the differences using an optimisation routine (subject to the constraint of N_{kt} being positive). This gives $N_{kt} = 14.4$, and the two sets of undrained strength data are compared in Figure 7.27. OCR is directly determined from the CPT data using Equation 7.41. The oedometer test data processing is identical to that described in Example 7.1, and is compared to the CPTU data in Figure 7.27, showing good agreement.

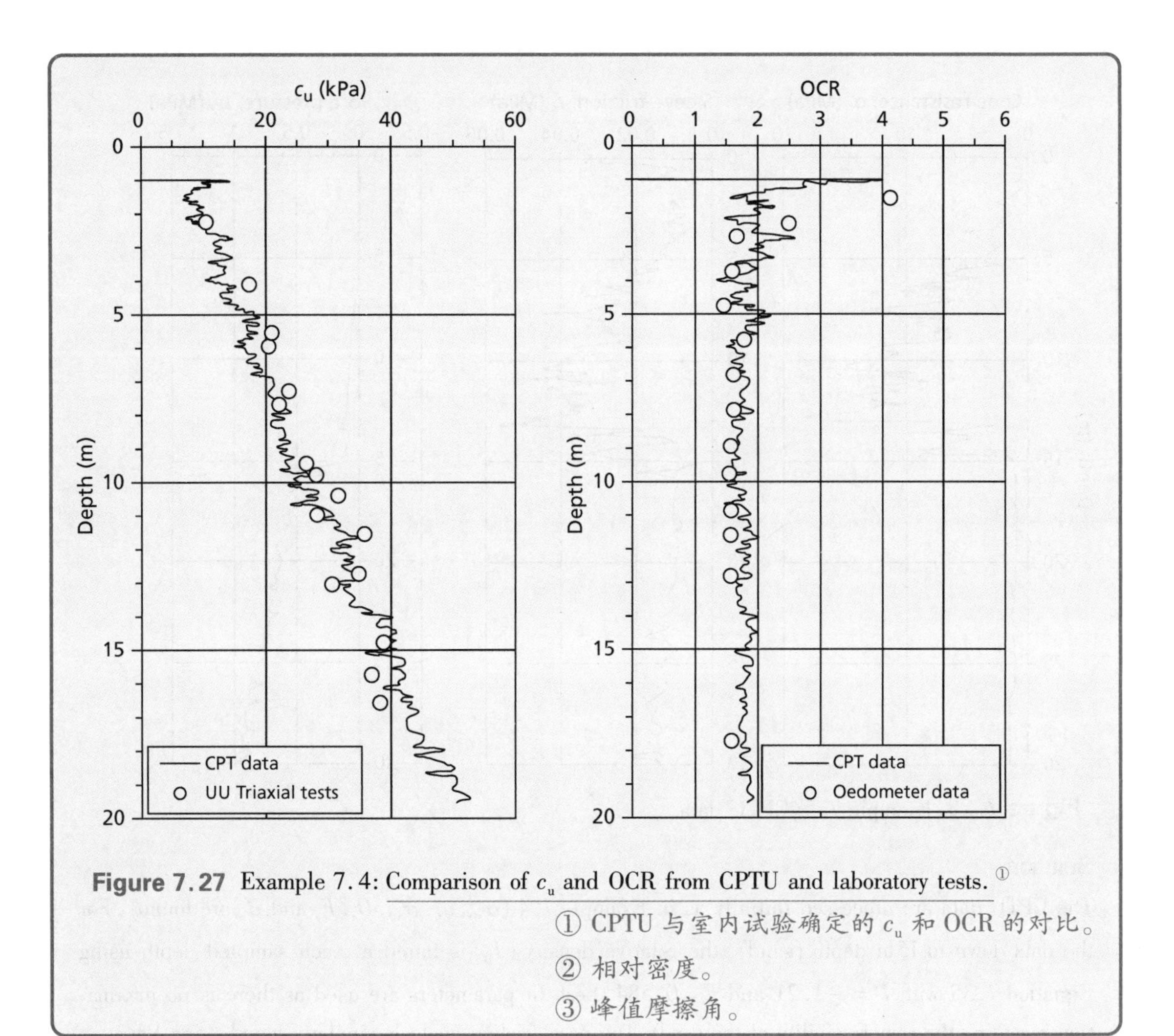

Figure 7.27 Example 7.4: Comparison of c_u and OCR from CPTU and laboratory tests.[①]

① CPTU 与室内试验确定的 c_u 和 OCR 的对比。

② 相对密度。

③ 峰值摩擦角。

Example 7.5

Figure 7.28 shows CPTU data for a site in Canada. SPT test data at the same site are presented in Table 7.3. Both sets of data are provided electronically on the Companion Website. The site has been identified, both from borehole logs and the CPTU record, as consisting of 15m of silica sand, overlying soft clay. The unit weight of both soils has been estimated at $\gamma \approx 17\text{kN/m}^3$ and the water table is at 2m BGL. Determine the relative density[②] and peak friction angle[③] of the sand layer and the undrained shear strength of the clay layer using both datasets.

Table 7.3 Example 7.5: SPT data

Depth (m)	1.3	3.4	5.1	6.7	8.2	9.8	11.3	12.8	14.3	15.2	17.3	19.0	21.2	22.6	24.0
N_{60}	2	3	6	17	15	24	19	17	6	4	4	4	6	6	5

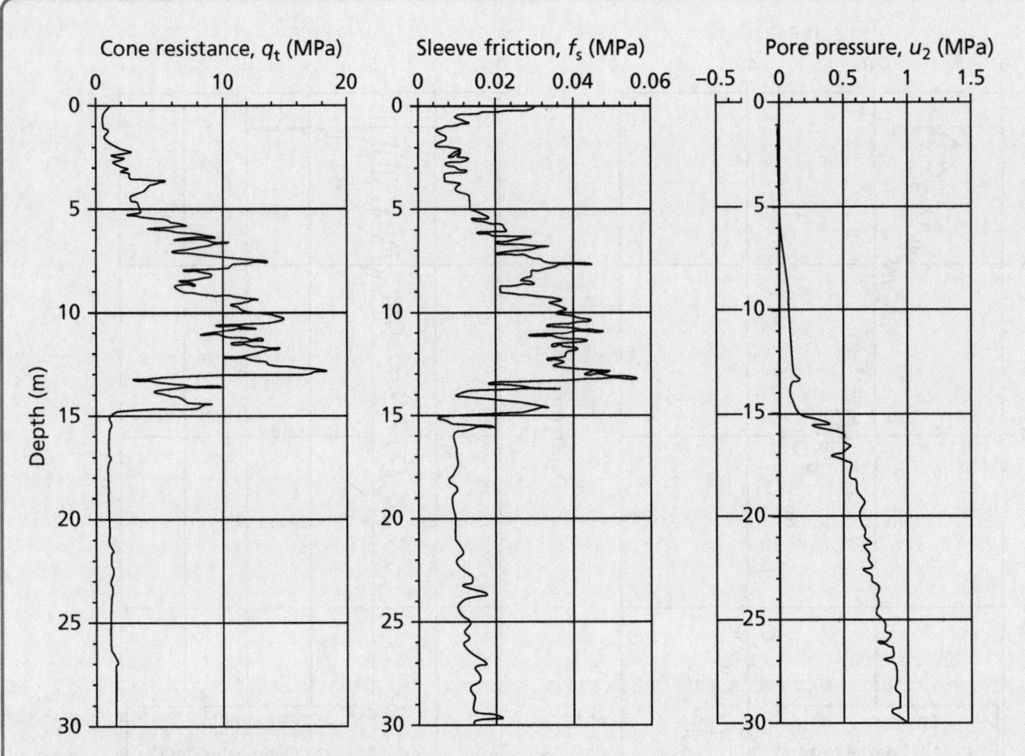

Figure 7.28 Example 7.5: CPTU data.

Solution

The CPTU data are processed initially as in Example 7.4 (σ_{v0}, u_0, σ'_{v0}, Q_t, F_r and B_q are found). For the data down to 15m depth (sand), the relative density (I_D) is found at each sampled depth using Equation 7.35 with $D = -1.21$ and $E = 0.584$ (best-fit parameters are used as there is no information regarding the compressibility of the sand). The peak friction angle is similarly found using Equation 7.36. Below 15-m depth (in the clay) c_u is determined at each sampled depth, where N_{kt} at each depth is found from B_q using Equation 7.39. For the SPT data, σ_{v0}, u_0 and σ'_{v0} are first determined at each test depth. The values of σ'_{v0} are then used to determine correction factors (C_N) at each depth using Equation 7.3, with $A = 200$ and $B = 100$ (in the absence of any more detailed information regarding grading). These values are used to determine corrected blowcounts $(N_1)_{60}$, from which relative density is approximated for tests down to 15-m depth using $(N_1)_{60}/I_D^2 = 60$, followed by determination of ϕ'_{max} values from Figure 7.4. For test points below 15m which are in the clay, undrained shear strengths are determined directly from the normalised blowcounts (N_{60}) using $c_u/N_{60} \approx 10$ (i. e. assuming a soft, insensitive non-fissured clay in the absence of any more detailed information). The derived parameters from the CPTU and SPT data are compared in Figure 7.29, and show reasonable agreement. The SPT data slightly underpredicts ϕ'_{max} compared to the CPTU (though the trend is similar), suggesting that the sand is overconsolidated (see Figure 7.4).

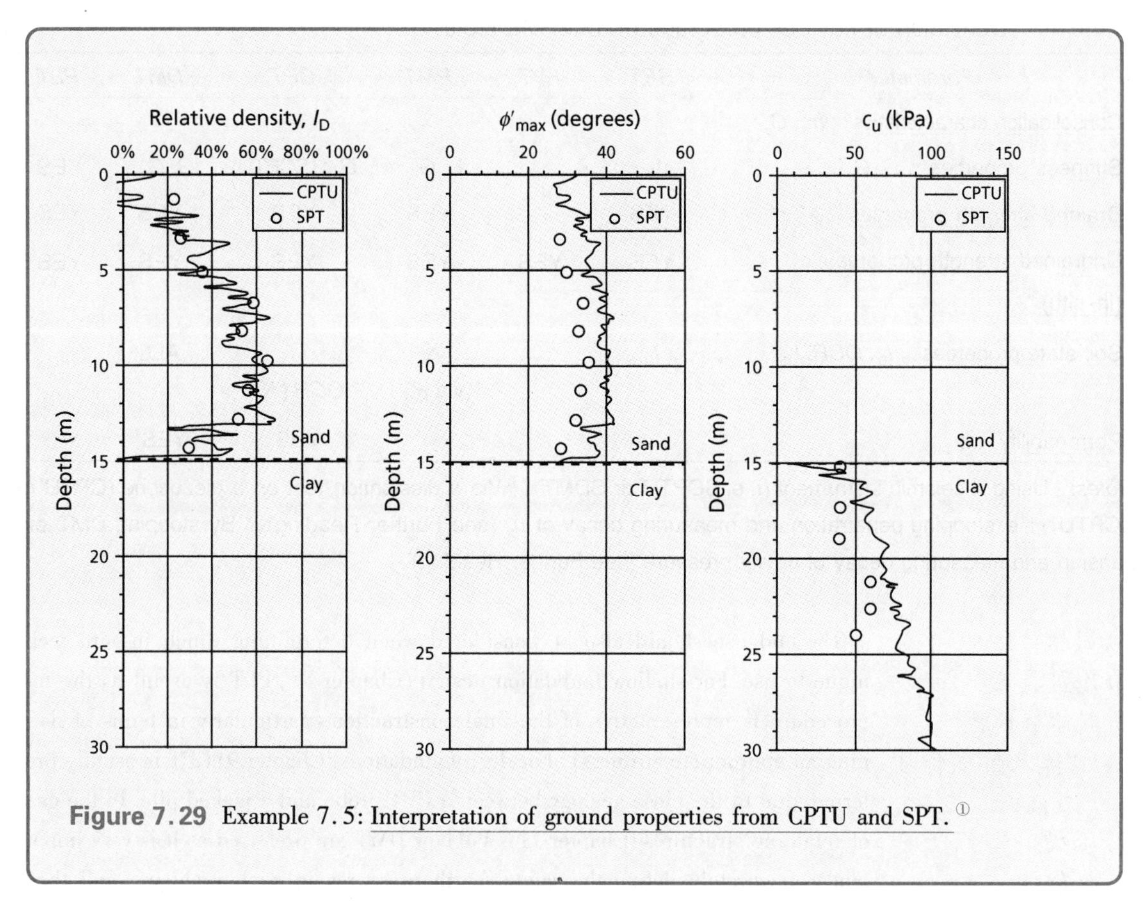

Figure 7.29 Example 7.5: Interpretation of ground properties from CPTU and SPT.[1]

① 由 CPTU 与 SPT 数据判读场地的特性。

7.6 Selection of in-situ test method(s)

Section 6.4 described the parameters that could be determined from the various laboratory tests to aid in the design of a programme of ground investigation. The same can be done for the in-situ tests described in this and the previous chapters. Table 7.4 summarises the mechanical characteristics[2] which can be obtained from each type of in-situ test, including those which were mentioned in Section 7.1, but which have not been described in detail (DMT, PLT). It will be shown in Part 2 of this book (Chapters 8-13 inclusive) that modern design approaches require both stiffness and strength parameters[3] to verify that an appropriate level of performance will be achieved in a rigorous way. This is in contrast to older 'traditional' approaches which relied on strength parameters only and applied highly empirical global factors of safety to ensure adequate performance. The prevalence of 'traditional' approaches until recently explains the popularity of the SPT, as it can determine the necessary strength parameters while being simple, quick and cheap. It is expected that over the coming years CPT and PMT will become more popular in general use as they can provide reliable data on both the strength and stiffness of soil.

② 力学特性。

③ 刚度和强度参数。

Table 7.4 Derivation of key soil properties from in-situ tests

Parameter	*SPT*	*FVT*	*PMT*	*CPT*	*DMT*	*PLT*
Consolidation characteristics①: m_v, C_c						
Stiffness properties②: G, G_0			G	G_0* (SCPTU)	G, G_0*	YES
Drained strength properties③: ϕ', c'	YES		YES	YES	YES	YES
Undrained strength properties: c_u (in-situ)④	YES	YES	YES	YES	YES	YES
Soil state properties⑤: I_D, OCR, K_0	I_D		K_0 (via σ_{h0})	I_D, OCR (K_0)	ALL	
Permeability⑥: k				YES†	YES‡	

Notes: * Using a seismic instrument (i. e. SCPTU or SDMT). † Via a dissipation test on a piezocone (CPTU or SCPTU) - i. e. stopping penetration and measuring decay of u_2 (see Further Reading). ‡ By stopping DMT expansion and measuring decay of cavity pressure (see Further Reading).

① 固结特性。
② 刚度特性。
③ 排水强度特性。
④ 不排水强度特性（原位）。
⑤ 土体状态特性。
⑥ 渗透性。

The end use should also be considered when determining which in-situ technique to use. For shallow foundation design (Chapter 8), PLT is useful as the test procedure is representative of the final construction (particularly in terms of defining an appropriate stiffness). For deep foundations (Chapter 9) CPT is usually preferred, due to the close analogy between a CPT probe and a jacked pile. In the case of retaining structures (Chapter 11), PMT or DMT are preferred as it is very important to accurately define the lateral earth pressures in such problems, and these tests are most reliable for this.

Summary

1 In-situ testing can be a valuable tool for evaluating the constitutive properties of the ground. Generally, a much larger body of soil is influenced during such tests, which can be advantageous over laboratory tests on small samples in certain soils (e. g. fissured clays). The tests also remove many of the issues associated with sampling, though attention must be paid instead to the disturbance of the soil that might occur during installation of the test device. The data collected from in-situ tests complements (rather than replaces) laboratory testing and the use of empirical correlations (Chapter 5). The use of in-situ testing can dramatically reduce the amount of sampling and laboratory testing required, provided it is calibrated against at least a small body of high-quality laboratory data. It can therefore be invaluable for the cost-effective ground investigation of large sites.

2 The four principle in-situ tests that are conducted in practice are (arguably) the Standard Penetration Test (SPT), Field Vane Test (FVT), Pressuremeter Test (PMT) and Cone Penetration Test (CPT). The first two of these are the simplest and cheapest tests, and are used to determine the strength characteristics of coarse-grained soil (SPT) and soft fine-grained soils (FVT); the SPT may also be used in stiffer fine-grained soils with greater caution. The latter two tests (PMT and CPT) represent modern devices making use of miniaturised sensors and computer control/data logging to measure multiple parameters, providing more detailed data and increasing their range of applicability. These tests are applicable to both coarse-and fine-grained soils, and can be used to reliably determine both strength and stiffness characteristics via theoretical models (PMT) or empirical correlations (CPT).

3 It has been demonstrated, through application to real soil test data, that spreadsheets are a useful tool for processing and interpreting a detailed suite of in-situ test data; moreover, they are essential for processing data from computerised PMT and CPT devices, which provide digital output. Digital exemplars have been provided for the worked examples in this chapter on the Companion Website, utilising data from all four of the tests discussed.

Problems

7.1 Table 7.5 presents corrected SPT blowcounts for a site consisting of 5m of silt overlying a thick deposit of clean silica sand. These data are provided electronically on the Companion Website. The saturated unit weight of both soils is $\gamma \approx 16\text{kN/m}^3$ and the water table is 1.6m below ground level (BGL). Determine the average relative density and peak friction angle of the sand between 10-20m BGL.

Table 7.5 Problem 7.1

Depth (m)	N_{60} *(blows)*	*Depth (m)*	N_{60} *(blows)*	*Depth (m)*	N_{60} *(blows)*
1.32	4	11.27	23	21.06	35
2.50	10	12.29	22	22.21	27
3.29	7	13.39	30	23.16	28
4.30	2	14.34	29	24.32	24
5.34	8	15.20	19	25.20	30
6.44	11	16.34	9	26.08	30
7.31	10	17.33	30	27.10	30
8.41	12	18.28	30	28.12	32
9.29	18	19.23	34	29.25	11
10.40	24	20.25	30	30.22	22

7.2 Figure 7.30 presents the results of a self-boring pressuremeter test undertaken at a depth of 10.4m

BGL in the sand deposit described in Problem 7.1. The data are provided electronically on the Companion Website. Determine the following parameters: σ_{h0}, u_0, σ'_{h0}, ϕ'_{max}, ψ. Find also the value of G from the unload-reload loop.

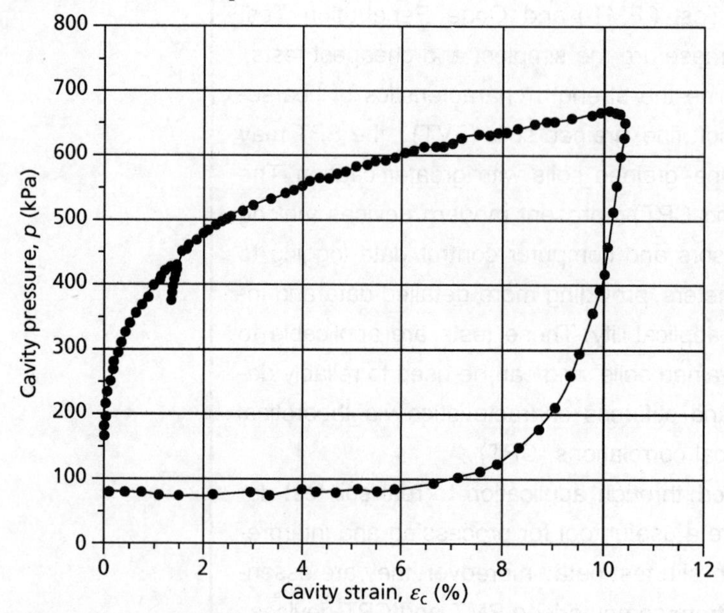

Figure 7.30 Problem 7.2.

7.3 Figure 7.31 presents the results of a CPT test undertaken in the sand deposit described in Problems 7.1 and 7.2. The data are provided electronically on the Companion Website. Determine the variation of relative density and peak friction angle with depth, and compare the results to those from the SPT tests in Problem 7.1 (the SPT data are also provided electronically).

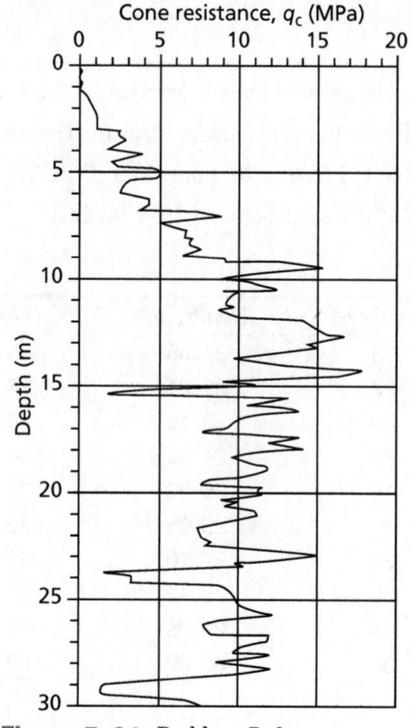

Figure 7.31 Problem 7.3.

7.4 Figure 7.32 presents the results of a self-boring pressuremeter test undertaken deep within a deposit of clay. The data are provided electronically on the Companion Website. Determine the in-situ horizontal total stress and the undrained shear strength of the clay at the test depth. Is the unload-reload loop appropriately sized?

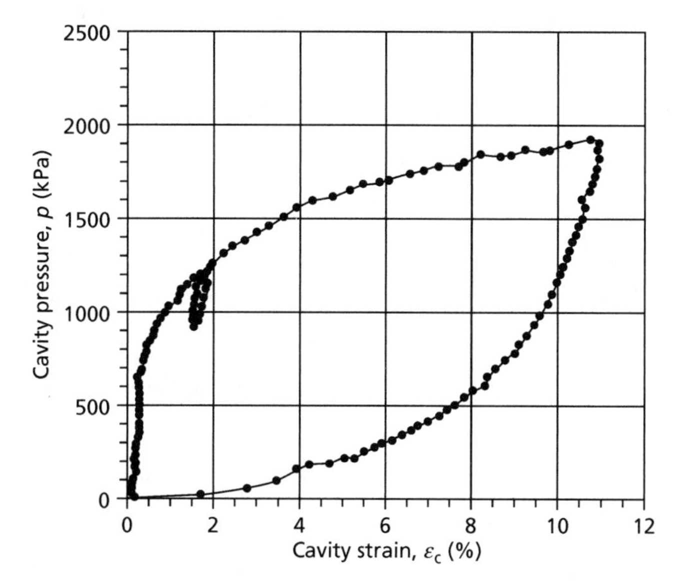

Figure 7.32 Problem 7.4.

7.5 Figure 7.33 presents the results of a CPTU test undertaken in the Gault clay shown in Figure 5.38 (a). The data are provided electronically on the Companion Website, along with data from a series of SBPM tests and SPT tests conducted in the same material. The clay has a unit weight of $\gamma \approx 19\mathrm{kN/m^3}$ above and below the water table, which is at a depth of 1 m BGL.

a Using the SBPM data as a reference, determine the value of N_{kt} appropriate for obtaining c_u from the CPTU data. What does this suggest about the clay deposit?

b Using your answer to (a) or otherwise, determine the variation of c_u with depth from the SPT data.

c Determine the variation of K_0 with depth using both the SBPM and CPTU data.

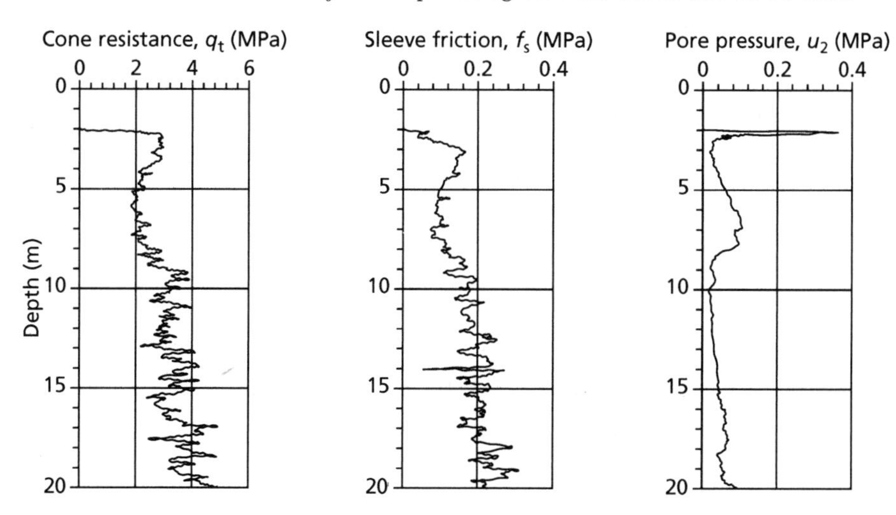

Figure 7.33 Problem 7.5.

References

Aas, G., Lacasse, S., Lunne, T. and Høeg, K. (1986) Use of in situ tests for foundation design on clay, in *Proceedings of the ASCE Speciality Conference In-situ '86: Use of In-situ Tests in Geotechnical Engineering, Blacksburg, VA*, pp. 1 – 30.

ASTM D1586 (2008) *Standard Test Method for Standard Penetration Test (SPT) and Split-Barrel Sampling of Soils*, American Society for Testing and Materials, West Conshohocken, PA.

ASTM D2573 (2008) *Standard Test Method for Field Vane Shear Test in Cohesive Soil*, American Society for Testing and Materials, West Conshohocken, PA.

ASTM D4719 (2007) *Standard Test Method for Prebored Pressuremeter Testing in Soils*, American Society for Testing and Materials, West Conshohocken, PA.

ASTM D5778 (2007) *Standard Test Method for Electronic Friction Cone and Piezocone Penetration Testing of Soils*, American Society for Testing and Materials, West Conshohocken, PA.

Azzouz, A., Baligh, M. and Ladd, C. C. (1983) Corrected field vane strength for embankment design, *Journal of Geotechnical and Geoenvironmental Engineering*, **109** (5), 730 – 734.

Bjerrum, L. (1973) Problems of soil mechanics and construction on soft clays, in *Proceedings of the 8th International Conference on SMFE, Moscow*, Vol. 3, pp. 111 – 159.

BS 1377 (1990) *Methods of test for soils for civil engineering purposes, Part 9: In-situ tests*, British Standards Institution, UK.

BS EN ISO 22476 (2005) *Geotechnical Investigation and Testing – Field Testing*, British Standards Institution, UK.

Clayton, C. R. I. (1995) The Standard Penetration Test (SPT): methods and use, *CIRIA Report 143*, CIRIA, London.

Dalton, J. C. P. and Hawkins, P. G. (1982) Fields of stress – some measurements of the in – situ stress in a meadow in the Cambridgeshire countryside. *Ground Engineering*, **15**, 15 – 23.

EC7 – 2 (2007) Eurocode 7: Geotechnical design – Part 2: Ground investigation and testing, BS EN 1997 – 2: 2007, British Standards Institution, London.

Gibson, R. E. and Anderson, W. F. (1961) In-situ measurement of soil properties with the pressuremeter, *Civil Engineering and Public Works Review*, **56**, 615 – 618.

Hughes, J. M. O., Wroth, C. P. and Windle, D. (1977) Pressuremeter tests in sands, *Géotechnique*, **27** (4), 455 – 477.

Jamiolkowski, M., LoPresti, D. C. F. and Manassero M. (2001) Evaluation of relative density and shear strength of sands from Cone Penetration Test and Flat Dilatometer Test, *Soil Behavior and Soft Ground Construction, Geotechnical Special Publication 119*, American Society of Civil Engineers, Reston, VA, pp. 201 – 238.

Karlsrud, K., Lunne, T. and Brattlien, K. (1996). Improved CPTU correlations based on block samples, in

Proceedings of the Nordic Geotechnical Conference, Reykjavik, Vol 1, pp. 195 – 201.

Kjekstad, O. , Lunne, T. and Clausen, C. J. F. (1978) Comparison between in situ cone resistance and laboratory strength for overconsolidated North Sea clays. *Marine Geotechnology*, **3**, 23 – 36.

Kulhawy, F. H. and Mayne, P. W. (1990) , *Manual on Estimating Soil Properties for Foundation Design*, Report EPRI EL – 6800, Electric Power Research Institute, Palo Alto, CA.

Lunne, T. and Kleven, A. (1981) Role of CPT in North Sea foundation engineering. *Session at the ASCE National Convention: Cone Penetration testing and materials*, St Louis, American Society of Civil Engineers, Reston, VA, 76 – 107.

Lunne, T. , Lacasse, S. and Rad, N. S. (1989) SPT, CPT, pressuremeter testing and recent developments on in situ testing of soils, in *General Report, Proceedings of the 12th International Conference of SMFE, Rio de Janeiro*, Vol. 4, pp. 2339 – 2403.

Mayne, P. W. (2007) *Cone penetration testing: a synthesis of highway practice*. NCHRP Synthesis Report 368, Transportation Research Board, Washington DC.

Mayne, P. W. and Mitchell, J. K. (1988) Profiling of overconsolidation ratio in clays by field vane. *Canadian Geotechnical Journal*, **25** (1) , 150 – 157.

Mayne, P. W. and Campanella, R. G. (2005) Versatile site characterization by seismic piezocone, in *Proceedings of the 16th International Conference on SMFE*, Osaka, Vol. 2, pp. 721 – 724.

Palmer, A. C. (1972) Undrained plane strain expansion of a cylindrical cavity in clay: a simple interpretation of the pressuremeter test, *Géotechnique*, **22** (3) , 451 – 457.

Powell, J. J. M. and Quarterman, R. S. T. (1988) The Interpretation of Cone Penetration Tests in Clays with Particular Reference to Rate Effects, *Penetration Testing* 1988, *Orlando*, **2**, 903 – 909.

Schmertmann, J. H. (1979) Statics of SPT. *Journal of the Geotechnical Division, Proceedings of the ASCE*, **105** (GT5) , 655 – 670.

Skempton, A. W. (1986) Standard penetration test procedures and the effects in sands of overburden pressure, relative density, particle size, ageing and overconsolidation, *Géotechnique*, **36** (3) , 425 – 447.

Stroud, M. A. (1989) The Standard Penetration Test – its application and interpretation, in *Proceedings of the ICE Conference on Penetration Testing in the UK*, Thomas Telford, London.

Terzaghi, K. and Peck, R. B. (1967) *Soil Mechanics in Engineering Practice* (2nd edn) , Wiley, New York.

Wroth, C. P. (1984) The interpretation of in-situ soil tests, *Géotechnique*, **34** (4) , 449 – 489.

Further reading

Clayton, C. R. I. , Matthews, M. C. and Simons, N. E. (1995) *Site Investigation* (2nd edn) , Blackwell, London.

A comprehensive book on site investigation which contains much useful practical advice and guidance on in-situ testing.

Lunne, T. , Robertson, P. K. and Powell, J. J. M. (1997) *Cone Penetration Testing in Geotechnical Practice*, E & FN Spon, London.

A comprehensive reference on all aspects of the CPT, including apparatus, test set-up, test procedure, and interpretation (of a wider range of parameters than covered herein). Includes a wealth of data from real sites all over the world, and includes interpretation in more challenging ground conditions (e. g. frozen ground, volcanic soil).

Marchetti, S. (1980) In-situ tests by flat dilatometer, *Journal of the Geotechnical Engineering Division, Proceedings of the ASCE*, **106** (GT3), 299 – 321.

This paper (by the test developer) describes the Marchetti flat dilatometer, the device most commonly used for the DMT, and its use in deriving soil properties.

Powell, J. J. M. and Uglow, I. M. (1988) The interpretation of the Marchetti Dilatometer Test in UK clays, *Proceedings of the ICE Conference on Penetration Testing in the UK*, Thomas Telford, London.

A companion to the previous item of further reading, describing the properties that can be reliably derived from the DMT, correlations for use in practice and an idea of the scatter associated with such observations.

Part 2

Applications in geotechnical engineering

Chapter 8

Shallow foundations[1]

① 浅基础。

Learning outcomes

After working through the material in this chapter, you should be able to:

1 Understand the working principles behind shallow foundations;

2 Solve simple foundation capacity problems using Terzaghi's bearing capacity[2] equation and/or limit analysis techniques;

3 Design shallow and deep foundation elements which are subjected to combined loading (vertical, horizontal, moment), using limit analysis techniques (ULS) and elastic solutions (SLS), within a limit-state design framework.

② 太沙基地基承载力。

8.1 Introduction

A foundation is that part of a structure which transmits loads directly to the underlying soil, a process known as **soil-structure interaction**.[3] This is shown indicatively in Figure 8.1 (a). To perform in a satisfactory way, the foundation must be designed to meet two principal performance requirements (known as **limit states**[4]), namely:

1 such that its capacity or resistance is sufficient to support the loads (actions) applied (i. e. so that it doesn't collapse[5]);

2 to avoid excessive deformation[6] under these applied loads, which might damage the supported structure or lead to a loss of function.

These criteria are shown schematically in Figure 8. 1 (b). **Ultimate limit states (ULS)**[7] are those involving the collapse or instability of the structure as a whole, or the failure of one of its components (point 1 above). **Serviceability limit states (SLS)**[8] are those involving excessive deformation, leading to damage or loss of function. Both ultimate and serviceability limit states must always be considered in design. The philosophy of limit states is the basis of Eurocode 7 (EC7, BSI 2004), a standard specifying all of the situations which must be considered in design in the UK and the rest of Europe. Similar standards are beginning to appear in other

③ 基础为结构的一部分，将荷载传递到下覆土层，这一过程也被称为土体-结构相互作用。

④ 极限状态。

⑤ 破坏。

⑥ 过大变形。

⑦ 承载力极限状态。

⑧ 正常使用极限状态。

parts of the world (e. g. GeoCode 21 in Japan). In the United States, the limit state design philosophy is termed Load and Resistance Factor Design (LRFD) ①.

This chapter will focus on the basic behaviour and design of shallow foundations under vertical loading. If a soil stratum near the surface is capable of adequately supporting the structural loads it is possible to use either **footings**② or a **raft**③, these being referred to in general as **shallow foundations**④. A footing is a relatively small slab giving independent support to part of the structure. A footing supporting a single column is referred to as an individual footing or **pad**⑤; one supporting a closely spaced group of columns is referred to as a combined footing, and one that supports a load-bearing wall as a **strip footing**⑥. A raft is a relatively large single slab, usually stiffened with cross members, supporting the structure as a whole. The resistance of a shallow foundation is quantified by its **bearing resistance**⑦ (a limiting load) or **bearing capacity**⑧ (a limiting pressure), the determination of which will be addressed in Sections 8. 2-8. 4. Bearing capacity is directly related to the shear strength of soil and therefore requires the strength properties ϕ', c' or c_u, depending on whether drained or undrained conditions are maintained, respectively.

① 荷载与抗力系数设计方法。
② 基础。
③ 筏基。
④ 浅基础。
⑤ 垫层。
⑥ 条形基础。
⑦ 承载抗力。
⑧ 承载能力。
⑨ 环境作用（如雪荷载）。
⑩ 活（可变）荷载（如层位荷载）。
⑪ 恒荷载（上层结构自重）。
⑫ 基底压力。
⑬ 作用力。
⑭ 承载抗力。
⑮ 荷载作用下基础的性状。
⑯ 工作荷载。
⑰ 极限沉降。
⑱ 实际沉降。
⑲ 与浅基础设计相关的概念。
⑳ 竖向荷载作用下土-结构相互作用。
㉑ 基础性能与极限状态设计。

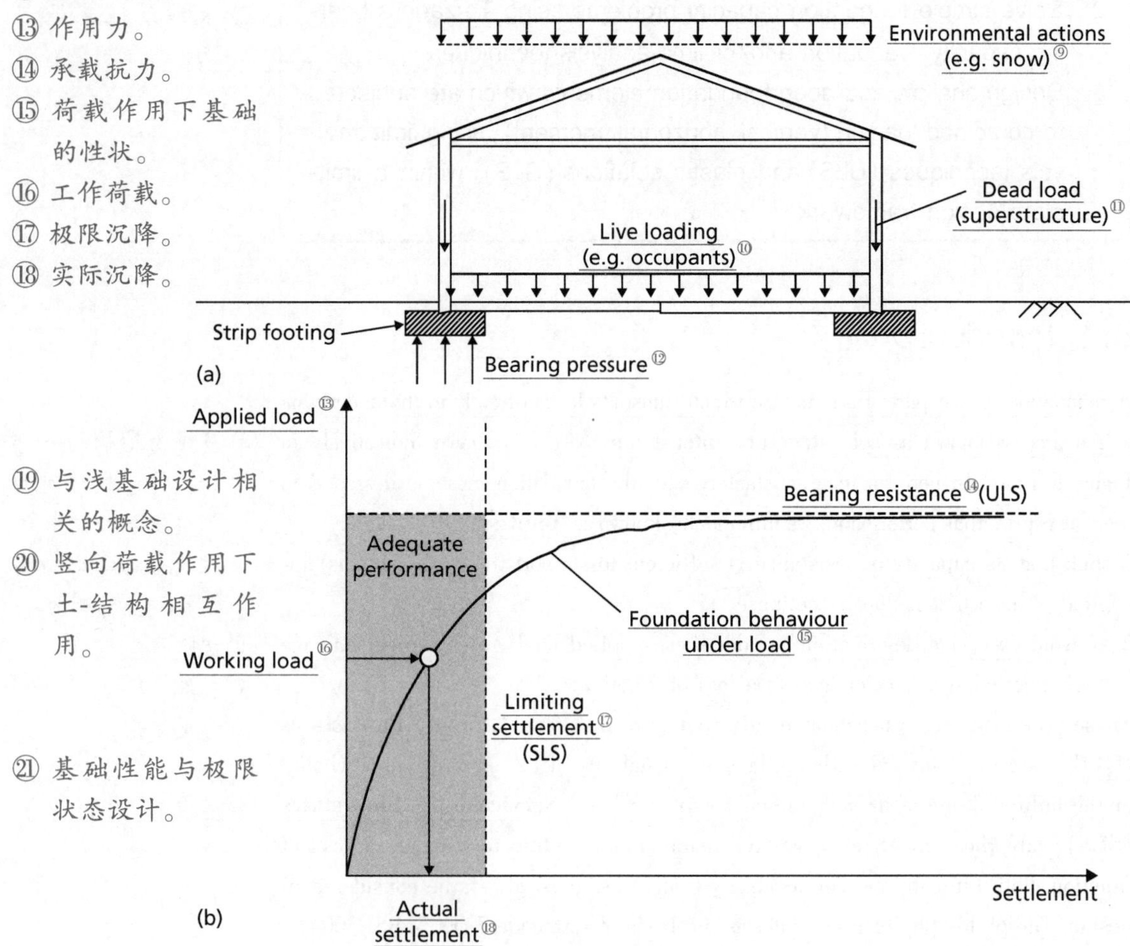

Figure 8.1 Concepts related to shallow foundation design⑲: (a) soil-structure interaction under vertical actions⑳, (b) foundation performance and limit state design㉑.

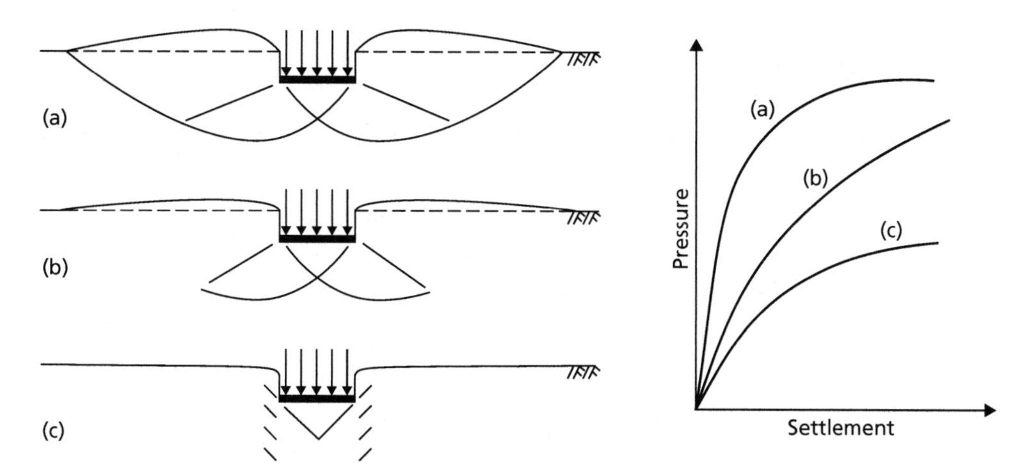

Figure 8.2 Modes of failure: (a) general shear, (b) local shear, and (c) punching shear.[①]

① 破坏模式：(a) 整体剪切破坏，(b) 局部剪切破坏，(c) 冲剪破坏。

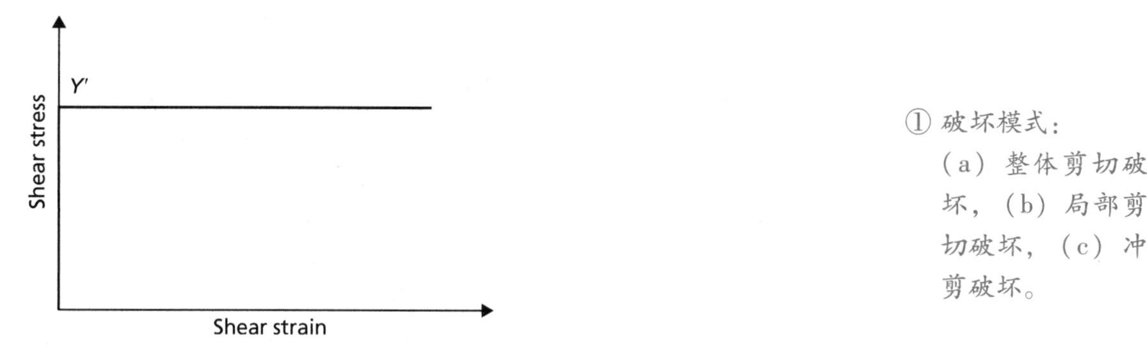

Figure 8.3 Idealised stress-strain relationship in a perfectly plastic material.

When individual shallow foundation elements are used as part of a foundation system (e.g. Figure 8.1 (a), consisting of two separate strips supporting the structure) they may additionally carry horizontal loads[②] and or moments, e.g. induced by wind loading (environmental action); this will be addressed in Section 8.5.

If the soil near the surface is incapable of adequately supporting the structural loads, **piles**[③], or other forms of **deep foundations**[④] such as piers[⑤] or caissons[⑥], are used to transmit the applied loads to suitable soil (or rock) at greater depth where the effective stresses (and hence shear strength, as described in Chapter 5) are larger. Deep foundations are addressed in Chapter 9.

In addition to being located within an adequate bearing stratum, a shallow foundation should be below the depth which is subjected to frost action[⑦] (around 0.5m in the UK) and, where appropriate, the depth to which seasonal swelling and shrinkage of the soil takes place. Consideration must also be given to the problems arising from excavating below the water table if it is necessary to locate foundations below this level. The choice of foundation level may also be influenced by the possibility of future excavations for services close to the structure, and by the effect of construction, particularly excavation, on existing structures and services.

② 水平荷载。

③ 桩。

④ 深基础。

⑤ 墩。

⑥ 沉井。

⑦ 冻融作用。

① 承载力与极限分析。

8.2 Bearing capacity and limit analysis[①]

Bearing capacity (q_f) is defined as the pressure which would cause shear failure of the supporting soil immediately below and adjacent to a foundation.

② 三种不同的破坏模式。
③ 整体剪切破坏。
④ 塑性平衡状态。

Three distinct modes of failure[②] have been identified, and these are illustrated in Figure 8.2; they will be described with reference to a strip footing. In the case of **general shear failure**[③], continuous failure surfaces develop between the edges of the footing and the ground surface, as shown in Figure 8.2. As the pressure is increased towards the value q_f a state of plastic equilibrium[④] is reached initially in the soil around the edges of the footing, which subsequently spreads downwards and outwards. Ultimately, the state of plastic equilibrium is fully developed throughout the soil above the failure surfaces. Heave of the ground surface occurs on both sides of the footing, although in many cases the final slip movement occurs only on one side, accompanied by tilting of the footing, as the footing will not be perfectly level and will hence be biased to fail towards one side. This mode of failure is typical of soils of low compressibility (i. e. dense coarse-grained or stiff fine-grained soils), and the pressure-settlement curve is of the general form shown in Figure 8.2, the ultimate bearing capacity being well defined. In the mode of **local shear failure**[⑤] there is significant compression of the soil under the footing, and only partial development of the state of plastic equilibrium. The failure surfaces, therefore, do not reach the ground surface and only slight heaving occurs. Tilting of the foundation would not be expected. Local shear failure is associated with soils of high compressibility and, as indicated in Figure 8.2, is characterised by the occurrence of relatively large settlements (which would be unacceptable in practice) and the fact that the ultimate bearing capacity is not clearly defined. **Punching shear failure**[⑥] occurs when there is relatively high compression of the soil under the footing, accompanied by shearing in the vertical direction around the edges of the footing. There is no heaving of the ground surface[⑦] away from the edges, and no tilting of the footing. Relatively large settlements are also a characteristic of this mode, and again the ultimate bearing capacity is not well defined. Punching shear failure will also occur in a soil of low compressibility if the foundation is located at considerable depth. In general, the mode of failure depends on the compressibility of the soil and the depth of the foundation relative to its breadth.[⑧]

⑤ 局部剪切破坏。
⑥ 冲剪破坏。
⑦ 地表隆起。
⑧ 总体而言，破坏的模式取决于土的压缩性和基础的深宽比。
⑨ 理想刚塑性理论。

The bearing capacity problem can be considered in terms of plasticity theory. It is assumed that the stress-strain behavior of the soil can be represented by the rigid-perfectly plastic idealisation[⑨], shown in Figure 8.3, in which both yielding and shear failure occur at the same state of stress: unrestricted plastic flow takes place at this stress level. A soil mass is said to be in a state of plastic equilibrium if the shear stress at every point within the mass reaches the value represented by point Y′.

⑩ 不稳定机构。

Plastic collapse occurs after the state of plastic equilibrium has been reached in part of a soil mass, resulting in the formation of an unstable mechanism[⑩]: that part of the soil mass slips relative to the rest of the mass. The applied load system, including body forces, for this condition is referred to as the collapse load. Determination of the collapse load is achieved using the limit theorems of plasticity (also known as limit analysis[1]) to calculate **lower** and **upper bounds**[2] to the true collapse load. In certain cases the theorems produce the same result, which would then be the exact value of the collapse load. The limit theorems can be stated as follows.

① 极限分析。
② 上限与下限解的值。

Lower bound (LB) theorem[3]

③ 下限定理。

If a state of stress can be found which at no point exceeds the failure criterion for the soil and is in equilibrium with a system of external loads (which includes the self-weight of the soil), then collapse cannot occur; the external load system thus constitutes a lower bound to the true collapse load (because a more efficient stress distribution may exist, which would be in equilibrium with higher external loads).

Upper bound (UB) theorem[4]

④ 上限定理。

If a **kinematically admissible** mechanism[5] of plastic collapse is postulated and if, in an increment of displacement, the work done by a system of external loads is equal to the dissipation of energy by the internal stresses, then collapse must occur; the external load system thus constitutes an upper bound to the true collapse load (because a more efficient mechanism may exist resulting in collapse under lower external loads).

⑤ 运动许可的机构。

In the upper bound approach, a mechanism of plastic collapse is formed by choosing a slip surface and the work done by the external forces is equated to the loss of energy by the stresses acting along the slip surface, without consideration of equilibrium. The chosen collapse mechanism[6] is not necessarily the true mechanism, but it must be kinematically admissible- i. e. the motion of the sliding soil mass must remain continuous[7] and be compatible with any boundary restrictions[8].

⑥ 破坏机构。
⑦ 连续。
⑧ 边界条件。

8.3 Bearing capacity in undrained materials[9]

⑨ 不排水条件下地基的承载力。

Analysis using the upper bound theorem

Upper bound approach, mechanism UB-1

It can be shown that for undrained conditions the failure mechanism within the soil mass should consist of slip lines[10] which are either straight lines or circular arcs (or a combination of the two). A simple mechanism consisting of three sliding blocks of soil is shown in Figure 8.4 for a strip footing under pure vertical loading.

⑩ 滑移线。

If the foundation inputs work[11] to the system by moving vertically downwards with a velocity v, the blocks will have to move as shown in Figure 8.4 to form a

⑪ 输入功。

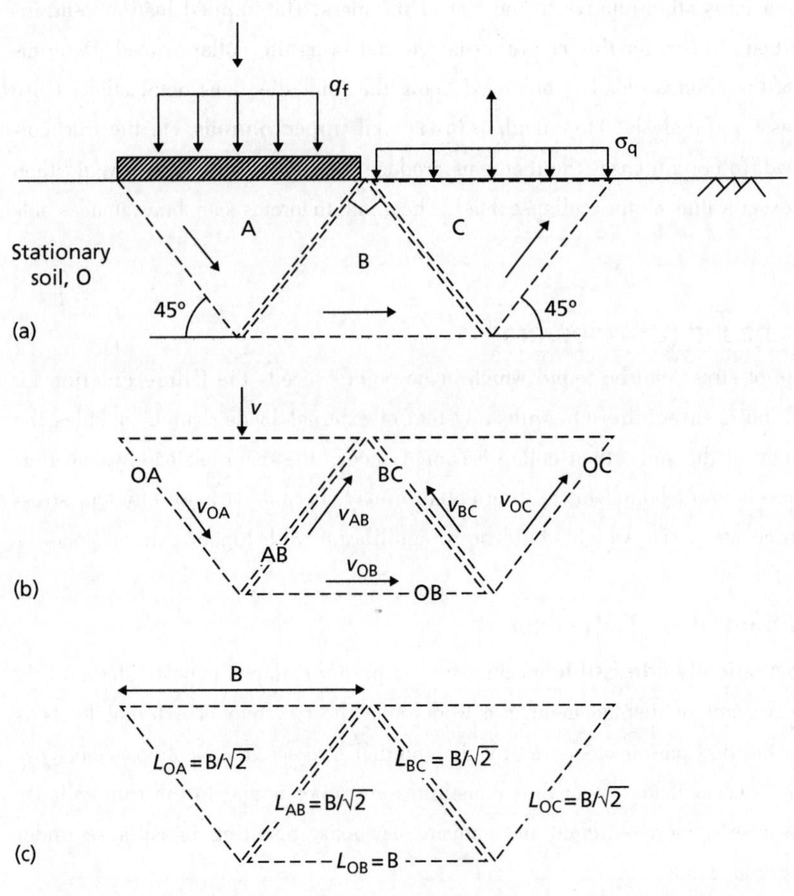

Figure 8.4 (a) Simple proposed mechanism, UB-1, (b) slip velocities, (c) dimensions.

mechanism and therefore be kinematically admissible. This movement generates relative velocities between each slipping block and its neighbours, which will result in energy dissipation along all of the slip lines shown (OA, OB, OC, AB and BC).[①] To determine the relative velocities along the slip lines a **hodograph** (velocity diagram)[②] is drawn, the construction procedure for which is shown in Figure 8.5. Starting with the known vertical displacement of the footing (v) gives point f (foundation) on the hodograph as shown in Figure 8.5 (a). Block A must move relative to the stationary soil at an angle of 45°; the vertical component of this motion must be equal to v so the footing and soil remain in contact. Two construction lines may be added to the hodograph as shown in Figure 8.5 (a) to represent these two limiting conditions. The crossing point of these two lines fixes the position of point a in the hodograph and therefore, the velocity v_{OA}. Block B moves horizontally with respect to O and at 45° relative to A. Adding two construction lines for these conditions fixes the position of b as shown in Figure 8.5 (b) and therefore the velocities v_{OB} and v_{BA}. Block C moves at 45° from points o and b as shown in Figure 8.5 (c) which fixes the position of c and therefore the relative velocities v_{OC} and v_{CB}. Basic trigonometry may then be used on the final hodograph to determine the lengths of

① 这种运动将使每一滑块与其相邻滑块产生相对速度，从而沿着滑移线（OA，OB，OC，AB 和 BC）引起能量的耗散。
② 速度矢量图。

the lines which represent the relative velocities in the mechanism in terms of the known foundation movement v, as shown in Figure 8.5 (d).

The energy dissipated (E_i) due to shearing at relative velocity v_i along a slip line of length L_i is given by:

$$E_i = \tau_f \cdot L_i \cdot v_i \qquad (8.1)$$

energy being force multiplied by velocity (stress multiplied by length is force per metre length of the slip plane into the page). The strength τ_f is used as the shear stress along the slip line, as the soil is at plastic failure along this line by definition (i. e. $\tau = \tau_f$). The total energy dissipated in the soil can then be found by summing E_i for all slip lines, as shown in Table 8.1.

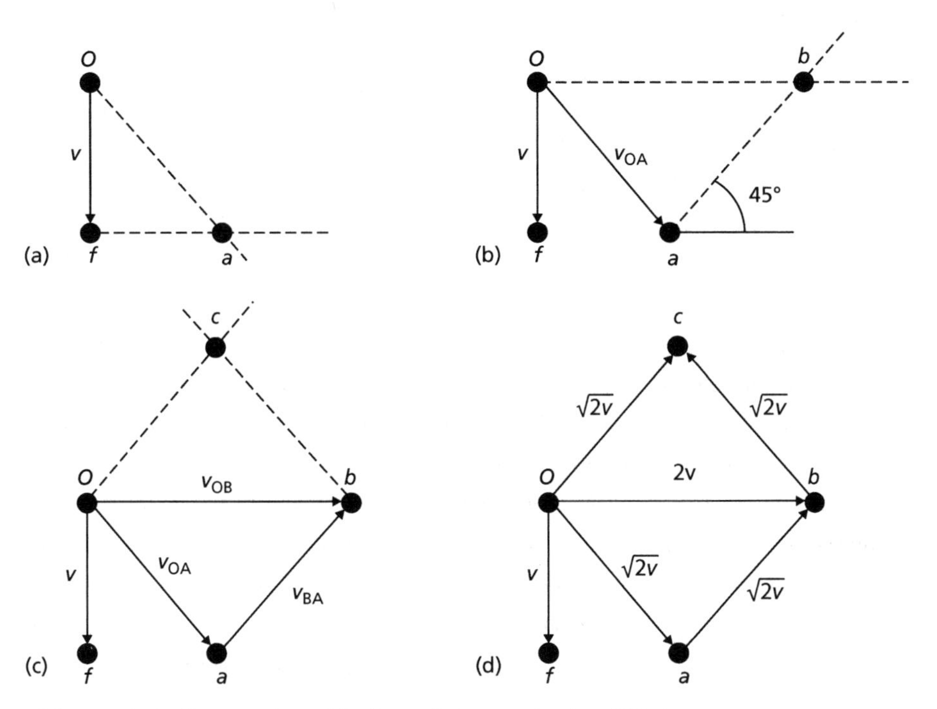

Figure 8.5 Construction of hodograph for mechanism UB-1.

Table 8.1 Energy dissipated within the soil mass in mechanism UB-1[1]

Slip line	Stress, τ_f	Length, L_i	Relative velocity, v_i	Energy dissipated, E_i
OA	c_u	$B/\sqrt{2}$	$\sqrt{2}v$	$c_u Bv$
OB	c_u	B	$2v$	$2c_u Bv$
OC	c_u	$B/\sqrt{2}$	$\sqrt{2}v$	$c_u Bv$
AB	c_u	$B/\sqrt{2}$	$\sqrt{2}v$	$c_u Bv$
BC	c_u	$B/\sqrt{2}$	$\sqrt{2}v$	$c_u Bv$
			Total Energy, ΣE_i =	$6c_u Bv$

① 机构 UB-1 中土体的能量耗散。

Work may be input to the system by the foundation moving downwards, where the work input is positive as the forcing and velocity are in the same direction. If there is a surcharge pressure[2] σ_q acting around the foundation as shown in Figure

② 超载。

8.4, this will be moved upwards as a result of the vertical component of the motion of block C. As the force and velocity are in opposite directions (the surcharge is being moved upwards against gravity), this will do negative work (effectively dissipate energy). The work done W_i by a pressure q_i acting over an area per unit length B_i moving at velocity v_i is given by

$$W_i = q_i \cdot B_i \cdot v_i \tag{8.2}$$

The total work done can then be found by summing W_i for all components, as shown in Table 8.2.

① 塑性破坏。

By the upper bound theorem, if the system is at plastic collapse[①], the work done by the external loads/pressures must be equal to the energy dissipated within the soil, i. e.

$$\sum W_i = \sum E_i \tag{8.3}$$

Therefore, for mechanism UB-1, inserting the values from Tables 8.1 and 8.2 gives the bearing capacity q_f:

$$(q_f - \sigma_q) Bv = 6c_u Bv$$

$$q_f = 6c_u + \sigma_q \tag{8.4}$$

Upper bound approach, mechanism UB-2

② 扇形剪切区。

The mechanism UB-1 requires sharp changes in the direction of movement of the blocks to translate the downwards motion beneath the footing into upwards motion of the adjacent soil. A more efficient mechanism (i. e. one that dissipates less energy) replaces block B in Figure 8.4 with a number of smaller wedges as shown in Figure 8.6 (a). These wedges describe a circular arc of radius R between the rigid blocks A and C, as shown in the figure, known as a **shear fan**[②]. The hodograph for this mechanism is shown in Figure 8.6 (d) - blocks A and C will move in the same direction and by the same magnitude as in Figure 8.5; the velocity around the edge of the circular arc will be constant as the soil in this region rotates about the point X.

③ 外部压力做功。

Table 8.2 Work done by the external pressures[③], mechanism UB-1

Component	*Pressure, p_i*	*Area, B_i*	*Relative velocity, v_i*	*Work done, W_i*
Footing pressure	q_f	B	v	$q_f Bv$
Surcharge	σ_q	B	$-v$	$-\sigma_q Bv$
			Total work done, ΣW_i =	$(q_f - \sigma_q) Bv$

The energy dissipated due to shearing between wedge i (of internal angle $\delta\theta$) and the stationary soil can then be found using Equation 8.1 where the length along the slip plane is $R\delta\theta$ as shown in Figure 8.6 (b) and (c):

$$E_i = c_u \cdot (R\delta\theta) \cdot v_{fan} \tag{8.5}$$

The energy dissipated due to the shearing occurring between each wedge and the

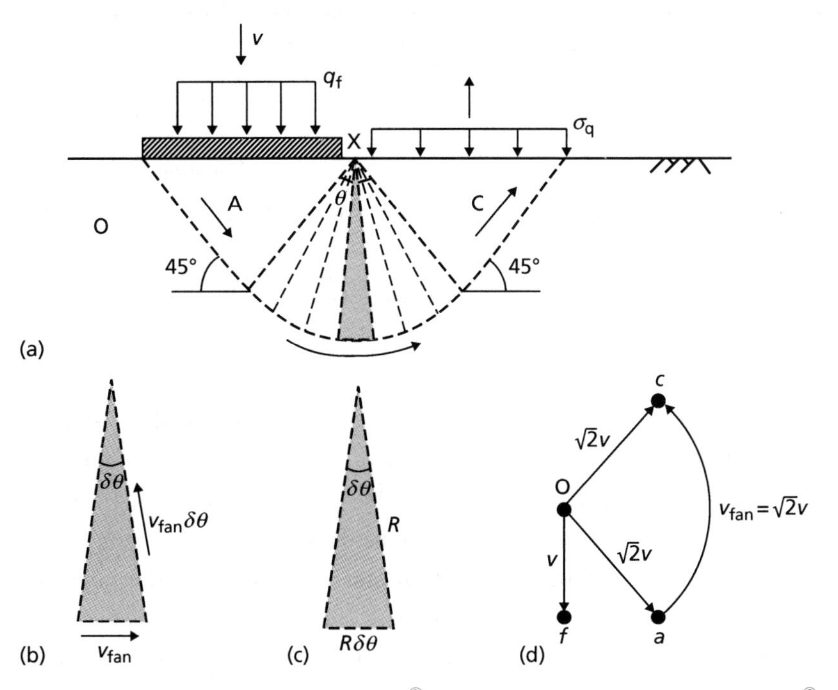

① 改进机构。
② 楔形体 i 的滑动速度。
③ 楔形体 i 的几何构造。
④ 速度矢量图。

Figure 8.6 (a) Refined mechanism① UB-2, (b) slip velocities on wedge i②, (c) geometry of wedge i③, (d) hodograph④.

next can be found similarly, with the length along the slip line being R and the relative velocity from the hodograph $v\delta\phi$:⑤

⑤ 由于每个楔形体之间的剪切而引发的能量耗散可以通过沿滑移线的长度 R 和速度矢量图中的相对速度 $v\delta\phi$ 近似得到。

$$E_{i,j} = c_u \cdot R \cdot (v_{fan}\delta\theta) \tag{8.6}$$

The total amount of energy dissipated within this zone is then given by summing the components from Equations 8.5 and 8.6 across all wedges. If the wedge angle $\delta\theta$ is made vanishingly small, this summation becomes an integral over the full internal angle of the zone (θ):

$$E_{fan} = \sum_i (E_i + E_{i,j}) = \sum_i 2c_u R v_{fan}\delta\theta = \int_0^\theta 2c_u R v_{fan}\delta\theta = 2c_u R v_{fan}\theta \tag{8.7}$$

The energy component E_{fan} replaces the terms along slip lines OB, AB and BC in mechanism UB-1. If blocks A and C still move in the same directions as before, the wedge angle $\theta = \pi/2$ (90°), $v_{fan} = v_{OA} = v_{OC} = \sqrt{2v}$ and $R = B/\sqrt{2}$. The energy dissipated in UB-2 is therefore as shown in Table 8.3.

⑥ 机构 UB-2 中土体的能量耗散。

Table 8.3 Energy dissipated within the soil mass in mechanism UB-2⑥

Slip line	*Stress*, τ_f	*Length*, L_i	*Relative velocity*, v_i	*Energy dissipated*, E_i
OA	c_u	$B/\sqrt{2}$	$\sqrt{2}v$	$c_u Bv$
Fan zone ($\theta = \pi/2$)	c_u	$R = B/\sqrt{2}$	$v_{fan} = \sqrt{2}v$	$\pi c_u Bv$ (from Eq. 8.7)
OC	c_u	$B/\sqrt{2}$	$\sqrt{2}v$	$c_u Bv$
			Total energy, ΣE_i =	$(2+\pi) c_u Bv$

The work dissipated in mechanism UB-2 is the same as for UB-1 (values in Table 8.2) such that applying Equation 8.3 yields:

$$(q_f - \sigma_q)Bv = (2+\pi)c_u Bv$$

$$q_f = (2+\pi)c_u + \sigma_q \tag{8.8}$$

The bearing pressure in Equation 8.8 is lower than for UB-1 (Equation 8.4), so UB-2 represents a better estimate of the true collapse load by the upper bound theorem.

① 利用下限定理分析。

Analysis using the lower bound theorem[①]

Lower bound approach, stress state LB-1

② 在下限解中，平衡与屈服条件的满足并没有考虑变形的模式。

③ 大主应力。

In the lower bound approach, the conditions of equilibrium and yield are satisfied without consideration of the mode of deformation.[②] For undrained conditions the yield criterion is represented by $\tau_f = c_u$ at all points within the soil mass. The simplest possible stress field satisfying equilibrium that may be drawn for a strip footing is shown in Figure 8.7 (a). Beneath the foundation (zone 1), the major principal stress[③] (σ_1) will be vertical. In the soil to either side of this (zone 2), the major principal stress will be horizontal. The minor principal stresses (σ_3) in each zone will be perpendicular to the major principal stresses. These two distinct zones are separated by a single frictionless stress discontinuity permitting the rotation of the major principal stress direction.

④ 无摩阻型应力不连续线。

⑤ 简化的应力场。

⑥ 莫尔圆。

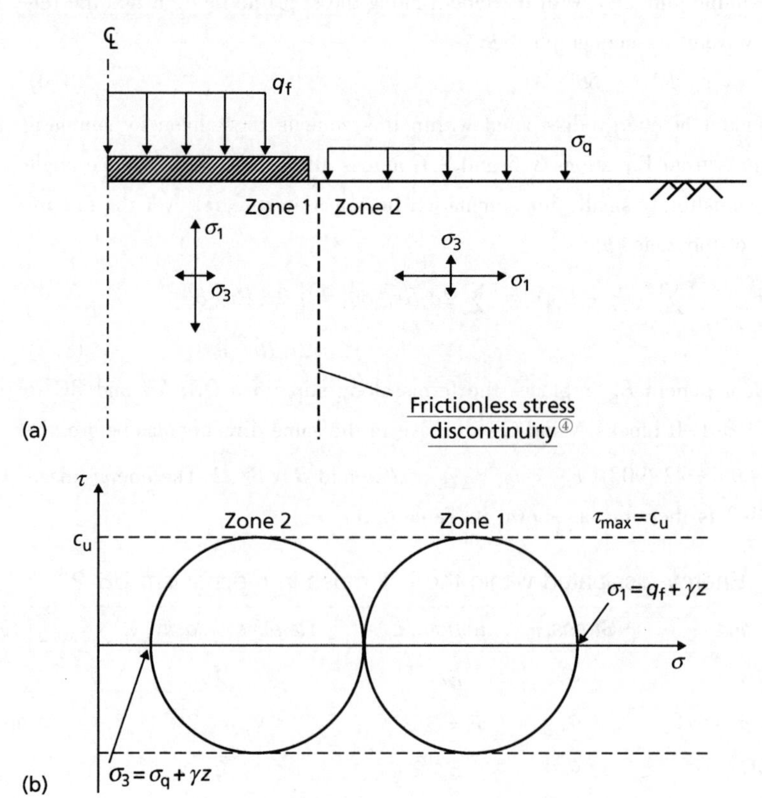

Figure 8.7 (a) Simple proposed stress state[⑤] LB-1, (b) Mohr circles[⑥].

Mohr circles[①] (see Chapter 5) may be drawn for each zone of soil as shown in Figure 8.7 (b). In order for the soil to be in equilibrium, σ_1 in zone 2 must be equal to σ_3 in zone 1[②]. This requirement causes the circles to just touch, as shown in Figure 8.7 (b). The major principal stress at any point within zone 1 is

$$\sigma_1 = q_f + \gamma z \tag{8.9}$$

i. e. the total vertical stress due to the weight of the soil (γz) plus the applied footing pressure (q_f). In zone two, the minor principal stress is similarly

$$\sigma_3 = \sigma_q + \gamma z \tag{8.10}$$

If the soil is undrained with shear strength c_u and the soil is everywhere in a state of plastic yielding, the diameter of each circle is $2c_u$.[③] Therefore at the point where the circles meet

$$q_f + \gamma z - 2c_u = \sigma_q + \gamma z + 2c_u$$

$$\therefore q_f = 4c_u + \sigma_q \tag{8.11}$$

① 莫尔圆。

② 为了满足土体平衡，在区域 2 的 σ_1 必须等于区域 1 的 σ_3。

③ 如果土体的抗剪强度为 c_u 且处于不排水条件下，土体任何一处都处于屈服阶段，则每一个圆的直径都是 $2c_u$。

Lower bound approach, stress state LB-2

As for the upper bound UB-1, the sharp change[④] in the stress field across the single discontinuity in stress state LB-1 is only a crude representation of the actual stress field within the ground. A more realistic stress state can be found by considering a series of frictional stress discontinuities along which a significant proportion of the soil strength can be mobilised, forming a **fan zone**[⑤] which gradually rotate the major principal stress from vertical beneath the footing to horizontal outside. This is shown in Figure 8.8 (a).

The change in direction of the major principal stresses across a frictional discontinuity[⑥] depends on the frictional strength along the discontinuity (τ_d, Figure 8.8 (b)). The Mohr circles representing the stress states in the zones either side of a discontinuity are shown in Figure 8.8 (c). The mean stress[⑦] in each zone is represented by s (see Chapter 5). As with mechanism LB-1 the circles will touch, but at a point where $\tau = \tau_d$ as shown in the figure. This defines the relative position of the two circles, i. e. the difference $s_A - s_B$. In crossing the discontinuity the major principal stress will rotate by an amount $\delta\theta$[⑧] (Figure 8.8 (b)):

$$\delta\theta = \frac{\pi}{2} - \Delta \tag{8.12}$$

As the radius of the Mohr circles are c_u, from Figure 8.8 (c)

$$s_A - s_B = 2c_u \cos\Delta \tag{8.13}$$

Equation 8.12 may then be substituted into Equation 8.13 for Δ. Then, in the limit[⑨] as $s_A - s_B \to \delta s'$, $\sin\delta\theta \to \delta\theta$

$$\begin{aligned} \delta s &= 2c_u \cos\left(\frac{\pi}{2} + \delta\theta\right) \\ &= 2c_u \sin\delta\theta \\ &= 2c_u \delta\theta \end{aligned} \tag{8.14}$$

④ 突变。

⑤ 扇形区。

⑥ 摩阻型不连续界面。

⑦ 平均应力。

⑧ 主应力穿过不连续线将旋转 $\delta\theta$。

⑨ 取极限时。

① 摩阻型应力不连续线。
② 改进的应力场。
③ 跨过摩阻型应力不连续面时主应力的旋转。
④ 莫尔圆。

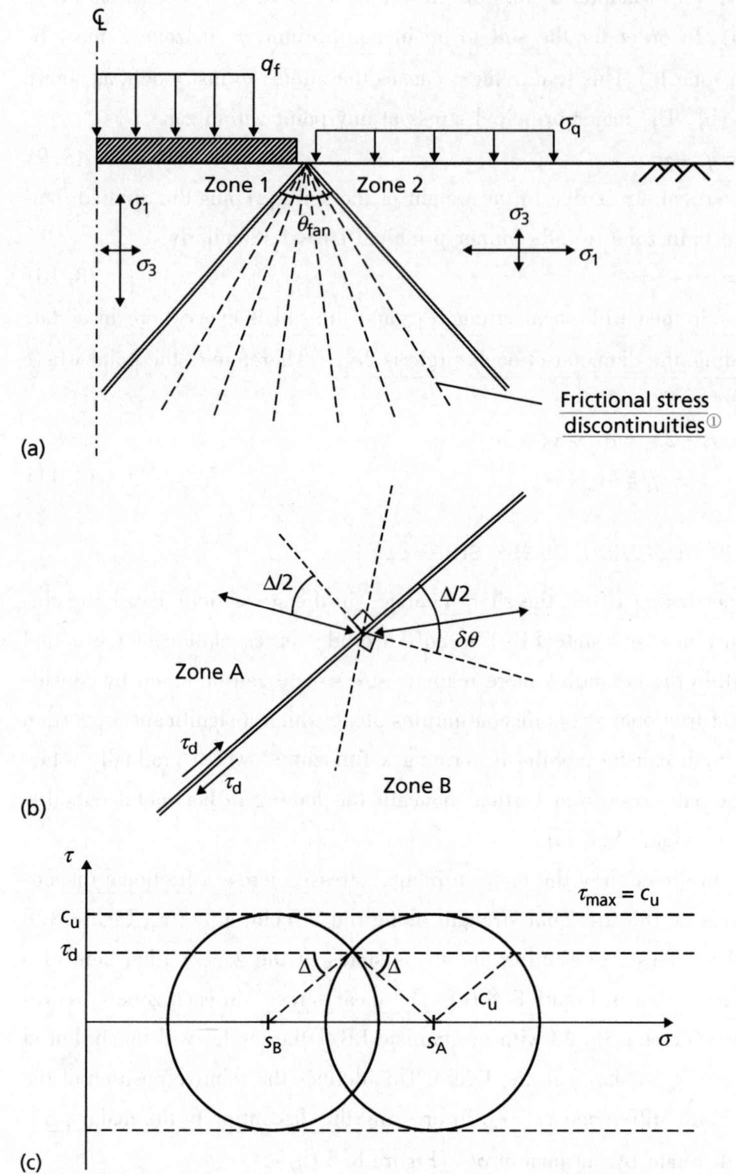

Figure 8.8 (a) Refined stress state② LB-2, (b) principal stress rotation across a frictional stress discontinuity③, (c) Mohr circles④.

For a fan zone of frictional stress discontinuities subtending an angle θ, Equation 8.14 may be integrated from zone 1 to zone 2 in Figure 8.8 (a) across the fan angle θ_{fan}, i.e.

$$\int_{s_2}^{s_1} \delta s = \int_0^{\theta_{fan}} 2c_u \delta\theta$$

$$s_1 - s_2 = 2c_u\theta_{fan} \tag{8.15}$$

For the shallow footing problem shown in Figure 8.9, σ_1 in zone 1 is still given by Equation 8.9 and σ_3 in zone 2 is still given by Equation 8.10. Therefore the principal stress rotation required in the fan is $\theta_{fan} = \pi/2$ (90°) giving, from Equation 8.15,

$$(q_f + \gamma z - c_u) - (\sigma_q + \gamma z + c_u) = 2c_u \frac{\pi}{2}$$

$$\therefore q_f = (2 + \pi) c_u + \sigma_q \tag{8.16}$$

The bearing pressure in Equation 8.16 is higher than for LB-1 (Equation 8.11), so LB-2 represents a better estimate of the true collapse load by the lower bound theorem.①

① 式（8.16）中的承载压力大于LB-1中的［式（8.11）］，LB-2利用下限法对真实极限荷载做出了更好的估算。

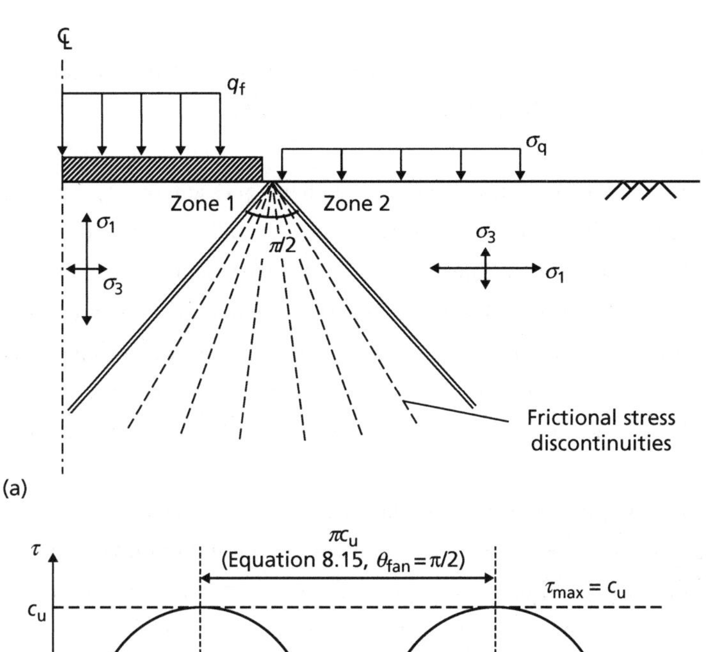

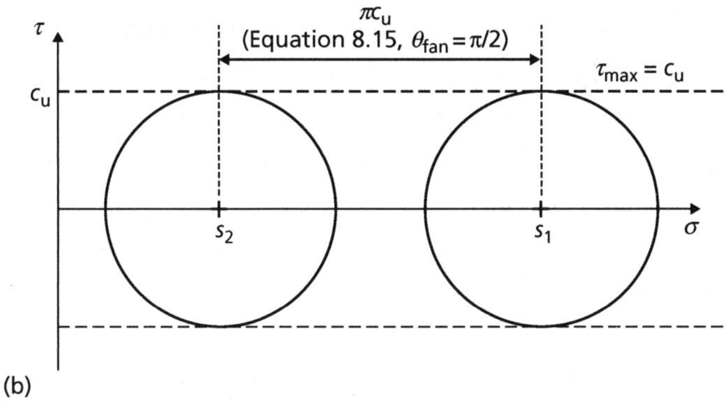

Figure 8.9 Stress state LB-2 for shallow foundation on undrained soil.

Combining upper and lower bounds to obtain the true collapse load②

Ignoring the surcharge pressure σ_q (which is the same in all of the expressions derived so far), the bearing capacity from UB-1 is $6c_u$ while that from LB-1 is $4c_u$. These form upper and lower bounds to the true collapse load, so that $4c_u \leqslant q_f \leqslant 6c_u$. However comparing the refined analyses, UB-2 and LB-2 both give the same value of $q_f = (2 + \pi) c_u = 5.14c_u$, so by the upper and lower bound theorems, this must be the exact solution③ - i. e. LB-2 represents the stress state just as mechanism UB-2 is formed.

② 结合上、下限法得到真实极限荷载。

③ 真实解。

For this problem it was possible with fairly minimal effort to determine the exact solution. In solving any generalised problem using limit analysis, it may be necessary to try many different mechanisms and stress states to determine the bearing capacity exactly (or if not exactly, with only a very narrow range between the best upper and lower bound solutions). Computers can be used to automate this optimisation process. Links to suitable software which is available for both commercial and academic use are provided on the Companion Website.

① 承载力系数。

Bearing capacity factors[①]

Comparing Equations 8.8 and 8.16, the bearing capacity of a shallow foundation on an undrained material may be written in a generalised form as

$$q_f = s_c N_c c_u + \sigma_q \qquad (8.17)$$

Where, for the case of a footing surrounded by a surcharge pressure σ_q, $N_c = 5.14$. N_c is the **bearing capacity factor** for a strip foundation under undrained conditions ($\tau_f = c_u$). The parameter s_c in Equation 8.17 is a shape factor[②] ($s_c = 1.0$ for a strip foundation). In principle, UB and LB analyses similar to those presented above may be conducted for various other cases (e.g. a footing near to a slope or on layered soils). In many cases, however, such analyses have been conducted and the results published as design charts for selecting the value of N_c for use in Equation 8.17.

② 基础形状系数。

Foundations are not normally located on the surface of a soil mass, but are embedded at a depth d below the surface. The soil above the **founding plane** (the level of the underside of the foundation)[③] is considered as a surcharge imposing a uniform pressure $\sigma_q = \gamma d$ on the horizontal plane at foundation level. This assumes that the shear strength of the soil between the surface and depth d is neglected. This is a reasonable assumption provided that d is not greater than the breadth of the foundation B. The soil above foundation level is normally weaker, especially if backfilled, than the soil at greater depth.

③ 基础下表面。

Skempton (1951) presented values of N_c for embedded strip foundations[④] in undrained soil as a function of d based on empirical evidence, which are given in Figure 8.10; also included are values suggested by Salgado *et al.* (2004), which are described by

④ 有埋深的条形基础。

$$N_c = (2 + \pi)\left(1 + 0.27\sqrt{\frac{d}{B}}\right). \qquad (8.18)$$

For a general rectangular footing[⑤] of dimensions $B \times L$ (where $B < L$), Eurocode 7 recommends that the shape factor s_c in Equation 8.17 is given by:

⑤ 矩形基础。

$$s_c = 1 + 0.2\frac{B}{L} \qquad (8.19)$$

Equations 8.18 and 8.19 are compared to Skempton's data for the extreme cases of strip ($B/L = 0$) and square footings ($B/L = 1$) in Figure 8.10. N_c for circular footings may be obtained by taking the square values. In practice, N_c is normally limited at a value of 9.0 for very deeply embedded square or circular foundations. Values of N_c obtained from Figure 8.10 may be used for stratified deposits[⑥], provided the value of c_u for a particular stratum is not greater than nor less than the average

⑥ 层状地基。

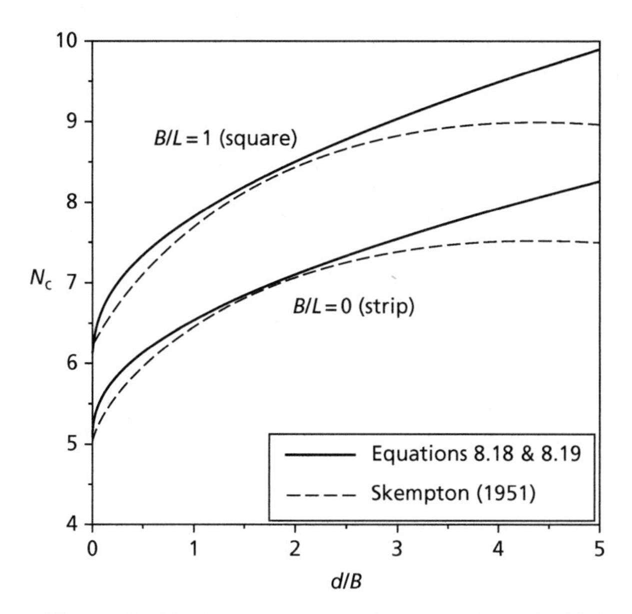

Figure 8.10 Bearing capacity factors N_c for embedded foundations in undrained soil.[①]

① 不排水土中有埋深基础的承载力系数 N_c。

value for all strata within the significant depth by more than 50% of that average value.

For layered soils, Merifield *et al.* (1999) presented upper and lower bound values of N_c for strip footings resting on a two- layer cohesive soil as a function of the thickness H of the upper layer of soil of strength c_{u1} which overlies a deep deposit of material of strength c_{u2}. Proposed design values of N_c for this case are given in Figure 8.11 (a) which are valid if the undrained shear strength of the upper layer is used in Equation 8.17 (i. e. $c_u = c_{u1}$). Subsequently, Merifield and Nguyen (2006) conducted further analyses for square footings with $B/L = 1.0$. The resulting shape factors they obtained are shown in Figure 8.11 (b).

If a shallow foundation is built close to a slope, its bearing capacity may be dramatically reduced.[②] This is a common case for transport infrastructure (e. g. a road or railway line) which is situated on an embankment. These types of construction are commonly very long, and so will always behave as strip foundations. Georgiadis (2010) presented charts for N_c for strip foundations set back from the crest of a slope of angle β by a multiple λ of the foundation width. These are based on upper bound analyses in which an optimal failure mechanism was found giving the lowest upper bound. For this case, it was important to include both ' local' failure mechanisms (bearing capacity failure of the foundation alone) and ' global' mechanisms (failure of the whole slope including the foundation).[③] Slope stability[④] is discussed in greater detail in Chapter 12. From Figure 8.12, the presence of a nearby slope reduces N_c (and hence also the bearing capacity). If the foundation is set far enough back from the crest of the slope ($\lambda > 2B$), then the slope will have no effect on the bearing capacity and $N_c = 2 + \pi$ as for level ground.

It is common for undrained strength to vary with depth, rather than be uniform (constant with depth). Davis and Booker (1973) conducted upper and lower bound

② 如果建造的浅基础靠近斜坡，其地基承载力会明显下降。

③ 对于此案例，同时考虑到“局部”破坏机理（基础承载力不足导致破坏）和“整体”机理（包括基础的整个边坡破坏）是非常重要的。

④ 边坡稳定。

plasticity analyses for soil with a linear variation of undrained shear strength with depth z below the founding plane, i. e.

$$c_u(z) = c_{u0} + Cz \tag{8.20}$$

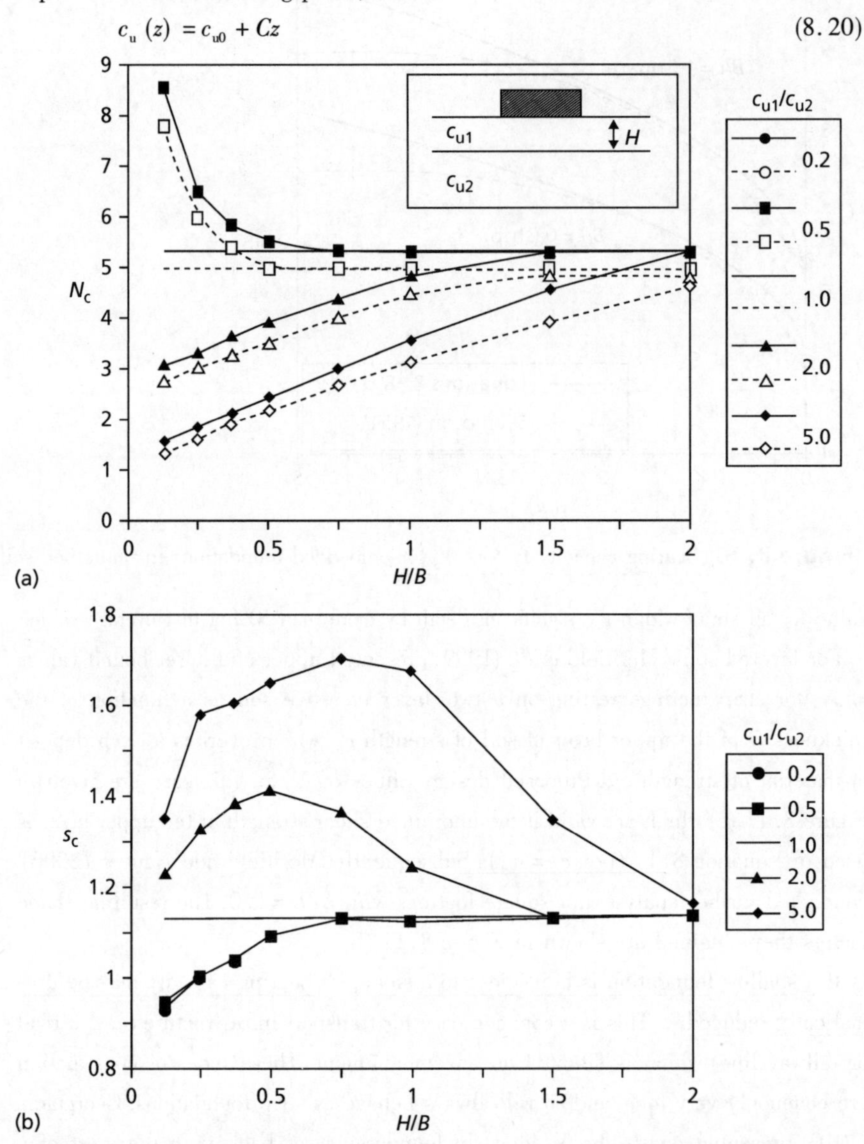

Figure 8. 11 (a) Bearing capacity factors N_c for strip foundations of width B on layered undrained soils (after Merifield *et al.*, 1999), solid lines- UB, dashed lines- LB, (b) shape factors s_c (after Merifield and Nguyen, 2006).

where c_{u0} is the undrained shear strength at the founding plane ($z = 0$) and C is the gradient of the c_u-z relationship. The bearing capacity is expressed in a different form compared to Equation 8. 17, as

$$q_f = \left[(2+\pi)c_{u0} + \frac{CB}{4}\right]F_z \tag{8.21}$$

The parameter F_z is read from Figure 8. 13. If the ratio of $CB/c_{u0} \leqslant 20$, the value of F_z may be read using the left side of the figure; if $CB/c_{u0} \geqslant 20$, it is more convenient to express the ratio as c_{u0}/CB and use the right side of the figure. For the special case of $C=0$ and $c_{u0}=c_u$ (uniform strength with depth), $CB/c_{u0}=0$, $F_z=1$ such that Equation 8.21 reduces to $q_f=5.14c_u$ as before.

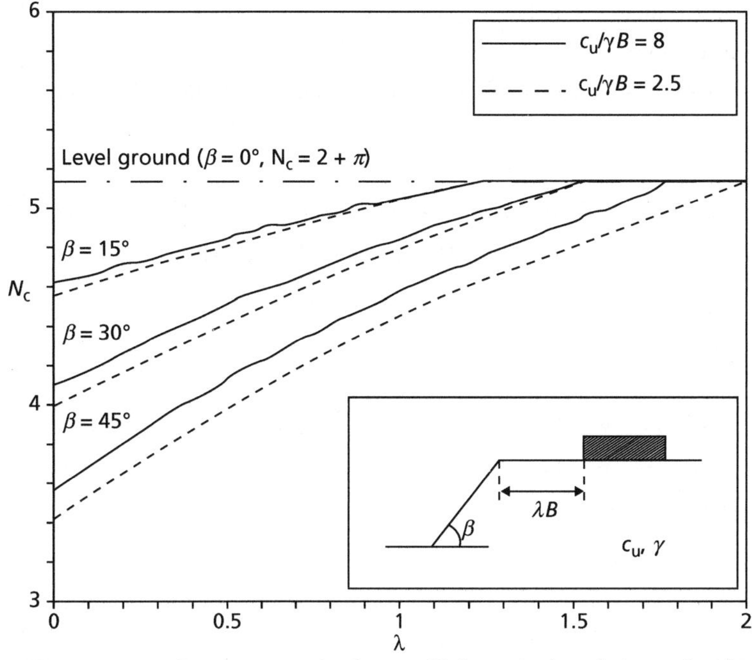

Figure 8.12 Bearing capacity factors N_c for strip foundations of width B at the crest of a slope of undrained soil[①] (after Georgiadis, 2010).

① 不排水条件下，位于斜坡顶端宽度为 B 的条形基础承载力系数 N_c。

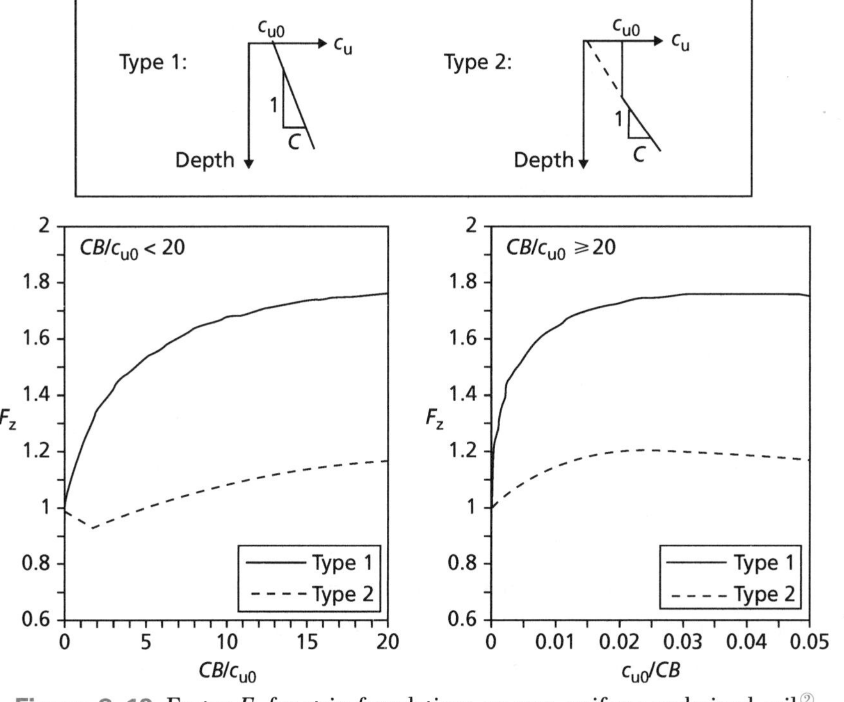

Figure 8.13 Factor F_z for strip foundations on non-uniform undrained soil[②] (after Davis and Booker, 1973).

② 不排水条件下非均质土中条形基础的承载力修正系数 F_z。

Example 8.1

A strip foundation 2.0m wide is located at a depth of 2.0m in a stiff clay of saturated unit weight $21kN/m^3$. The undrained shear strength is uniform with depth, with $c_u = 120kPa$. Determine the undrained bearing capacity of the foundation under the following conditions:

a the foundation is constructed in level ground;

b a cutting at a gradient of 1 : 2 is subsequently made adjacent to the foundation, with the crest 1.5m from the edge of the foundation.

Solution

For case (a) $d/B = 2.0/2.0 = 1$, so, from Figure 8.10a, $N_c = 6.4$ (using Skempton's values). As the footing is a strip, $s_c = 1.0$. The surcharge pressure $\sigma_q = \gamma d = 21 \times 2.0 = 42kPa$. Therefore:

$$
\begin{aligned}
q_f &= s_c N_c c_u + \sigma_q \\
&= (1.0 \cdot 6.4 \cdot 120) + 42 \\
&= 810kPa
\end{aligned}
$$

For case (b), a 1 : 2 slope has an angle $\beta = \tan^{-1}(1/2) = 26.6°$. The parameter $c_u/\gamma B = 120/(21 \times 2.0) = 2.9$ and $\lambda = 1.5/2.0 = 0.75$. Interpolation between lines in Figure 8.12 is then used to find $N_c = 4.4$. The shape factor is the same as before, giving:

$$
\begin{aligned}
q_f &= s_c N_c c_u + \sigma_q \\
&= (1.0 \cdot 4.4 \cdot 120) + 42 \\
&= 570kPa
\end{aligned}
$$

Construction of the slope will therefore reduce the bearing capacity of the foundation. It should be noted that the actual bearing capacity in case (b) is likely to be higher as the value of N_c is uncorrected for the depth of embedment.

8.4 Bearing capacity in drained materials[①]

① 排水条件下地基的承载力。

Analysis using the upper bound theorem[②]

② 利用上限定理分析。
③ 对数螺旋线。
④ 剪胀性。
⑤ 正交原理。
⑥ 相关联流动法则。

It can be shown that for drained conditions the slip surfaces within a kinematically admissible failure mechanism should consist of either straight lines or curves of a specific form known as logarithmic spirals[③] (or a combination of the two). The conditions on a slip surface at any point will be analogous to a direct shear test (as described in Section 5.4). This is shown schematically in Figure 8.14 for a cohesionless soil ($c' = 0$). Referring to Figure 5.16, all drained materials will exhibit some amount of dilatancy[④] during shear (quantified by the angle of dilation, ψ). In general, all soils will have $\psi \leqslant \phi'$. However, for the special case of $\psi = \phi'$ the direction of movement will be perpendicular to the resultant force (R_s) on the shear plane, so, by Equation 8.1, there is no energy dissipated in shearing along the slip line (i.e. there is no movement in the direction of the resultant force). This condition is known as **the normality principle**[⑤]. This special case of $\psi = \phi'$ represents an **associative flow rule**[⑥], and the use of this flow rule considerably simplifies limit analysis in drained materials.

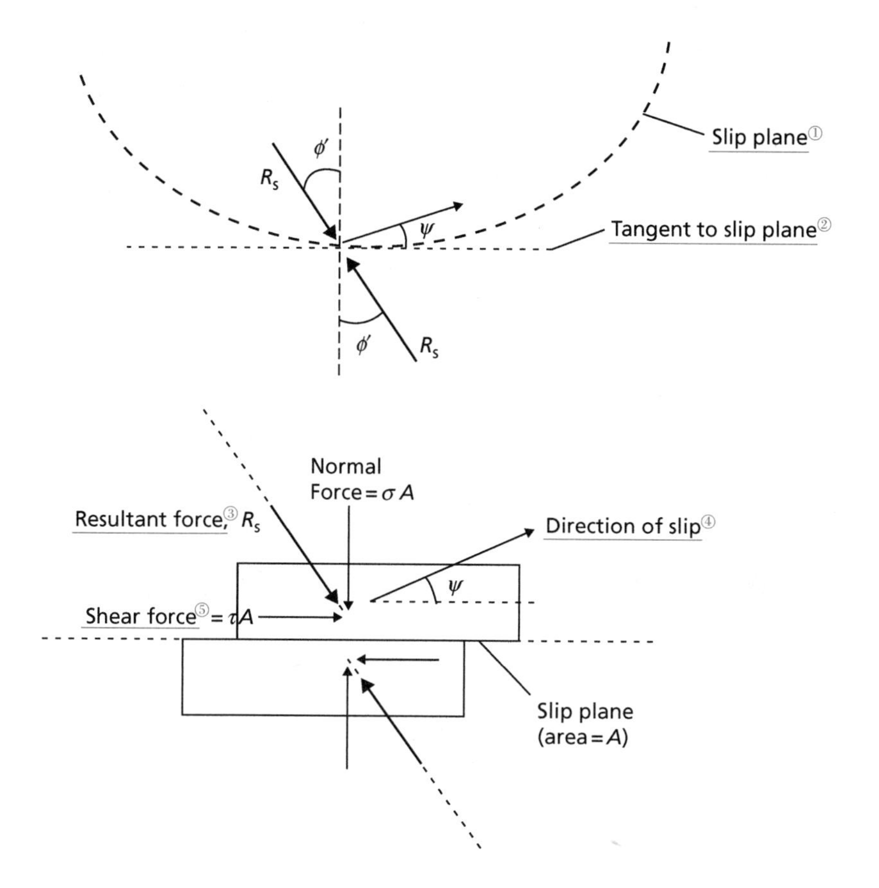

① 滑移面。
② 正切于滑移面。
③ 合力。
④ 滑移方向。
⑤ 剪力。
⑥ 排水条件下土中沿滑移面的受力情况。

Figure 8.14 Conditions along a slip plane in drained material.[6]

Figure 8.15 (a) shows a failure mechanism in a weightless[7] cohesionless soil ($\gamma = c' = 0$) with a friction angle of ϕ'. This is similar to UB-2 (Figure 8.6), but with a logarithmic spiral[8] replacing the circular arc for the fan shear zone B. Beneath the footing, a rigid block will form with slip surfaces at an angle of $\pi/4 + \phi'/2$ to the horizontal as shown. It should be noted that this is a general value for the internal wedge angles in any soil; the internal wedge angles of $45°(\pi/4)$ used earlier for the undrained case arise from the fact that $\phi' = \phi_u = 0$ in undrained materials. These angles fix the lengths of the slip planes OA and AB. To determine the geometry of the rest of the mechanism, an equation describing the logarithmic spiral must first be found. Figure 8.15 (b) shows the change in geometry between two points on the slip line relative to the centre of rotation of the shear zone, from which it can be seen that if $d\theta$ is small

⑦ 无自重。
⑧ 对数螺旋线。

$$\tan\psi = \frac{dr}{rd\theta} \tag{8.22}$$

Rearranging Equation 8.22, it may be integrated from an initial radius r_0 at $\theta = 0$ to a general radius r at θ:

$$\int_{r_0}^{r} \frac{dr}{r} = \int_{0}^{\theta} \tan\psi d\theta$$

$$\ln\left(\frac{r}{r_0}\right) = \theta\tan\psi$$

$$r = r_0 e^{\theta \tan\psi} \tag{8.23}$$

Equation 8.23 can then be applied to the mechanism in Figure 8.15 (a), where $r_0 = L_{AB}$, $r = L_{BC}$ and $\theta = \pi/2$. For associative flow[①], $\psi = \phi'$. Length L_{AB} is found from trigonometry with the known foundation width B and wedge angles $\pi/4 + \phi'/2$, i.e.

① 相关流动法则。

$$L_{AB} = \frac{B}{2\cos\left(\frac{\pi}{4} + \frac{\phi'}{2}\right)}$$

$$L_{BC} = \frac{Be^{\frac{\pi}{2}\tan\phi'}}{2\cos\left(\frac{\pi}{4} + \frac{\phi'}{2}\right)} = L_{OC}$$

② 滑移面。
③ 瞬时滑移方向。
④ 排水条件下土中的上限机构。
⑤ 机构的几何关系。
⑥ 对数螺旋线的几何构造。
⑦ 速度矢量图。

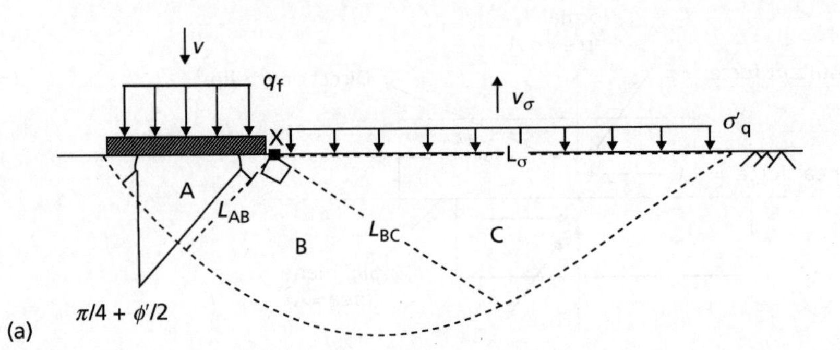

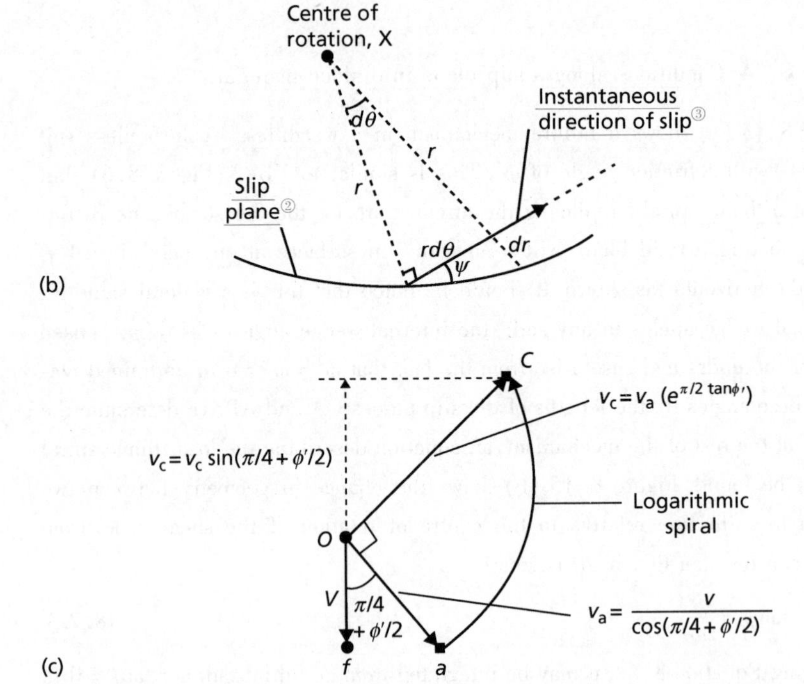

Figure 8.15 Upper bound mechanism in drained soil[④]: (a) geometry of mechanism[⑤], (b) geometry of logarithmic spiral[⑥], (c) hodograph[⑦].

The area per unit length over which the surcharge acts on the mechanism L_σ can then be found from trigonometry

$$L_{\sigma} = Be^{\frac{\pi}{2}\tan\phi'} \cdot \tan\left(\frac{\pi}{4} + \frac{\phi'}{2}\right) \tag{8.24}$$

The hodograph for the mechanism is shown in Figure 8.15 (c). This is similar to the hodograph for the undrained case (UB-2) shown in Figure 8.6 (d); in the drained case, however, the curved line joining points a and c is a logarithmic spiral rather than a circular arc (so the magnitude of v_c can be found from v_a using Equation 8.23, interchanging radii for velocity magnitudes).

As a result of the normality principle, there is no energy dissipated by shearing within the soil mass so $\Sigma E_i = 0$.[1] As for the undrained case, the footing and surcharge pressures still do work and the computations for the drained case are shown in Table 8.4.

Applying Equation 8.3 then gives

$$\sum W_i = 0$$

$$q_f Bv - \sigma'_q Bve^{\pi\tan\phi'}\tan^2\left(\frac{\pi}{4} + \frac{\phi'}{2}\right) = 0$$

$$q_f = \left[e^{\pi\tan\phi'}\tan^2\left(\frac{\pi}{4} + \frac{\phi'}{2}\right)\right]\sigma'_q \tag{8.25}$$

① 由于采用正交流动法则，土体内部的剪切将不会发生能量耗散，即$\sum E_i = 0$。

Analysis using the lower bound theorem[2]

The proposed stress state considered is the same as LB-2 for the undrained case i.e. considering a fan zone of frictional stress discontinuities.[3] This is shown in Figure 8.16 (a). The change in direction of the major principal stresses across a frictional discontinuity depends on the frictional strength along the discontinuity as before (τ_d, Figure 8.16 (b)). However the envelope bounding the Mohr circles[4] in zones 1 and 2 is now of the form $\tau_f = \sigma'\tan\phi'$ for the drained case, and $\tau_d = \sigma'_d \tan \phi'_{mob}$, where ϕ'_{mob} is the mobilised friction angle[5] along the discontinuity. This is shown in Figure 8.16 (c). The mean effective stress in each zone is represented by s'.

As with mechanism LB-2 the circles will touch at a point where $\tau = \tau_d$ as shown in the figure defining the relative position of the two circles. In crossing the discontinuity, the major principal stress will rotate by an amount $\delta\theta$. From Figure 8.16 (b) it can be determined that

$$\delta\theta = \frac{\pi}{2} - \Delta \tag{8.26}$$

Considering the shear strength at the crossing point of the two Mohr circles in Figure 8.16 (c)

② 利用下限定理分析。

③ 建议考虑的应力状态与LB-2的不排水案例一样，如呈扇形的摩阻力不连续面。

④ 莫尔圆的外包络线。

⑤ 发挥的摩擦角。

Table 8.4 Work done by the external pressures, mechanism UB-1

Component	*Pressure*	*Area, B_i*	*Relative velocity, v_i*	*Work done, W_i*
Footing pressure	q_f	B	v	$q_f Bv$
Surcharge	σ'_q	$Be^{(\pi/2\tan\phi')} \times \tan(\pi/4 + \phi'/2)$	$v_2 = -ve^{(\pi/2\tan\phi')} \times \tan(\pi/4 + \phi'/2)$	$-\sigma'_q Bve^{(\pi\tan\phi')} \times \tan^2(\pi/4 + \phi'/2)$

① 摩阻型应力不连续面。
② 不连续面上的应力状态。
③ 应力状态。
④ 摩阻型应力不连续面上的主应力旋转。
⑤ 莫尔圆。

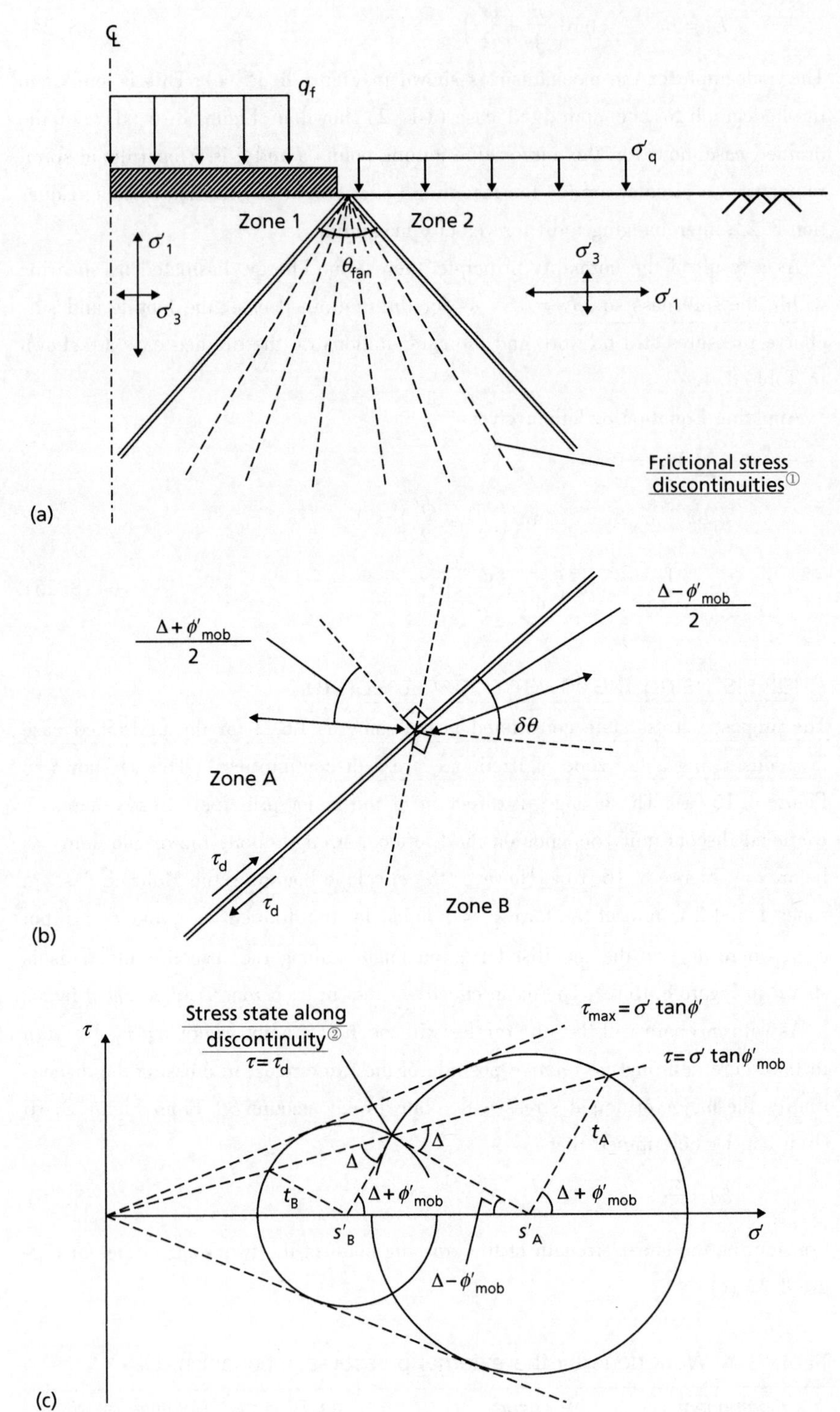

Figure 8. 16 (a) Stress state③, (b) principal stress rotation across a frictional stress discontinuity④, (c) Mohr circles⑤.

$$\tau_d = t_B \sin(\Delta - \phi'_{mob}) = t_A \sin(\Delta + \phi'_{mob})$$

$$\frac{t_B}{t_A} = \frac{\sin(\Delta + \phi'_{mob})}{\sin(\Delta - \phi'_{mob})} \quad (8.27)$$

The radii of the Mohr circles (t_A and t_B) can also be described by $t = s'\sin\phi'$ for a cohesionless soil[①] as shown in Figure 5.6. This condition means that $s'_B/s'_A = t_B/t_A$. Applying this and substituting Equation 8.26, Equation 8.27 becomes

① 无黏性土的莫尔圆半径（t_A 和 t_B）也可表示为 $t = s'\sin\phi'$。

$$\frac{s'_A}{s'_B} = \frac{s'_B + \delta s'}{s'_B} = \frac{\cos(\phi'_{mob} - \delta\theta)}{\cos(\phi'_{mob} + \delta\theta)} \quad (8.28)$$

Setting $s'_B = s'$, as the strength of the discontinuity approaches the strength of the soil ($\phi'_{mob} \to \phi'$) Equation 8.28 can be simplified to

$$1 + \frac{\delta s'}{s'} = 1 + 2\delta\theta\tan\phi'$$

$$\frac{\delta s'}{s'} = 2\delta\theta\tan\phi' \quad (8.29)$$

for small $\delta\theta$. For a fan zone of frictional stress discontinuities subtending an angle θ_{fan}, Equation 8.29 may be integrated from zone 1 to zone 2, i.e.

$$\int_{s'_2}^{s'_1} \frac{\delta s'}{s'} = \int_0^{\theta_{fan}} 2\tan\phi' \delta\theta$$

$$\frac{s'_1}{s'_2} = e^{2\theta_{fan}\tan\phi'} \quad (8.30)$$

From Figure 8.16 (c), $s'_1 = q_f - s'_1 \sin\phi'$ in zone 1 and $s'_2 = \sigma'_q + s'_2 \sin\phi'$ in zone 2. The principal stress rotation required in the fan is $\theta_{fan} = \pi/2$ (90°) giving from Equation 8.30

$$\frac{q_f}{(1+\sin\phi')} \cdot \frac{(1-\sin\phi')}{\sigma'_q} = e^{\pi\tan\phi'}$$

$$\therefore q_f = \left[\frac{(1+\sin\phi')}{(1-\sin\phi')} e^{\pi\tan\phi'}\right]\sigma'_q \quad (8.31)$$

Bearing capacity factors[②]

② 承载力系数。

The upper and lower bound solutions to the bearing capacity of a strip foundation on a weightless cohesionless soil (Equations 8.25 and 8.31 respectively) give the same answer for q_f as it can be shown mathematically that

$$\frac{(1+\sin\phi')}{(1-\sin\phi')} = \tan^2\left(\frac{\pi}{4} + \frac{\phi'}{2}\right)$$

As for the undrained case, the bearing capacity may be written in a generalised form as

$$q_f = s_q N_q \sigma'_q$$

where N_q is a bearing capacity factor relating to surcharge applied around a foundation under drained conditions and s_q is a shape factor.[③] For the undrained case, it was shown that inclusion of the soil unit weight γ in the lower bound analyses did not influence the magnitude of q_f. This is because the amount of soil moving downwards with gravity beneath the foundation was equal to the amount of soil moving upwards against gravity, so that there is no net work done due to the soil weight.

③ N_q 是与排水条件下基础周围附加外力有关的承载力系数，s_q 为形状系数。

The same is not true for the mechanism shown in Figure 8. 15, where the size of the upward moving block beneath the surcharge (doing negative work) is greater than the downward moving block beneath the footing (doing positive work). In this case, therefore, there will be an additional amount of resistance due to the additional net negative work input as a result of the self-weight. Any cohesion c' in the soil will also increase the bearing capacity. As a result, the bearing capacity in drained materials[①] is usually expressed by

① 排水条件下土体的承载能力。

$$q_f = s_q N_q \sigma'_q + \frac{1}{2}\gamma B s_\gamma N_\gamma + s_c N_c c' \tag{8.32}$$

where N_γ is the bearing capacity factor relating to self-weight, N_c is the factor relating to cohesion, and s_γ and s_c are further shape factors. Values of N_q were found previously by limit analysis, and are given in closed-form[②] by:

② 解析解。

$$N_q = \frac{(1+\sin\phi')}{(1-\sin\phi')}e^{\pi\tan\phi'} \tag{8.33}$$

Parameter N_c can be similarly derived for soil with non-zero c' to give

$$N_c = \frac{N_q - 1}{\tan\phi'} \tag{8.34}$$

The final bearing capacity factor, N_γ, is difficult to determine analytically, and is influenced by the roughness of the footing-soil interface[③] (Kumar and Kouzer, 2007). Furthermore, the dilation of soil (and hence the representative value of ϕ' and the degree of associativity) is controlled by the average effective stress beneath the footing ($0.5\gamma B$ in Equation 8. 32) as described in Section 5. 4, such that there is also a size effect[④] on the value of N_γ (Zhu *et al.*, 2001). Salgado (2008) recommended that for an associative soil with a rough footing-soil interface,

③ 土与基础底面的粗糙度。

④ 尺寸效应。

$$N_\gamma = (N_q - 1)\tan(1.32\phi') \tag{8.35}$$

In EC7 the following expression is proposed:

$$N_\gamma = 2(N_q - 1)\tan\phi' \tag{8.36}$$

Values of N_q, N_c and N_γ are plotted in terms of ϕ' in Figure 8. 17. It should be noted that the from Equation 8. 34, $N_c \rightarrow 2+\pi$ as $\phi' \rightarrow 0$, which matches the value found in Section 8. 3 for undrained conditions.

Lyamin *et al.* (2007) present the shape factors for rectangular foundations derived from rigorous limit analyses. Their results for s_q are shown in Figure 8. 18 (a), where it will be seen that s_q varies with ϕ' and B/L. Also shown in this figure are values using the expression recommended in EC7:

$$s_q = 1 + \frac{B}{L}\sin\phi' \tag{8.37}$$

It can be seen from Figure 8. 18 (a) that for low values of ϕ' (typical of the drained strength of fine-grained soils) Equation 8. 37 overpredicts s_q, which would result in an overestimation of bearing capacity; for higher values of $\phi' > 30°$ (typical of coarse-grained soils) Equation 8. 37 underpredicts s_q, giving a conservative estimate of bearing capacity. Once s_q has been found, it can be shown analytically that

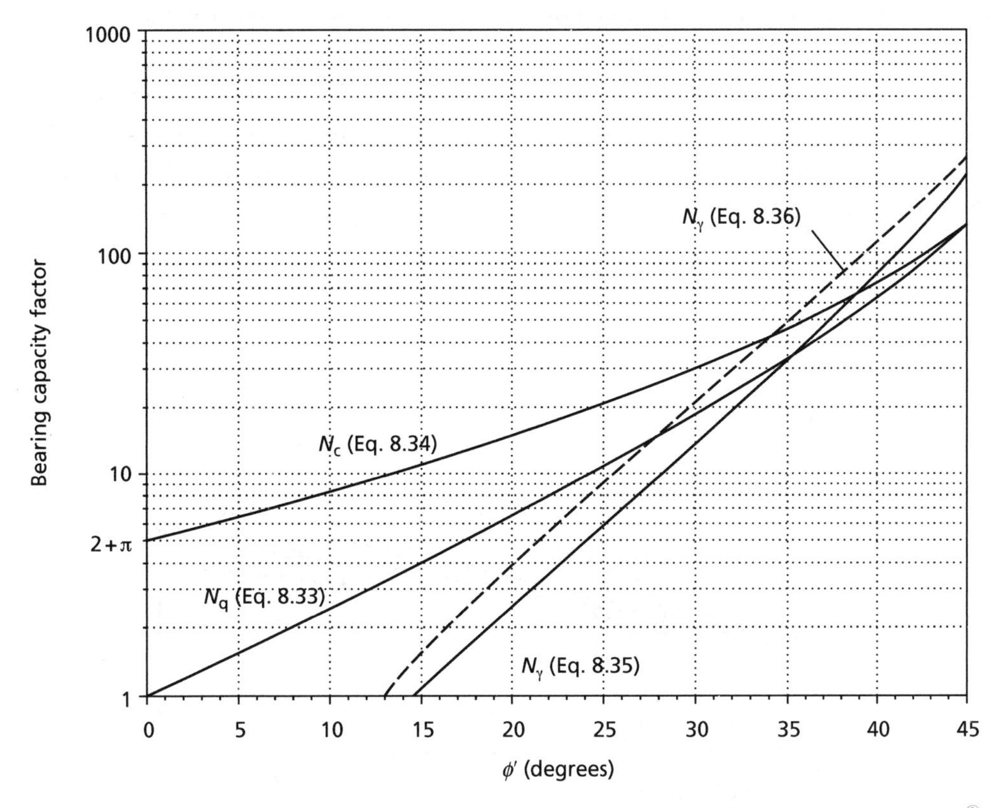

Figure 8.17 Bearing capacity factors for shallow foundations under drained conditions.[①]

① 排水条件下浅基础的承载力系数。

$$s_c = \frac{s_q N_q - 1}{N_q - 1} \tag{8.38}$$

Equation 8.38 is the approach recommended in EC7 for finding s_c.

Data for s_γ, also from Lyamin *et al.* (2007) are shown in Figure 8.18 (b), which can similarly be seen to vary with $\phi'$and B/L. Eurocode 7 recommends the use of the following expression:

$$s_\gamma = 1 - 0.3\frac{B}{L} \tag{8.39}$$

which can be seen, from Figure 8.18 (b), to form a lower bound to the data. Therefore, for $\phi' > 20°$, the use of Equation 8.39 will always give a conservative estimate of the bearing capacity of a rectangular foundation.

Depth factors d_c, d_q and d_γ can also be applied to the terms in Equation 8.32 for cases where the soil above the founding plane has non- negligible strength (depth effects for undrained soils were previously considered in Figure 8.10). These are a function of d/B, and recommended values may be found in Lyamin *et al.* (2007). However, they should only be used if it is certain that the shear strength of the soil above foundation level is, and will remain, equal (or almost equal) to that below foundation level.[②] Indeed, EC7 does not recommend the use of depth factors[③] (i.e. $d_c = d_q = d_\gamma = 1$).

② 但是，其只适于基础之上的土与基础之下的土的抗剪强度相等或近似相等的情况。

③ 深度系数。

For foundations under working load, the maximum shear strain within the supporting soil will normally be less than that required to develop peak shear strength in dense sands or stiff clays, as strains must be low enough to ensure that the settlement of the foundation is acceptable (Sections 8.6-8.8). The allowable bearing capacity or the design bearing resistance should be calculated, therefore, using the peak strength parameters corresponding to the appropriate stress levels.① It should be recognised, however, that the results of bearing capacity calculations are very sensitive to the values assumed for the shear strength parameters, especially the

① 允许承载力或设计承载力应利用与适当应力水平相关的峰值强度参数计算。

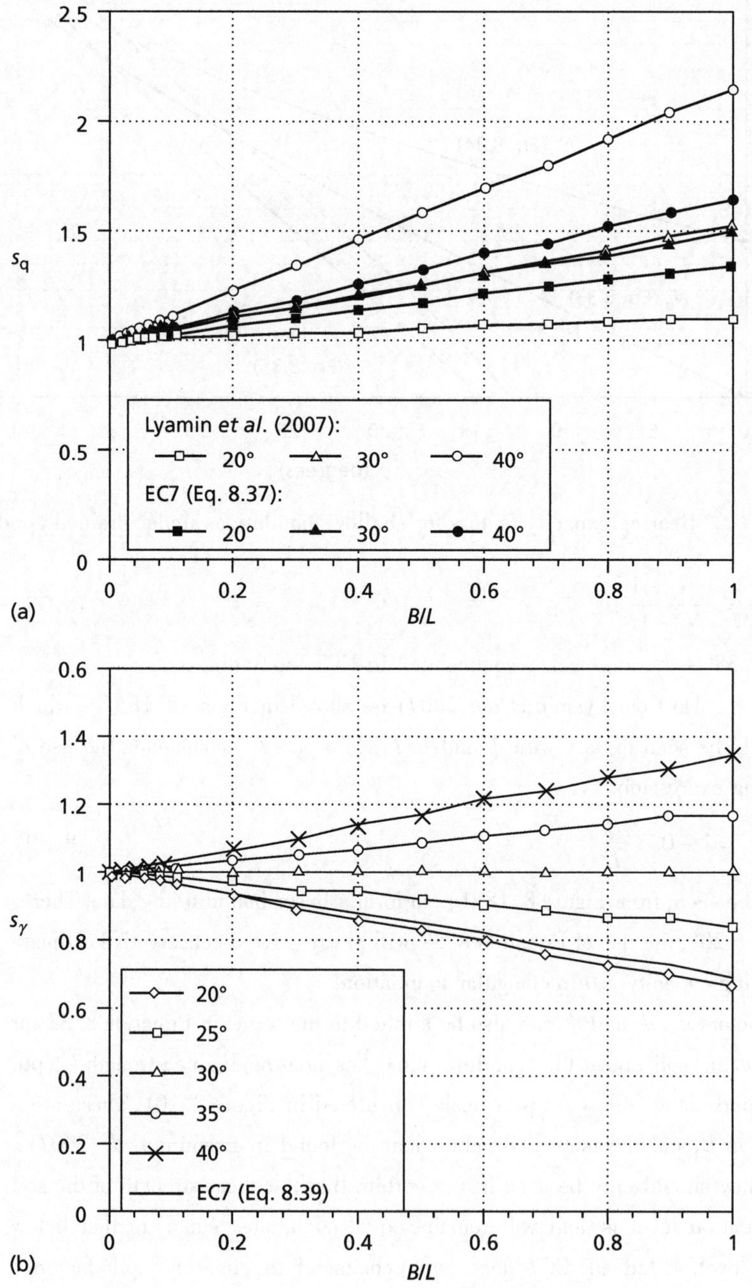

② 排水条件下浅基础的形状参数。

Figure 8.18 Shape factors for shallow foundations under drained conditions②: (a) s_q, (b) s_γ.

higher values of ϕ'. Due consideration must therefore be given to the probable degree of accuracy of the parameters[1].

① 参数准确性的可靠度。

It is vital that the appropriate values of unit weight are used in the bearing capacity equation[2]. In an effective stress analysis, three different situations should be considered:

② 在承载力的式中选用恰当的重度值非常重要。

i If the water table is well below the founding plane, the bulk (total) unit weight (γ) is used in the first and second terms of Equation 8.32.

ii If the water table is at the founding plane, the **effective (buoyant) unit weight**[3] ($\gamma' = \gamma - \gamma_w$) must be used in the second term (which represents the resistance due to the weight of the soil below foundation level), the bulk unit weight being used in the first term (resistance due to surcharge).

③ 有效重度（浮重度）。

iii If the water table is at the ground surface or above (e.g. for soil beneath lakes/rivers or seabed soil), the effective unit weight must be used in both the first and second terms.

Example 8.2

A footing 2.25 × 2.25m is located at a depth of 1.5m in a sand for which $c' = 0$ and $\phi' = 38°$. Determine the bearing resistance (a) if the water table is well below foundation level and (b) if the water table is at the surface. The unit weight of the sand above the water table is 18kN/m³; the saturated unit weight is 20kN/m³.

Solution

For $\phi' = 38°$, the bearing capacity factors are $N_q = 49$ (Equation 8.33) and $N_\gamma = 75$ (Equation 8.36). The footing is square ($B/L = 1$), so the shape factors are $s_q = 1.62$ (Equation 8.37) and $s_\gamma = 0.70$ (Equation 8.39). The values of s_q and s_γ are both conservative (as $\phi' > 30°$). As $c' = 0$ in this case there is no need to compute N_c and s_c. For case (a), when the water table is well below the founding plane:

$$\begin{aligned} q_f &= s_q N_q \gamma d + \frac{1}{2}\gamma B s_\gamma N_\gamma \\ &= (1.62 \cdot 49 \cdot 18 \cdot 1.5) + (0.5 \cdot 18 \cdot 2.25 \cdot 0.70 \cdot 75) \\ &= 3206\text{kPa} \end{aligned}$$

When the water table is at the surface, the ultimate bearing capacity is given by

$$\begin{aligned} q_f &= s_q N_q \gamma' d + \frac{1}{2}\gamma' B s_\gamma N_\gamma \\ &= [1.62 \cdot 49 \cdot (20 - 9.81) \cdot 1.5] + [0.5 \cdot (20 - 9.81) \cdot 2.25 \cdot 0.70 \cdot 75] \\ &= 1815\text{kPa} \end{aligned}$$

Comparing these two results, it can be seen that changing the hydraulic conditions (pore pressures) within the ground has a significant effect on the bearing capacity.

8.5 Shallow foundations under combined loading[4]

④ 组合荷载下的浅基础。

Most foundations are subjected to a horizontal component of loading (H) in addition to the vertical actions (V) considered in Chapter 8. If this is relatively small in relation to the vertical component, it need not be considered in design-e. g. the typical wind loading on a building structure can normally be carried safely by a foundation designed satisfactorily to carry the vertical actions. However, if H (and/or any applied moment, M) is relatively large (e. g. a tall building under hurricane wind loading), the overall stability of the foundation under the combination of actions must be verified.

① 竖向力、水平力、弯矩。

For a foundation loaded by actions V, H and M[①] (usually abbreviated to V-H-M), the following limit states must be met:

ULS-1 The resultant vertical action on the foundation V must not exceed the bearing resistance of the supporting soil;

ULS-2 Sliding must not occur between the base of the wall and the underlying soil due to the resultant lateral action, H;

ULS-3 Overturning of the wall must not occur due to the resultant moment action, M; SLS The resulting foundation movements due to any settlement, horizontal displacement and rotation must not cause undue distress or loss-of-function in the supported structure.

Limit analysis techniques may be applied to determine stability at the ULS and elastic solutions may be used to determine foundation movements at the SLS, similar to those undertaken in Chapter 8 for pure vertical loading.

② 基础稳定性的极限分析法。

Foundation stability from limit analysis (ULS)[②]

Before the general case of V-H-M loading is considered, the stability of foundations under simpler combinations of V-H loading will be considered initially to introduce the key concepts. This builds on the lower bound limit analysis techniques from Chapter 8.

③ 土与基础面的附加剪应力 $\tau_f = H/A_f$。

The addition of horizontal load H to a conventional vertical loading problem ($q_f = V/A_f$) will induce an additional shear stress $\tau_f = H/A_f$ at the soil footing interface[③] as shown in Figure 8.19 (a). It is assumed in this analysis that the footing is perfectly rough. This will serve to rotate the major principal stress direction in zone 1. For an undrained material, this rotation will be $\theta = \Delta/2$ from the vertical (Figure 8.19(b)). The stress conditions in zone 2 are unchanged from those shown in Figure 8.9 (a). Therefore, the overall rotation of principal stresses across the fan zone is now $\theta_{fan} = \pi/2 - \Delta/2$, such that from Equation 8.15

$$s_1 - s_2 = c_u (\pi - \Delta) \tag{8.40}$$

From Figure 8.19 (b),

$$\sin\Delta = \frac{\tau_f}{c_u} = \frac{H}{A_f c_u} \tag{8.41}$$

In zone 2, $s_2 = \sigma_q + \gamma z + c_u$ as in Section 8.3 (unchanged), while in zone 1, $s_1 =$

$q_f + \gamma z - c_u \cos\Delta$ from Figure 8.19 (b). Substituting these relationships into Equation 8.40 and rearranging gives

$$q_f = \frac{V}{A_f} = c_u(1 + \pi - \Delta + \cos\Delta) + \sigma_q$$

$$= c_u N_c + \sigma_q \tag{8.42}$$

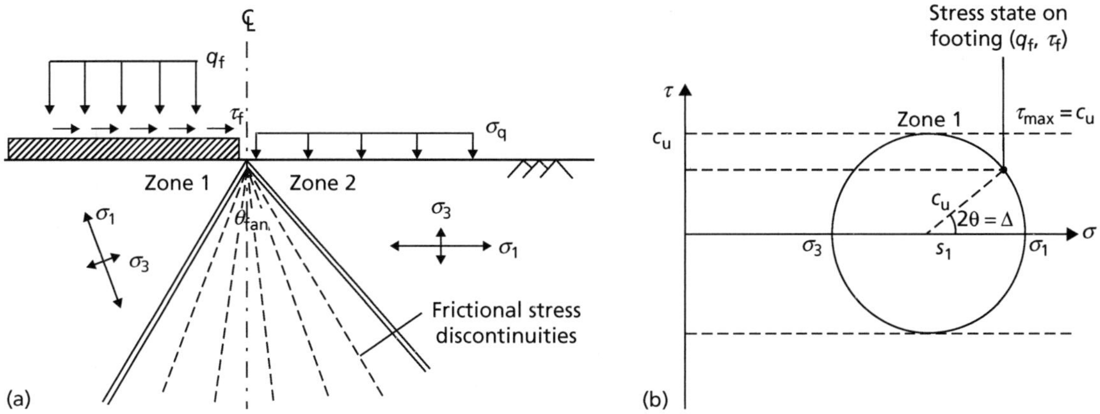

Figure 8.19 (a) Stress state for V-H loading, undrained soil, (b) Mohr circle in zone 1.

For all possible values of H ($0 \leqslant H/A_f c_u \leqslant 1$), Δ can be found from Equation 8.41 and $V/A_f c_u$ (= bearing capacity factor N_c) found from Equation 8.42. These are plotted in Figure 8.20 for the case of no surcharge ($\sigma_q = 0$). When $H/A_f c_u = 0$ (i.e. purely vertical load), $\Delta = 0$ and $V/A_f c_u = 2 + \pi$ (see Equation 8.16); when $H/A_f c_u = 1$, the shear stress $\tau_f = c_u$ and the footing will slide horizontally, irrespective of the value of V. The resulting curve represents the **yield surface**[①] for the foundation under V-H loading. Combinations of V and H which lie within the yield surface will be stable, while those lying outside the yield surface will be unstable (i.e. result in plastic collapse). If $V \gg H$, collapse will be predominantly in bearing (vertical, ULS-1); if $V \ll H$, collapse will be predominantly by sliding (horizontal translation, ULS-2). For intermediate states, a combined mechanism resulting in significant vertical and horizontal components will occur.

① 破坏包络面。

In Eurocode 7 and many other design specifications worldwide, a different approach is adopted. This consists of applying an additional **inclination factor**[②] i_c to the standard bearing capacity equation (Equation 8.17), where

② 倾斜系数。

$$i_c = \frac{1}{2}\left(1 + \sqrt{1 - \frac{H}{A_f c_u}}\right) \tag{8.43}$$

For the case of no surcharge and a strip footing ($s_c = 1$), from Equation 8.17

$$q_f = \frac{V}{A_f} = i_c(2 + \pi)c_u$$

$$\frac{V}{A_f c_u} = \frac{(2 + \pi)}{2}\left(1 + \sqrt{1 - \frac{H}{A_f c_u}}\right) \tag{8.44}$$

Equation 8.44 is also plotted in Figure 8.40 where it is practically indistinguishable from the rigorous plasticity solution represented by Equation 8.42.

① 竖向-水平荷载作用下条形基础在不排水土中的破坏包络面。

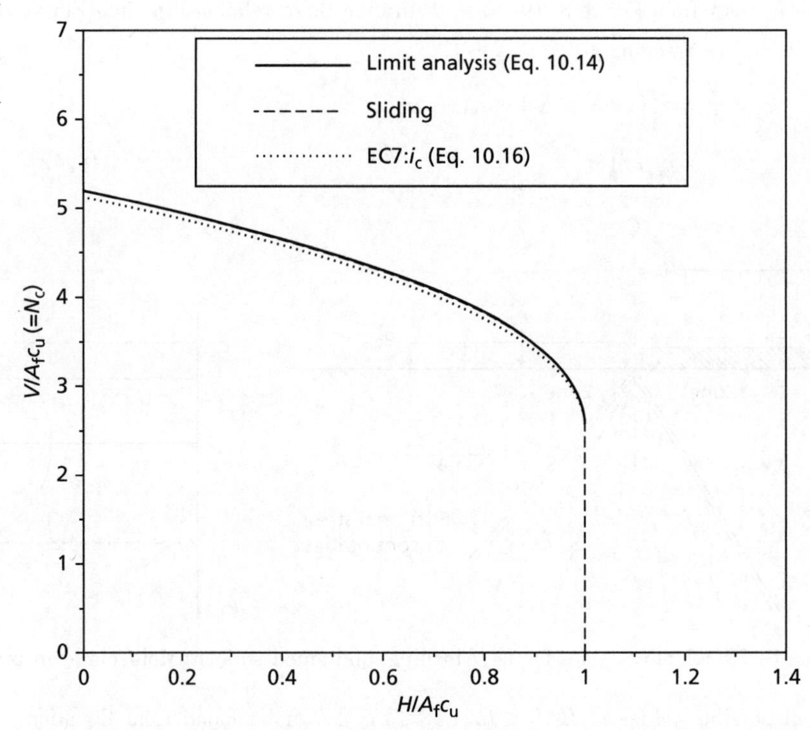

Figure 8.20 Yield surface for a strip foundation on undrained soil under V-H loading.[①]

Gourvenec (2007) presented a yield surface for the general case of V-H-M loading on undrained soil, where

$$\left[\frac{1.29\,\frac{H}{R}}{0.25-\left(\frac{V}{R}-0.5\right)^2}\right]^2+\left[\frac{2.01\,\frac{M}{BR}}{\frac{V}{R}-\left(\frac{V}{R}\right)^2}\right]^2=1 \tag{8.45}$$

In Equation 8.45, R is the vertical resistance of the foundation under pure vertical loading, ($H = M = 0$), as found in Chapter 8, and B is the breadth of the foundation. As $R = (2+\pi)A_f c_u$, Equations 8.42 and 8.44 can be rewritten in terms of V/R and H/R by substituting for $A_f c_u$. Figure 8.21 (a) compares the lower bound solution, Eurocode 7 approach and the full yield surface (Equation 8.45) for the case of V-H loading ($M = 0$). Using this alternative normalisation, it can be seen that the maximum horizontal action which can be sustained is approximately 19% of the vertical resistance.

② 当 $M\neq0$ 时，破坏包络面为三维曲面。

When $M\neq0$, the yield surface becomes a three-dimensional surface[②] (a function of V, H and M). Figure 8.21 (b) shows contours of V/R for combinations of H and M under general loading for use in ULS design. The presence of moments allows for overturning (rotation) of the foundation when $M \gg V, H$. The yield surface represented in Figure 8.21 (b) assumes that tension cannot be sustained along the soil-footing interface, i. e. the foundation will **uplift**[③] if the overturning effect is strong. Provided that the combination of V, H and M lies within the yield surface, the founda-

③ 上浮，隆起。

tion will not fail in bearing, sliding or overturning such that ULS-1-ULS-3 will all be satisfied (and can be checked simultaneously), showing the power of the yield surface concept.

A lower bound analysis may also be conducted for a foundation on a weightless drained material[①] (Figure 8.22 (a)). Here, $H/V=\tau_f/q_f=\tan\beta$. From Figure 8.22 (b), the rotation of the major principal stress direction in zone 1 is $\theta=(\Delta+\beta)/2$ from the vertical. The stress conditions in zone 2 are unchanged from those shown in Figure 8.16 (a). Therefore, the overall rotation of principal stresses across the fan zone is now $\theta_{fan}=\pi/2-(\Delta+\beta)/2$, such that from Equation 8.30

① 无自重排水材料。

$$\frac{s_1'}{s_2'}=e^{(\pi-\Delta-\beta)\tan\phi'} \tag{8.46}$$

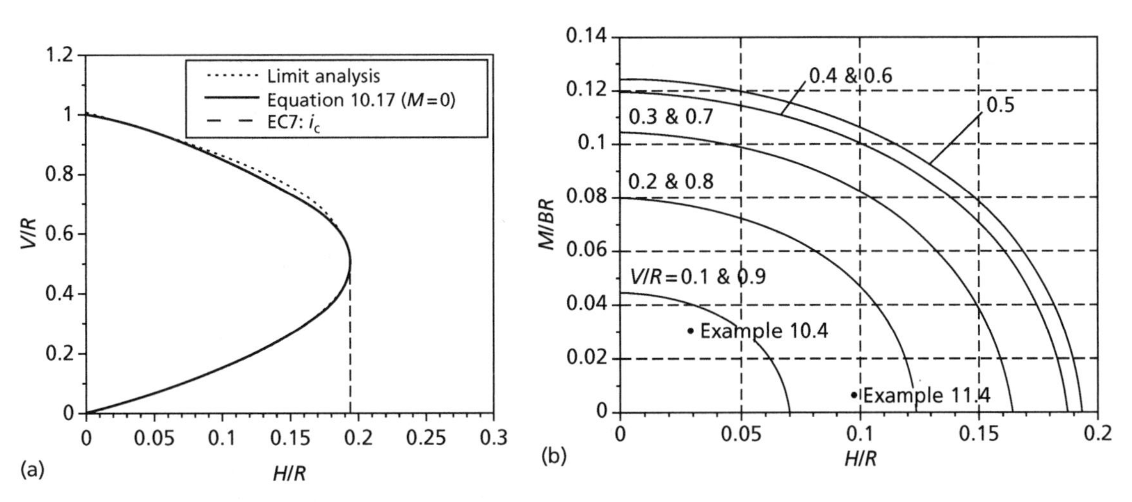

Figure 8.21 Yield surfaces for a strip foundation on undrained soil under (a) V-H loading; (b) V-H-M loading.

From Figure 8.22 (b),

$$\sin\Delta=\frac{\sin\beta}{\sin\phi'} \tag{8.47}$$

In zone 2, $s_2'=\sigma_q+s_2'\sin\phi'$ as in Section 8.4 (unchanged), while in zone 1, $s_1'=q_f-s_1'\sin\phi'\cos(\Delta+\beta)$ from Figure 8.22 (b). Substituting these relationships into Equation 10.18 and rearranging gives

$$q_f=\frac{V}{A_f}=\left[\frac{1+\sin\phi'\cos(\Delta+\beta)}{1-\sin\phi'}\right]e^{(\pi-\Delta-\beta)\tan\phi'}\cdot\sigma_q$$

$$=N_q\sigma_q \tag{8.48}$$

Values of β may be found for any combination of V and H, from which Δ may be found from Equation 10.19 and values of N_q from Equation 10.20. These are plotted in Figure 10.15. If the footing is perfectly rough ($\delta'=\phi'$) then sliding will occur if $H/V\geqslant\tan\phi'$[②].

② 如果基底完全粗糙（$\delta'=\phi'$），当$H/V\geqslant\tan\phi'$时发生滑动。

As for the undrained case, Eurocode 7 and many other design specifications worldwide adopt an alternative approach. This consists of applying an additional inclination factor[③] i_q to the standard bearing capacity equation (Equation 8.31),

③ 附加倾斜系数。

where

$$i_q = \left(1 - \frac{H}{V}\right)^2 \tag{8.49}$$

For the case of a strip footing on cohesionless soil ($c' = 0$), from Equation 8.31

$$N_q = \frac{q_f}{\sigma_q} = i_q\left[\frac{(1+\sin\phi')}{(1-\sin\phi')}e^{\pi\tan\phi'}\right] \tag{8.50}$$

Equation 8.50 is also plotted in Figure 8.23 where it is practically indistinguishable from the rigorous plasticity solution represented by Equation 8.48, though it should be noted that Equation 8.49 is only valid when $H/V \leqslant \tan\phi'$ to account for sliding.

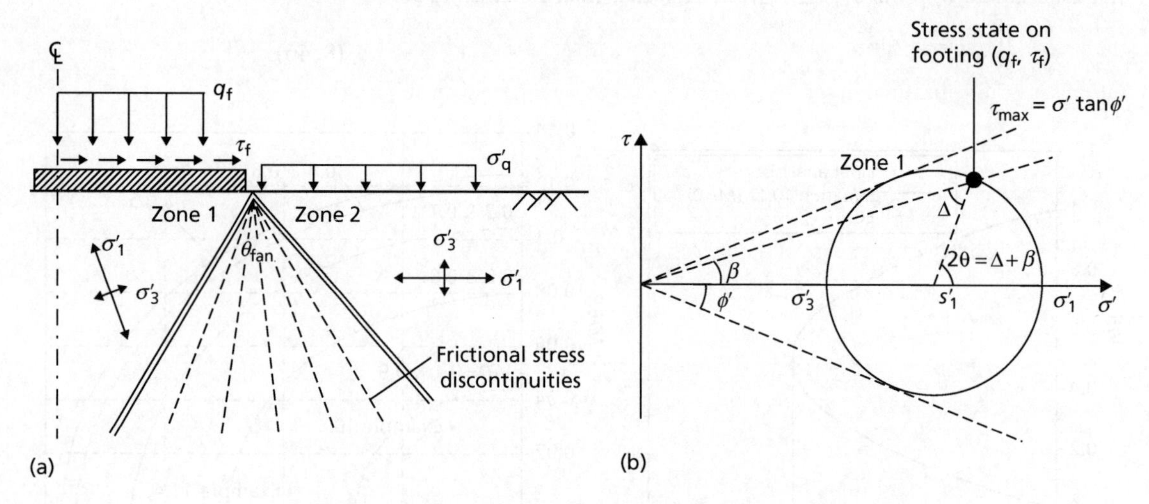

Figure 8.22 (a) Stress state for V-H loading, drained soil, (b) Mohr circle in zone 1.

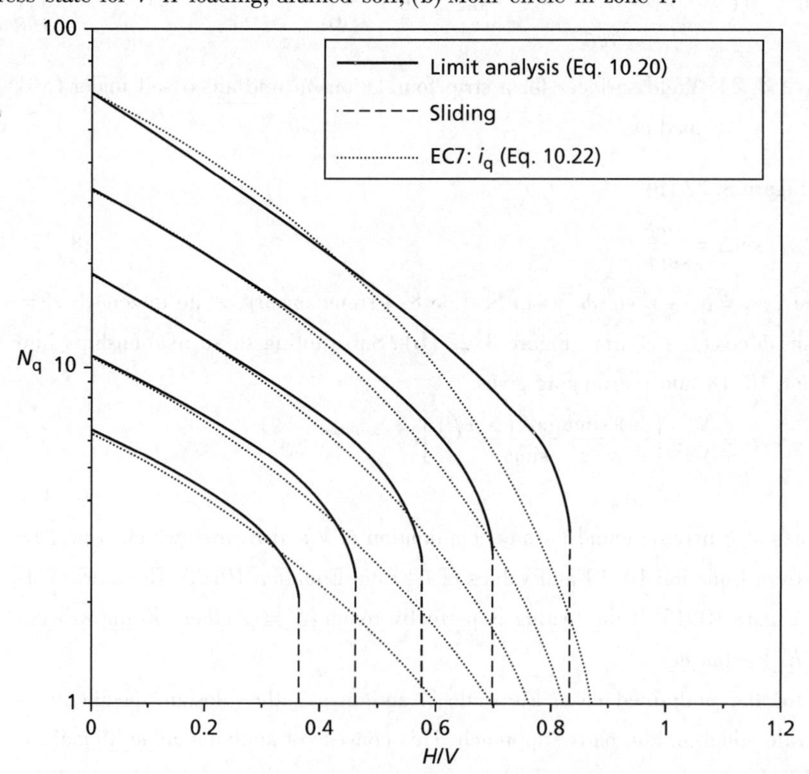

Figure 8.23 N_q for a strip foundation on drained soil under V-H loading.

Butterfield and Gottardi (1994) presented a yield surface for the general case of V-H-M loading on drained soil, where

$$\frac{3.70\left(\frac{H}{R}\right)^{2}-2.42\left(\frac{H}{R}\right)\left(\frac{M}{BR}\right)+8.16\left(\frac{M}{BR}\right)^{2}}{\left[\frac{V}{R}\left(1-\frac{V}{R}\right)\right]^{2}}=1 \tag{8.51}$$

Recognising that V/R at any value of H/V is the value of N_q from Equation 8.50 divided by the value of N_q at $H/V=0$ (Equation 8.31), and that $H/R=(H/V)\times(V/R)$, Equations 8.48 and 8.50 can be expressed in terms of V/R and H/R for the case of $M=0$. Figure 10.16 (a) compares the lower bound solution, Eurocode 7 approach and the full yield surface for the case of V-H loading ($M=0$). Using this alternative normalisation, it can be seen that the maximum horizontal thrust which can be sustained is approximately 13% of the vertical resistance, lower than for the undrained case.①

When $M\neq 0$ the yield surface becomes a three dimensional surface, as before. Figure 8.24 (b) shows contours of V/R for combinations of H and M under general loading for use in ULS design from Equation 8.51. For the drained case, Eurocode 7 is also able to account for moment effects through the use of a reduced footing width② $B'=B-2e_m$ in Equation 8.31, where e_m is the eccentricity③ of the vertical load from the centre of the footing which creates a moment of magnitude M, i.e. $e_m=M/V$. For a strip footing, the footing-soil contact area is therefore B' per metre length under V-H-M loading and

$$q_f=\frac{V}{B'}=i_q N_q \sigma_q \tag{8.52}$$

Under pure vertical loading V (where $V=R$ at bearing capacity failure),

$$q_f=\frac{R}{B}=N_q \sigma_q \tag{8.53}$$

Dividing Equation 8.52 by 8.53 and substituting for i_q (Equation 8.49), B' and e_m gives

$$\frac{V}{R}=i_q\left(\frac{B'}{B}\right)=\left(1-\frac{H}{V}\right)^{2}\left(1-\frac{2M}{BV}\right) \tag{8.54}$$

Equation 8.54 may be plotted out as a yield surface for comparison with Equation 8.51, from which it can be seen that for low values of $V/R\leqslant 0.3$ Equation 8.54 will provide an unconservative estimation of foundation stability. For all values of V/R, Equation 8.54 also overestimates the capacity at low H/R (i.e. where overturning is the predominant failure mechanism), making it less suitable for checking ULS-3.

① 排水条件下，基础能承受的最大水平推力约为最大竖向承载力的13%，且小于不排水条件的值。

② 修正的基础宽度。

③ 偏心距。

Foundation displacement from elastic solutions (SLS)④

④ 基础位移的弹性分析方法。

If the combination of actions V-H-M applied to a shallow foundation lies within the yield surface, it is still necessary to check that the foundation displacements under the applied actions are tolerable (SLS). Whilst in Chapter 8 settlement s (vertical displacement) was the action effect associated with the action V, under multi-axial

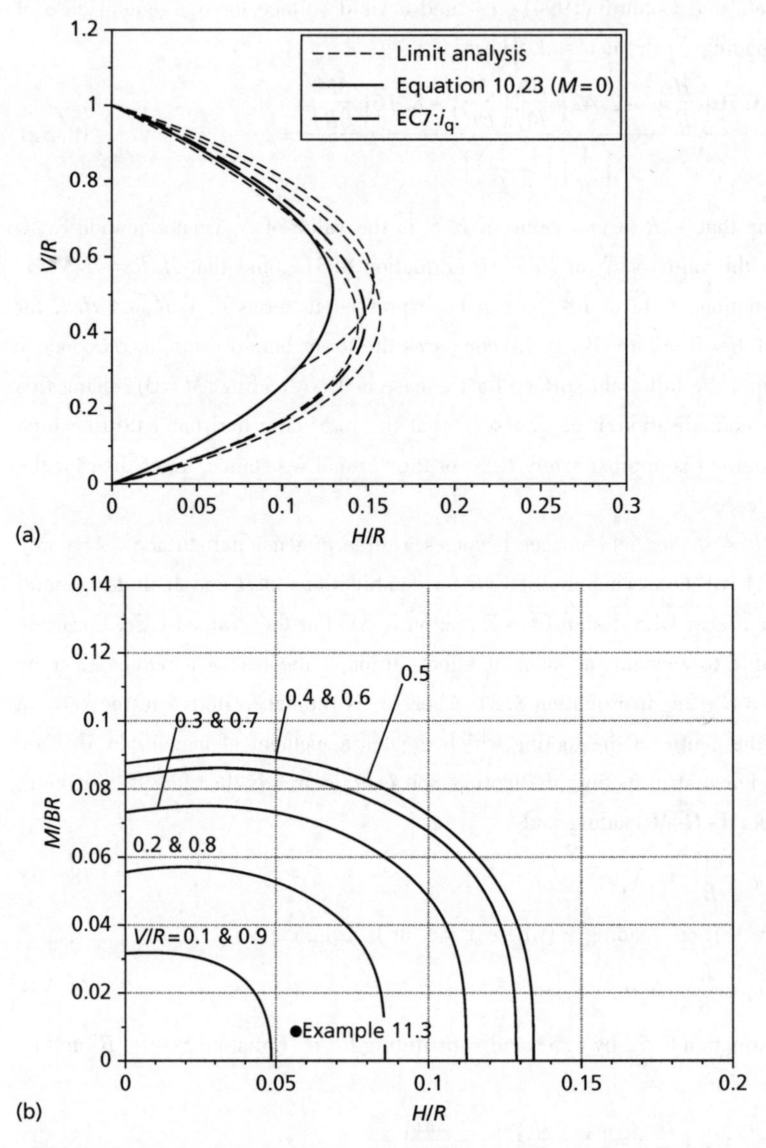

Figure 8.24 Yield surfaces for a strip foundation on drained soil under (a) V-H loading, (b) V-H-M loading.

loading the footing may additionally displace horizontally by h (under the action of H) and rotate by θ (under the action of M). The relationship between the action and the action effect in each case may be related by elastic stiffness $K_v = V/s$ (vertical stiffness①), $K_h = H/h$ (horizontal stiffness②) and $K_\theta = M/\theta$ (rotational stiffness③).

The elastic solution for vertical settlement given as $s = \frac{qB}{E}(1-\nu^2)I_s$ may be re-expressed as a vertical stiffness K_v, by recognising that the bearing pressure $q = V/BL$. It is common to express foundation stiffness in terms of shear modulus G, rather than Young's Modulus E, so that consideration of undrained or drained conditions only involves changing the value of ν④. By using Equation 5.6, then,

① 竖向刚度。
② 水平向刚度。
③ 转动刚度。
④ 排水或不排水条件的转换只需通过改变泊松比 ν 就可实现。

$s = \frac{qB}{E}(1 - \nu^2) I_s$ may be rearranged as:

$$K_v = \frac{V}{s} = \left(\frac{2L}{I_s}\right)\frac{G}{(1-\nu)} \tag{8.55}$$

The horizontal stiffness of a shallow foundation was derived by Barkan (1962) as

$$K_h = \frac{H}{h} = 2G(1+\nu) F_h \sqrt{BL} \tag{8.56}$$

where F_h is a function of L/B as shown in Figure 8.25. The rotational stiffness of a shallow foundation was derived by Gorbunov-Possadov and Serebrajanyi (1961) as

$$K_\theta = \frac{M}{\theta} = \frac{G}{1-\nu} F_\theta B L^2 \tag{8.57}$$

where F_θ is also a function of L/B as shown in Figure 8.25. The three foregoing equations 8.55- 8.57 assume that there is no coupling between the different terms.[①]

① 式（8.55）—式（8.57）假设各分项之间没有相互影响。

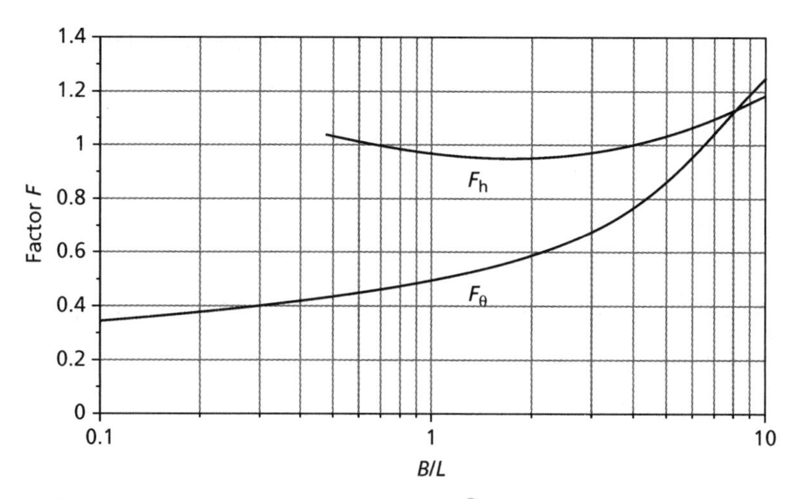

Figure 8.25 Non- dimensional factors[②] F_h and F_θ for foundation stiffness determination.

② 无量纲系数。

Example 8.3

An offshore wind turbine is to be installed on a square gravity-base foundation as shown in Figure 8.26(a gravity base is a large shallow foundation). The subsoil is clay with $c_u = 20\text{kPa}$ (constant with depth). The weight of the turbine structure is 2.6MN, and the gravity base is neutrally buoyant (i. e. it is hollow, with its weight balancing the resultant uplift force form the water pressure). Determine the required width of the foundation to satisfy ULS to DA1a if the horizontal (environmental) loading is 20% of the vertical load and acts at the level of the seabed.

Solution

If the horizontal loading is a variable unfavourable action ($\gamma_A = 1.50$) which acts rapidly (undrained) and the vertical load a permanent unfavourable action ($\gamma_A = 1.35$), then the ratio of the design loads is

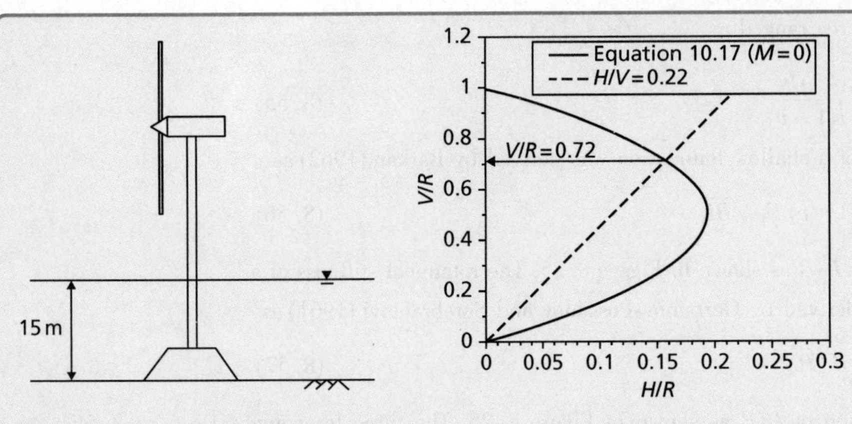

Figure 8.26 Example 8.3.

$$\frac{H}{V}=0.2\cdot\frac{1.50}{1.35}=0.22$$

This plots as a straight line with gradient $1/0.22=4.5$ on Figure 8.21 (a), as shown in Figure 8.26. This intersects the yield surface at $V/R=0.72$. The design load $V=1.35\times 2.6=3.51\text{MN}$ (neutral buoyancy means that the gravity base does not apply any net load). The vertical resistance R is given by Equation 8.17 as

$$\frac{R}{B^2}=\frac{s_c N_c\left(\frac{c_u}{\gamma_{cu}}\right)}{\gamma_{Rv}}$$

where $s_c=1.2$ (Equation 8.19), $N_c=5.14$, $c_u=20\text{kPa}$ and $\gamma_{cu}=\gamma_{Rv}=1.00$ for DA1a. Substituting these values gives $R=123.4B^2$ (kN). Then, for stability,

$$\frac{V}{R}\leqslant 0.72 \quad \therefore 3510\leqslant 0.72\cdot 123.4B^2$$

$$B\geqslant 6.29\text{m}$$

Example 8.4

The gravity base for the wind turbine in Example 8.3 was subsequently constructed $15\times 15\text{m}$ square. If the horizontal action now acts at sea level, determine whether the foundation satisfies the ULS to EC7 DA1a. If the seabed soil has $E_u/c_u=500$, determine the displacements of the foundation under the applied actions.

Solution

At ULS, the vertical design load $V=3.51\text{MN}$ as before. The horizontal design load $H=1.50\times 0.2\times 2.6=0.78\text{MN}$. As this acts at sea level (15m above the founding plane), $M=15\times H=11.7\text{MNm}$. The design resistance is now

$$R=\frac{s_c N_c\left(\frac{c_u}{\gamma_{cu}}\right)}{\gamma_{Rv}}B^2=\frac{1.2\cdot 5.14\cdot\left(\frac{20}{1.00}\right)}{1.00}\cdot 15^2=27.8\text{MN}$$

The normalised parameters for use with Figure 8.21 (b) are then: $H/R=0.78/27.8=0.03$, $M/BR=11.7/(15\times 27.8)=0.03$. The point defined by these parameters is plotted in Figure 8.21 (b), which shows that $0.07<V/R<0.93$. As $V/R=3.51/27.8=0.13$, the foundation does satisfy the ULS.

The undrained Young's Modulus $E_u = 500 \times 20 = 10\text{MPa}$, so that, from Equation 5.6,

$$G = \frac{E_u}{2(1+v_u)} = \frac{10}{2(1+0.5)} = 3.3\text{MPa}$$

Treating the gravity base as rigid, $I_s = 0.82$, from Equation 8.55,

$$\frac{V}{s} = \left(\frac{2L}{I_s}\right)\frac{G}{(1-v)} \quad \therefore s = \frac{2.60}{\left(\frac{2 \cdot 15}{0.82}\right)\frac{3.3}{(1-0.5)}} = 0.0107\text{m}$$

noting that V is now the characteristic load for SLS calculations. From Figure 8.25, $F_h = 0.95$ and $F_\theta = 0.5$, so that from Equations 8.56 and 8.57,

$$\frac{H}{h} = 2G(1+v)F_h\sqrt{BL} \quad \therefore h = \frac{(0.2 \cdot 2.6)}{2 \cdot 3.3 \cdot 1.5 \cdot 0.95 \cdot \sqrt{15 \cdot 15}} = 0.0037\text{m}$$

$$\frac{M}{\theta} = \frac{G}{1-v}F_\theta BL^2 \quad \therefore \theta = \frac{(0.2 \cdot 2.6 \cdot 15)}{\left(\frac{3.3}{1-0.5}\right) \cdot 0.5 \cdot 15^3} = 7.0 \times 10^{-4}\text{radians}$$

The gravity base will therefore displace vertically by 10.7 mm and horizontally by 3.7 mm, and rotate by 0.04°, under the applied actions.

Summary

1 The application of load to a shallow foundation induces stresses within the underlying soil mass generating shear. As the foundation is loaded it will settle. If the shear stress reaches a condition of plastic equilibrium and a compatible failure mechanism can be formed, then the footing will suffer bearing capacity failure (the settlement will become infinite).

2 The condition of plastic failure within a soil mass may be analysed using limit analysis. Upper bound techniques involve postulating a compatible failure mechanism and performing an energy balance based on the movement within the mechanism. Lower bound techniques involve postulating a stress field which is in equilibrium with the applied external load. The true failure load will lie between the upper and lower bound solutions- if the upper and lower bounds are equal, then the true solution has been found. These methods have been applied to both undrained and drained soil conditions to determine bearing capacity. More advanced solutions given in the literature have extended these approaches to more complex ground conditions, to determine bearing capacity under more realistic conditions for use in foundation design.

3 The vertical bearing resistance (and therefore the carrying capacity) of shallow foundations is reduced if significant horizontal and moment actions are present. Limit analysis can be used to derive a yield surface for use in ULS design, which can efficiently check the overall stability of the foundation (i. e. against bearing capacity failure, sliding and overturning).

Problems

8.1 A strip footing 2m wide is founded at a depth of 1m in a stiff clay of saturated unit weight 21kN/m^3, the water table being at ground level. Determine the bearing capacity of the foundation (a) when c_u = 105kPa and $\phi_u = 0$, and (b) when $c' = 10$kPa and $\phi' = 28°$.

8.2 Determine the allowable design load on a footing 4.50 × 2.25m at a depth of 3.50m in a stiff clay to EC7 DA1a. The saturated unit weight of the clay is 20kN/m^3 and the characteristic shear strength parameters are $c_u = 135$kPa and $\phi_u = 0$.

8.3 A footing 2.5 × 2.5m carries a pressure of 400kPa at a depth of 1m in a sand. The saturated unit weight of the sand is 20kN/m^3 and the unit weight above the water table is 17kN/m^3. The design shear strength parameters are $c' = 0$ and $\phi' = 40°$. Determine the bearing capacity of the footing for the following cases:

a the water table is 5m below ground level;

b the water table is 1m below ground level;

c the water table is at ground level and there is seepage vertically upwards under a hydraulic gradient of 0.2.

8.4 A shallow tunnel, formed from pre-fabricated concrete box sections ($\gamma = 23.5\text{kN/m}^3$), is to be installed in water-bearing ground. The tunnel has exterior dimensions of 15m deep by 30m wide, and has walls, roof and ceiling 1.5m thick. It will be installed such that the top surface of the box is level with the ground surface. During construction the water table is drawn-down to below the underside of the box, but this will be relaxed once construction is complete.

a Determine the depth of the water table at which the structure will start to float, to EC7 (assuming that the walls of the box are smooth).

b If tension piles are installed beneath the box, determine the design resistance that the piling must be able to provide.

References

Barkan, D. D. (1962) *Dynamic Bases and Foundations,* McGraw-Hill Book Company, New York, NY.

Butterfield, R. and Gottardi, G. (1994) A complete three-dimensional failure envelope for shallow footings on sand, *Geotchnique*, **44**(1), 181 – 184.

Davis, E. H. and Booker, J. R. (1973) The effect of increasing strength with depth on the bearing capacity of clays, *Géotechnique*, **23** (4), 551 – 563.

EC7 - 1 (2004) *Eurocode 7: Geotechnical design - Part 1: General rules, BS EN 1997 - 1: 2004*, British Standards Institution, London.

Georgiadis, K. (2010) An upper-bound solution for the undrained bearing capacity of strip footings at the top of a slope, *Géotechnique*, **60** (10), 801 - 806.

Giroud, J. P. (1972) Settlement of rectangular foundation on soil layer, *Journal of the ASCE*, **98** (SM1), 149 - 54.

Gorbunov-Possadov, M. I. and Serebrajanyi, R. V. (1961) Design of structures upon elastic foundations, in *Proceedings of the 5th International Conference on Soil Mechanics and Foundation Engineering*, Vol. 1, pp. 643 - 648.

Gourvenec, S. (2007) Shape effects on the capacity of rectangular footings under general loading, *Geotechnique*, **57**(8), 637 - 646.

Kumar, J. and Kouzer, K. M. (2007) Effect of footing roughness on bearing capacity factor N_γ, *Journal of Geotechnical and Geoenvironmental Engineering*, **133** (5), 502 - 511.

Lyamin, A. V., Salgado, R., Sloan, S. W. andPrezzi, M. (2007) Two-and three-dimensional bearing capacity of footings in sand, *Géotechnique*, **57** (8), 647 - 662.

Merifield, R. S. and Nguyen, V. Q. (2006) Two-and three-dimensional bearing-capacity solutions for footings on two-layered clays, *Geomechanics and Geoengineering: An International Journal*, **1** (2), 151 - 162.

Merifield, R. S., Sloan, S. W. and Yu, H. S. (1999) Rigorous plasticity solutions for the bearing capacity of two-layered clays, *Géotechnique*, **49** (4), 471 - 490.

Salgado, R. (2008). *The Engineering of Foundations*, McGraw-Hill, New York, NY.

Salgado, R., Lyamin, A. V., Sloan, S. W. and Yu, H. S. (2004) Two-and three-dimensional bearing capacity of foundations in clay, *Géotechnique*, **54** (5), 297 - 306.

Skempton, A. W. (1951) The bearing capacity of clays, *Proceedings of the Building Research Congress*, Vol. 1, pp. 180 - 189.

Zhu, F., Clark, J. I. and Philips, R. (2001) Scale effect of strip and circular footings resting on dense sand, *Journal of Geotechnical and Geoenvironmental Engineering*, **127** (7), 613 - 620.

Further reading

Frank, R., Bauduin, C., Driscoll, R., Kavvadas, M., Krebs Ovesen, N., Orr, T. and Schuppener, B. (2004) *Designers' Guide to EN 1997 - 1 Eurocode 7: Geotechnical Design - General rules*, Thomas Telford, London.

This book provides a guide to limit state design of a range of constructions (including shallow foundations) using Eurocode 7 from a designer's perspective and provides a useful companion to the Eurocodes when conducting design. It is easy to read and has plenty of worked examples.

Chapter 9

Retaining structures[1]

① 挡土结构。

Learning outcomes

After working through the material in this chapter, you should be able to:

1 Use limit analysis and limit equilibrium techniques[2] to determine the limiting lateral earth pressures[3] acting on retaining structures;

2 Determine in-situ lateral stresses based on fundamental soil properties and understand how limiting earth pressures are mobilised from these values by relative soil-structure movement;

3 Determine the lateral stresses induced on a retaining structure due to external loads and construction procedures;

4 Design a gravity retaining structure[4], an embedded wall[5], a braced excavation[6] or a reinforced soil retaining structure[7] within a limit-state design framework (Eurocode 7).

② 极限平衡法。
③ 侧向土压力。
④ 重力式挡土结构。
⑤ 埋置式挡土墙。
⑥ 支撑开挖。
⑦ 加筋土挡土结构。

9.1 Introduction

It is often necessary in geotechnical engineering to retain masses of soil (Figure 9.1). Such applications may be permanent, e.g.

- retaining unstable soil next to a road or railway,
- raising a section of ground with minimal land-take,
- creation of underground space;

or temporary, e.g.

- creating an excavation to install service pipes/cables or to repair existing services.

In permanent applications, a structural element is usually used to support the retained mass of soil. This will typically be either a **gravity retaining wall**[8], which keeps the retained soil stable due to its mass, or a **flexible retaining wall**[9] which resists soil movement by bending. In both cases it is essential to determine the

⑧ 重力式挡土墙。
⑨ 柔性挡土墙。

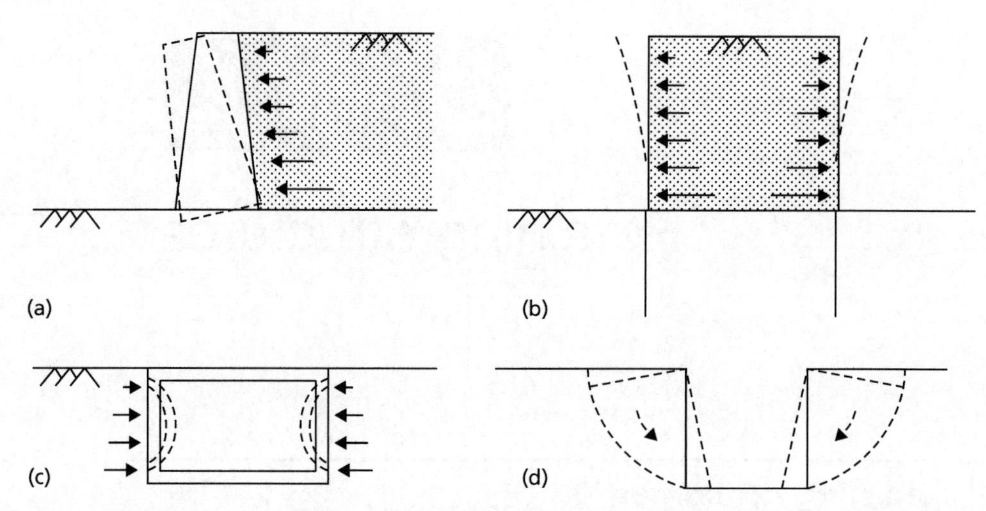

Figure 9.1 Some applications of retained soil: (a) restraint of unstable soil mass, (b) creation of elevated ground, (c) creation of underground space, (d) temporary excavations.

magnitude and distribution of lateral pressure between the soil mass and the adjoining retaining structure to check the stability of a gravity wall against sliding and overturning[①] or to undertake the structural design of a flexible retaining wall. As with foundations, such permanent structures must be designed to satisfy both ultimate and serviceability limit states. These will be discussed in greater detail in Section 9.4 and 9.7. Soil may also be retained by reinforcing the soil mass[②] itself; this will be described in Section 9.11.

① 验算重力式挡土墙的抗滑稳定性和抗倾覆能力。
② 增强土体。

Excavations may be self-supporting if undrained strength can be mobilised, and in these cases lateral earth-pressure theory may be used to determine the maximum depth to which such excavations can safely be made (an ultimate limiting state). Supported excavations will be discussed in greater detail in Section 9.9, while unsupported excavations will be discussed in Chapter 10.

Section 9.2 introduces the basic theories of lateral earth pressure using limit analysis techniques[③], as in Chapter 8. Rigorous lower bound solutions will be derived for both undrained and drained conditions. As previously, it is assumed that the stress – strain behaviour of the soil can be represented by the rigid – perfectly plastic idealisation, shown in Figure 8.3. Conditions of plane strain[④] are also assumed (as for strip footings), i. e. strains in the longitudinal direction of the structure are assumed to be zero due to the length of most retaining structures.

③ 极限分析技术。
④ 平面应变。

9.2 Limiting earth pressures[⑤] from limit analysis Limiting lateral earth pressures

⑤ 极限土压力。

Figure 9.2 (a) shows the stress conditions in soil on either side of an embedded retaining wall, where major principal stresses are defined by σ_1, σ_1' (total and effective, respectively) and minor principal stresses are defined by σ_3, σ_3'. If the wall

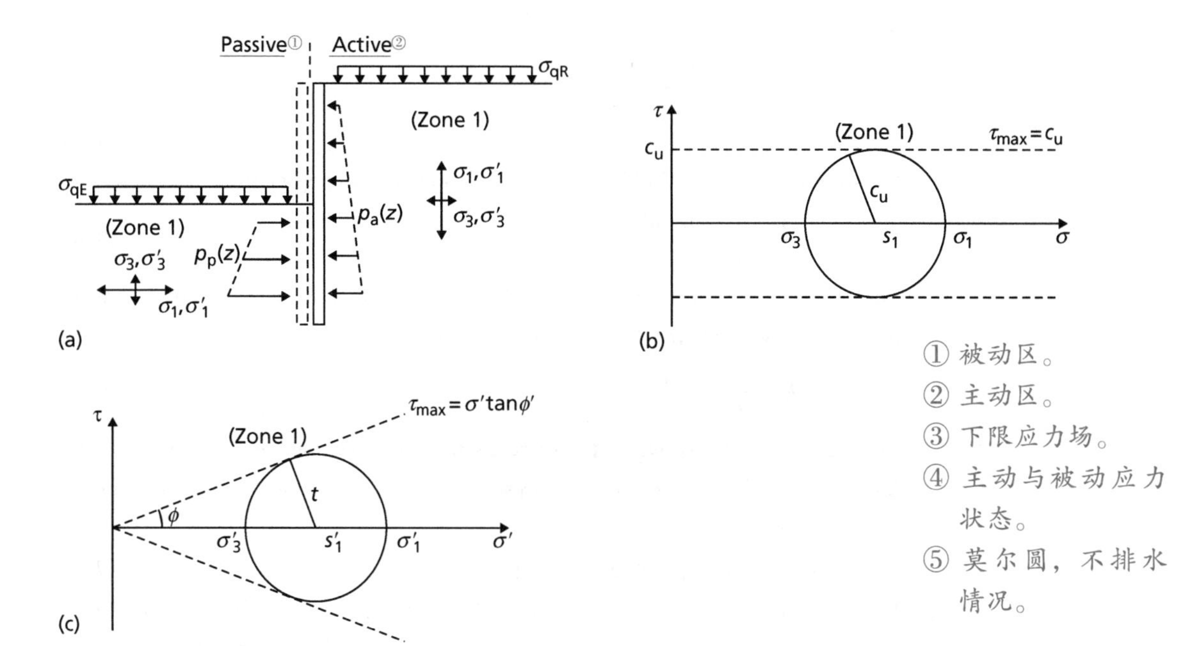

① 被动区。
② 主动区。
③ 下限应力场。
④ 主动与被动应力状态。
⑤ 莫尔圆，不排水情况。

Figure 9.2 Lower bound stress field[3]: (a) stress conditions under active and passive conditions[4], (b) Mohr circle, undrained case[5], (c) Mohr circle, drained case[6].

⑥ 莫尔圆，排水情况。

were to fail by moving horizontally (translating) in the direction shown, the horizontal stresses within the **retained soil**[7] behind the wall will reduce. If the movement is large enough, the value of horizontal stress decreases to a minimum value such that a state of plastic equilibrium develops for which the major principal total and effective stresses are vertical. This is known as the **active condition**[8]. In the soil on the other side of the wall there will be lateral compression of the soil as the wall displaces, resulting in an increase in the horizontal stresses until a state of plastic equilibrium is reached, such that the major principal total and effective stresses are horizontal. This is known as the **passive condition**[9]. Mohr circles at the point of failure are shown in Figure 9.2 (b) for undrained soil and Figure 9.2 (c) for drained material.

⑦ 被围护土。
⑧ 主动条件。
⑨ 被动条件。

In the active case for an undrained cohesive material at failure, $\sigma_1(z) = \sigma_v(z) = \gamma z + \sigma_{qR}$ (total vertical stress) and σ_3 is the horizontal stress in the soil, which by equilibrium must also act on the wall. Therefore, $\sigma_3 = \sigma_h = p_a$, where p_a is the **active earth pressure**[10]. From Figure 9.2 (b),

⑩ 主动土压力。

$$\sigma_3 = \sigma_1 - 2c_u$$

$$P_a(z) = \sigma_v(z) - 2c_u \tag{9.1}$$

In the passive case, $\sigma_3(z) = \sigma_v(z) = \gamma z + \sigma_{qE}$ and $\sigma_1 = p_p$, where p_p is the **passive earth pressure**[11]. From Figure 9.2 (b),

⑪ 被动土压力。

$$\sigma_1 = \sigma_3 + 2c_u$$

$$P_p(z) = \sigma_v(z) + 2c_u \tag{9.2}$$

In the active case for a drained cohesionless material, $\sigma'_1(z) = \sigma'_v(z)$ (effective vertical stress) and $\sigma'_3(z) = \sigma'_h(z)$ is the horizontal effective stress in the soil. From Figure 9. 2 (c),

$$\sin \phi' = \frac{t}{s_1'} = \frac{\sigma_1' - \sigma_3'}{\sigma_1' + \sigma_3'} \tag{9.3}$$

Substituting for σ_1' and σ_3' in Equation 9. 3 and rearranging gives

$$\frac{\phi_h'}{\phi_v'} = \frac{1 - \sin \phi'}{1 + \sin \phi'} \tag{9.4}$$

The ratio σ_h'/σ_v' is termed the earth pressure coefficient, K (see Chapter 7). As Equation 9. 4 was derived for active conditions,

$$K_a = \frac{1 - \sin \phi'}{1 + \sin \phi'} \tag{9.5}$$

where K_a is the **active earth pressure coefficient**[①]. The active earth pressure acting on the wall ($p_a = \sigma_h$, a total stress) is then found using Terzaghi's Principle ($\sigma_h = \sigma_h' + u$)

① 主动土压力系数。

$$p_a(z) = \sigma_h(z) = K_a \sigma_v'(z) + u(z) \tag{9.6}$$

In the passive case for a drained material, $\sigma_3'(z) = \sigma_v'(z)$ (effective vertical stress) and $\sigma_1'(z) = \sigma_h'(z)$ is the horizontal effective stress in the soil. Substituting for σ_1' and σ_3' in Equation 9. 3 and rearranging gives

$$\frac{\phi_h'}{\phi_v'} = \frac{1 + \sin \phi'}{1 - \sin \phi'} = K_p \tag{9.7}$$

where K_p is the **passive earth pressure coefficient**[②]. The passive earth pressure acting on the wall is then

② 被动土压力系数。

$$p_p(z) = \sigma_h(z) = K_p \sigma_v'(z) + u(z) \tag{9.8}$$

Rankine's theory of earth pressure (general ϕ', c' material)

Rankine developed a lower bound solution based on the **Method of Characteristics** for the case of a soil with general strength parameters c' and ϕ'[③]. The Mohr circle representing the state of stress at failure in a two-dimensional element is shown in Figure 9. 3, the relevant shear strength parameters being denoted by c' and ϕ'. Shear failure occurs along a plane at an angle of $45° + \phi'/2$ to the major principal plane[④]. If the soil mass as a whole is stressed such that the principal stresses at every point are in the same directions then, theoretically, there will be a network of failure planes (known as a slip line field) equally inclined to the principal planes, as shown in Figure 9. 3. The two sets of planes are termed α-and β-characteristics, from where the method gets its name. It should be appreciated that the state of plastic equilibrium can be developed only if sufficient deformation of the soil mass can take place (see Section 9. 3).

③ 针对土体的通用强度参数 c' 和 ϕ'，朗肯基于特征线法提出了下限求解法。

④ 剪切破坏发生在与大主应力平面呈（$45° + \phi'/2$）夹角的平面内。

A semi-infinite mass of soil with a horizontal surface is considered as before, having a vertical boundary formed by a smooth wall surface extending to semi-infinite depth, as represented in Figure 9. 4 (a). Referring to Figure 9. 3,

$$\sin\phi' = \frac{(\sigma_1' - \sigma_3')}{(\sigma_1' + \sigma_3' + 2c'\cot\phi')}$$

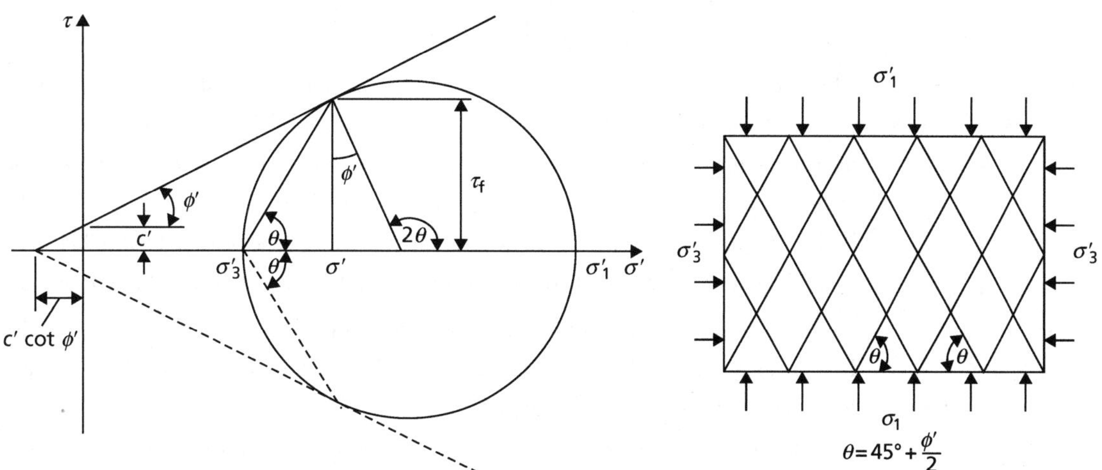

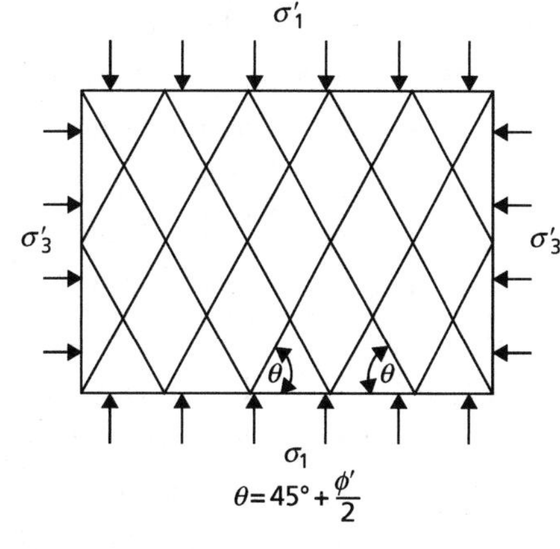

Figure 9.3 State of plastic equilibrium[①].

① 塑性平衡状态。

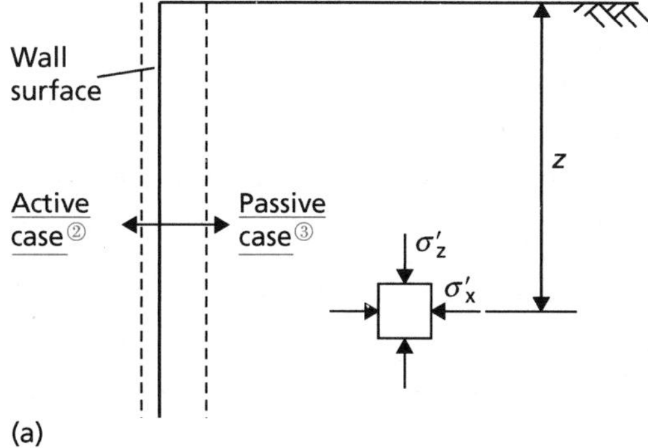

② 主动情况下墙的位移。

③ 被动情况下墙的位移。

④ 朗肯主动状态。

⑤ 朗肯被动状态。

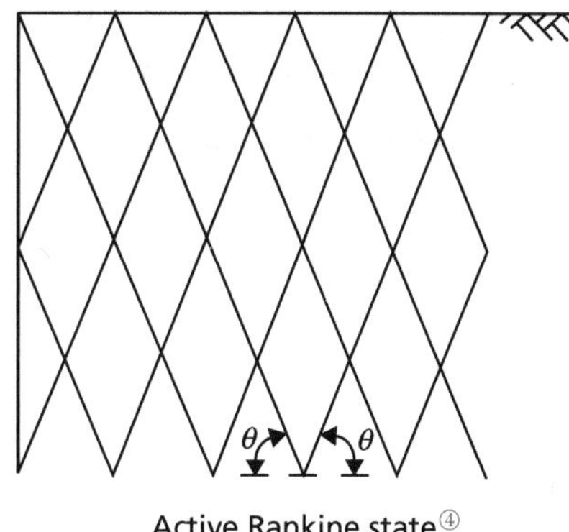

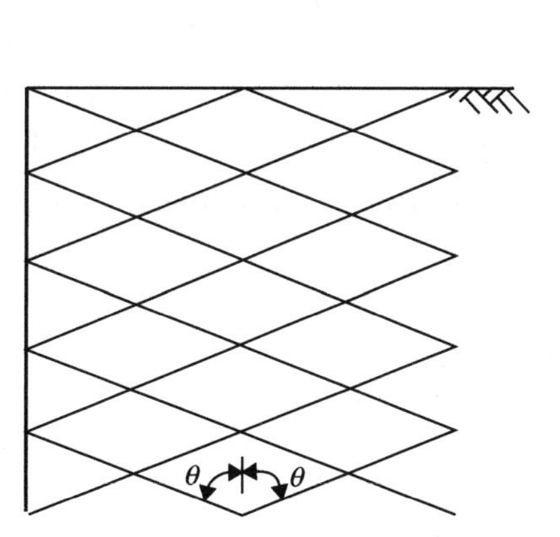

Figure 9.4 Active and passive Rankine states.

$$\therefore \sigma_3'(1+\sin\phi') = \sigma_1'(1-\sin\phi') - 2c'\cos\phi' \tag{9.9}$$

$$\therefore \sigma_3' = \sigma_1'\left(\frac{1-\sin\phi'}{1+\sin\phi'}\right) - 2c'\left(\frac{\sqrt{(1-\sin^2\phi')}}{1+\sin\phi'}\right)$$

① 当水平应力等于土的主动压力时，土体处在朗肯主动状态，土体中将出现两组与水平面（最大主应力面）呈夹角$\theta=(45^\circ+\phi'/2)$的破坏面。

② 太沙基原理。

③ 当水平应力等于土的被动压力时，土体处于朗肯被动状态，土体中将出现两组与竖直面（最大主应力面）呈夹角$\theta=(45^\circ+\phi'/2)$的破坏面。

④ 注：此处英文版原书错误，应为"K_p"。

$$\therefore\ \sigma'_3=\sigma'_1\left(\frac{1-\sin\phi'}{1+\sin\phi'}\right)-2c'\left(\sqrt{\frac{1-\sin^2\phi'}{1+\sin\phi'}}\right)$$

Alternatively, $\tan^2(45^\circ-\phi'/2)$ can be substituted for $(1-\sin\phi')/(1+\sin\phi')$.

As before, in the active case $\sigma'_1=\sigma'_v$ and $\sigma'_3=\sigma'_h$. When the horizontal stress becomes equal to the active pressure the soil is said to be in the active Rankine state, there being two sets of failure planes each inclined at $\theta=45^\circ+\phi'/2$ to the horizontal (the direction of the major principal plane)[①] as shown in Figure 9.4 (b). Using Equation 11.5 and Terzaghi's Principle[②], Equation 9.9 may be written as

$$p_a(z)=\sigma_h(z)=K_a\sigma'_v(z)-2c'\sqrt{K_a}+u(z) \tag{9.10}$$

If $c'=0$, Equation 9.10 reduces to Equation 9.6; if $\phi'=0$ and $c'=c_u$, $K_a=1$ and Equation 9.10 reduces to Equation 9.1.

In the passive case $\sigma'_1=\sigma'_h$ and $\sigma'_3=\sigma'_v$. When the horizontal stress becomes equal to the passive pressure the soil is said to be in the passive Rankine state, there being two sets of failure planes each inclined at $\theta=45^\circ+\phi'/2$ to the vertical (the direction of the major principal plane)[③] as shown in Figure 9.4 (c). Rearranging Equation 9.9 gives

$$\sigma'_1=\sigma'_3\left(\frac{1+\sin\phi'}{1-\sin\phi'}\right)+2c'\left(\sqrt{\frac{1+\sin\phi'}{1-\sin\phi'}}\right) \tag{9.11}$$

Using Equation 9.5 and Terzaghi's Principle, Equation 9.11 may be written as

$$p_p(z)=\sigma_h(z)=K_p\sigma'_v(z)+2c'\sqrt{K_p}+u(z) \tag{9.12}$$

If $c'=0$, Equation 9.12 reduces to Equation 9.8; if $\phi'=0$ and $c'=c_u$, K_a[④] $=1$ and Equation 9.12 reduces to Equation 9.2.

Example 9.1

The soil conditions adjacent to a sheet pile wall are given in Figure 9.5, a surcharge pressure of 50kPa being carried on the surface behind the wall. For soil 1, a sand above the water table, $c'=0$, $\phi'=38^\circ$ and $\gamma=18\text{kN/m}^3$. For soil 2, a saturated clay, $c'=10\text{kPa}$, $\phi'=28^\circ$ and $\gamma_{sat}=20\text{kN/m}^3$. Plot the distributions of active pressure behind the wall and passive pressure in front of the wall.

Solution

For soil 1,

$$K_a=\frac{1-\sin 38^\circ}{1+\sin 38^\circ}=0.24,\quad K_p=\frac{1}{0.24}=4.17$$

⑤ 被动土压力。

⑥ 主动土压力。

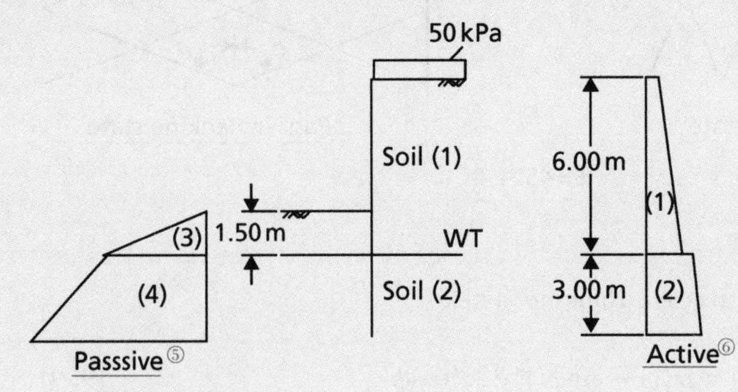

Figure 9.5 Example 9.1.

For soil 2,

$$K_a = \frac{1 - \sin 28^\circ}{1 + \sin 28^\circ} = 0.36, \quad K_p = \frac{1}{0.36} = 2.78$$

The pressures in soil 1 are calculated using $K_a = 0.24$, $K_p = 4.17$ and $\gamma = 18\text{kN/m}^3$. Soil 1 is then considered as a surcharge of (18 × 6) kPa on soil 2, in addition to the surface surcharge. The pressures in soil 2 are calculated using $K_a = 0.36$, $K_p = 2.78$ and $\gamma' = (20 - 9.8) = 10\text{kN/m}^3$ (see Table 9.1). The active and passive pressure distributions are shown in Figure 9.5. In addition, there is equal hydrostatic pressure on each side of the wall below the water table.

Table 9.1 Example 9.1

Soil	Depth[1] (m)	Pressure[2] (kPa)	
Active pressure:[3]			
(1)	0	0.24 × 50	= 12.0
(1)	6	(0.24 × 50) + (0.24 × 18 × 6) = 12.0 + 25.9	= 37.9
(2)	6	0.36[50 + (18 × 6)] − (2 × 10 × $\sqrt{0.36}$) = 56.9 − 12.0	= 44.9
(2)	9	0.36[50 + (18 × 6)] − (2 × 10 × $\sqrt{0.36}$) + (0.36 × 10.2 × 3) = 56.9 − 12.0 + 11.0	= 55.9
Passive pressure:[4]			
(3)	0	0	
(3)	1.5	4.17 × 18 × 1.5	= 112.6
(4)	1.5	(2.78 × 18 × 1.5) + (2 × 10 × $\sqrt{2.78}$) = 75.1 + 33.3	= 108.4
(4)	4.5	(2.78 × 18 × 1.5) + (2 × 10 × $\sqrt{2.78}$) + (2.78 × 10.2 × 3) = 75.1 + 33.3 + 85.1	= 193.5

Effect of wall properties (roughness, batter angle)[5]

In most practical cases the wall will not be smooth such that shear stresses may be generated along the soil – wall interface[6], which may also not be vertical but slope at an angle w to the vertical. This additional shear will cause a rotation of the principal stresses close to the wall, while in the soil further away the major principal stresses will still be vertical (active case) or horizontal (passive case) as before. In order to ensure equilibrium throughout the soil mass, frictional stress discontinuities[7] (see Sections 8.3 and 8.4) must be used to rotate the principal stresses between zone 1 and zone 2 as shown in Figure 9.6.

The amount by which the principal stresses rotate[8] depends on the magnitude of the shear stress that can be developed along the soil – wall interface, τ_w (the interface shear strength). In undrained materials $\tau_w = \alpha c_u$ is assumed, while in drained materials $\tau_w = \sigma' \tan \delta'$ is used.

In an undrained material, the stress conditions in zone 1 are still represented by the Mohr circle shown in Figure 9.2 (b). The Mohr circle for zone 2 is shown in Figure 9.7 (a) for the active case when the soil in zone 2 is at plastic failure. The major principal stress σ_1 acts on a plane which is rotated by $2\theta = \pi - \Delta_2$ from the

① 土体深度。
② 压力。
③ 主动土压力。
④ 被动土压力。
⑤ 挡土墙特性的影响（粗糙度，倾角）。
⑥ 土-墙接触面。
⑦ 摩阻型应力不连续面。
⑧ 主应力旋转。

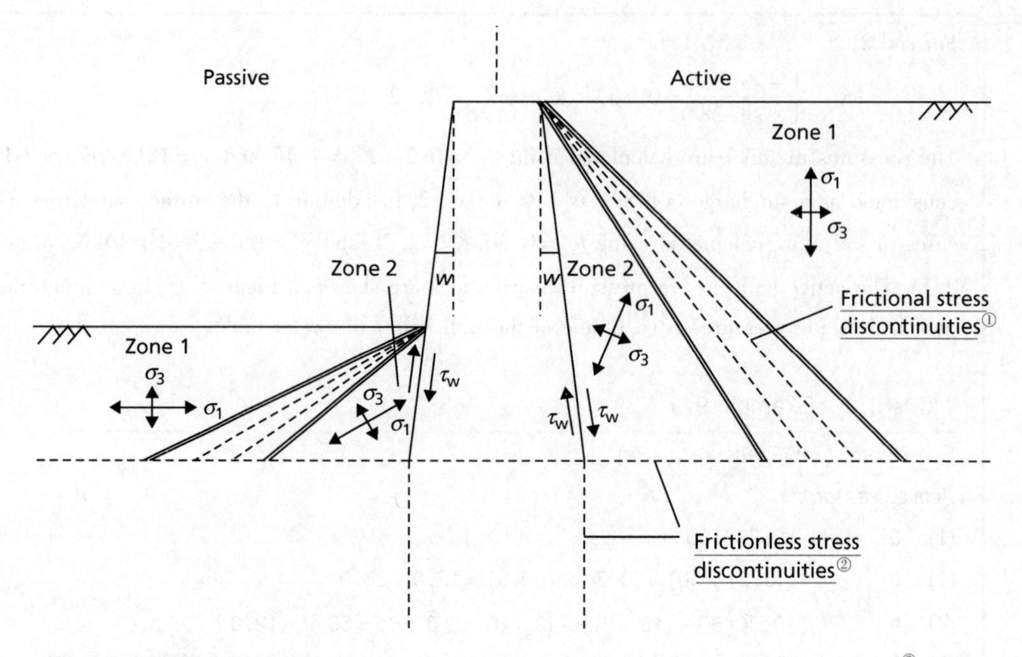

Figure 9.6 Rotation of principal stresses due to wall roughness and batter angle[3] (only total stresses shown).

① 摩阻型应力不连续面。
② 无摩阻型应力不连续面。
③ 由于墙面粗糙与墙身倾斜而产生的主应力旋转。
④ 主动土压力。
⑤ 莫尔圆。
⑥ 作用在挡土墙上的法向总应力。

stress state on the wall, which is itself at an angle w to the vertical. As σ_1 in zone 1 is vertical, the rotation in principal stress direction from zone 1 to 2 is $\theta_{fan} = \Delta_2/2 - w$ from Figure 9.7 (a). The magnitude of s_1 reduces moving from zone 1 to zone 2, so, from Equation 8.15,

$$s_1 - s_2 = c_u(\Delta_2 - 2w) \tag{9.13}$$

In zone 1 (from Figure 9.2 (b)),

$$s_1 = \sigma_v - c_u \tag{9.14}$$

In zone 2, the total stress acting normal to the wall (the active earth pressure)[4] from Figure 9.7 (a) is given by

$$p_a = s_2 - c_u \cos \Delta_2 \tag{9.15}$$

The Mohr circle[5] for zone 2 is shown in Figure 9.7 (b) for the passive case when the soil in zone 2 is at undrained plastic failure. The stress state on the wall represents the stresses in the vertical direction. The major principal stress σ_1 acts on a plane which is rotated by $2\theta = \Delta_2$ from the stress state on the wall, which is itself at an angle of w to the vertical. As σ_1 in zone 1 was horizontal, the rotation in principal stress direction is $\theta_{fan} = \Delta_2/2 - w$. The magnitude of s_1 increases moving from zone 1 to zone 2, so, from Equation 8.15,

$$s_2 - s_1 = c_u(\Delta_2 - 2w) \tag{9.16}$$

In zone 1 (from Figure 9.2 (b)),

$$s_1 = \sigma_v + c_u \tag{9.17}$$

In zone 2, the total stress acting normal to the wall[6] (the active earth pressure) from Figure 9.7 (b) is given by

$$p_p = s_2 + c_u \cos \Delta_2 \tag{9.18}$$

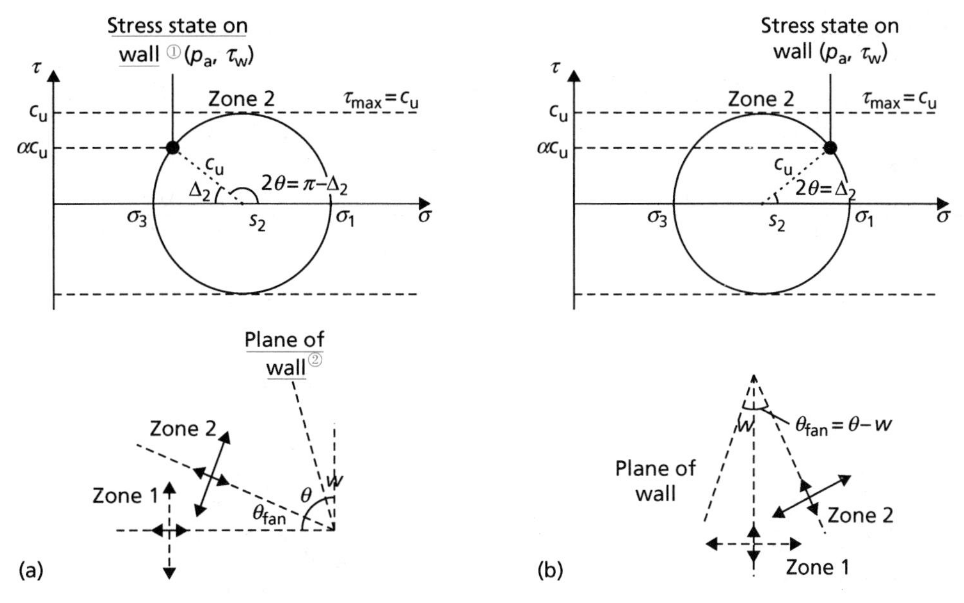

Figure 9.7 Mohr circles for zone 2 soil (adjacent to wall) under undrained conditions: (a) active case, (b) passive case ③.

① 挡土墙的应力状态。

② 挡土墙平面。

③区域 2 土体(靠近挡土墙) 在不排水条件下的莫尔圆：(a) 主动状态；(b) 被动状态。

From Figure 9.7, it is clear that

$$\sin \Delta_2 = \frac{\tau_w}{c_u} = \alpha \tag{9.19}$$

To determine the active or passive earth pressures acting on a given wall, the following procedure should be followed:

1 Find the mean stress in zone 1 from Equations 9.14 or 9.17 for active or passive conditions respectively;
2 Determine Δ_2 from Equation 9.19;
3 Find the mean stress in zone 2 from Equation 9.13 or 9.16;
4 Evaluate earth pressures using Equation 9.15 or 9.18.

It should be noted that for the special case of a smooth vertical wall ④, $\alpha = w = 0$, so $\Delta_2 = 0$ (Equation 8.47) and Equations 9.15 and 9.19 reduce to Equations 9.1 and 9.2 respectively.

④ 光滑竖直挡土墙。

The Mohr circle for zone 2 in drained soil is shown in Figure 9.8 (a) for the active case when the soil in zone 2 is at plastic failure. The major principal stress σ'_1 acts on a plane which is rotated by $2\theta = \pi - (\Delta_2 - \delta')$ from the stress state on the wall, which is itself at an angle w to the vertical. The stress conditions in zone 1 are still represented by the Mohr circle shown in Figure 9.2 (b). As σ'_1 in zone 1 is vertical, the rotation in principal stress direction is $\theta_{fan} = (\Delta_2 - \delta')/2 - w$ from Figure 9.8 (a). The magnitude of s'_1 reduces moving from zone 1 to zone 2, so, from Equation 8.30,

$$\frac{s'_1}{s'_2} = e^{(\Delta_2 - \delta' - 2w)\tan\phi'} \tag{9.20}$$

In zone 1 (from Figure 9.2 (b)),

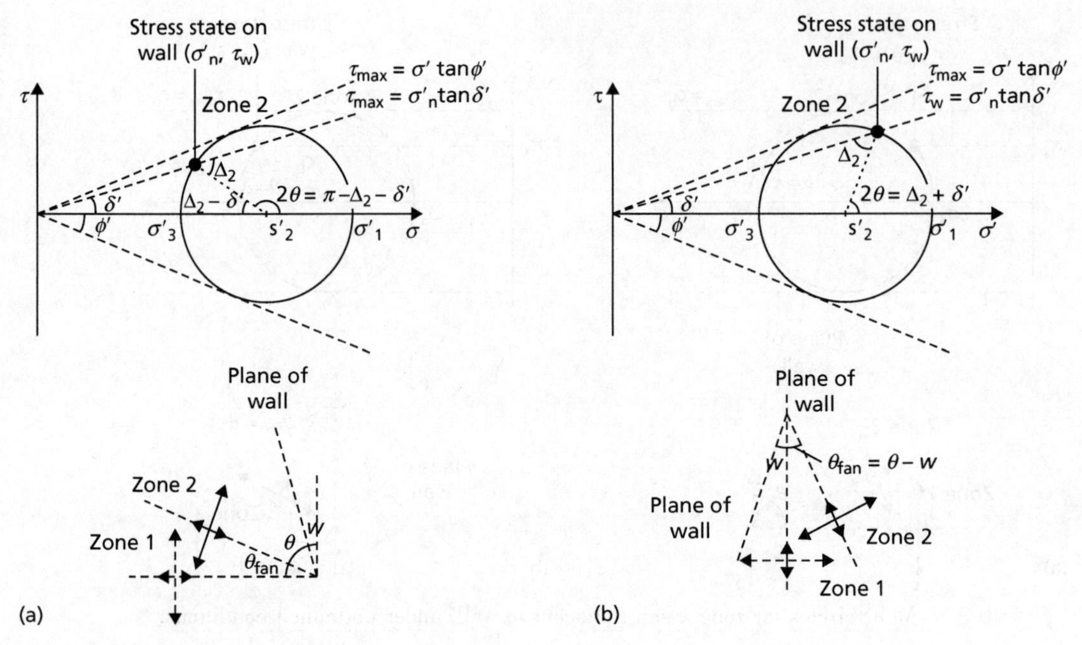

Figure 9.8 Mohr circles for zone 2 soil (adjacent to wall) under drained conditions: (a) active case, (b) passive case[①].

$$s_1' = \frac{\sigma_v'}{1 + \sin \phi'} \tag{9.21}$$

In zone 2, the effective stress acting normal to the wall[②] (σ_n') from Figure 9.8 (a) is given by

$$\begin{aligned}\sigma_n' &= s_2' - s_2' \sin \phi' \cos (\Delta_2 - \delta') \\ &= s_2'[1 - \sin \phi' \cos (\Delta_2 - \delta')]\end{aligned} \tag{9.22}$$

Then, defining the active earth pressure coefficient in terms of the normal rather than horizontal effective stress[③] (as the wall is battered by w)

$$\begin{aligned}K_a &= \frac{\sigma_n'}{\sigma_v'} = \frac{s_2'}{s_1'} \cdot \frac{1 - \sin \phi' \cos (\Delta_2 - \delta')}{1 + \sin \phi'} \\ &= \frac{1 - \sin \phi' \cos (\Delta_2 - \delta')}{1 + \sin \phi'} e^{-(\Delta_2 - \delta' - 2w)\tan \phi'}\end{aligned} \tag{9.23}$$

The Mohr circle for zone 2 is shown in Figure 9.8 (b) for the passive case when the soil in zone 2 is at plastic failure. The major principal stress σ_1' acts on a plane which is rotated by $2\theta = \Delta 2 + \delta'$ from the stress state on the wall, which is itself at an angle of w to the vertical[④]. As σ_1' in zone 1 is horizontal, the rotation in principal stress direction is $\theta_{fan} = (\Delta_2 + \delta')/2 - w$. The magnitude of s_1' increases moving from zone 1 to zone 2, so, from Equation 8.30,

$$\frac{s_2'}{s_1'} = e^{(-\Delta_2 + \delta' - 2w)\tan \phi'} \tag{9.24}$$

In zone 1 (from Figure 9.2 (b)),

$$s_1' = \frac{\sigma_v'}{1 - \sin \phi'} \tag{9.25}$$

① 区域2土体（靠近挡土墙）在排水条件下的莫尔圆：(a) 主动状态；(b) 被动状态。

② 沿法向作用在挡土墙上的有效应力。

③ 土压力系数由墙上的法向有效应力而不是水平向有效应力来定义。

④ 最大主应力σ_1'作用在与挡土墙上应力状态旋转$2\theta = \Delta_2 + \delta'$的平面上，这一平面与垂直平面夹角为$w$。

In zone 2, the effective stress acting normal to the wall from Figure 9.8 (b) is given by

$$\sigma'_n = s'_2 + s'_2 \sin\phi' \cos(\Delta_2 + \delta')$$
$$= s'_2[1 + \sin\phi' \cos(\Delta_2 + \delta')] \quad (9.26)$$

Then, defining the passive earth pressure coefficient in terms of the normal rather than horizontal effective stress (as the wall is battered by w)

$$K_p = \frac{\sigma'_n}{\sigma'_v} = \frac{s'_2}{s'_1} \cdot \frac{1 + \sin\phi' \cos(\Delta_2 + \delta')}{1 - \sin\phi'}$$
$$= \frac{1 + \sin\phi' \cos(\Delta_2 + \delta')}{1 - \sin\phi'} e^{(\Delta_2 + \delta' - 2w)\tan\phi'} \quad (9.27)$$

In both Equations 9.23 and 9.27

$$\sin\Delta_2 = \frac{\sin\delta'}{\sin\phi'} \quad (9.28)$$

To determine the active or passive earth pressures for a given wall, the following procedure should be followed[①]:

1 Find the vertical effective stresses σ'_v in zone 1[②];
2 Determine Δ_2 from Equation 9.28[③];
3 Find the earth pressure coefficient from Equation 9.23 or 9.27[④];
4 Evaluate earth pressures using Equation 9.6 or 9.8[⑤].

It should be noted that for the special case of a smooth vertical wall, $\delta' = w = 0$, so $\Delta_2 = 0$ (Equation 9.28) and Equations 9.23 and 9.27 reduce to Equations 9.5 and 9.7 respectively.

Sloping retained soil[⑥]

In many cases the retained soil behind the wall may not be level but may slope upwards at an angle β to the horizontal, as shown in Figure 9.9 (a). In this case, the major principal stress in the retained soil in zone 1 will no longer be vertical as the slope will induce a permanent static shear stress[⑦] within the soil mass, rotating the principal stress direction. The stress conditions (normal and shear stresses) on a plane parallel to the surface at any depth may be found by considering equilibrium as shown in Figure 9.9 (b). The component of the block's self-weight W, acting normal to the inclined plane per metre length of the soil mass, is then

$$W = \gamma z A \cos\beta \quad (9.29)$$

Considering equilibrium parallel to the plane,

$$\tau_{mob} = \frac{W}{A}\sin\beta = \gamma z \cos^2\beta \sin\beta \quad (9.30)$$

Considering equilibrium perpendicular to the plane,

$$\sigma' = \sigma - u = \frac{W}{A}\cos\beta - u = \gamma z \cos^2\beta - u \quad (9.31)$$

where u is the pore pressure on the plane. The stress conditions in the retained soil may be represented by the stress ratio τ_{mob}/σ' given by

① 按如下步骤确定给定挡土墙上的主动或被动土压力。
② 在区域1中找到竖向有效应力σ'_v。
③ 通过式(9.28)确定Δ_2。
④ 通过式(9.23)或式(9.27)确定土压力系数。
⑤ 通过式(9.6)或式(9.8)估算土压力。
⑥ 挡土墙后土体倾斜情况。
⑦ 静态(初始)剪应力。

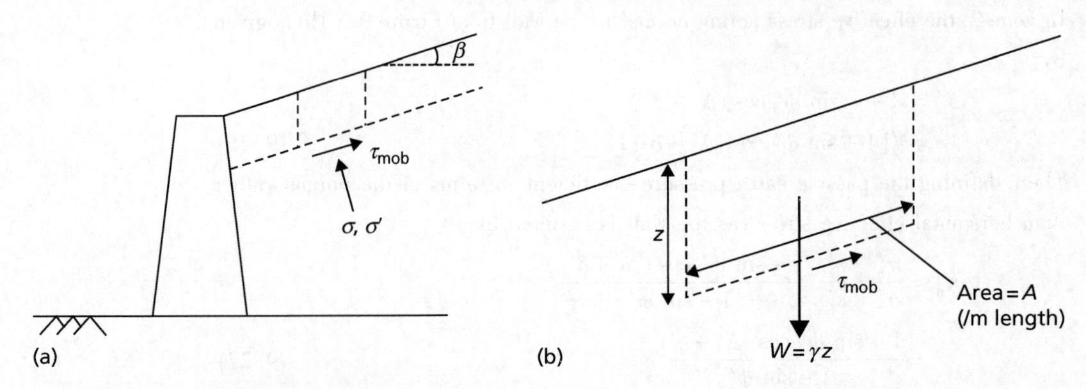

Figure 9. 9 Equilibrium of sloping retained soil.

$$\frac{\tau_{mob}}{\sigma'} = \frac{\gamma z\cos\beta\sin\beta}{\gamma z\cos^2\beta - u} = \tan\phi'_{mob} \tag{9.32a}$$

Equation 9. 32 applies for drained conditions. For undrained conditions where only total stresses are required,

$$\frac{\tau_{mob}}{\sigma'} = \frac{\gamma z\cos\beta\sin\beta}{\gamma z\cos^2\beta} = \tan\beta \tag{9.32b}$$

The Mohr circle for undrained conditions in zone 1 in the inclined case is shown in Figure 9. 10 (a). It can be seen that the major principal stress σ_1 is rotated $2\theta = \Delta_1$ from the stress state on the inclined plane in the retained soil, which is itself at an angle of β to the horizontal. The resultant rotation of the principal stresses in zone 1 compared to the level ground case is $\theta = \Delta_1/2 - \beta$. As a result, the value of θ_{fan} in Equation 9. 13 should be modified to

$$\theta_{fan} = \left(\frac{\Delta_2}{2} - w\right) - \left(\frac{\Delta_1}{2} - \beta\right) \tag{9.33}$$

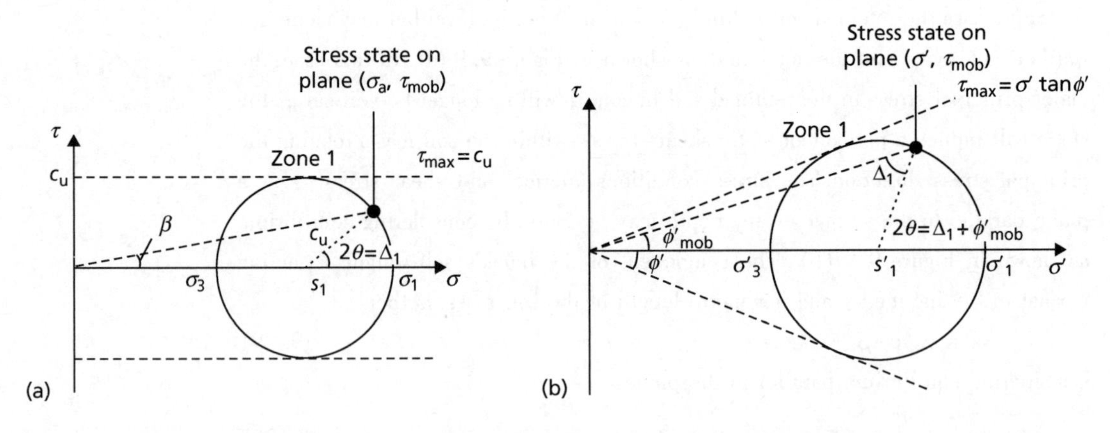

Figure 9. 10 Mohr circles for zone 1 soil under active conditions: (a) undrained case, (b) drained case[①].

① 区域1土体在主动状态下的莫尔圆：(a)不排水条件；(b)排水条件。

In zone 1 (from Figure 9. 10 (a)),

$$s_1 = \sigma' - c_u\cos\Delta_1 \tag{9.34}$$

which replaces Equation 9. 14. The difference in mean stress between zones 1 and 2 (active case) for the general case of sloping backfill retained by a battered rough wall① is then given by

$$s_1 - s_2 = c_u\,(\Delta_2 - 2w - \Delta_1 + 2\beta) \tag{9.35}$$

① 墙身倾斜、墙后填土倾斜的情况。

where

$$\sin\Delta_1 = \frac{\tau_{mob}}{c_u} \tag{9.36}$$

Equation 9.35 replaces Equation 9.13 for the case of sloping backfill and undrained conditions; the solution procedure remains unchanged.

The Mohr circle for drained conditions in zone 1 is shown in Figure 9.10 (b). It can be seen that the major principal effective stress σ'_1 is rotated $2\theta = \Delta_1 + \phi'_{mob}$ from the stress state on the inclined plane in the retained soil, which is itself at an angle of β to the horizontal. The resultant rotation of the principal stresses② in zone 1 compared to the level ground case is then $\theta = (\Delta_1 + \phi'_{mob})/2 - \beta$. As a result, the value of θ_{fan} in Equation 9.20 should be modified to

② 主应力旋转角度。

$$\theta_{fan} = \left(\frac{\Delta_2 - \delta}{2} - w\right) - \left(\frac{\Delta_1 + \phi'_{mob}}{2} - \beta\right) \tag{9.37}$$

From Figure 9.10 (b) it can also be seen that

$$\begin{aligned}\sigma' &= s'_1 + s'_1\sin\phi'\cos(\Delta_1 + \beta)\\ &= s'_1[1 + \sin\phi'\cos(\Delta_1 + \beta)]\end{aligned} \tag{9.38}$$

so that the active earth pressure coefficient for the general case of sloping backfill retained by a battered rough wall is given by

$$K_a = \frac{\sigma'_n}{\sigma'} = \frac{1 - \sin\phi'\cos(\Delta_2 - \delta')}{1 + \sin\phi'\cos(\Delta_1 + \beta)} e^{-(\Delta_2 - \delta' - 2w - \Delta_1 - \phi'_{mob} + 2\beta)\tan\phi'} \tag{9.39}$$

where

$$\sin\Delta_1 = \frac{\sin\phi'_{mob}}{\sin\phi'} \tag{9.40}$$

Equation 9.39 replaces Equation 9.23 for the case of sloping backfill and drained conditions; the solution procedure and all other equations remain unchanged.

9.3 Earth pressure at rest③

③ 静止土压力。

It has been shown that active pressure is associated with lateral expansion④ of the soil at failure and is a minimum value; passive pressure is associated with lateral compression⑤ of the soil at failure and is a maximum value. The active and passive values are therefore referred to as **limit pressures**⑥. If the lateral strain in the soil is zero, the corresponding lateral pressure, p'_0, acting on a retaining structure is called the **earth pressure at-rest**⑦, and is usually expressed in terms of effective stress by the equation

④ 侧向伸展。

⑤ 侧向压缩。

⑥ 极限压力。

⑦ 静止土压力。

$$p'_0 = K_0\sigma'_v \tag{9.41}$$

where K_0 is the coefficient of earth pressure at-rest⑧ in terms of effective stress. In the absence of a retaining structure p'_0 is the horizontal effective stress in the ground, σ'_h, such that Equation 9.41 becomes a restatement of Equation 7.42.

⑧ 静止土压力系数。

Since the at-rest condition does not involve failure of the soil, the Mohr circle representing the vertical and horizontal stresses does not touch the failure envelope and the horizontal stress cannot be determined analytically through limit analysis. The value of K_0, however, can be determined experimentally by means of a triaxial test in which the axial stress and the all-round pressure[①] are increased simultaneously such that the lateral strain in the specimen is maintained at zero[②]; this will generally require the use of a stress-path cell[③].

① 围压。
② 试样的侧向应变为零。
③ 应力路径三轴仪。

The value of K_0 may also be estimated from in-situ test data, notably using the pressuremeter (PMT) or CPT. Of these methods, the PMT is most reliable as the required parameters are directly measured, while the CPT relies on empirical correlation. Using the PMT, the total in-situ horizontal stress σ_{h0} in any type of soil is determined from the lift-off pressure as shown in Figures 7.14 and 7.16. The in-situ vertical total stress σ_{v0} is then determined from bulk unit weight[④] (from disturbed samples taken from the borehole), and pore pressures (u_0) are determined from the observed depth of the water table in the borehole or other piezometric measurements (see Chapter 6). Then

④ 天然重度。

$$K_0 = \frac{\sigma'_{h0}}{\sigma'_{v0}} = \frac{\sigma_{h0} - u_0}{\sigma_{v0} - u_0} \tag{9.42}$$

From CPT data, K_0 is determined empirically from the normalised cone tip resistance using Equation 7.43 (note: this only applies to fine-grained soils).

For normally consolidated soils[⑤], the value of K_0 can also be related approximately to the strength parameter ϕ' by the following formula proposed by Jaky (1944):

⑤ 正常固结土。

$$K_{0,NC} = 1 - \sin\phi' \tag{9.43a}$$

For overconsolidated soils the value of K_0 depends on the stress history and can be greater than unity, a proportion of the at-rest pressure developed during initial consolidation being retained in the soil when the effective vertical stress is subsequently reduced. Mayne and Kulhawy (1982) proposed the following correlation for overconsolidated soils[⑥] during expansion (but not recompression):

⑥ 超固结土。

$$K_0 = (1 - \sin\phi') \cdot \mathrm{OCR}^{\sin\phi'} \tag{9.43b}$$

In Eurocode 7 it is proposed that

$$K_0 = (1 - \sin\phi') \cdot \sqrt{\mathrm{OCR}} \tag{9.43c}$$

Values of K_0 from Equation 9.43 are shown in Figure 9.11, where they are compared with values for a range of soils derived from laboratory tests collated by Pipatpongsa *et al.* (2007), Mayne (2007) and Mayne and Kulhawy (1982). While the analytical expressions do fit the data, it should be noted that the scatter suggests that there is likely to be a significant amount of uncertainty[⑦] associated with this parameter.

⑦ 较大的不确定性。

Generally, for any condition intermediate to the active and passive states, the value of the lateral stress is unknown[⑧]. Figure 9.12 shows the form of the relationship between strain and the lateral pressure coefficient. The exact relationship depends on the initial value of K_0 and on whether excavation or backfilling is involved in construction of the retained soil mass. The strain required to mobilise the passive pressure is considerably greater than that required to mobilise the active pressure[⑨].

⑧ 总体上来说，对于处在主动与被动状态之间的状态，侧向应力值是未知的。
⑨ 达到被动状态需要的应变远大于达到主动状态的应变。

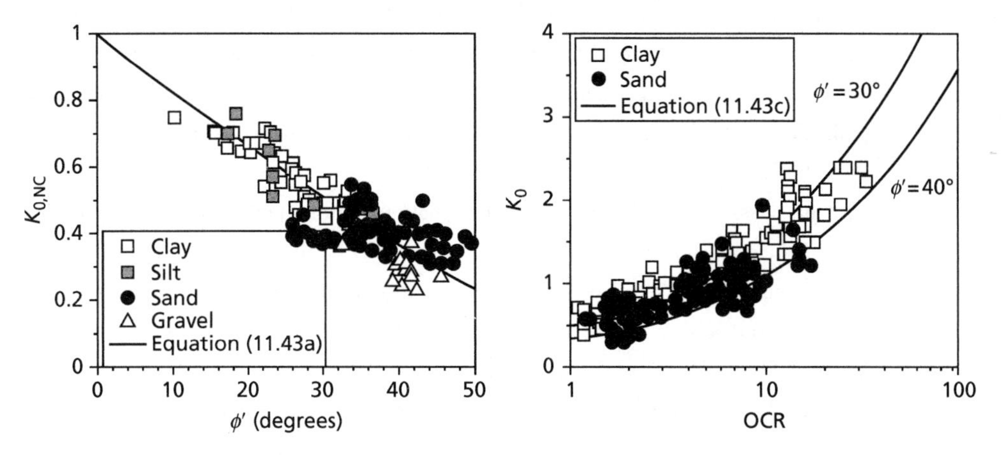

Figure 9.11 Estimation of K_0 from ϕ' and OCR, and comparison to in-situ test data.

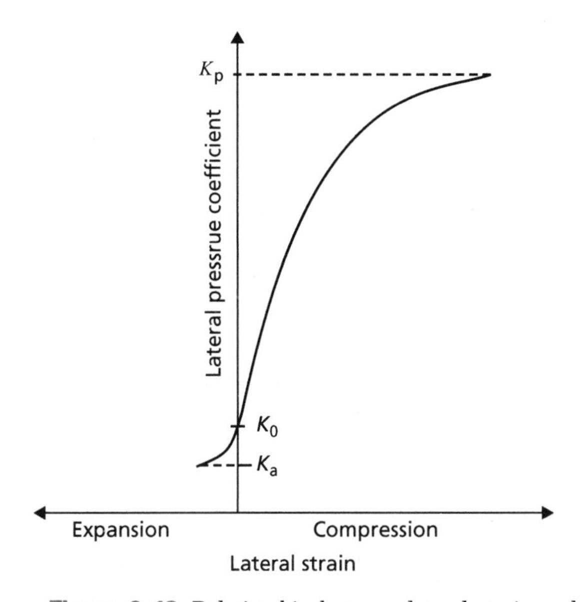

Figure 9.12 Relationship between lateral strain and lateral pressure coefficient.[①]

① 侧向应变与侧向土压力系数的关系。

In the lower bound limit analyses presented in Section 9.2, the entire soil mass was subjected to lateral expansion (active case) or compression (passive case). However, the movement of a retaining wall of finite dimensions cannot develop the active or passive state in the soil mass as a whole. The active state, for example, would be developed only within a wedge of soil between the wall and a failure plane passing through the lower end of the wall and at an angle of $45° + \phi'/2$ to the horizontal, as shown in Figure 9.13 (a); the remainder of the soil mass would not reach a state of plastic equilibrium. A specific (minimum) value of lateral strain would be necessary for the development of the active state within this wedge. A uniform strain within the wedge would be produced by a rotational movement[②] (A′B) of the wall, away from the soil, about its lower end, and a deformation of this type, of sufficient magnitude, constitutes the minimum deformation requirement for the development of the active state. Any deformation configuration enveloping A′B,

② 转动。

for example, a uniform translational movement A′B′, would also result in the development of the active state. If the deformation of the wall were not to satisfy the minimum deformation requirement, the soil adjacent to the wall would not reach a state of plastic equilibrium and the lateral pressure would be between the active and at-rest values.

In the passive case, the minimum deformation requirement is a rotational movement of the wall, about its lower end, into the soil. If this movement were of sufficient magnitude, the passive state would be developed within a wedge of soil between the wall and a failure plane at an angle of $45° + \phi'/2$ to the vertical, as shown in Figure 9. 13 (b). In practice, however, only part of the potential passive resistance would normally be mobilised. The relatively large deformation necessary for the full development of passive resistance would be unacceptable, with the result that the pressure under working conditions would be between the at-rest and passive values, as indicated in Figure 9. 12 (and consequently providing a factor of safety against passive failure).

① 被动土压力完全发挥。

Experimental evidence indicates that the mobilisation of full passive resistance① requires a wall movement of the order of 2% –4% of embedded depth in the case of dense sands and of the order of 10% –15% in the case of loose sands. The corresponding percentages for the mobilisation of active pressure are of the order of 0. 25 and 1%, respectively.

② 重力式挡土结构。

9. 4 Gravity retaining structures②

③ 重力式挡土墙的稳定性是由于墙体的自重，也有可能由墙趾前部的被动阻力而得到加强。

The stability of gravity (or freestanding) walls is due to the self-weight of the wall, perhaps aided by passive resistance developed in front of the toe of the wall③. The traditional gravity wall (Figure 9. 14 (a)), constructed of masonry or mass concrete, is uneconomic because the material is used only for its dead weight. Reinforced concrete cantilever walls (Figure 9. 14 (b)) can be more economic because the backfill itself, acting on the base, is employed to provide most of the required dead weight. Other types of gravity structure include gabion and crib walls (Figures 9. 14 (c) and (d)). Gabions are cages of steel mesh, rectangular in plan and elevation, filled with particles generally of cobble size, the units being used as the building blocks of a gravity structure. Cribs are open structures assembled from precast concrete or timber members and enclosing coarse-grained fill, the structure and fill acting as a composite unit to form a gravity wall.

Limit state design

④ 最终（承载力）极限状态。

At the ultimate limit state (ULS)④ failure will occur with the retained soil under active conditions as the wall moves towards the excavation, as the generation of passive pressures (higher than K_0, Figure 9. 12) would require additional forcing towards the retained soil to induce slip in this direction. A gravity retaining wall is more complex than the uniaxially loaded foundations of Chapters 8, with vertical

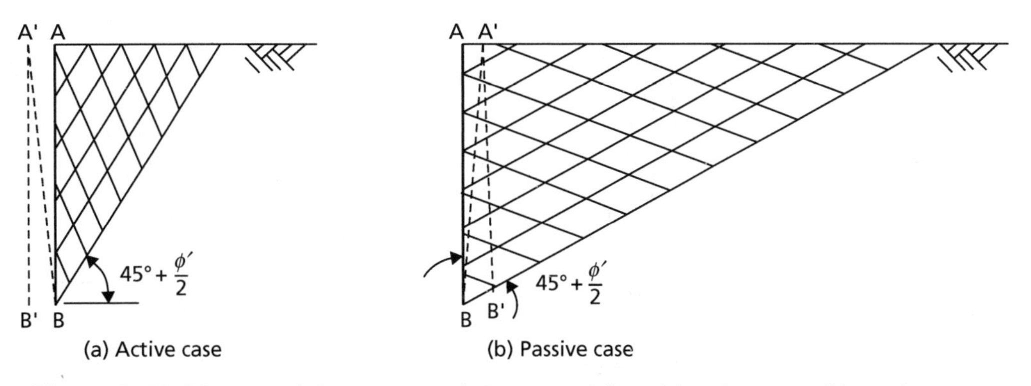

Figure 9.13 Minimum deformation conditions to mobilise: (a) active state, (b) passive state.

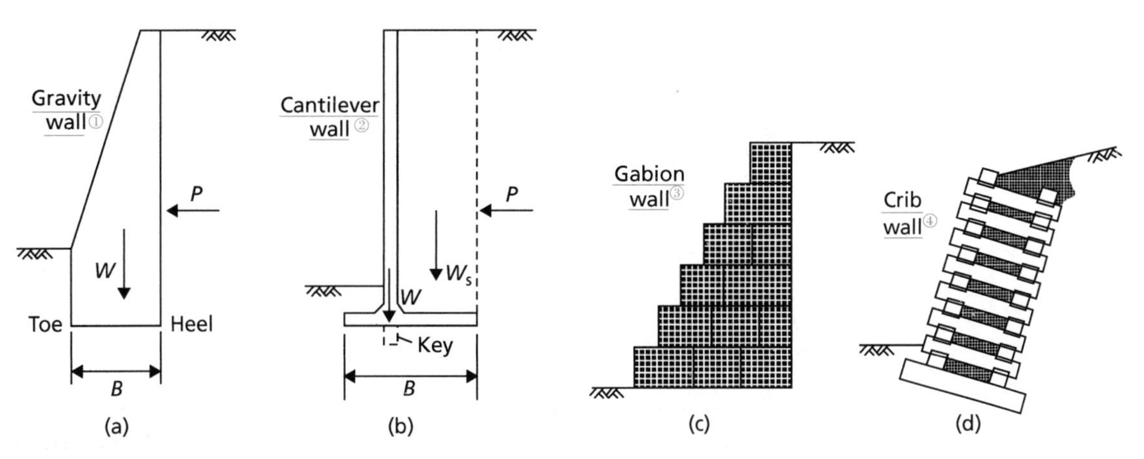

Figure 9.14 Gravity retaining structures⑤.

① 重力式挡土墙。
② 悬臂式挡土墙。
③ 格宾挡土墙。
④ 框架挡土墙。
⑤ 重力式挡土结构。
⑥ 地基的极限承载力。
⑦ 滑移。
⑧ 倾覆。
⑨ 深层滑动面。
⑩ 不利渗流的影响。

actions (wall dead load), horizontal actions (active earth pressures), potential seepage effects and uneven ground levels. Ultimate limit states which must be considered in wall design are shown schematically in Figure 9.15 and described as follows:

- ULS-1 Base pressure applied by the wall must not exceed the ultimate bearing capacity of the supporting soil⑥;
- ULS-2 Sliding⑦ between the base of the wall and the underlying soil due to the lateral earth pressures;
- ULS-3 Overturning⑧ of the wall due to horizontal earth pressure forces when the retained soil mass becomes unstable (active failure);
- ULS-4 The development of a deep slip surface⑨ which envelops the structure as a whole (analysed using methods which will be described in Chapter 10);
- ULS-5 Adverse seepage effects⑩ around the wall, internal erosion or leakage through the wall: consideration should be given to the consequences of the failure of drainage systems to operate as intended (Chapter 2);

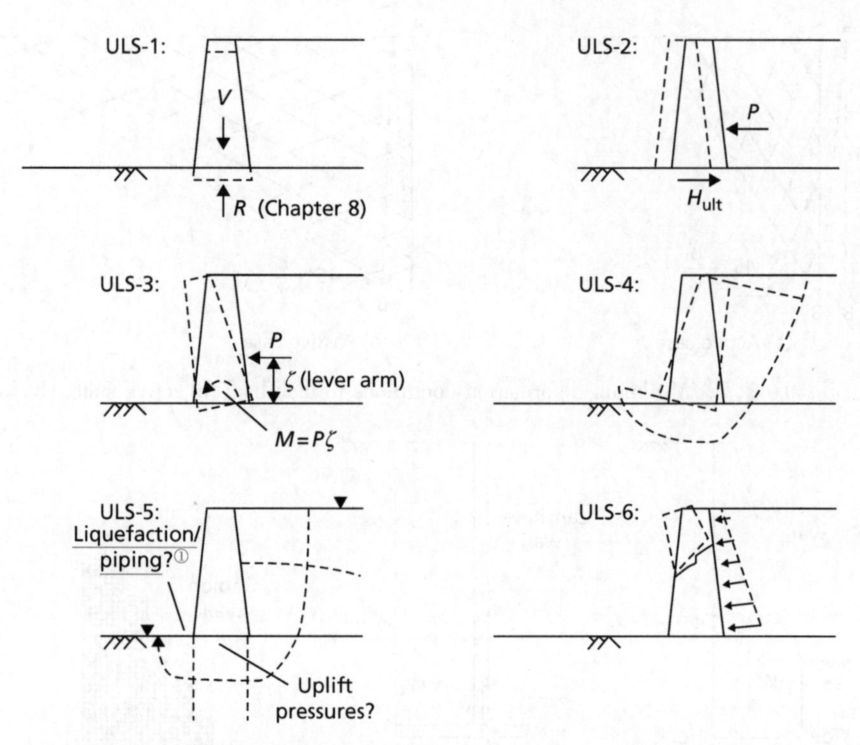

Figure 9.15 Failure modes for gravity retaining structures at ULS②.

① 液化/管涌。

② 最终极限状态下重力式挡土结构的破坏模式。

ULS-6	Structural failure③ of any element of the wall or combined soil/structure failure.

③ 结构破坏。

Serviceability criteria must also be met at the serviceability limit state (SLS)④ as follows:

④ 正常使用极限状态。

SLS-1	Soil and wall deformations must not cause adverse effects on the wall itself or on adjacent structures and services;
SLS-2	Excessive deformations⑤ of the wall structure under the applied earth pressures must be avoided (typically this is only significant in the case of slender cantilever walls, Figure 9.14b, which will be considered separately later in this chapter).

⑤ 过量变形。

The first step in design is to determine all the actions/earth pressures acting on the wall from which the resultant thrusts can be determined as described later in this section. Soil and water levels should represent the most unfavourable conditions conceivable in practice. Allowance must be made for the possibility of future (planned or unplanned) excavation in front of the wall known as **overdig**⑥, a minimum depth of 0.5m being recommended; accordingly, passive resistance in front of the wall is normally neglected.

⑥ 超挖。

In terms of ULS design, this chapter will chiefly focus on limit states ULS-1 to ULS-3. Once the earth pressure thrusts⑦ are known, the resultant vertical, horizon-

⑦ 土推力。

tal and moment components acting on the base of the wall (V, H and M, respectively) are obtained. ULS-1 to ULS-3 can then be checked using the yield surface concepts, treating the gravity wall as a shallow foundation under combined loading[①]. ULS-2 requires an additional check if the wall – soil interface is not perfectly rough. In this case, the resultant active earth pressure force from the retained soil represents the action that is inducing failure. With passive resistance in front of the wall neglected, the resistance to sliding per metre length of wall (H_{ult}) comes from interface friction along the base of the wall[②]. For a rough wall sliding on undrained soil the design resistance is

① 组合荷载。

② 墙底的界面摩阻力。

$$H_{ult} = \frac{\alpha c_u B}{\gamma_{Rh}} \tag{9.44a}$$

while for sliding on drained soil is

$$H_{ult} = \frac{V \tan \delta'}{\gamma_{Rh}} \tag{9.44b}$$

where α and δ' represent the adhesion and interface friction angle respectively, as before. H_{ult} is a resistance, so parameter γ_{Rh} is a partial resistance factor for sliding, analogous to γ_{Rv} for the bearing resistance of shallow foundations. In Eurocode 7, the normative value of $\gamma_{Rh} = 1.00$ for sets R1 and R3 and 1.10 for set R2. Then, to satisfy ULS-2,

$$H \leqslant H_{ult} \tag{9.45}$$

Of the remaining ULS conditions, ULS-4 will be considered in Chapter 10 (stability of unsupported soil masses), as in this case the failure bypasses the wall completely (Figure 9.15). The effects of seepage on retaining wall stability have been partially considered in earlier chapters (ULS-5), and ULS-6 is not considered herein as it relates solely to the structural strength of the wall itself under the lateral earth pressures, which must be determined using structural/continuum mechanics principles. It should be noted that if seepage is occurring, then the effects of seepage should also be accounted for in the other ULS failure modes[③] (e.g. uplift pressures[④] will reduce the normal contact stress in ULS-2, reducing sliding resistance).

③ 注意到如果渗流发生，则极限状态破坏模式分析时需要考虑渗流效应。

④ 上浮力。

Resultant thrust

In order to check the overall stability of a retaining structure, the active and passive earth pressure distributions are integrated over the height of the wall to determine the **resultant thrust**[⑤] (a force per unit length of wall), which is used to define the horizontal action. If the retaining structure is smooth, the resultant thrust will act normal to the wall (this may not be horizontal if the wall has a battered back). For the case of retained undrained soil having uniform bulk density γ (so that $\sigma = \gamma z \cos^2 \beta$), combining Equations 9.35, 9.15 and 9.34 gives

⑤ 推力的合力。

$$P_a = \gamma z \cos^2 \beta - c_u (1 + \Delta_2 + \cos \Delta_2 - 2w - \Delta_1 - 2\beta) \tag{9.46a}$$

while combining Equations 9.16, 9.17 and 9.18 gives

$$P_a = \gamma z + c_u (1 + \Delta_2 + \cos \Delta_2 - 2w) \tag{9.46b}$$

Equation 9.46 is linear in z and these total pressure distributions are shown in Figure 9. 16. In the active case, the value of p_a is zero at a particular depth z_0. From Equation 9. 46a, with $p_a = 0$,

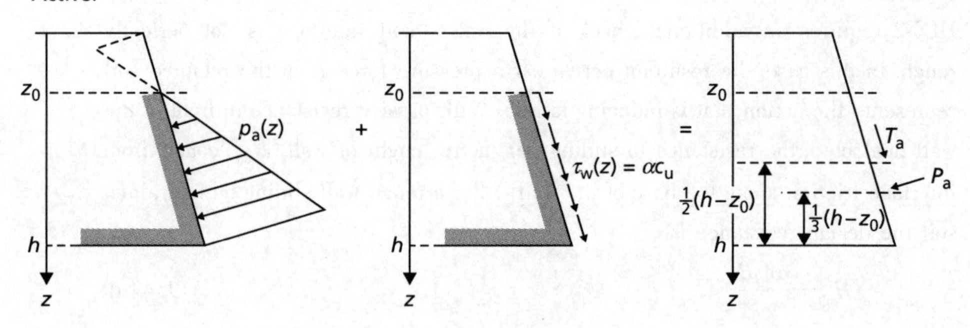

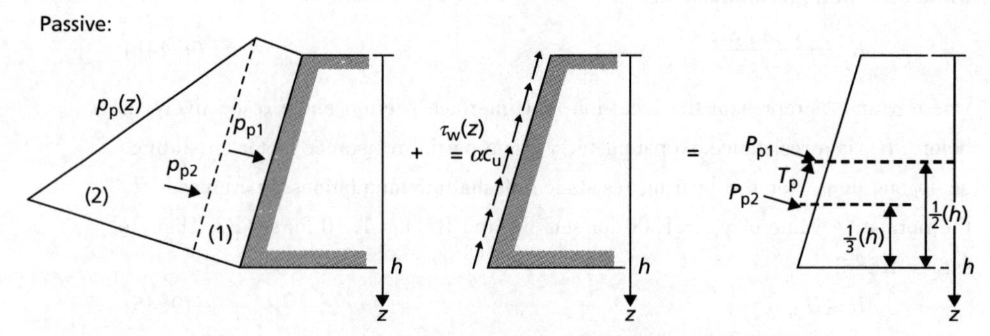

Figure 9. 16 Pressure distributions and resultant thrusts: undrained soil.

$$z_0 = \frac{c_u\,(1 + \Delta_2 + \cos\,\Delta_2 - 2w - \Delta_1 - 2\beta)}{\gamma} \tag{9.47}$$

① 拉力。

② 裂缝。

③ 主动土压力的合力。

④ 牵引力。

⑤ 界面剪力。

⑥ 被动土压力的合力。

This means that in the active case the soil is in a state of tension[①] between the surface and depth z_0. In practice, however, this tension cannot be relied upon to act on the wall, since cracks[②] are likely to develop within the tension zone and the part of the pressure distribution diagram above depth z_0 should be neglected. The total active thrust[③] (P_a) acting normal to a wall of height h with a batter angle of w is then

$$P_a = \int_{z_0}^{h} p_a \frac{dz}{\cos\,w} \tag{9.48}$$

The force P_a acts at a distance of $\frac{1}{3}(h - z_0)$ above the bottom of the wall. If the wall is rough there is additionally a traction force[④] (T_a) due to the interface shearing, acting downwards along the wall surface at the same point of

$$T_a = \int_{z_0}^{h} \tau_w \frac{dz}{\cos\,w} \tag{9.49}$$

where $\tau_w = \alpha c_u$. Equation 9. 49 neglects any interface shear[⑤] in the tension zone above z_0 (i. e. it is assumed that a crack opens between the soil and the wall in this zone), so the force acts at a distance of $\frac{1}{2}(h - z_0)$ above the bottom of the wall.

In the passive case p_p is always positive so that the total passive thrust[⑥] (P_p) acting normal to a wall of height h with a batter angle of w is

$$P_p = \int_{0}^{h} p_p \frac{dz}{\cos\,w} \tag{9.50}$$

The two components of P_p act at distances of $\frac{1}{3}h$ and $\frac{1}{2}h$, respectively, above the bottom of the wall surface, as shown in Figure 9.16. If the wall is rough there is additionally a traction force[①] (T_p) acting upwards along the wall surface of

$$T_p = \int_0^h \tau_w \frac{dz}{\cos w} \tag{9.51}$$

In the case of a drained cohesionless material ($c' = 0$), the lateral pressures acting on the retaining structure should be separated into the component due to the effective stress in the soil ($p' = K\sigma'$) and the component due to any pore water pressure[②] (u). The effective thrust[③] (P') acting normal to a wall of height h with a batter angle of w is

$$P'_a = \int_0^h K_a \sigma' \frac{dz}{\cos w} \text{ (active)} \tag{9.52a}$$

$$P'_p = \int_0^h K_p \sigma' \frac{dz}{\cos w} \text{ (passive)} \tag{9.52b}$$

and resultant pore water pressure thrust[④] (U)

$$U = \int_0^h u \frac{dz}{\cos w} \tag{9.53}$$

If the wall is rough, there will be an additional interface shear force acting downwards along the wall surface on the active side and upwards along the wall surface on the passive side[⑤]. This is found using Equation 9.51 on both active and passive sides (as there is no tension zone[⑥]); however, the interface shear strength is $\tau_w = \sigma' \tan \delta'$. The normal and shear components of the thrust P' and T are often combined into a single resultant force which acts at an angle of δ' to the wall normal, as shown in Figure 9.17.

① 牵引力。
② 孔隙水压力。
③ 有效推力。
④ 孔隙水压力产生的合推力。
⑤ 如果墙面粗糙，则墙体的主动面上会有向下作用的附加界面剪力和被动面上向上作用的附加界面剪力。
⑥ 张拉区。
⑦ 压力分布与推力合力：排水条件。

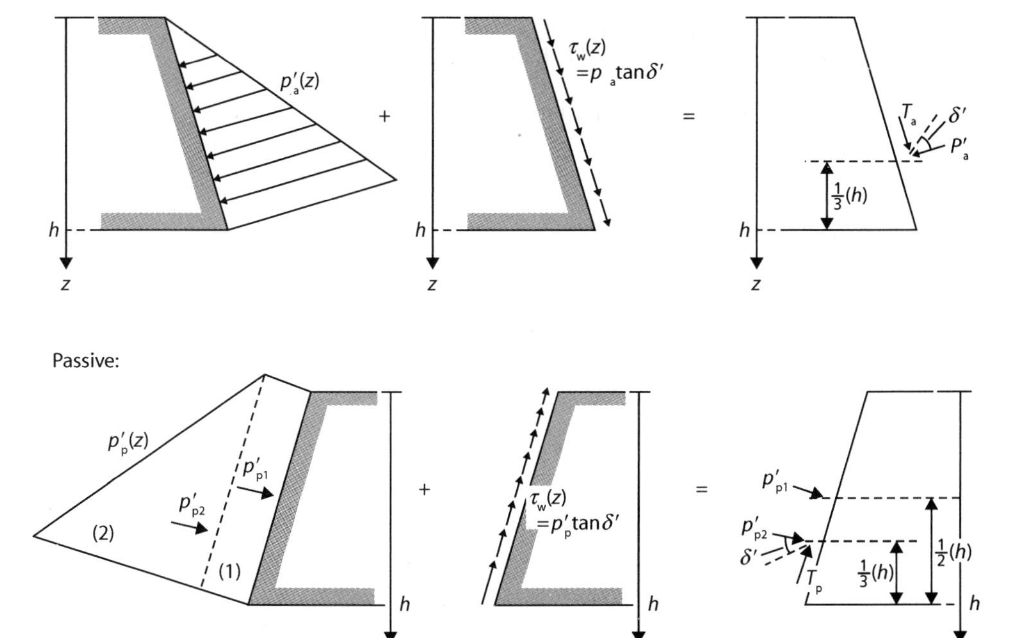

Figure 9.17 Pressure distributions and resultant thrusts: drained soil.[⑦]

Example 9.2

a Calculate the total active thrust on a smooth vertical wall 5m high retaining a sand of unit weight 17kN/m^3 for which $\phi' = 35°$ and $c' = 0$; the surface of the sand is horizontal and the water table is below the bottom of the wall.

b Determine the thrust on the wall if the water table rises to a level 2m below the surface of the sand. The saturated unit weight① of the sand is 20kN/m^3. ① 饱和重度。

Solution

a $w = \delta' = \beta = 0$, so using either Equation 9.39, 9.23 or 9.5 gives

$$K_a = \frac{1 - \sin\phi'}{1 + \sin\phi'} = 0.27$$

With the water table below the bottom of the wall, $\sigma' = \sigma'_v = \gamma_{dry} z$, so, from Equation 9.52a,

$$\begin{aligned} P'_a &= \int_0^h K_a \gamma_{dry} z \mathrm{d}z \\ &= \frac{1}{2} K_a \gamma_{dry} h^2 \\ &= \frac{1}{2} \cdot 0.27 \cdot 17 \cdot 5^2 \\ &= 57.5\text{kN/m} \end{aligned}$$

b The pressure distribution on the wall is now as shown in Figure 9.18, including hydrostatic pressure② on the lower 3m of the wall. The thrust components are: ② 静水压力。

$$(1)\ P'_a = \int_0^2 K_a \gamma_{dry} z \mathrm{d}z = \frac{1}{2} \cdot 0.27 \cdot 17 \cdot 2^2 = 9.2\text{kN/m}$$

$$(2)\ P'_a = 0.27 \cdot 17 \cdot 2 \cdot 3 = 27.6\text{kN/m}$$

$$(3)\ P'_a = \int_{2-2}^{5-2} K_a(\gamma_{sat} z - \gamma_w z)\,\mathrm{d}z = \frac{1}{2} \cdot 0.27 \cdot (20 - 9.81) \cdot 3^2 = 12.4\text{kN/m}$$

$$(4)\ U_a = \int_{2-2}^{5-2} \gamma_w z \mathrm{d}z = \frac{1}{2} \cdot 9.81 \cdot 3^2 = 44.1\text{kN/m}$$

Summing the four components of thrust gives a total thrust = 93.3kN/m.

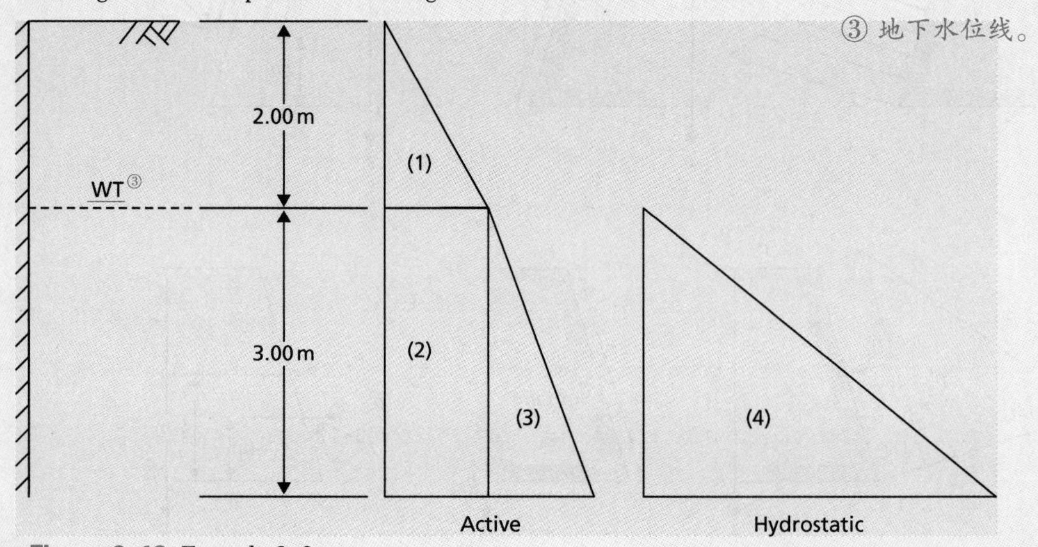

③ 地下水位线。

Figure 9.18 Example 9.2.

Example 9.3

Details of a cantilever retaining wall[1] are shown in Figure 9.19, the water table being below the base of the wall. The unit weight of the backfill is 17kN/m^3 and a surcharge pressure of 10kPa acts on the surface. Characteristic values of the shear strength parameters for the backfill are $c' = 0$ and $\phi' = 36°$. The angle of friction between the base and the foundation soil is 27° (i.e. $\delta' = 0.75\phi'$). Is the design of the wall satisfactory at the ultimate limit state to EC7, DA1b?

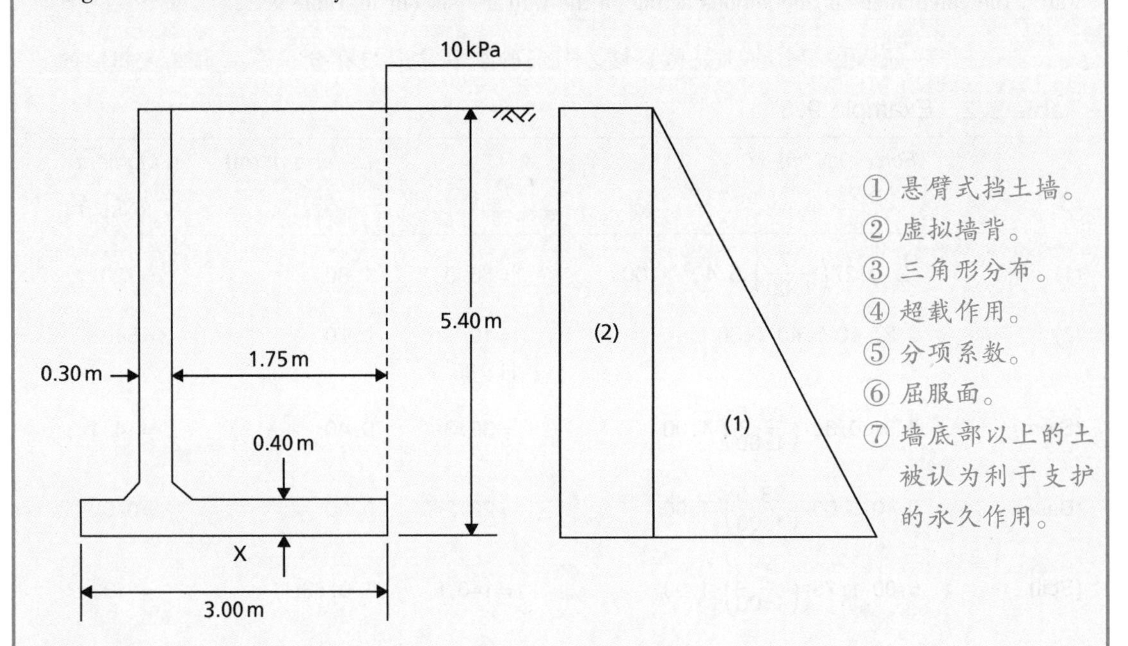

Figure 9.19 Example 9.3.

① 悬臂式挡土墙。
② 虚拟墙背。
③ 三角形分布。
④ 超载作用。
⑤ 分项系数。
⑥ 屈服面。
⑦ 墙底部以上的土被认为利于支护的永久作用。

Solution

The unit weight of concrete will be taken as 23.5kN/m^3. The active thrust from the retained soil acts on the vertical plane through the soil shown by the dotted line in Figure 9.19 (also known as **the virtual back**[2]), so $\delta' = \phi'$ and $w = 0$ on this interface. The value of K_a can then be found from Equation 9.39 with $w = \beta = \Delta_1 = \phi'_{mob} = 0$, or from Equation 9.23 (as the retained soil is not sloping). For DA1b, the design value of $\phi' = \tan^{-1}(\tan 36°/1.25) = 30°$. Therefore $\delta' = 30°$ ($\pi/6$) and $\Delta_2 = \pi/2$, giving $K_a = 0.27$. The soil is dry, so $\sigma' = \sigma'_v = \gamma z + \sigma_q$, giving $p_a = K_a\,\sigma'_v$ (Equation 9.6). Then, from Equation 9.53 $U_a = 0$ and from Equation 9.52

$$p_a = p'_a = \int_0^h K_a(\gamma z + \sigma_q)\,dz = \frac{1}{2}K_a\gamma z h^2 + K_a\sigma_q h$$

The first term in P_a relates to (1), which has a triangular distribution[3] with depth, and will therefore act at $h/3$ above the base of the wall. The second term relates to the surcharge (2) which is uniform with depth, so that this component acts at $h/2$ above the base of the wall. Action (1) is a permanent unfavourable action (partial factor = 1.00), while the surcharge effect[4] (2) is a variable unfavourable action (partial factor[5] = 1.30). For using the yield surface[6] in Figure 8.24b, the vertical forces (downwards positive) due to the weight of the wall (stem + base) and soil above the base are considered as permanent favourable actions[7] for $V/R < 0.5$ (where sliding and overturning dominate) and permanent unfavourable actions for $V/R \geqslant 0.5$ (when bearing failure dominates). For DA1b this does not affect the value of the partial factors, which in both cases are 1.00. The vertical (downwards) traction force $T_a = P'_a \tan\delta'$ from the interface shear on the virtual back is factored similarly①. The moments induced by the vertical and horizontal actions are determined about the centre of the footing base (point X in Figure 9.19), with anticlockwise moments positive (these act to overturn the wall). The calculations of the actions acting on the wall are set out in Table 9.2.

① 由虚拟背部上界面剪切力引起的垂直牵引力在分项系数上也是相似的。

Table 9.2 Example 9.3

	Force (kN/m)		Lever arm (m)	Moment (kNm/m)
(1)	$\frac{1}{2}\cdot 0.27\left(\frac{17}{1.00}\right)\cdot 5.40^2\cdot 1.00$	=66.9	1.80	=120.4
(2)	$0.27\cdot 10\cdot 5.40\cdot 1.30$	=19.0 H=85.9	2.70	=51.3
(Stem)	$5.00\cdot 0.30\cdot\left(\frac{23.5}{1.00}\right)\cdot 1.00$	=35.3	0.40	=14.1
(Base)	$0.40\cdot 3.00\cdot\left(\frac{23.5}{1.00}\right)\cdot 1.00$	=28.2	0.00	=0.0
(Soil)	$5.00\cdot 1.75\cdot\left(\frac{17}{1.00}\right)\cdot 1.00$	=148.8	−0.63	= −93.7
(Virtual back)	(66.9 + 19.0) tan 30	=49.6	−1.50	= −74.4
		V=261.9		M=17.7

The bearing resistance of the wall is calculated assuming that the failure occurs in front of the wall (worst case), so, from Equation 8.32,

$$R = \frac{q_f A}{\gamma_R} = \frac{\frac{1}{2}\gamma B^2 N_\gamma}{\gamma_R} = \frac{\frac{1}{2}\cdot\left(\frac{17}{1.00}\right)\cdot 3^2\cdot 20.1}{1.00} = 1.54\text{MN/m}$$

where N_γ is from Equation 8.36. Therefore, $H/R = 0.056$ and $M/BR = 0.004$. This point is plotted, from where it can be seen that, to lie within the yield surface, $0.12 < V/R < 0.88$ (interpolating between contours). In this case, $V/R = 0.11$, so ULS is not satisfied. Checking ULS-2 explicitly (Equations 9.44b and 9.45) $H_{ult} = V\tan\delta' = 261.9 \times \tan 30 = 151.2$kN/m, which is greater than H so the wall does not slide. Also, $V \ll R$, so ULS-1 is also satisfied. It is the overturning that is therefore causing the problem.

If the wall is required to be 5.40 m tall, the width of the base would need to be extended (into the retained soil). This would increase the value of V and reduce the value of M (by increasing the restoring moment from the soil above the base). It should be noted that passive resistance in front of the wall has been neglected to allow for unplanned excavation.

Example 9.4

Details of a mass concrete gravity-retaining wall are shown in Figure 9.20, the unit weight of the wall material being 23.5kN/m^3. The unit weight of the dry retained soil is 18kN/m^3, and the characteristic shear strength parameters are $c' = 0$ and $\phi' = 33°$. The value of δ' between wall and retained soil and between wall and foundation soil is 26°. The wall is founded on uniform clay with $c_u = 120$kPa and $\alpha = 0.8$ between the base of the wall and the clay. Is the design of the wall satisfactory at the ultimate limit state to EC7 DA1a?

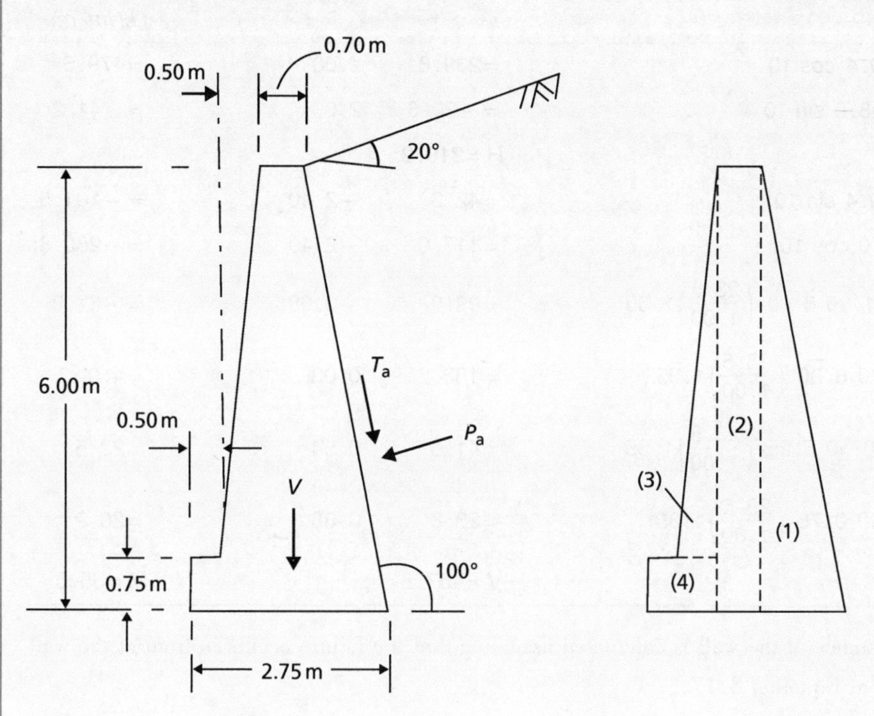

Figure 9.20 Example 9.4.

Solution

As the back of the wall and the soil surface are both inclined, the value of K_a will be calculated from Equation 9.39. The design values of the angles of shearing resistance in this equation are $\phi' = 33°$, $\delta' = 26°$ for DA1a, $w = 100 - 90 = 10°$, $\Delta_2 = 53.6°$ (Equation 9.28), $\beta = 20°$, $\phi'_{mob} = \beta = 20°$(Equation 9.32), and $\Delta_1 = 38.9°$(Equation 9.40), giving $K_a = 0.46$. From Equation 9.53$U_a = 0$, and from Equations 9.52 and 9.31

$$P_a = P'_a = \int_0^h K_a \sigma' \frac{dz}{\cos w} = \int_0^h \frac{K_a \gamma z \cos^2 \beta}{\cos w} dz = \frac{K_a \gamma \cos^2 \beta}{2\cos w} h^2$$

This force acts normal to the back of the wall at a vertical distance of $h/3$ above the base of the wall and is a permanent unfavourable action (partial factor = 1.35). The (downwards) interface friction force along the back of the wall is given by

$$T_a = \int_0^h (K_a \sigma') \tan \delta' \frac{dz}{\cos w} = P'_a \tan \delta' = \frac{K_a \gamma z \cos^2 \beta}{2\cos w} = h^2 \tan \delta'$$

This is similarly a permanent unfavourable action. Evaluating these two expressions and applying partial factors gives design values of the actions: $P_a = 180.4$kN/m, $T_a = 88.0$kN/m. These actions are shown in Figure 9.20. The vertical action due to the dead weight of the retaining wall is determined by splitting the wall into smaller, simpler sections as shown in Figure 9.20. This is also a permanent unfavourable action. Moments are considered about the centre of the base of the wall (anticlockwise positive), the calculations being set out in Table 9.3.

Table 9.3 Example 9.4

	Force[1] (kN/m)		Lever arm[2] (m)	Moment[3] (kNm/m)
$P_a \cos 10°$	180.4 cos 10	=239.8	2.00	=479.6
$T_a \sin 10°$	−88.0 sin 10	= −20.6	2.00	= −41.2
		H =219.2		
$P_a \sin 10°$	180.4 sin 10	=42.3	−2.40	= −101.5
$T_a \cos 10°$	88.0 cos 10	=117.0	−2.40	= −280.8
Wall (1)	$\frac{1}{2} 1.05 \cdot 6.00 \cdot \left(\frac{23.5}{1.00}\right) 1.35$	=99.9	−0.68	= −67.9
(2)	$0.70 \cdot 6.00 \cdot \left(\frac{23.5}{1.00}\right) 1.35$	=133.2	0.03	=4.0
(3)	$\frac{1}{2} 0.50 \cdot 5.25 \cdot \left(\frac{23.5}{1.00}\right) 1.35$	=41.6	0.54	=22.5
(4)	$1.00 \cdot 0.75 \cdot \left(\frac{23.5}{1.00}\right) 1.35$	=23.8	0.88	=20.9
		V =457.8		M =35.6

The bearing resistance of the wall is calculated assuming that the failure occurs in front of the wall (worst case), so, from Equation 8.17,

$$R = \frac{q_f A}{\gamma_R} = \frac{N_c c_u B}{\gamma_R} = \frac{5.14 \cdot \left(\frac{120}{1.00}\right) \cdot 2.75}{1.00} = 1.70\text{MN/m}$$

where $N_c = 5.14$. Therefore, $H/R = 0.129$ and $M/BR = 0.008$. This point is plotted in Figure 8.21b, from where it can be seen that to lie within the yield surface $0.21 < V/R < 0.79$. In this case, $V/R = 0.25$, so ULS is satisfied. A further check should be made for sliding (ULS-2) in this case, as the yield surface approach assumes perfect bonding between the wall and the clay (i.e. $\alpha = 1$). From Equations 9.44a and 9.45, $H_{ult} = (\alpha c_u B)/\gamma_{Rh} = (0.8 \times 120 \times 2.75)/1.00 = 264\text{kN/m}$, which is greater than H so the wall will not slide.

9.5 Coulomb's theory of earth pressure[4]

While the rigorous limit analysis solutions[5] presented in Section 9.2 are applicable to a wide range of retaining structure problems, they are by no means the only way of determining lateral earth pressures. An alternative method of analysis, known as **limit equilibrium**[6], involves consideration of the stability, as a whole, of the

① 力。
② 力臂。
③ 力矩。
④ 库仑土压力理论。
⑤ 精确的极限分析解答。
⑥ 极限平衡法。

wedge of soil between a retaining wall and a trial failure plane. The force between the wedge and the wall surface is determined by considering the equilibrium of forces acting on the wedge when it is on the point of sliding either up or down the failure plane, i. e. when the wedge is in a condition of limiting equilibrium. Friction between the wall and the adjacent soil can be taken into account as for the limit analyses in Section 9. 2, represented by $\tau_w = \alpha c_u$ in undrained soil and $\tau_w = \sigma' \tan \delta'$ in drained soil. The method was first developed for retaining structures by Coulomb (1776), and is popular for use in design.

The limit equilibrium theory is now interpreted as an upper bound plasticity solution[①] (although analysis is based on force equilibrium and not on the work – energy balance[②] defined in Chapter 8), collapse of the soil mass above the chosen failure plane occurring as the wall moves away from or into the soil. Thus, in general, the theory underestimates the total active thrust and overestimates the total passive resistance[③] (i. e. provides upper bounds to the true collapse load).

① 塑性上限分析法。
② 功-能平衡。
③ 该理论低估了总的主动土压力，高估了总的被动土压力。

Active case

Figure 9. 21 (a) shows the forces acting on the soil wedge between a wall surface AB, inclined at angle x to the horizontal, and a trial failure plane BC, at angle θ_p to the horizontal. The soil surface AC is inclined at angle β to the horizontal. The shear strength parameter c' is initially taken as zero, though this limitation can subsequently be relaxed for a general ϕ', c' material. For the failure condition, the soil wedge[④] is in equilibrium under its own weight (W), the reaction to the force (P) between the soil and the wall, and the resultant reaction (R_s) on the failure plane. Because the soil wedge tends to move down the plane BC at failure, the reaction P

④ 土楔。
⑤ 库仑理论：$c' = 0$ 的主动情况：(a) 土楔几何示意图，(b) 作用力多边形。

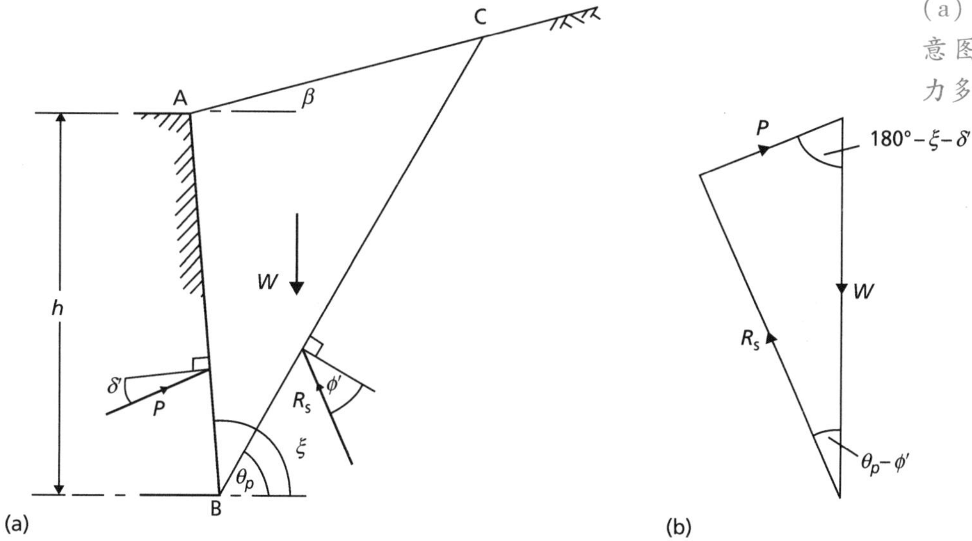

Figure 9. 21 Coulomb theory: active case with $c' = 0$: (a) wedge geometry, (b) force polygon[⑤].

acts at angle δ' below the normal to the wall. (If the wall were to settle more than the backfill, the reaction P would act at angle δ' above the normal.) At failure, when the shear strength of the soil has been fully mobilised, the direction of R_s is at an angle ϕ' below the normal to the failure plane. The directions of all three forces, and the magnitude of W, are known, and therefore a triangle of forces (Figure 9.21 (b)) can be drawn and the magnitude of P determined for the trial in question.

A number of trial failure planes would have to be selected to obtain the maximum value of P, which would be the total active thrust on the wall[①]. However, using the sine rule, P can be expressed in terms of W and the angles in the triangle of forces. Then the maximum value of P, corresponding to a particular value of θ_p, is given by $\partial P/\partial\theta = 0$. This leads to the following solution for P'_a in dry soil:

① 通过选择一系列的潜在破裂面试算所获得的 P 的最大值，就是作用在墙上的总主动土压力。

$$P'_a = \frac{1}{2}K_a\gamma H^2 \tag{9.54}$$

where

$$K_a = \left[\frac{\dfrac{\sin(\zeta-\phi')}{\sin\zeta}}{\sqrt{\sin(\zeta+\delta')}+\sqrt{\dfrac{\sin(\phi'+\delta')\sin(\phi'-\beta)}{\sin(\zeta-\beta)}}}\right]^2 \tag{9.55}$$

The point of application of the total active thrust is not given by the Coulomb theory but is assumed to act at a distance of ⅓h above the base of the wall as considered previously. Had Equation 9.55 been used to determine the values of K_a in Examples 9.3 and 9.4, values of 0.30 and 0.48 would have been obtained respectively, which compare favourably with the values from lower bound limit analysis of 0.27 and 0.46.

The analysis can be extended to dry soil cases in which the shear strength parameter c' is greater than zero. It is assumed that tension cracks[②] may extend to a depth z_0, the trial failure plane (at angle θ_p to the horizontal) extending from the heel of the wall to the bottom of the tension zone, as shown in Figure 9.22. The forces acting on the soil wedge at failure are then as follows:

② 张拉裂缝。

1 the weight of the wedge (W);
2 the reaction (P) between the wall and the soil, acting at angle δ' below the normal[③];
3 the force due to the constant component of shearing resistance on the wall ($T_a = \tau_w \times$ EB);
4 the reaction (R_s) on the failure plane, acting at angle ϕ' below the normal;
5 the force on the failure plane due to the constant component of shear strength ($C = c' \times$ BC).

③ 土与挡土墙之间的作用力，作用方向为墙背的法向向下 δ'。

The directions of all five forces are known together with the magnitudes of W, T_a and C, and therefore the value of P can be determined from the force diagram for the trial failure plane. Again, a number of trial failure planes would be selected to obtain the maximum value of P. The resultant thrust is then expressed as

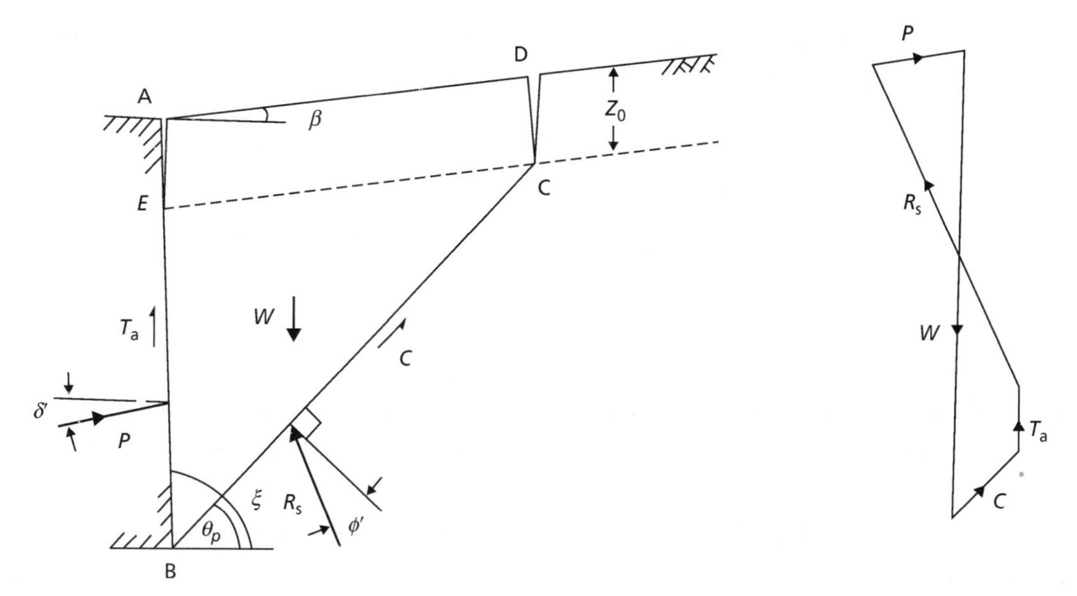

Figure 9.22 Coulomb theory: active case with $c' > 0$.

$$P'_a = \frac{1}{2}K_a\gamma h^2 - 2K_{ac}c'h \tag{9.56}$$

where

$$K_{ac} = 2\sqrt{K_a\left(1 + \frac{\tau_w}{c'}\right)} \tag{9.57}$$

If the soil is saturated, there will be additional forces acting due to the pore water①. Under these conditions it is best to determine the resultant active thrust using a force diagram that includes the resultant pore water thrust U_a acting on the slip plane. This general technique will then allow for the determination of the resultant thrust on a retaining structure under either static or seepage conditions (see Example 9.5).

① 由于孔隙水产生的附加作用力。

Passive case

In the passive case, the reaction P acts at angle δ' above the normal to the wall surface (or δ' below the normal if the wall were to settle more than the adjacent soil) and the reaction R_s at angle ϕ' above the normal to the failure plane. In the triangle of forces, the angle between W and P is $180° - \xi + \delta'$ and the angle between W and R_s is $\theta_p + \phi'$. The total passive resistance, equal to the minimum value of P, is given by

$$P'_p = \frac{1}{2}K_a\gamma H^2 \tag{9.58}$$

where

$$K_p = \left[\frac{\dfrac{\sin(\xi + \phi')}{\sin\xi}}{\sqrt{\sin(\xi + \delta')} - \sqrt{\dfrac{\sin(\phi' + \delta')\sin(\phi' + \beta)}{\sin(\xi - \beta)}}}\right] \tag{9.59}$$

Care must be exercised when using Equation 9.59 as it is known to overestimate passive resistance, seriously so for the higher values of ϕ', representing an error on the unsafe side.②

② 运用式（9.59）时需注意，它将高估被动土压力，特别是ϕ'较高时，将导致偏不安全的误差。

As before, the analysis can be extended to dry soil cases in which the shear strength parameter c' is greater than zero, giving

$$P_p = \frac{1}{2}K_p\gamma h^2 - 2K_{pc}c'h \tag{9.60}$$

where

$$K_{pc} = 2\sqrt{K_p\left(1 + \frac{\tau_w}{c'}\right)} \tag{9.61}$$

Example 9.5

Details of a retaining structure, with a vertical drain[①] adjacent to the back surface, are shown in Figure 9.23(a), the saturated unit weight of the retained soil being 20kN/m^3. The design strength parameters for the soil are $c' = 0$, $\phi' = 38°$ and $\delta' = 15°$. Assuming an active failure plane at $\theta_p = 45° + \phi'/2$ to the horizontal (see Figure 9.13), determine the total horizontal thrust on the wall under the following conditions:

① 竖向排水。

a when the backfill becomes fully saturated due to continuous rainfall, with steady-state seepage towards the drain[②];

② 由于连续降雨，回填土完全饱和，伴随稳定渗流的发生。

b if the vertical drain were replaced by an inclined drain below the failure plane at 45° to the horizontal[③];

③ 如果竖向排水被倾斜排水取代，并且倾斜排水发生在与破坏面呈向下45°角的方向。

c if there were no drainage system behind the wall[④] (i. e. the drains in (a) or (b) become blocked).

④ 如果挡墙后无任何排水系统。

Solution

This problem demonstrates how the effects of seepage[⑤] may be accounted for in earth pressure problems. In each case resultant forces due to the pore water must be determined. This may be achieved by drawing a flow net[⑥] (Chapter 2); however, only the pore water pressures are required in order to determine the resultant pore water thrust. The results of this approach are shown below.

⑤ 渗流作用。 ⑥ 流网。

a The equipotentials[⑦] for seepage towards the vertical drain from the spreadsheet tool are shown in Figure 9.23 (a). Since the permeability of the drain must be considerably greater than that of the backfill, the drain remains unsaturated and the pore water pressure at every point within the drain is zero (atmospheric[⑧]). Thus, at every point on the boundary between the drain and the backfill, total head is equal to elevation head[⑨]. The equipotentials, therefore, must intersect this boundary at points spaced at equal vertical intervals Δh as shown. The boundary itself is neither a flow line nor an equipotential.

⑦ 等势线。 ⑧ 环境大气压。 ⑨ 位置水头。

The values of total head and elevation head may be determined at the points of intersection of the equipotentials with the failure plane. The pore water pressures at these points are then determined using Equation 2.1 and integrated numerically along the slip plane to give the pore water thrust on the slip plane, $U = 36.8\text{kN/m}$. The pore water forces on the other two boundaries of the soil wedge are zero.

The total weight (W) of the soil wedge is now calculated, i. e.

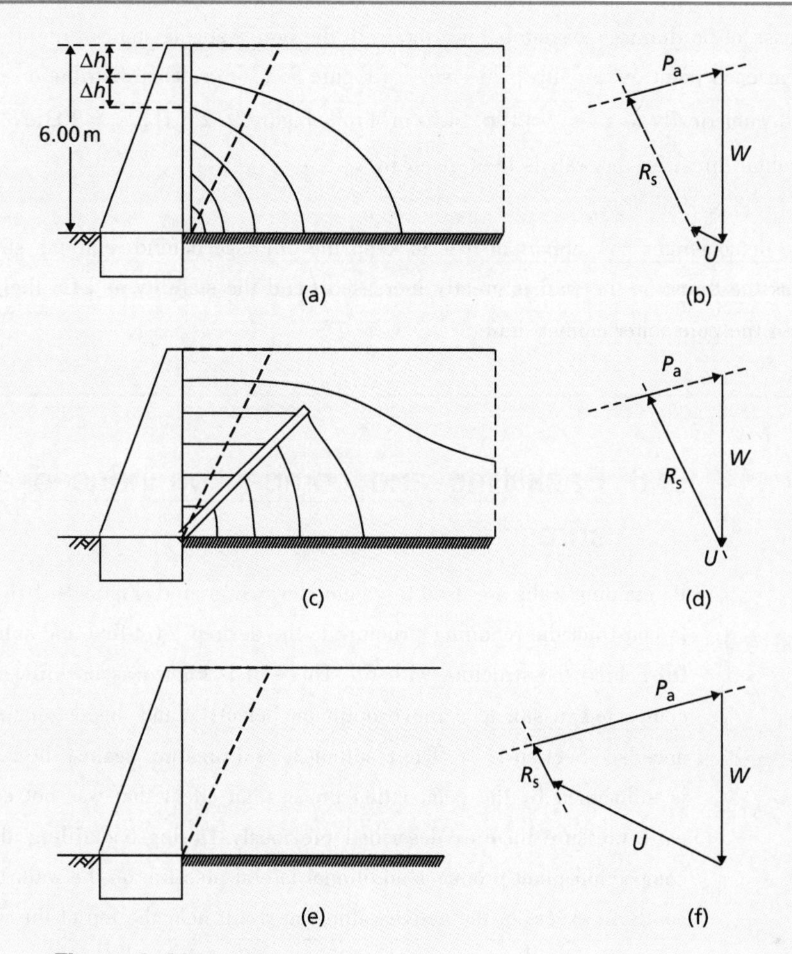

Figure 9.23 Example 9.5.

$$W = \frac{1}{2} \cdot 6.00 \cdot \left[\frac{6.00}{\tan(45 + 19)^\circ} \right] \cdot 20 = 176\text{kN/m}$$

The forces acting on the wedge are shown in Figure 9.23(b). Since the directions of the four forces are known, together with the magnitudes of W and U, the force polygon can be drawn to scale as shown, from which P_a = 104kN/m can be measured from the diagram. The horizontal thrust on the wall is then given by

$$P_a \cos \delta' = 101\text{kN/m}$$

Other failure surfaces would have to be chosen in order that the maximum value of total active thrust can be determined (most critical case).

b For the inclined drain shown in Figure 9.23(c), the equipotentials above the drain are horizontal, the total head[①] being equal to the elevation head in this zone. Thus at every point on the failure plane the pore water pressure is zero, and hence $U = 0$ also. This form of drain is preferable to the vertical drain. In this case, from Figure 9.23(d), P_a = 80kN/m such that the horizontal thrust on the wall is then given by

$$P_a \cos \delta' = 77\text{kN/m}$$

① 总水头。

c For the case of no drainage system behind the wall, the pore water is static[①], i. e. the pore water pressure at each point on the slip plane is $\gamma_w z$ (Figure 9. 23(e)). This distribution can again be integrated numerically to give $U = 196.5\text{kN/m}$. From Figure 9. 23(f), $P_a = 203\text{kN/m}$ such that the horizontal thrust on the wall is then given by

$$P_a \cos \delta' = 196\text{kN/m}$$

This example demonstrates how important it is to keep the drainage behind retaining structures well maintained, as the thrust on the wall is greatly increased (and the stability at ULS therefore greatly reduced) when the pore water cannot drain[②].

① 静孔隙水压力。
② 这一例子说明了保持挡土墙后良好排水的重要性。当孔隙水无法排出时，作用在挡土墙上的推力增加，即边坡的稳定性急剧下降。
③ 回填及压实引起的土压力。
④ 一般通过现场压实来达到最优密实度。
⑤ 作用于挡土墙上的残余侧向土压力。
⑥ 碾压设备。
⑦ 滚轮。
⑧ 离心力。

9.6 Backfilling and compaction-induced earth pressures[③]

If retaining walls are used to create elevated ground (Figure 9. 1 (b)), it is common to construct the retaining structure to the desired level first and subsequently **backfill** behind the structure with fill. This soil is known as backfill, and is commonly compacted in-situ to achieve optimum density[④] (and, hence, engineering performance; see Section 1. 7). The resultant lateral pressure against the retaining structure is influenced by the compaction process, an effect that was not considered in the earth pressure theories described previously. During backfilling, the weight of the compaction plant produces additional lateral pressure on the wall. Pressures significantly in excess of the active value can result near the top of the wall, especially if it is restrained by propping during compaction. As each layer is compacted, the soil adjacent to the wall is pushed downwards against the frictional resistance on the wall surface (τ_w). When the compaction plant is removed the potential rebound of the soil is restricted by wall friction, thus inhibiting reduction of the additional lateral pressure. Also, the lateral strains induced by compaction have a significant plastic component which is irrecoverable. Thus, there is a residual lateral pressure on the wall[⑤]. A simple analytical method of estimating this residual lateral pressure has been proposed by Ingold (1979).

Compaction of backfill behind a retaining wall is normally effected by rolling. The compaction plant[⑥] can be represented approximately by a line load equal to the weight of the roller[⑦]. If a vibratory roller is employed, the centrifugal force[⑧] due to the vibrating mechanism should be added to the static weight. Referring to Figure 9. 24, the stresses at point X due to a line load of Q per unit length on the surface are as follows:

$$\sigma_z = \frac{2Q}{\pi} \frac{z^3}{(x^2 + z^2)^2} \tag{9.62}$$

$$\sigma_x = \frac{2Q}{\pi} \frac{x^2 z}{(x^2 + z^2)^2} \tag{9.63}$$

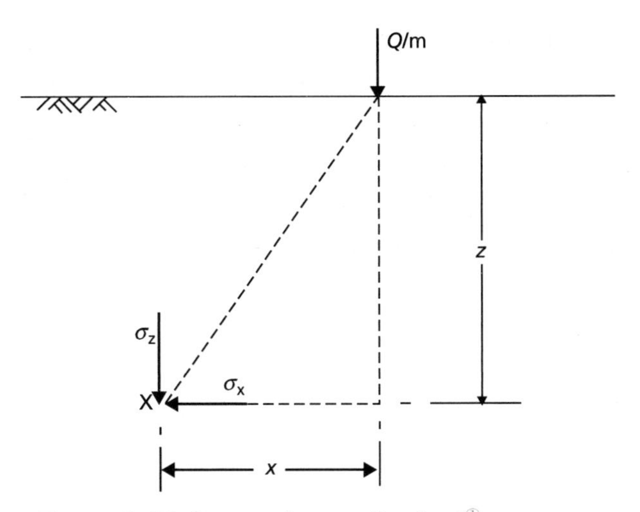

Figure 9.24 Stresses due to a line load[①].

① 线荷载引起的应力。

$$\tau_{xz} = \frac{2Q}{\pi} \frac{xz^2}{(x^2 + z^2)^2} \tag{9.64}$$

From Equation 9.62, the vertical stress immediately below the line load is

$$\sigma_z = \frac{2Q}{\pi z}$$

Then the lateral pressure on the wall at depth z is given by

$$p_c = K_a (\gamma z + \sigma_z)$$

When the stress σ_z is removed, the lateral stress may not revert to the original value ($K_a\gamma z$). At shallow depth the residual lateral pressure could be high enough, relative to the vertical stress γz, to cause passive failure in the soil. Therefore, assuming there is no reduction in lateral stress on removal of the compaction plant, the maximum (or critical) depth (z_c) to which failure could occur is given by

$$p_c = K_p \gamma z_c$$

Thus

$$K_a (\gamma z_c + \sigma_z) = \frac{1}{K_a} \gamma z_c$$

If it is assumed that γz_c is negligible compared to σ_z, then

$$z_c = \frac{K_a^2 \sigma_z}{\gamma} = \frac{K_a^2 2Q}{\gamma \pi z_c}$$

Therefore

$$z_c = K_a \sqrt{\frac{2Q}{\pi \gamma}}$$

The maximum value of lateral pressure (p_{max}) occurs at the critical depth[②], therefore (again neglecting γz_c)

② 临界深度。

$$p_{max} = \frac{2QK_a}{\pi z_c} = \sqrt{\frac{2Q\gamma}{\pi}} \tag{9.65}$$

① 假设达到最大压力 p_{max}，并且保持不变，在接下来的每一层中，可画一条垂直线当作临界深度之下的应力包络线。

Backfill is normally placed and compacted in layers. Assuming that the pressure p_{max} is reached, and remains, in each successive layer, a vertical line can be drawn as a pressure envelope below the critical depth[①]. Thus, the distribution shown in Figure 9. 25 represents a conservative basis for design. However, at a depth z_a the active pressure will exceed the value p_{max}. The depth z_a, being the limiting depth of the vertical envelope, is obtained from the equation

$$K_a \gamma z_a = \sqrt{\frac{2Q\gamma}{\pi}}$$

Thus

$$z_a = \frac{1}{K_a}\sqrt{\frac{2Q}{\pi\gamma}} \tag{9.66}$$

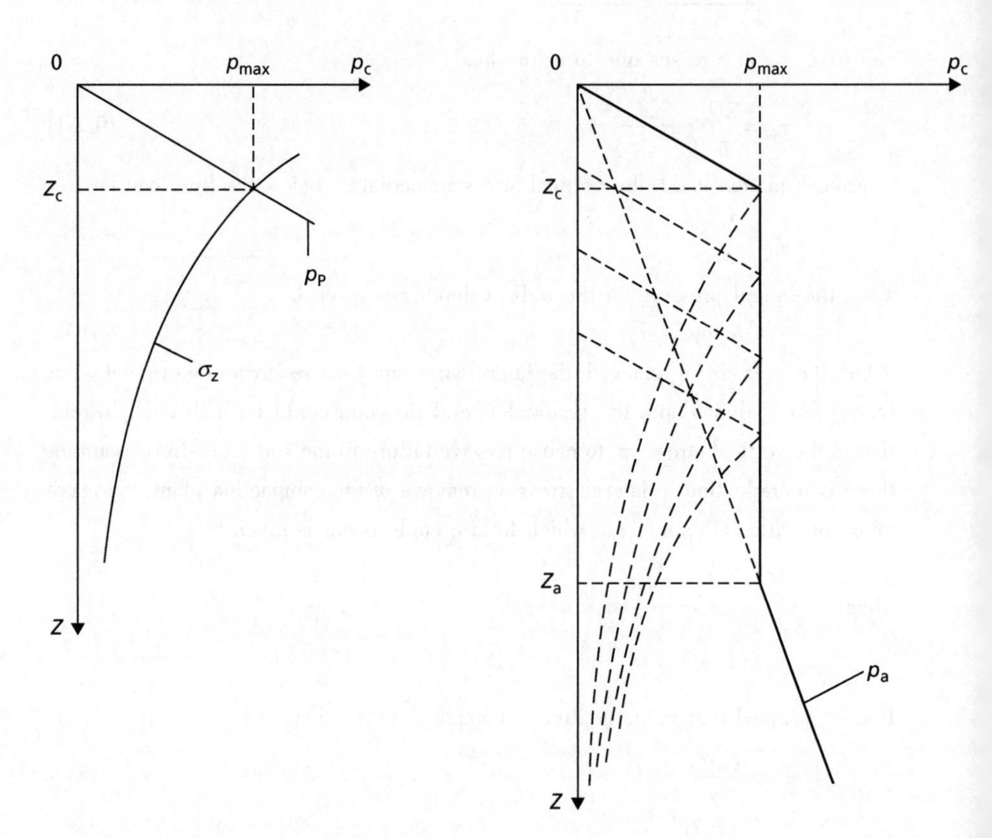

② 压实引起的压力。

Figure 9.25 Compaction-induced pressure[②].

③ 嵌入式挡土墙。

9.7 Embedded walls[③]

Cantilever walls

④ 钢板桩。

Walls of this type are mainly of steel sheet piling[④], and are used only when the retained height of soil is relatively low. In sands and gravels these walls may be used as permanent structures, but in general they are used only for temporary support. The stability of the wall is due entirely to passive resistance mobilised in front of the wall[①]. The principal limit states considered previously for gravity retaining structures are then replaced by:

① 挡土墙的稳定性取决于挡土墙前产生的被动阻力。

ULS-1	Horizontal translation of the wall[②];
ULS-2	Rotation of the wall[③];
ULS-3	Bearing failure of the wall (acting as a pile) under its own self weight and interface shear forces from the retained soil[④].

② 挡土墙平动。

③ 挡土墙转动。

④ 墙体自重和土-墙界面剪切作用下导致的挡土墙承载破坏。

Limit states ULS-4 to ULS-6 listed for gravity walls (Section 9. 4) should also be considered, as should serviceability limit states SLS-1 and SLS-2. The mode of failure of an embedded wall is by rotation about a point O near the lower end of the wall, as shown in Figure 9. 26 (a). Consequently, passive resistance acts in front of the wall above O and behind the wall below O, as shown in Figure 9. 26 (b),

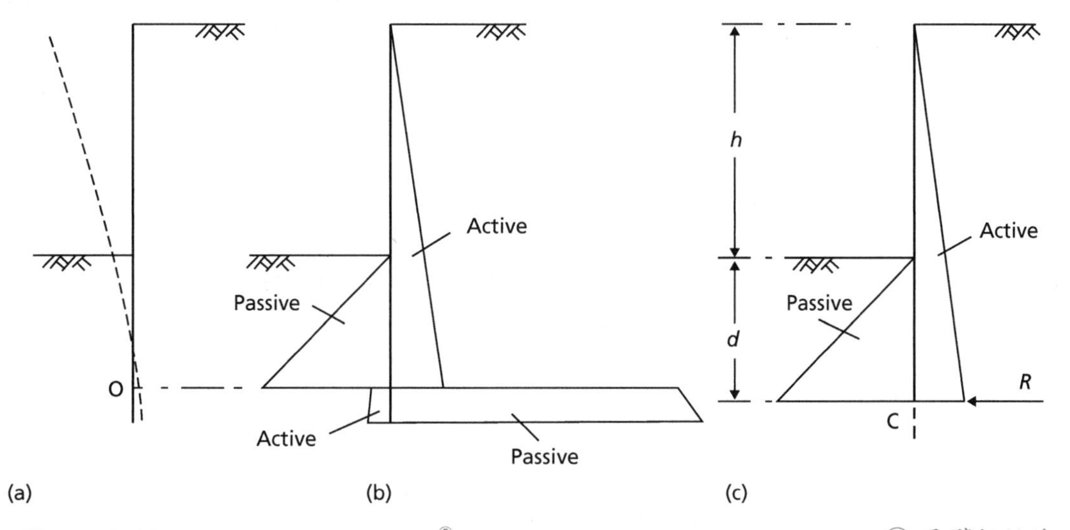

Figure 9.26 Cantilever sheet pile wall[⑤].

⑤ 悬臂板桩墙。

thus providing a fixing moment. However, this pressure distribution is an idealisation, as there is unlikely to be a complete change in passive resistance from the front to the back of the wall at point O[⑥]. To allow for over-excavation, it is recommended that the soil level in front of the wall should be reduced by 10% of the retained height, subject to a maximum of 0. 5m (overdig).

⑥ 但是，由于不可能在挡土墙 O 点处被动阻力发生完全翻转，所以这是理想状态的压力分布。

Design is generally based on the simplification shown in Figure 9. 26 (c), it being assumed that the net passive resistance below point O is represented by a concentrated force R acting at a point C, slightly below O, at depth d below the lower soil surface. ULS-1 is verified by ensuring that the resultant passive thrust is greater than the resultant active thrust. ULS-2 is verified by ensuring that the (clockwise) restoring moment about O due to the resultant passive thrust of the soil in front of the wall is greater than the (anticlockwise) driving moment due to the resultant active thrust of the soil behind the wall. Given that a cantilever wall is flexible, the check of the internal structural stability and performance of the wall (ULS-6 and SLS-2) is of greater importance than for gravity structures, so it is typical in design to determine the shear force and bending moment distributions in the wall for subsequent structural checks (which are beyond the scope of this book).

① 拉锚或支撑式挡墙。

Anchored and propped walls[①]

Generally, structures of this type are either of steel sheet piling or reinforced concrete diaphragm walls, the construction of which is described in Section 9.10; however, secant piling may also be used to form a wall which would be anchored or propped. Additional support to embedded walls is provided by a row of **tie-backs (anchors)**[②] or props near the top of the wall, as illustrated in Figure 9.27 (a). Tie-backs are normally high-tensile steel cables or rods, anchored in the soil some distance behind the wall. Walls of this type are used extensively in the support of deep excavations and in waterfront construction. In the case of sheet pile walls there are two basic modes of construction. Excavated walls are constructed by driving a row of sheet piling, followed by excavation or dredging to the required depth in front of the wall. Backfilled walls[③] are constructed by partial driving, followed by backfilling to the required height behind the piling (see Section 9.6). In the case of diaphragm walls[④], excavation takes place in front of the wall after it has been cast in-situ. Stability is due to the passive resistance developed in front of the wall together with the supporting forces in the ties or props.

② 拉锚。

③ 回填式挡土墙。

④ 地下连续墙。

The limit states to be considered include those listed above for cantilever walls. In addition, the following two limit states must be considered:

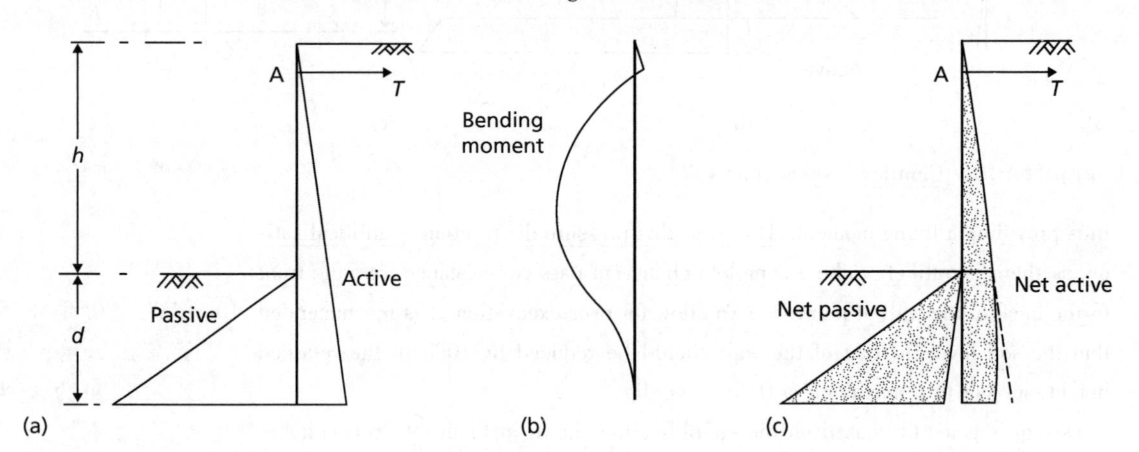

Figure 9.27 Anchored sheet pile wall: free earth support method.

⑤ 锚杆从土中拔出产生的破坏，这是由土-锚接触强度失效引起的。

ULS-7 Failure of anchors/ties by pull-out from the soil[⑤] (anchored walls with ties in tension only). This is essentially a check of the soil – tie interface strength.

ULS-8 Structural failure of ties/props[1]. For props which are loaded in compression, failure would be by buckling[2]; in anchors tie failure would be by fracture. This is essentially a check of the structural strength of the tie/prop itself.

① 支撑、锚杆的结构破坏。
② 屈曲。

An additional serviceability limit state is

SLS-3 Yield of ties/props should be minimal.

These additional limiting states are described in greater detail in Section 9. 8.

The consequences of not meeting these limiting states can be severe. On 20 April 2004, a section of a 33m deep propped excavation collapsed in Singapore (Figure 9. 28). The failure of the wall led to collapse of a zone of soil 150 × 100m in plan and 30m deep, making the adjacent Nicoll Highway impassable for 8 months after the collapse. Of the construction workers who were in the excavation at the time of the collapse, four were killed and a further three injured. The causes of the collapse were ultimately identified as underestimation of the thrusts acting on the wall[3] (ULS-1/2) coupled with inadequate structural design of the prop-wall connection[4] (ULS-8). Excessive deformation of the wall was observed prior to collapse, but was not acted upon.

③ 低估挡土墙上的作用力。
④ 支撑与墙的接头的结构设计不足。

Figure 9.28 Nicoll Highway collapse, Singapore.

Free earth support analysis for tied/propped walls

It is assumed that the depth of embedment below excavation level is insufficient to produce fixity at the lower end of the wall. Thus, the wall is free to rotate at its low-

① 弯矩。
② 转动极限状态。

er end, the bending moment[①] diagram being of the form shown in Figure 9. 27 (b). To satisfy the rotation limit state[②] ULS-2, the sum of the restoring moments ($\sum M_R$ factored as resistances) about the anchor or prop must be greater than or equal to the sum of the overturning moments ($\sum M_A$ factored as actions), i. e.

$$\sum M_A \leqslant \sum M_R \tag{9.67}$$

③ 水平力平衡。

ULS-1 is then verified by considering equilibrium of the horizontal forces[③]. For this limiting state the active thrusts behind the wall are still the actions (causing the wall to move outwards), while the prop/tie force and passive thrusts are resistances. This gives the minimum resistance that the prop/tie must be able to provide to satisfy ULS-1. These prop forces are then the characteristic actions in the tie/prop for verification of ULS-7 and ULS-8. The net earth pressure distribution[④] on the wall using the final propping/tie system is then determined under working load conditions[⑤] (all partial factors = 1. 00) to provide the characteristic loadings for verification of the wall's structural stability (ULS-6). Finally, if appropriate, the vertical forces on the wall are calculated and checked, it being a requirement that the downward force (e. g. the component of the force in an inclined tie-back) should not exceed the (upward) frictional resistance available between the wall and the soil on the passive side minus the (downward) frictional force on the active side (ULS-3).

④ 净土压力分布。
⑤ 工作荷载情况。

⑥ 自由端分析法。
⑦ 恢复力矩。

When applying free earth support analysis[⑥] in ULS design, the active (effective) earth pressures which result in overturning moments are considered as actions and factored accordingly (as for gravity retaining structures). The passive earth pressures acting in front of the wall which result in restoring moments[⑦] are treated as resistances. The partial factors for earth pressure resistance $\gamma_{Re} = 1.00$ for sets R1 and R3, and 1. 40 for set R2. These factors are included on the EC7 quick reference sheet on the Companion Website. The net pore pressure distribution on an embedded wall will always be an overturning moment, and is therefore treated as an action.

It should be realised that full passive resistance is only developed under conditions of limiting equilibrium, i. e. when the structure is at the pint of failure. Under working conditions, analytical and experimental work has indicated that the distribution of lateral pressure is likely to be of the form shown in Figure 9. 29, with passive resistance being fully mobilised close to the lower surface. The extra depth of embedment required to provide adequate safety results in a partial fixing moment at the lower end of the wall and, consequently, a lower maximum bending moment than the value under limiting equilibrium or collapse conditions. In view of the uncertainty regarding the pressure distributions under working conditions, it is recommended that bending moments and tie or prop force under limiting equilibrium conditions should be used in the structural design of the wall[⑧]. The tie or prop force thus calculated should be increased by 25% to allow for possible redistribution of pressure due to arching[⑨] (see below). Bending moments should be calculated on the same basis in the case of cantilever walls.

⑧ 推荐采用在极限平衡条件下的弯矩、锚或支撑的力用于挡土墙的结构设计。
⑨ 土拱效应。

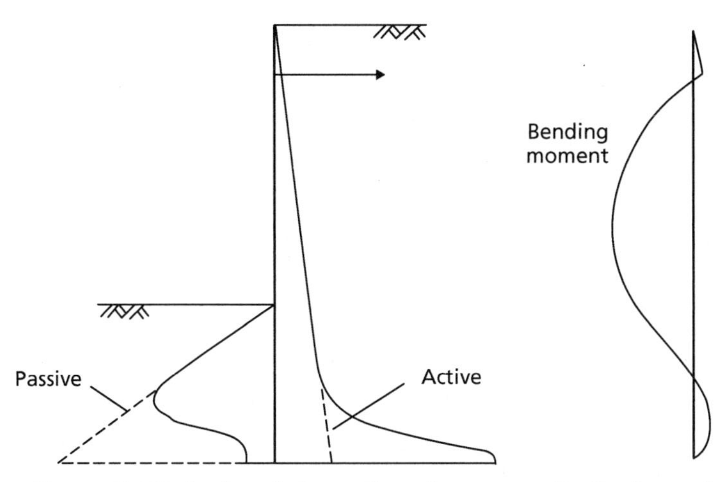

Figure 9.29 Anchored sheet pile wall: pressure distribution under working conditions.

The behaviour of an anchored wall is also influenced by its degree of flexibility or stiffness. In the case of flexible sheet pile walls①, experimental and analytical results indicate that redistributions of lateral pressure take place. The pressures on the yielding parts of the wall (between the tie and excavation level) are reduced and those on the unyielding parts (in the vicinity of the tie and below excavation level) are increased with respect to the theoretical values, as illustrated in Figure 9.30. These redistributions of lateral pressure are the result of the phenomenon known as **arching**. No such redistributions take place in the case of stiff walls, such as concrete diaphragm walls (Section 9.10).

① 柔性板桩墙。
② 锚杆。
③ 土压力分布。
④ 自由端法。
⑤ 试验。

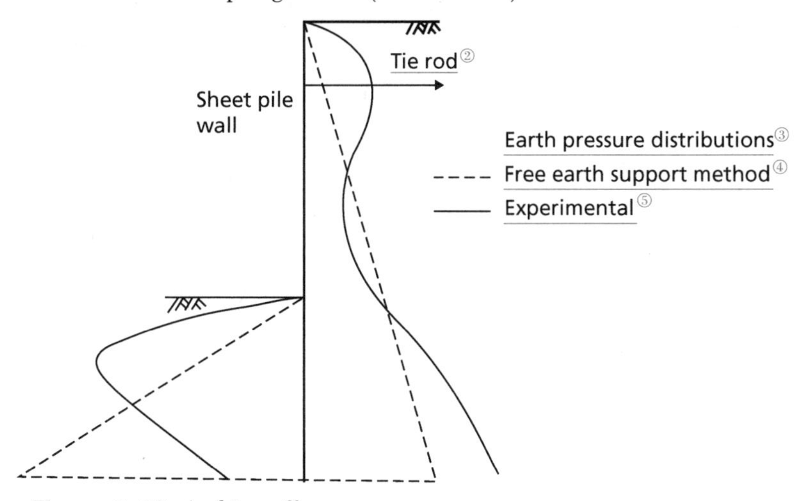

Figure 9.30 Arching effects.

Arching was defined by Terzaghi(1943) in the following way:

> If one part of the support of a soil mass yields while the remainder stays in place, the soil adjoining the yielding part moves out of its original position between adjacent stationary soil masses. The relative movement within the soil is opposed by shearing resistance within the zone of contact between the yielding and stationary masses. Since the shearing resistance tends to keep the yielding mass in its original position, the pressure on the yielding part⑥ of the support

⑥ 屈服部分。

① 静止部分。

② 压力由屈服土体传递到邻近的不屈服土体称为土拱效应。

is reduced and the pressure on the stationary parts[1] is increased. This transfer of pressure from a yielding part to adjacent non-yielding parts of a soil mass is called the arching effect[2]. Arching also occurs when one part of a support yields more than the adjacent parts.

③ 土拱效应在砂土中比粉土和黏土中显著，并且密砂的土拱效应比松砂的显著。

The conditions for arching are present in anchored sheet pile walls when they deflect. If yield of an anchor takes place (ULS-7), arching effects are reduced to an extent depending on the amount of yielding. On the passive side of the wall, the pressure is increased just below excavation level as a result of larger deflections into the soil. In the case of backfilled walls, arching is only partly effective until the fill is above tie level. Arching effects are much greater in sands than in silts or clays, and are greater in dense sands than in loose sands[3].

Generally, redistributions of earth pressure result in lower bending moments than those obtained from the free earth support method of analysis; the greater the flexibility of the wall, the greater the moment reduction. However, for stiff walls, such as diaphragm walls, formed by excavation in soils having a high K_0 value (in the range 1 – 2), such as overconsolidated clays, Potts and Fourie (1984, 1985) showed that both maximum bending moment and prop force could be significantly higher than those obtained using the free earth support method.

④ 孔隙水压力分布。

⑤ 嵌入式挡土墙通常用有效应力法来分析。

Pore water pressure distribution[4]

Embedded walls are normally analysed in terms of effective stresses[5]. Care is therefore required in deciding on the appropriate distribution of pore water pressure. Several different situations are illustrated in Figure 9. 31.

If the water table levels are the same on both sides of the wall, the pore water pressure distributions will be hydrostatic and will balance (Figure 9. 31 (a)); they can thus be eliminated from the calculations, as the resultant thrusts and moments due to the pore water on either side of the wall will balance.

⑥ 假设墙前墙后两个水位的总水头沿挡土墙土体均匀地消散。

If the water table levels are different and if steady seepage conditions have developed and are maintained, the distributions on the two sides of the wall will be unbalanced. The pressure distributions on each side of the wall may be combined into a single net pressure distribution because no earth pressure coefficient is involved. The net distribution on the back of the wall could be determined from a flow net, as illustrated in Example 2. 2, or using the FDM spreadsheet accompanying Chapter 2. However, in many situations an approximate distribution, ABC in Figure 9. 31 (b), can be obtained by assuming that the total head is dissipated uniformly along the back and front wall surfaces between the two water table levels[6]. The maximum net pressure occurs opposite the lower water table level and, referring to Figure 9. 31 (b), is given by

$$u_C = \frac{2ba}{2b + a}\gamma_w \tag{9.68}$$

In general, this approximate method will underestimate net water pressure, espe-

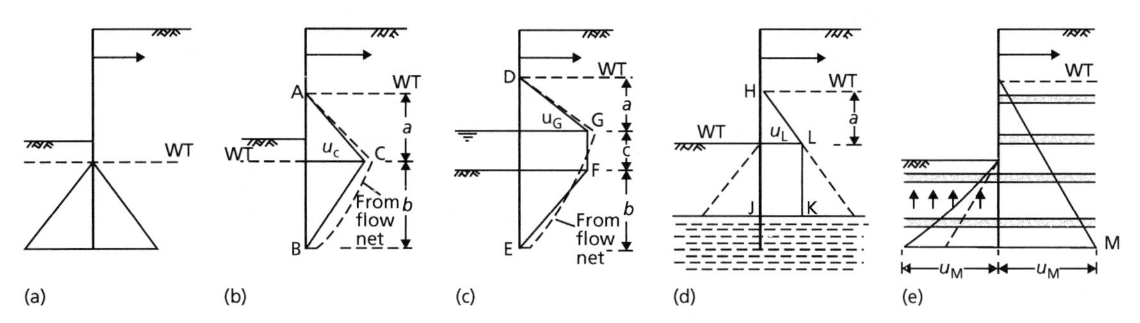

Figure 9.31 Various pore water pressure distributions.

cially if the bottom of the wall is relatively close to the lower boundary of the flow region (i. e. if there are large differences in the sizes of curvilinear squares in the flow net approaching the bottom of the wall). The approximation should not be used in the case of a narrow excavation between two lines of sheet piling where curvilinear squares are relatively small (and seepage pressure relatively high) approaching the base of the excavation. In these cases, the FDM spreadsheet or a flow net should be used.

In Figure 9.31 (c), a depth of water is shown in front of the wall, the water level being below that of the water table behind the wall. In this case the approximate distribution DEFG should be used in appropriate cases, the net pressure at G being given by

$$u_G = \frac{(2b + c)a}{2b + c + a}\gamma_w \tag{9.69}$$

If $c = 0$, then Equation 11.69 reduces to Equation 11.68.

A wall constructed mainly in a soil of relatively high permeability but penetrating a layer of low permeability layer (typically clay) is shown in Figure 9.31 (d). This may be done to reduce the amount of seepage underneath the wall by using the low permeability soil as a natural barrier to seepage[①]. If undrained conditions apply within the clay, the pore water pressure in the overlying soil would be hydrostatic and the net pressure distribution would be HJKL as shown, where

① 该方法可以通过利用低渗透性土当作天然的挡水屏障，从而减弱墙底的渗流作用。

$$u_L = a\gamma_w \tag{9.70}$$

A wall constructed in a low permeability soil (e. g. clay) which contains thin layers or partings of high permeability material (e. g. sand or silt) is shown in Figure 9.31 (e). In this case, it should be assumed that the sand or silt allows water at hydrostatic pressure to reach the back surface of the wall. This implies pressure in excess of hydrostatic, and consequent upward seepage in front of the wall (see Section 3.7).

For short-term situations[②] for walls in clay (e. g. during and immediately after excavation), there exists the possibility of tension cracks[③] developing or fissures opening[④]. If such cracks or fissures fill with water, hydrostatic pressure should be assumed over the depth in question. The water in the cracks or fissures would also result in softening of the clay. Softening[⑤] would also occur near the soil surface in

② 短期情况。
③ 张拉裂缝。
④ 节理张开。
⑤ 软化。

① 有效应力分析方法。
② 稳态渗流。
③ 渗透压力。
④ 饱和重度。
⑤ 超挖。
⑥ 静水压力。
⑦ 力、力臂、弯矩的设计值（考虑分项系数）。

front of the wall as a result of stress relief on excavation. An effective stress analysis[①] would ensure a safe design in the event of rapid softening of the clay taking place, or if work were delayed during the temporary stage of construction.

Under conditions of steady seepage[②], use of the approximation that total head is dissipated uniformly along the wall has the consequent advantage that the seepage pressure is constant. For the conditions shown in Figure 9. 31 (b), for example, the seepage pressure[③] at any depth is

$$j = \frac{a}{2b + a}\gamma_w \tag{9.71}$$

The effective unit weight of the soil below the water table, therefore, would be increased to $\gamma' + j$ behind the wall, where seepage is downwards, and reduced to $\gamma' - j$ in front of the wall, where seepage is upwards. These values should be used in the calculation of effective active and passive pressures, respectively, if groundwater conditions are such that steady seepage is maintained. Thus, active pressures are increased and passive pressures are decreased relative to the corresponding static values.

Example 9.6

The sides of an excavation 2. 25m deep in sand are to be supported by a cantilever sheet pile wall, the water table being 1. 25m below the bottom of the excavation. The unit weight of the sand above the water table is 17kN/m^3, and below the water table the saturated unit weight[④] is 20kN/m^3. Characteristic strength parameters are $c' = 0$, $\phi' = 35°$ and $\delta' = 0$. Allowing for a surcharge pressure of 10kPa on the surface, determine the required depth of embedment of the piling to satisfy the rotational (ULS-2) and translational (ULS-1) limit states to Eurocode 7, DA1b.

Solution

Below the water table, the effective unit weight of the soil is $(20 - 9.81) = 10.2\text{kN/m}^3$. To allow for possible over-excavation[⑤] the soil level should be reduced by 10% of the retained height of 2. 25m, i. e. by 0. 225m. The depth of the excavation therefore becomes 2. 475m, say 2. 50m, and the water table will be 1. 00m below this level.

The design dimensions and the earth pressure diagrams are shown in Figure 9. 32. The distributions of hydrostatic pressure[⑥] on the two sides of the wall balance and can be eliminated from the calculations (there is no seepage occurring). For applying partial factors:

- the active earth pressures on the retained side (2 – 4 in Figure 9. 32) are unfavourable permanent actions (causing the wall to rotate or slide);
- the horizontal earth pressure due to the surcharge (1 in Figure 9. 32) is an unfavourable variable action;
- the passive earth pressures in front of the wall (5 – 7in Figure 9. 32) are treated as resistances (as they prevent the wall rotating or sliding) and factored by γ_{Re}.

The procedure is to check the rotational stability first by applying Equation 9. 67 about point C. The design value of ϕ' for DA1b is $\tan^{-1}(\tan 35°/1.25) = 29°$. The corresponding values of K_a and K_p are 0. 35 and 2. 88, respectively, using Equations 9. 5 and 9. 7, or 9. 23 and 9. 27. The design (factored) values of forces, lever arms and moments[⑦] are set out in Table 9. 4.

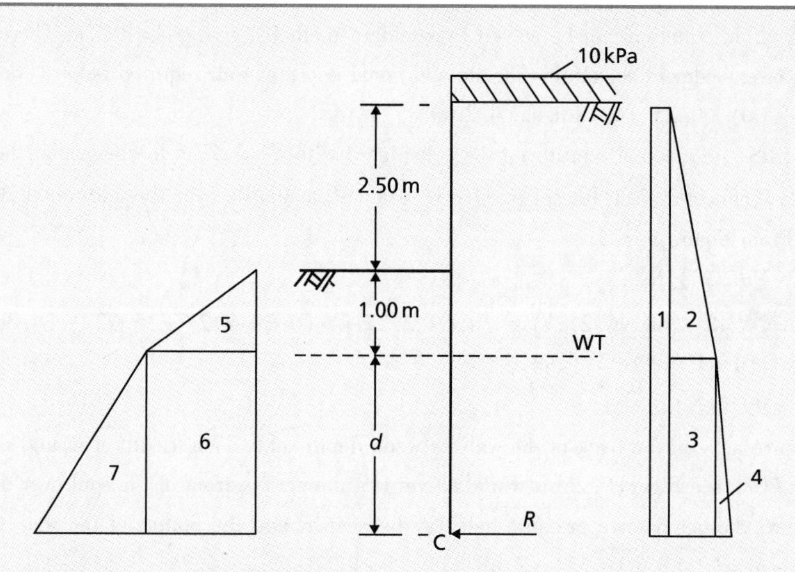

Figure 9.32 Example 9.6.

Table 9.4 Example 9.6

	Force (kN/m)	*Lever arm (m)*	*Moment (kNm/m)*
	Actions (H_A, M_A):		
(1)	$0.33 \cdot 10 \cdot (d+3.5) \cdot 1.30 = 4.29d + 15.02$	$\frac{d+3.5}{2}$	$2.15d^2 + 15.02d + 26.29$
(2)	$\frac{1}{2} \cdot 0.33 \cdot 17 \cdot 3.5^2 \cdot 1.00 = 34.36$	$d + \frac{3.5}{3}$	$34.36d + 40.09$
(3)	$0.33 \cdot 17 \cdot 3.5 \cdot d \cdot 1.00 = 19.64d$	$\frac{d}{2}$	$9.82d^2$
(4)	$\frac{1}{2} \cdot 0.33 \cdot 10.2 \cdot d^2 \cdot 1.00 = 1.68d^2$	$\frac{d}{3}$	$0.56d^3$
	Resistances (H_R, M_R):		
(5)	$\frac{\frac{1}{2} \cdot 2.88 \cdot 17 \cdot 1.0^2}{1.00} = 24.48$	$d + \frac{1.0}{3}$	$24.48d + 8.16$
(6)	$\frac{2.88 \cdot 17 \cdot 1 \cdot d}{1.00} = 48.96d$	$\frac{d}{2}$	$24.48d^2$
(7)	$\frac{\frac{1}{2} \cdot 2.88 \cdot 10.2 \cdot d^2}{1.00} = 14.69d^2$	$\frac{d}{3}$	$4.90d^3$

Applying Equation 9.67 to check ULS-2, the minimum depth of embedment[①] will just satisfy ULS, so:

① 最小埋置深度。

$$\sum M_A = \sum M_R$$

$$0.56d^3 + 11.97d^2 + 49.38d + 66.38 = 4.90d^3 + 24.48d^2 + 24.48d + 8.16$$

$$0 = 4.34d^3 + 12.51d^2 - 24.9d - 58.22$$

The resulting cubic equation can be solved by standard methods, giving $d = 2.27$m. Therefore, the required depth of embedment accounting for the additional depth of wall required below C and overdig = 1.2 (2.27 + 1.00) + 0.25 = 4.18m, say 4.20m.

To check ULS-1, horizontal equilibrium is considered with $d = 2.27$m to check that the R is sufficient for fixity, compared with the net passive resistance[①] available over the additional 20% embedment depth. From Figure 9.32, ① 净被动阻力。

$$R + \sum H_A = \sum H_R$$

$$R = 24.48 + (48.96 \cdot 2.27) + (14.69 \cdot 2.27^2) - (4.29 \cdot 2.27 + 15.02) - 34.36 - (19.64 \cdot 2.27) - (1.68 \cdot 2.27^2)$$
$$= 99.0\text{kN/m}$$

Passive pressure acts on the back of the wall between depths of 5.77m (depth of R) and the bottom of the wall at 6.70m (see Figure 9.26b), while active pressure acts in front of the wall over the same distance. Therefore, the net passive pressure half way between R and the bottom of the wall (6.24m) is

$$p'_p - p'_a = [2.88 \cdot 10 \cdot 6.24 + 2.88 \cdot 17 \cdot 3.5 + 2.88 \cdot 10.2 \cdot (6.24 - 3.5)] - [0.35 \cdot 17 \cdot 1 + 0.35 \cdot 10.2 \cdot (6.24 - 3.5)]$$
$$= 415.8\text{kPa}$$

The net passive resistance available over the additional embedded depth[②] ($P_{p,\,net}$) must then be greater than or equal to R to satisfy ULS-1: ② 附加埋置深度。

$$P_{p,\,net} = 415.8 \cdot (6.70 - 5.77)$$
$$= 386.7\text{kN/m}$$

$P_{p,\,net} > R$, so ULS-1 is satisfied.

The procedure detailed above could then be repeated for any other design approaches as required by altering the partial factors[③] applied in Table 9.4. ③ 分项系数。

Example 9.7

A propped cantilever wall supporting the sides of an excavation in stiff clay is shown in Figure 9.33. The saturated unit weight of the clay (above and below the water table) is 20kN/m^3. The design values of the active and passive coefficients of lateral earth pressure are 0.30 and 4.2, respectively. Assuming conditions of steady-state seepage, determine the required depth of embedment for the wall to be stable (use EC7 DA1b). Determine also the force in each prop.

Solution

The distributions of earth pressure and net pore water pressure (assuming uniform decrease of total head around the wall as shown in Figure 9.31b) are shown in Figure 9.33.

The maximum net water pressure at level D from Equation 9.68 is:

$$u_D = \frac{2 \cdot d \cdot (6.0 - 1.5)}{2 \cdot d + (6.0 - 1.5)} \cdot 9.81 = \frac{88.3d}{2d + 4.5}\text{kPa}$$

and the average seepage pressure[④] is ④ 平均渗流压力。

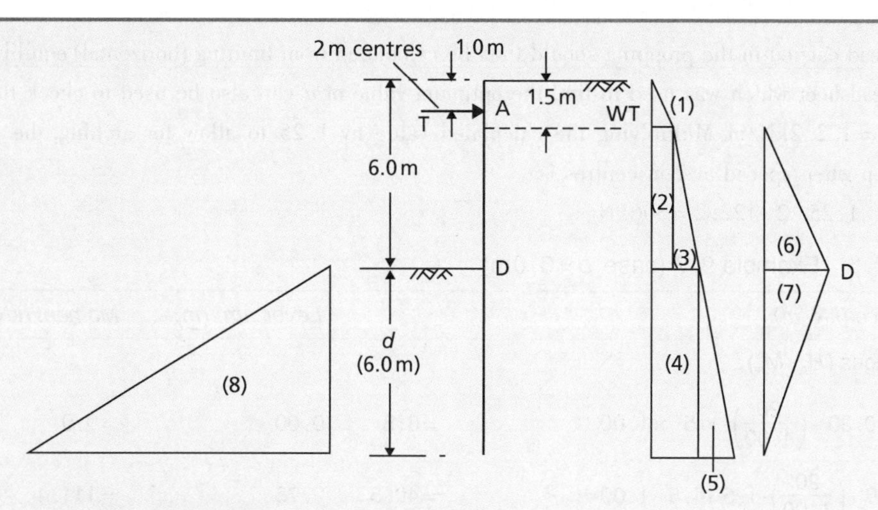

Figure 9.33 Example 9.7.

$$j = \frac{(6.0 - 1.5)}{2 \cdot d + (6.0 - 1.5)} \cdot 9.81 = \frac{44.1}{2d + 4.5}\text{kPa}$$

Thus, below the water table, active forces are calculated using an effective unit weight[①] of

$$(\gamma' + j) = 10.2 + \frac{44.1}{2d + 4.5}\text{kN/m}^3$$

① 有效重度。

and passive forces are calculated using an effective unit weight of

$$(\gamma' + j) = 10.2 - \frac{44.1}{2d + 4.5}\text{kN/m}^3$$

If the forces, lever arms and moments are expressed in terms of the unknown embedment depth d, complex algebraic expressions would result; thus it is preferable to assume a series of trial values of d and check ULS-2 for each. If the ULS is satisfied at the initial trial value of d, further trials should be made, reducing d until the ULS is just not satisfied. Equally, if the ULS is not satisfied on the first trial, further trials should increase d until ULS is just satisfied[②]. In either case, the final value at which ULS is just satisfied represents the minimum value of d. To avoid the unknown prop force, T, ULS-2 is verified by taking moments about point A through which T acts[③].

② 如果不能满足 ULS 状态，则在接下来的尝试中要增大 d 直到满足 ULS。 ③ 按支撑轴力 T 作用点 A 取矩分析。

Following this procedure, a trial value of $d = 6.0$m is first selected. Then, $u_D = 32.1$kPa, $(\gamma' + j) = 12.9\text{kN/m}^3$ and $(\gamma' - j) = 7.5\text{kN/m}^3$. The active thrusts generate disturbing moments and are factored as unfavourable permanent actions as before. The net pressure due to seepage also acts in this direction and is factored in the same way. The passive thrusts are factored as resistances. For DA1b, all of these factors are 1.00; they are, however, included in the calculations below for completeness. The calculations for $d = 6.0$m are then shown in Table 9.5, from which $\sum M_A = 3068.3\text{kNm/m} > \sum M_R = 5103.0$, satisfying ULS-2 and suggesting that d can be reduced to produce a more efficient design.

The calculations may be input into a spreadsheet in a table similar to the above, but as a function of d. An optimisation tool (e.g. Solver in MS Excel) can then be used to find the value of d which makes ULS-2 just satisfied. An example of this approach is provided on the Companion Website, from which $d = 4.57$m (for DA1b). The use of a spreadsheet makes it straightforward to consider other design approaches, as it is only necessary to change the partial factors appropriately and re-run the optimisation.

The load carried in the propping should then be calculated from limiting (horizontal) equilibrium[①]. The spreadsheet which was used to find the optimum value of d can also be used to check this, from which $T = 122.2\text{kN/m}$. Multiplying the calculated value by 1.25 to allow for arching, the force in each prop when spaced at 2-m centres is

$$1.25 \cdot 2 \cdot 122.2 = 306\text{kN}$$

Table 9.5 Example 9.7 (case $d = 6.0\text{m}$)

Force (kN/m)		Lever arm (m)	Moment (kNm/m)
Actions (H_A, M_A):			
(1) $\frac{1}{2} \cdot 0.30 \cdot \left(\frac{20}{1.00}\right) \cdot 1.5^2 \times 1.00$	=6.8	0.00	=0.0
(2) $0.30 \cdot \left(\frac{20}{1.00}\right) \cdot 1.5 \cdot 4.5 \cdot 1.00$	=40.5	2.75	=111.4
(3) $\frac{1}{2} \cdot 0.30 \cdot \left(\frac{12.9}{1.00}\right) \cdot 4.5^2 \times 1.00$	=39.2	3.50	=137.2
(4) $0.30 \cdot \left[\left(\frac{20}{1.00} \cdot 1.5\right) + \left(\frac{12.9}{1.00} \cdot 4.5\right)\right] \cdot 6.0 \cdot 1.00$	=158.2	8.00	=1265.6
(5) $\frac{1}{2} \cdot 0.30 \cdot \left(\frac{12.9}{1.00}\right) \cdot 6.0^2 \times 1.00$	=69.7	9.00	=627.3
(6) $\frac{1}{2} \cdot 32.1 \cdot 4.5 \cdot 1.00$	=72.2	3.50	=252.7
(7) $\frac{1}{2} \cdot 32.1 \cdot 6.0 \cdot 1.00$	=96.3	7.00	=674.1
Resistances (H_R, M_R):			
(8) $\dfrac{\frac{1}{2} \cdot 4.2 \cdot \left(\frac{7.5}{1.00}\right) \cdot 6.0^2}{1.00}$	=567.0	9.00	=5103.0

① 支撑结构上的荷载应通过水平方向上的极限平衡求得。

9.8 Ground anchorages[②]

② 岩土锚固。

Tie rods[③] are normally anchored in beams, plates or concrete blocks (known as dead-man anchors) some distance behind the wall (Figure 9.34 (a)). The tie rod force from a free earth support analysis (T) is resisted by the passive resistance developed in front of the anchor, reduced by the active pressure acting on the back. To avoid the possibility of progressive failure of a line of ties, it should be assumed that any single tie could fail either by fracture or by becoming detached and that its load could be redistributed safely to the two adjacent ties. Accordingly, it is recommended that a load factor of at least 2.0 should be applied to the tie rod force[④], in addition to any partial factors involved in the ULS checks. The following sections describe the calculation models for determining the pull-out resistance[⑤] of anchorages (T_f) for verification of ULS-7. The normative values of the partial factor

③ 锚杆。

④ 锚杆荷载系数至少为2.0。

⑤ 抗拔阻力。

which should be applied to this resistance according to Eurocode 7 are $\gamma_{Ra} = 1.10$ for sets R1 and R2, and 1.00 for set R3. These factors are included on the EC7 quick reference sheet on the Companion Website. For ULS-7 to be verified:

$$T \leqslant T_f \tag{9.72}$$

Plate anchors[①]

① 板锚。

If the width (b) of the anchor is not less than half the depth (d_a) from the surface to the bottom of the anchor, it can be assumed that passive resistance is developed over the depth d_a. The anchor must be located beyond the plane YZ (Figure 9.34 (a)) to ensure that the passive wedge of the anchor does not encroach on the active wedge behind the wall.

The equation of equilibrium for a ground anchor at failure (for ULS-7) is

$$T_f = \frac{1}{2}(K_p - K_a)\gamma d_a^2 l - K_a \sigma_q d_a l \tag{9.73}$$

where l = length of anchor per tie and σ_q = surface surcharge pressure.

Ground anchors[②]

② 地锚。

Tensioned cables, attached to the wall and anchored in a mass of cement grout[③] or grouted soil (Figure 9.34 (b)), are another means of support. These are known as **ground anchors**. A ground anchor normally consists of a high-tensile steel cable or bar, called the tendon, one end of which is held securely in the soil by a mass of cement grout or grouted soil; the other end of the tendon is anchored against a bearing plate on the structural unit to be supported. While the main application of ground anchors is in the construction of tie-backs for diaphragm or sheet pile walls, other applications are in the anchoring of any structure subjected to overturning, sliding or buoyancy, or in the provision of reaction for in-situ load tests. Ground anchors can be constructed in sands (including gravelly sands and silty sands) and stiff clays, and they can be used in situations where either temporary or permanent support is required[④].

③ 水泥浆。

④ 地锚可以建于砂土（含砾砂和粉砂）或黏土中，并可作为临时或永久支护。

The grouted length of tendon, through which force is transmitted to the surrounding soil, is called the **fixed anchor length**[⑤]. The length of tendon between the fixed anchor and the bearing plate is called the **free anchor length**[⑥]; no force is transmitted to the soil over this length. For temporary anchors, the tendon is normally greased and covered with plastic tape over the free anchor length. This allows for free movement of the tendon and gives protection against corrosion. For permanent anchors, the tendon is normally greased and sheathed with polythene under factory conditions; on site, the tendon is stripped and degreased over what will be the fixed anchor length.

⑤ 锚固长度。

⑥ 自由段长度。

The ultimate load which can be carried by an anchor depends on the soil resistance (principally skin friction) mobilised adjacent to the fixed anchor length. This, of course, assumes that there will be no prior failure at the grout-tendon interface,

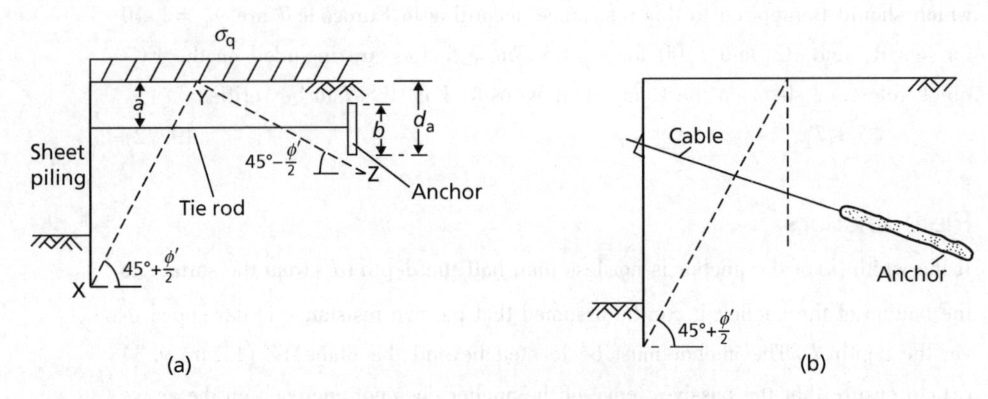

Figure 9.34 Anchorage types: (a) plate anchor, (b) ground anchor.

or of the tendon itself (i. e. ULS-7 will be achieved before ULS-8). Anchors are usually prestressed① in order to reduce the lateral displacement required to mobilise soil resistance, and to minimise ground movements in general. Each anchor is subjected to a test loading after installation, temporary anchors② usually being tested to 1.2 times the working load and permanent anchors③ to 1.5 times the working load. Finally, prestressing of the anchor takes place. Creep displacements under constant load will occur in ground anchors. A creep coefficient④, defined as the displacement per unit log time, can be determined by means of a load test.

① 施加预应力。
② 临时锚杆。
③ 永久锚杆。
④ 蠕变系数。

A comprehensive ground investigation is essential in any location where ground anchors are to be employed. The soil profile must be determined accurately, any variations in the level and thickness of strata being particularly important. In the case of sands, the particle size distribution should be determined in order that permeability and grout acceptability can be estimated.

Design of ground anchors in coarse-grained soils⑤

⑤ 粗粒土中的地锚设计。

In general, the sequence of construction is as follows. A cased borehole (diameter usually within the range 75 – 125mm) is advanced through the soil to the required depth. The tendon is then positioned in the hole, and cement grout is injected under pressure⑥ over the fixed anchor length as the casing is withdrawn. The grout penetrates the soil around the borehole, to an extent depending on the permeability of the soil and on the injection pressure, forming a zone of grouted soil, the diameter of which can be up to four times that of the borehole (Figure 9.35 (a)). Care must be taken to ensure that the injection pressure does not exceed the overburden pressure of the soil above the anchor, otherwise heaving or fissuring may result. When the grout has achieved adequate strength, the other end of the tendon is anchored against the bearing plate. The space between the sheathed tendon and the sides of the borehole, over the free anchor length, is normally filled with grout (under low pressure); this grout gives additional corrosion protection⑦ to the tendon.

⑥ 压力注浆。
⑦ 防腐蚀保护。

The ultimate resistance of an anchor to pull-out⑧ (ULS-7) is equal to the sum of the side resistance and the end resistance of the grouted mass. Considering the an-

⑧ 极限抗拔承载力。

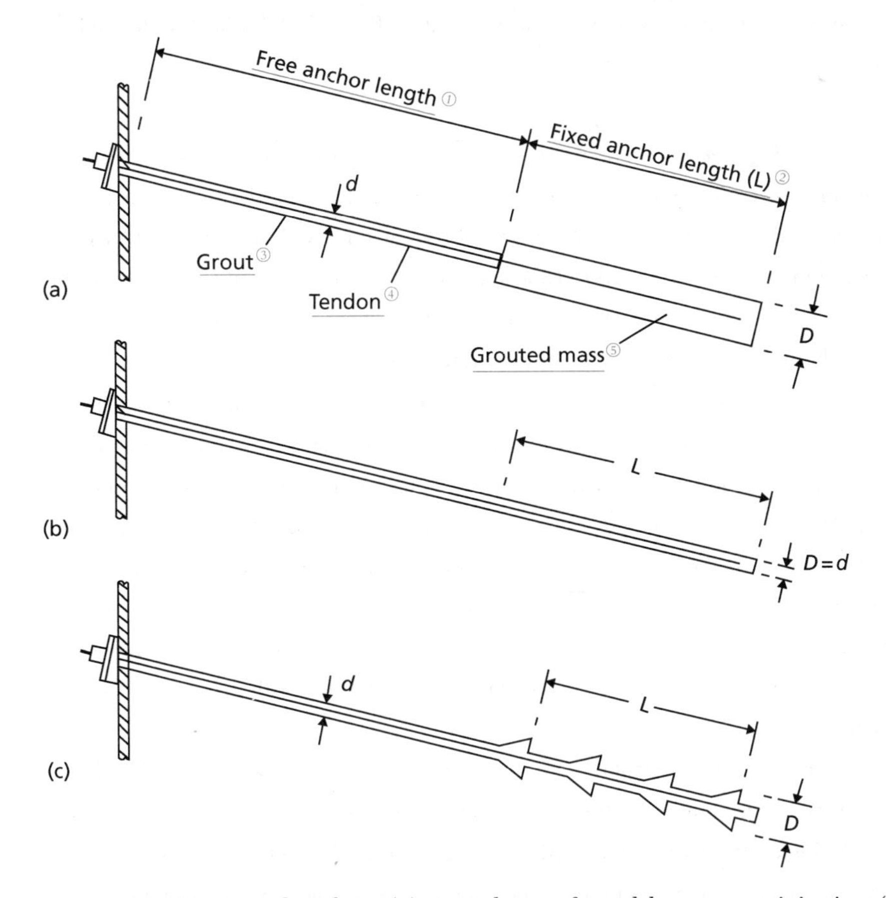

Figure 9.35 Ground anchors: (a) grouted mass formed by pressure injection, (b) grout cylinder, and (c) multiple under-reamed anchor.

① 自由段长度。
② 锚固段长度。
③ 水泥浆。
④ 钢筋束。
⑤ 注浆体。

chor to act as a pile, the following theoretical expression has been proposed:

$$T_f = A\sigma'_v \pi DL\tan\phi' + B\gamma' h\frac{\pi}{4}(D^2 - d^2) \qquad (9.74)$$

where T_f = pull-out capacity of anchor, A = ratio of normal pressure at interface to effective overburden pressure (essentially an earth pressure coefficient), σ'_v = effective overburden pressure adjacent to the fixed anchor and B = bearing capacity factor.

It was suggested that the value of A is normally within the range 1 – 2. The factor B is analogous to the bearing capacity factor N_q in the case of piles, and it was suggested that the ratio N_q/B is within the range 1.3 – 1.4, using the N_q values of Berezantzev *et al.* (1961) However, Equation 9.74 is unlikely to represent all the relevant factors in a complex problem. The ultimate resistance also depends on details of the installation technique⑥, and a number of semi-empirical formulae have been proposed by specialist contractors, suitable for use with their particular technique. An example of such a formula is

⑥ 施工工艺。

$$T_f = Ln\tan\phi' \qquad (9.75)$$

The value of the empirical factor n is normally within the range 400 – 600kN/m for coarse sands and gravels, and within the range 130 – 165kN/m for fine to medium sands.

① 细粒土中的地锚设计。

Design of ground anchors in fine-grained soils[①]

The simplest construction technique for anchors in stiff clays is to auger a hole to the required depth, position the tendon and grout the fixed anchor length using a tremie pipe (Figure 9. 35 (b)). However, such a technique would produce an anchor of relatively low capacity because the skin friction at the grout – clay interface would be unlikely to exceed $0.3c_u$ (i. e. $\alpha \leqslant 0.3$).

② 碎石注入技术。

Anchor capacity can be increased by the technique of gravel injection[②]. The augered hole is filled with pea gravel over the fixed anchor length, then a casing, fitted with a pointed shoe, is driven into the gravel, forcing it into the surrounding clay. The tendon is then positioned and grout is injected into the gravel as the casing is withdrawn (leaving the shoe behind). This technique results in an increase in the effective diameter of the fixed anchor (of the order of 50%) and an increase in side resistance: a value of $\alpha \approx 0.6$ can be expected. In addition, there will be some end resistance. The borehole is again filled with grout over the free anchor length.

③ 扩大端。

Another technique employs an expanding cutter to form a series of enlargements[③] (or under-reams) of the augered hole at close intervals over the fixed anchor length (Figure 9. 35 (c)); the cuttings are generally removed by flushing with water. The cable is then positioned and grouting takes place. A value of $\alpha \approx 1$ can normally be assumed along the cylindrical surface through the extremities of the enlargements.

The following design formula (analogous to Equation 9. 74) can be used for anchors in undrained soil conditions (at ULS-7):

$$T_f = \pi D L \alpha c_u + \frac{\pi}{4}(D^2 + d^2) c_u N_c \tag{9.76}$$

④ 表面摩擦系数。

⑤ 饱和重度。

where T_f = pull-out capacity of anchor, L = fixed anchor length, D = diameter of fixed anchor, d = diameter of borehole, α = skin friction coefficient[④] and N_c = bearing capacity factor (generally assumed to be 9). Resistance at the grout-clay interface along the free anchor length may also be taken into account.

Example 9.8

Details of an anchored sheet pile wall are given in Figure 9. 36, the design ground and water levels being as shown. The ties are spaced at 2.0m centres. Above the water table the unit weight of the soil is 17kN/m^3, and below the water table the saturated unit weight[⑤] is 20kN/m^3. Characteristic soil parameters are $c' = 0$, $\phi' = 36°$, and δ' is taken to be $1/2\phi'$. Determine the required depth of embedment and the minimum capacity of each tie to satisfy ULS to Eurocode 7 (DA1b). Design a continuous anchor to support the ties.

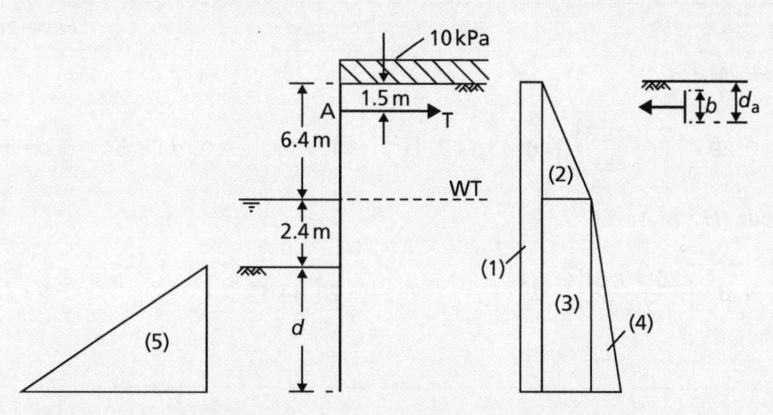

Figure 9.36 Example 9.8.

Solution

For DA1b, the design value of $\phi' = \tan^{-1}(\tan 36°/1.25) = 30°$. Hence (for $\delta' = 1/2\phi'$) $K_a = 0.29$ and $K_p = 4.6$. The lateral pressure diagrams are shown in Figure 9.36. The water levels on the two sides of the wall are equal, therefore the hydrostatic pressure distributions are in balance and can be eliminated from the calculations. The forces and their lever arms are set out in Table 9.6. Force (1) in Table 9.6 is multiplied by the partial factor 1.30, surcharge being a variable unfavourable action. The partial factor for all other forces, being permanent unfavourable actions, is 1.00.

Applying Equation 9.67 to check ULS-2, the minimum depth of embedment will just satisfy ULS, so:

$$\sum M_A = \sum M_R$$

$$0.99d^3 + 32.02d^2 + 312.6d + 893.12 = 15.63d^3 + 171.11d^2$$

$$0 = 14.64d^3 + 139.09d^2 - 312.6d - 893.12$$

The resulting cubic equation can be solved by standard methods, giving $d = 3.19$m. For ULS-1 to be satisfied,

$$\sum H_A = \sum H_R$$

$$1.48d^2 + 42.41d + 218.38 = 23.44d^2 + T$$

$$T = -21.96 \cdot (3.19)^2 + 42.41 \cdot (3.19) + 218.38$$

$$= 130.2 \text{kN/m}$$

Table 9.6 Example 9.8

Force (kN/m)		*Lever arm* (m)
Actions (H_A, M_A):		
(1) $1.30 \cdot 0.29 \cdot 10 \cdot (d+8.8)$	$=3.77d+33.18$	$\frac{1}{2}d+2.9$
(2) $1.00 \cdot \frac{1}{2} \cdot 0.29 \cdot \left(\frac{17}{1.00}\right) \cdot 6.4^2$	$=100.97$	2.77
(3) $1.00 \cdot 0.29 \cdot \left(\frac{17}{1.00}\right) \cdot 6.4\ (d+2.4)$	$=31.55d+75.72$	$\frac{1}{2}d+6.1$

Force (kN/m)		*Lever arm* (m)
Actions (H_A, M_A):		
(4) $1.00 \cdot \frac{1}{2} \cdot 0.29 \cdot \left(\frac{20-9.81}{1.00}\right) \cdot 6.4\,(d+2.4)^2$	$=1.48d^2+7.09d+8.51$	$\frac{2}{3}d+6.5$
Resistances (H_R, M_R):		
(5) $1.00 \cdot \frac{1}{2} \cdot 4.6 \cdot \left(\frac{20-9.81}{1.00}\right) \cdot d^2$	$=23.44d^2$	$\frac{2}{3}d+7.3$
Tie	$=T$	0

Hence the force in each tie = 2 × 130. 2 = 260kN. The design load to be resisted by the anchor is 130. 2kN/m. Therefore, the minimum value of d_a is given from Equation 9. 73 as

$$\frac{T_f}{l} = \frac{1}{2}(K_p - K_a)\gamma d_a^2 - K_a \sigma_q d_a$$

$$130.2 = \frac{1}{2} \cdot (4.6 - 0.29) \cdot \left(\frac{17}{1.00}\right) \cdot d_a^2 - 0.29 \cdot 10 \cdot d_a$$

$$0 = 36.64 d_a^2 - 2.90 d_a - 130.2$$

$$\therefore d_a = 1.93\text{m}$$

Then the vertical dimension (b) of the anchor = 2 (1.93 − 1.5) = 0. 86m.

① 支撑开挖。

9. 9 Braced excavations[1]

② 支撑。

Sheet piling or timbering is normally used to support the sides of deep, narrow excavations, stability being maintained by means of struts[2] acting across the excavation, as shown in Figure 9. 37 (a). The piling is usually driven first, the struts being installed in stages as excavation proceeds. When the first row of struts is installed, the depth of excavation is small and no significant yielding of the soil mass will have taken place. As the depth of excavation increases, significant yielding of the soil occurs before strut installation but the first row of struts prevents yielding near the surface. Deformation of the wall, therefore, will be of the form shown in Figure 9. 37 (a), being negligible at the top and increasing with depth. Thus the deformation condition for active conditions in Figure 9. 13 is not satisfied and active earth pressures cannot be assumed to act on such walls. Failure of the soil will take place along a slip surface of the form shown in Figure 9. 37 (a), only the lower part of the soil wedge within this surface reaching a state of plastic equilibrium, the upper part remaining in a state of elastic equilibrium. The limit states outlined earlier for propped walls must be met in design; however, because active conditions are not mobilised[3], free-earth support analysis should not be used to determine the strut/prop forces for ULS-8. An alternative procedure is described below. Additionally, because propping generally allows excavation to be made to greater depths, there

③ 未达到主动状态。

will be a larger stress relief due to excavation which may result in heave of the soil at the bottom of the excavation[①] in fine-grained soils (also termed **basal heave**). This limit state (denoted ULS-9) is essentially a reverse bearing capacity problem (involving unloading rather than loading of the ground), and is also described below.

① 基底隆起。

Determination of strut forces[②]

② 支撑轴力的确定。

Failure of a braced wall is normally due to the initial failure of one of the struts (i. e. at ULS-8), resulting in the progressive failure of the whole system[③]. The forces in the individual struts may differ widely because they depend on such random factors as the force with which the struts are lodged home and the time between excavation and installation of struts. The usual design procedure for braced walls is semiempirical, being based on actual measurements of strut loads in excavations in sands and clays in a number of locations. For example, Figure 9.37 (b) shows the apparent distributions of earth pressure derived from load measurements in the struts at three sections of a braced excavation in a dense sand. Since it is essential that no individual strut should fail, the pressure distribution assumed in design is taken as the envelope covering all the random distributions obtained from field measurements. Such an envelope[④] should not be thought of as representing the actual distribution of earth pressure with depth but as a hypothetical pressure diagram from which the maximum likely characteristic strut loads can be obtained with some degree of confidence.

③ 整个系统的渐近破坏。

④ 包络线。

Based on 81 case studies in a range of soils in the UK, Twine and Roscoe (1999) presented the pressure envelopes shown in Figure 9.37 (c) and (d). For soft and firm clays an envelope of the form shown in Figure 9.37 (c) is proposed for flexible walls[⑤] (i. e. sheet pile walls and timber sheeting) and, tentatively, for stiff walls[⑥] (i. e. diaphragm and contiguous pile walls, see Section 9.10). The upper and lower pressure values are represented by $a\gamma h$ and $b\gamma h$, respectively, where γ is the total unit weight of the soil and h the depth of the excavation, including an al-

⑤ 柔性挡土墙。

⑥ 刚性挡土墙。

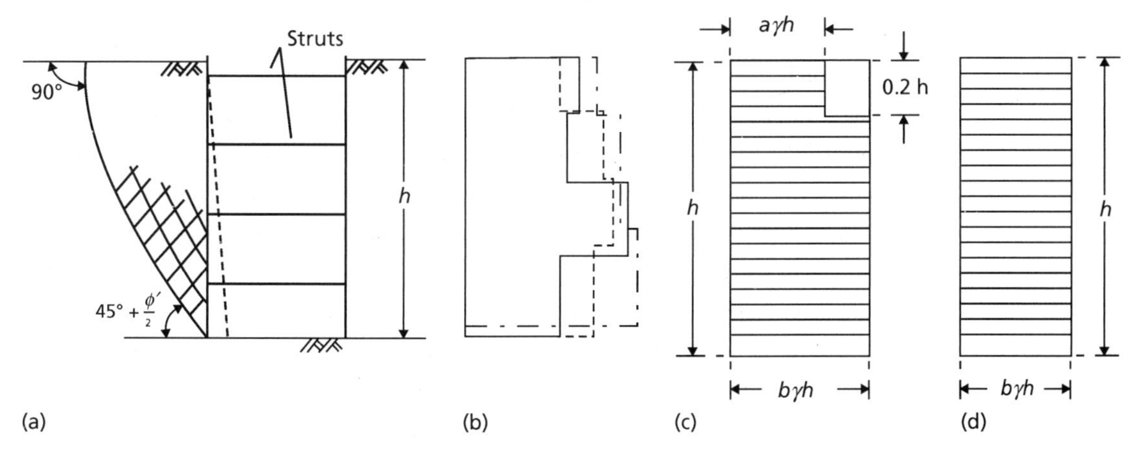

Figure 9.37 Earth pressure envelopes for braced excavations.

lowance for over-excavation. For soft clay with wall elements extending to the base of the excavation, $a = 0.65$ and $b = 0.50$. For soft clay with wall elements embedded below the base of the excavation, $a = 1.15$ and $b = 0.5$. For firm clay, the values of a and b are 0.3 and 0.2, respectively.

① 极硬黏土、硬黏土和粗粒土中应力包络线都是矩形。

The envelopes for stiff and very stiff clay and for coarse soils are rectangular① (Figure 9.37 (d)). For stiff and very stiff clays, the value of b for flexible walls is 0.3 and for stiff walls is 0.5. For coarse soils $b = 0.2$, but below the water table the pressure is $b\gamma' H$ (γ' being the buoyant unit weight) with hydrostatic pressure acting in addition.

② 开挖引起的超静负孔隙水压力的消散。

In clays, the envelopes take into account the increase in strut load which accompanies dissipation of the negative excess pore water pressures induced during excavation②. The envelopes are based on characteristic strut loads, the appropriate partial factor then being applied to give the design load for structural checks of the propping at ULS-8. The envelopes also allow for a nominal surface surcharge of 10kPa on the retained soil on either side of the excavation.

③ 基底隆起。

Basal heave③

④ 破坏机理。

Bearing capacity theory (Chapter 8) can be applied to the problem of base failure in braced excavations in fine-grained soils under undrained conditions (ULS-9). The application is limited to the analysis of cases in which the bracing is adequate to prevent significant lateral deformation of the soil adjacent to the excavation (i.e. when the other ULS conditions are satisfied). A simple failure mechanism④, originally proposed by Terzaghi (1943), is illustrated in Figure 9.38, the angle at a being 45° and bc being a circular arc in an undrained material; therefore, the length of ab is $(B/2)/\cos 45°$ (approximately $0.7B$).

Failure occurs when the shear strength of the soil is insufficient to resist the average shear stress resulting from the vertical pressure (p) on ac due to the weight of the soil ($0.7\gamma Bh$) plus any surcharge (σ_q) reduced by the shear strength on cd ($c_u h$). Thus,

$$p = \gamma h + \sigma_q - \frac{c_u h}{0.7B} \tag{9.77}$$

⑤ 地基承载力分析。

⑥ 上覆应力。

The problem is essentially that of a bearing capacity analysis⑤ in reverse, there being zero pressure at the bottom of the excavation and p representing the overburden pressure⑥. The shear strength available along the failure surface, acting in the opposite direction to that in the bearing capacity problem, can be expressed as $p_f = c_u N_c$ (Equation 8.17). Thus, for limiting equilibrium (i.e. to satisfy ULS-9),

$$p \leqslant p_f \tag{9.78}$$

Equation 9.78 may be solved for h to find the maximum depth of excavation for ULS-9 to be satisfied. In applying partial factors, p is the action, while p_f is the resistance, and these should be factored accordingly. If a firm stratum were to exist at depth D_f below the base of the excavation, where $D_f < 0.7B$, then D_f replaces $0.7B$ in Equation 9.77.

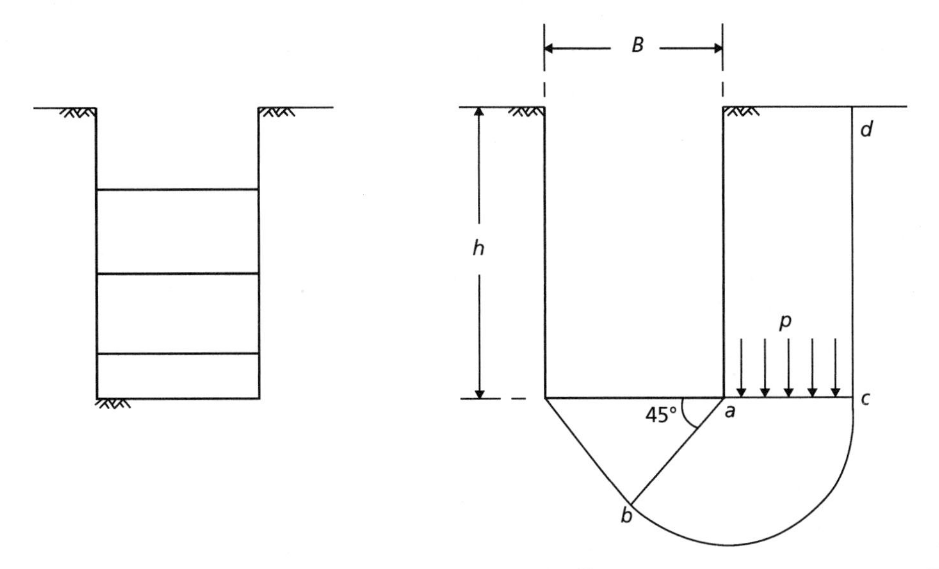

Figure 9.38 Base failure in a braced excavation.[①]

① 基坑支护中的基底破坏。

Based on observations of actual base failures in Oslo, Bjerrum and Eide (1956) concluded that Equation 9.77 gave reliable results only in the case of excavations with relatively low depth/breadth (h/B)[②] ratios. In the case of excavations with relatively large depth/breadth ratios, local failure occurred before shear failure on *cd* was fully mobilised up to surface level, such that

$$p = \gamma h + \sigma_q \quad (9.79)$$

i.e. applying Equation 8.17 with $q_f = 0$.

② 深宽比。

Where there is a possibility that the base of the excavation will fail by heaving, this should be analysed before the strut loads are considered. Due to basal heave and the inward deformation of the clay[③], there will be horizontal and vertical movement of the soil outside the excavation. Such movements may result in damage to adjacent structures and services, and should be monitored during excavation; advance warning of excessive movement or possible instability can thus be obtained.

③ 基底隆起导致黏土土体向坑内侧变形。

SLS design of braced excavations[④]

④ 正常使用极限状态下有支撑基坑的设计。

Diaphragm or piled walls[⑤] are commonly used in braced excavations. In general, the greater the flexibility of the wall system and the longer the time before struts or anchors are installed, the greater will be the movements outside the excavation[⑥]. Settlement of the retained soil beside the excavation (SLS-1) is usually critical when braced excavations are made in urban areas, where differential settlement behind the wall may affect adjacent structures or services. Analysis of the serviceability limit state for retaining structures is difficult, typically requiring complex numerical analyses using a large body of soil parameters and requiring extensive validation. It is therefore preferable in design to use empirical methods based on observations of wall movements in successful excavations[⑦]. Fortunately, databases of many tens of case histories now exist in a range of materials (for example Gaba *et al.* (2003); see

⑤ 地下连续墙或桩墙。

⑥ 墙体重度越大，且支撑或锚杆施加越晚，则基坑墙后土体的位移越大。

⑦ 采用基于开挖成功的墙体变形统计的经验方法。

Further reading), and limiting envelopes are summarised in Figure 9. 39. In this figure, x is the distance behind the excavation (normalised by the excavation depth, h) and s_g is the settlement of the ground surface (again, normalised by h).

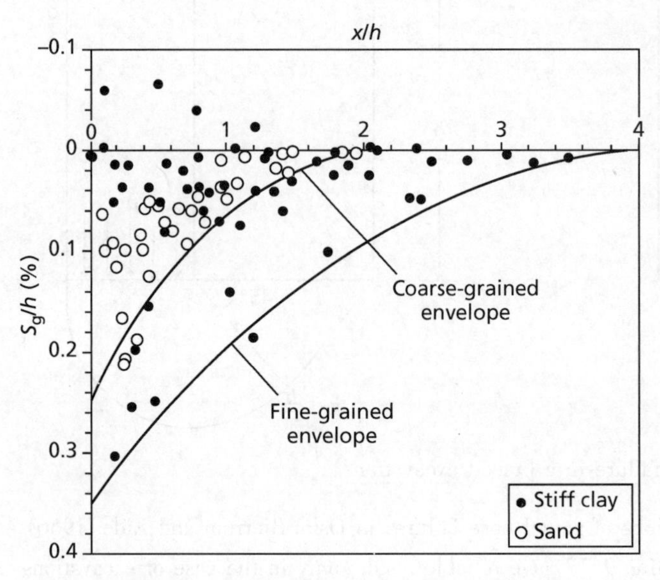

Figure 9. 39 Envelopes of ground settlement behind excavations.

The magnitude and distribution of the ground movements depend on the type of soil, the dimensions of the excavation, details of the construction procedure, and the standard of workmanship. Ground movements should be monitored during excavation and compared with the limits shown in Figure 9. 39 so that advance warning of excessive movement or possible instability① can be obtained. Assuming comparable construction techniques and workmanship, the magnitude of settlement adjacent to an excavation is likely to be relatively small in dense cohesionless soils but can be excessive in soft plastic clays.

① 过大变形或可能失稳破坏的提前预警。

9. 10 Diaphragm walls

A **diaphragm wall**② is a relatively thin reinforced concrete membrane cast in a trench, the sides of which are supported prior to casting by the hydrostatic pressure of a slurry of **bentonite**③ (a montmorillonite clay) in water. Stability of the trench during the excavation and casting phase is described in more detail in Section 10. 2. When mixed with water, bentonite readily disperses to form a colloidal suspension which exhibits thixotropic properties – i. e. it gels when left undisturbed but becomes fluid when agitated. The trench, the width of which is equal to that of the wall, is excavated progressively in suitable lengths (known as **panels**)④ from the ground surface, as shown in Figure 9. 40 (a), generally using a powerclosing clamshell grab; shallow concrete guide walls are normally constructed as an aid to excavation. The trench is filled with the bentonite slurry as excavation proceeds; excavation thus takes place through the slurry already in place. The excavation process

② 地下连续墙。

③ 膨润土。

④ 连续墙幅长。

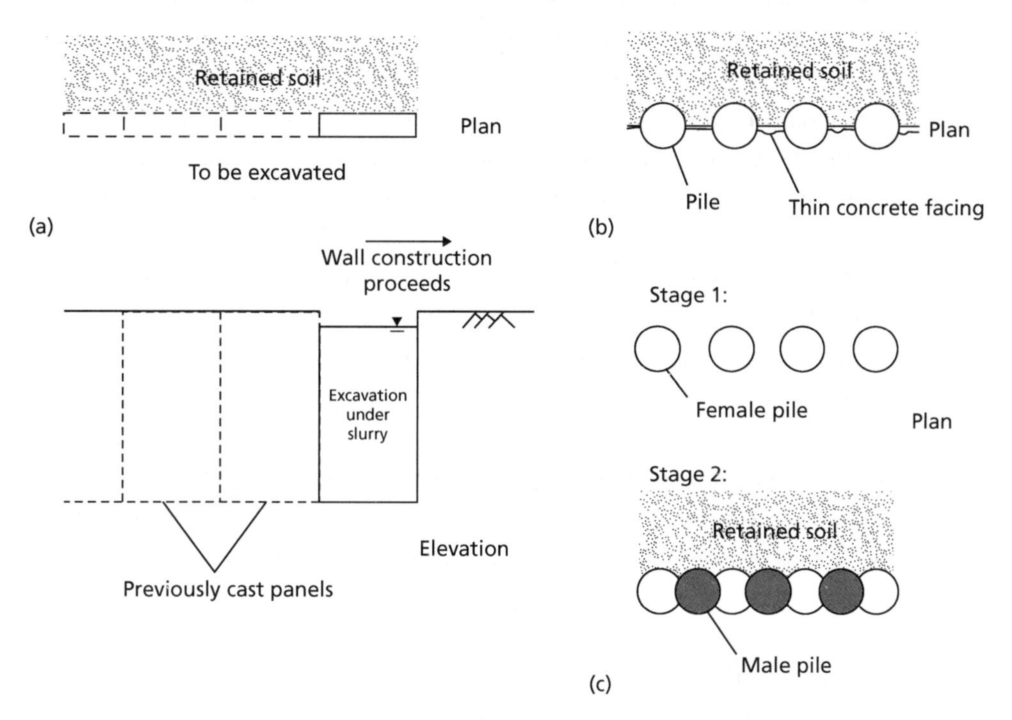

Figure 9.40 (a) Diaphragm wall, (b) contiguous pile wall, (c) secant pile wall.

turns the gel into a fluid, but the gel becomes reestablished when disturbance ceases. The slurry tends to become contaminated with soil and cement in the course of construction, but can be cleaned and re-used.

The bentonite particles form a skin of very low permeability①, known as the **filter cake**②, on the excavated soil faces. This occurs due to the fact that water filters from the slurry into the soil, leaving a layer of bentonite particles, a few millimetres thick, on the surface of the soil. Consequently, the full hydrostatic pressure of the slurry acts against the sides of the trench, enabling stability to be maintained. The filter cake will form only if the fluid pressure in the trench is greater than the pore water pressure in the soil; a high water table level can thus be a considerable impediment to diaphragm wall construction. In soils of low permeability, such as clays, there will be virtually no filtration of water into the soil and therefore no significant filter-cake formation will take place; however, total stress conditions apply, and slurry pressure will act against the clay. In soils of high permeability, such as sandy gravels, there may be excessive loss of bentonite into the soil, resulting in a layer of bentonite-impregnated soil and poor filter-cake formation. However, if a small quantity of fine sand (around 1%) is added to the slurry the sealing mechanism in soils of high permeability can be improved, with a considerable reduction in bentonite loss. Trench stability③ depends on the presence of an efficient seal on the soil surface; the higher the permeability of the soil, the more vital the efficiency of the seal becomes.

① 低渗透性的泥皮。
② 滤饼。
③ 沟槽的稳定性。

A slurry having a relatively high density is desirable from the points of view of trench stability, reduction of loss into soils of high permeability, and the retention of contaminating particles in suspension. On the other hand, a slurry of relatively low density will be displaced more cleanly from the soil surfaces and the reinforcement, and will be more easily pumped and decontaminated. The specification for the slurry must reflect a compromise between these conflicting requirements. Slurry specifications① are usually based on density, viscosity②, gel strength and pH.

① 泥浆要求。
② 黏度。

On completion of excavation, the reinforcement cage is positioned and the panel is filled with wet concrete using a **tremie pipe**③ which is dropped through the slurry to the base of the excavation. The wet concrete (which has a density of approximately twice that of the slurry) displaces the slurry upwards from the bottom of the trench, the tremie being raised in stages as the level of concrete rises. Once the wall (constructed as a series of individual panels keyed together) has been completed and the concrete has achieved adequate strength, the soil on one side of the wall can be excavated. It is usual for ground anchors or props to be installed at appropriate levels, as excavation proceeds, to tie the wall back into the retained soil (Sections 9.7 – 9.9). The method is very convenient for the construction of deep basements and underpasses, an important advantage being that the wall can be constructed close to adjoining structures, provided that the soil is moderately compact and ground deformations are tolerable. Diaphragm walls are often preferred to sheet pile walls because of their relative rigidity and their ability to be incorporated as part of the final structure.

③ 下料导管。

An alternative to a diaphragm wall is the **contiguous pile wall**④, in which a row of bored piles form the wall, relying on arching between piles to support the retained soil. A thin concrete (or other) facing may be applied to the soil between piles to prevent erosion⑤ which could lead to progressive failure. A **secant pile wall**⑥ is similar to a contiguous wall though two sets of piles are installed to overlap with each other, forming a continuous structure. These are normally achieved by installing an initial row of cast-in-situ bored concrete piles in the ground at a centre-to-centre spacing of less than one diameter. These are known as ‘female’ piles. A lower strength concrete (or preferably, a higher strength concrete that has slow strength gain) is used in these piles. Once the row is complete, further ‘male’ piles are installed in the spaces between piles. As there is less than one pile diameter of spacing available, boring of the second set of piles will partially drill through the existing piles (hence the need for the concrete to have a low strength at this time). As the wall is formed from a series of overlapping circular piles, the resulting wall will have a ribbed surface; it may therefore be necessary to provide additional facing on the wall if it is to be used as part of the final structure.

④ 连续桩墙。
⑤ 侵蚀。
⑥ 咬合桩。

The decision whether to use a triangular or a trapezoidal distribution of lateral pressure in the design of a diaphragm wall depends on the anticipated wall deformation⑦. A triangular distribution (Section 9.7) would probably be indicated in the case of a single row of tie-backs or props near the top of the wall. In the case of

⑦ 由预期的墙体变形，决定用三角形还是梯形分布的侧向压力来设计地下连续墙。

multiple rows of tie-backs or props over the height of the wall, the trapezoidal distributions shown in Figure 9.37 might be considered more appropriate.

9.11 Reinforced soil

Reinforced soil consists of a compacted soil mass within which tensile reinforcing elements, typically in the form of horizontal steel strips, are embedded. (Patents for the technique were taken out by Henri Vidal and the Reinforced Earth Company, the term **reinforced earth**[①] being their trademark.) Other forms of reinforcement include strips, rods, grids and meshes of metallic or polymeric materials, and sheets of **geotextiles**[②]. The mass is stabilised as a result of interaction between the soil and the elements, the lateral stresses within the soil being transferred to the elements which are thereby placed in tension. The soil used as fill material should be predominantly coarse-grained, and must be adequately drained to prevent it from becoming saturated. In coarse fills, the interaction is due to frictional forces which depend on the characteristics of the soil together with the type and surface texture[③] of the elements. In the cases of grid and mesh reinforcement, interaction is enhanced by interlocking[④] between the soil and the apertures in the material.

① 加筋土。

② 土工织物。

③ 表面特性。

④ 咬合作用。

In a reinforced-soil retaining structure (also referred to as a **composite wall**[⑤]), a facing is attached to the outside ends of the elements to prevent the soil spilling out at the edge of the mass and to satisfy aesthetic requirements[⑥]; the facing does not act as a retaining wall. The facing should be sufficiently flexible to withstand any deformation of the fill. The types of facing normally used are discrete precast concrete panels, full-height panels and pliant U-shaped sections aligned horizontally. The basic features of a reinforced-soil retaining wall are shown in Figure 9.41. Such structures possess considerable inherent flexibility, and consequently can withstand relatively large differential settlement[⑦]. The reinforced soil principle can also be employed in embankments, normally by means of geotextiles, and in slope

⑤ 复合墙体。

⑥ 满足美观要求。

⑦ 差异沉降。

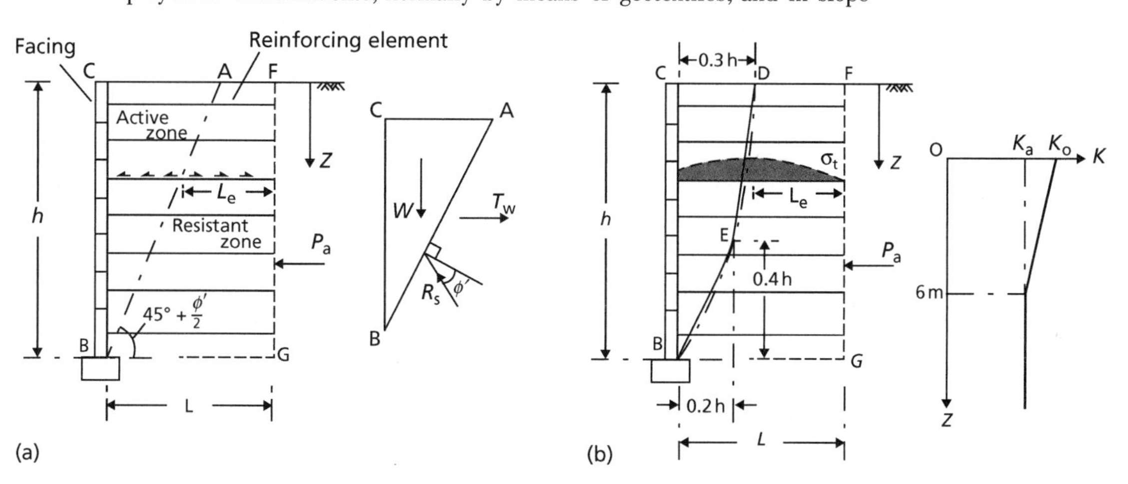

Figure 9.41 Reinforced soil-retaining structure: (a) tie-back wedge method, (b) coherent gravity method.

① 土钉。
② 外部和内部稳定性。

stabilisation by the insertion of steel rods – a technique known as soil nailing[①].

Both external and internal stability[②] must be considered in design. The external stability of a reinforced soil structure is usually analysed using a limit equilibrium approach (Section 9.5). The back of the wall should be taken as the vertical plane (FG) through the inner end of the lowest element, the total active thrust (P_a) due to the backfill behind this plane then being calculated as for a gravity wall. The ultimate limit states for external stability are:

③ 下卧层承载力模式破坏，造成结构倾斜。
④ 下卧层与加筋土间的滑动。
⑤ 深层滑动面。

> ULS-1: bearing resistance failure of the underlying soil, resulting in tilting of the structure[③];
> ULS-2: sliding between the reinforced fill and the underlying soil[④]; and
> ULS-3: development of a deep slip surface[⑤] which envelops the structure as a whole.

ULS-1 is verified using the methods described in Section 9.4 for gravity walls; the methods required for verification of ULS-3 will be described in Chapter 10. The remainder of this section will focus on the verification of ULS-2. Serviceability limit states are excessive values of settlement and wall deformation as for the other classes of retaining structure considered previously.

⑥ 筋体的受拉破坏。
⑦ 土与筋体间的滑动。

In considering the internal stability of the structure, the principal limit states are tensile failure of the elements[⑥] and slipping between the elements and the soil.[⑦] Tensile failure of one of the elements could lead to the progressive collapse of the whole structure (an ultimate limit state). Local slipping due to inadequate frictional resistance would result in a redistribution of tensile stress and the gradual deformation of the structure, not necessarily leading to collapse (i. e. a serviceability limit state).

Consider a reinforcing element at depth z below the surface of a soil mass. The tensile force in the element due to the transfer of lateral stress from the soil is given by

$$T = K\sigma_z S_x S_z \tag{9.80}$$

where K is the appropriate earth pressure coefficient at depth z, σ_z the vertical stress, S_x the horizontal spacing of the elements and S_z the vertical spacing. If the reinforcement consists of a continuous layer, such as a grid, then the value of S_x is unity, and T is the tensile force per unit length of wall. The vertical stress σ_z is due to the overburden pressure at depth z plus the stresses due to any surcharge loading and external bending moment (including that due to the total active thrust on the part of plane FG between the surface and depth z). The average vertical stress can be expressed as

$$\sigma_Z = \frac{V}{L - 2e}$$

where V is the vertical component of the resultant force at depth z, e the eccentricity of the force and L the length of the reinforcing element at that depth. Given the design tensile strength of the reinforcing material, the required cross-sectional area or thickness of the element can be obtained from the value of T. The addition of surcharge loading at the surface of the retained soil will cause an increase in vertical stress which can be calculated from elastic theory.

There are two procedures for the design of retaining structures. One approach is the **tie-back wedge method**[1] which is applicable to structures with reinforcement of relatively high extensibility, such as grids, meshes and geotextile sheets. This method is an extension of Coulomb's method, and considers the forces acting on a wedge of soil from which a force diagram can be drawn The active state is assumed to be reached throughout the soil mass because of the relatively large strains possible at the interface between the soil and the reinforcement; therefore the earth pressure coefficient in Equation 9.80 is taken as K_a at all depths and the failure surface at collapse will be a plane AB inclined at $45° + \phi'/2$ to the horizontal, as shown in Figure 9.41 (a), dividing the reinforced mass into an active zone[2] within which shear stresses on elements act outwards towards the face of the structure and a resistant zone[3] within which shear stresses act inwards. The frictional resistance available on the top and bottom surfaces of an element is then given by

① 锚固楔体法。

② 主动区。

③ 抗力作用区。

$$T_f = 2bL_e\sigma_z\tan\delta' \tag{9.81}$$

where b is the width of the element, L_e the length of the element in the resistant zone and δ' the angle of friction between soil and element. Slippage between elements and soil (referred to as **bond failure**[4]) will not occur if T_f is greater than or equal to T. The value of δ' can be determined by means of direct shear tests or full-scale pullout tests[5].

④ 黏结破坏。

⑤ 全尺寸抗拔试验。

The stability of the wedge ABC is checked in addition to the external and internal stability of the structure as a whole. The forces acting on the wedge, as shown in Figure 9.41 (a), are the weight of the wedge (W), the reaction on the failure plane (R) acting at angle ϕ' to the normal (being the resultant of the normal and shear forces), and the total tensile force resisted by all the reinforcing elements (T_w). The value of T_w can thus be determined. In effect, the force T_w replaces the reaction P of a retaining wall (as, for example, in Figure 9.21 (a)). Any external forces must be included in the analysis, in which case the inclination of the failure plane will not be equal to $45° + \phi'/2$ and a series of trial wedges must be analysed to obtain the maximum value of T_w. The design requirement is that the factored sum of the tensile forces in all the elements, calculated from Equation 9.81, must be greater than or equal to T_w to satisfy ULS-2.

The second design procedure is the **coherent gravity method**[6], due to Juran and Schlosser (1978), and is applicable to structures with elements of relatively low extensibility[7], such as steel strips. Experimental work indicated that the distribution of tensile stress[8] (σ_t) along such an element was of the general form shown in Figure 9.41 (b), the maximum value occurring not at the face of the structure but at a point within the reinforced soil, the position of this point varying with depth as indicated by the curve DB in Figure 9.41 (b). This curve again divides the reinforced mass into an active zone and a resistant zone, the method being based on the stability analysis of the active zone. The assumed mode of failure is that the reinfor-

⑥ 黏结重力法。

⑦ 低延性加筋体。

⑧ 拉应力分布形式。

cing elements fracture progressively at the points of maximum tensile stress and, consequently, that conditions of plastic equilibrium develop in a thin layer of soil along the path of fracture. The curve of maximum tensile stress therefore defines the potential failure surface. If it is assumed that the soil becomes perfectly plastic, the failure surface will be a logarithmic spiral①. The spiral is assumed to pass through the bottom of the facing and to intersect the surface of the fill at right angles, at a point approximately 0.3h behind the facing, as shown in Figure 9.41 (b). A simplified analysis can be made by assuming that the curve of maximum tensile stress is represented by the bilinear approximation DEB shown in Figure 9.41 (b), where CD = 0.3h. Equations 9.80 and 9.81 are then applied. The earth pressure coefficient in Equation 9.80 is assumed to be equal to K_0 (the at-rest coefficient)② at the top of the wall, reducing linearly to K_a at a depth of 6m. The addition of surface loading would result in the modification of the line of maximum tensile stress, and for this situation an amended bilinear approximation is proposed in BS 8006 (BSI, 1995), which provides valuable guidance on the deign of reinforced soil constructions to complement EC7.

① 对数螺旋。

② 静止土压力系数。

Summary

1 Lower bound limit analysis③ may be used to determine the limiting lateral earth pressures acting on retaining structures in homogeneous soil conditions either directly (in undrained materials) or via lateral earth pressure coefficients K_a and K_p (in drained material). If the retaining structure moves away from the retained soil at failure, active earth pressures will be generated, while larger passive earth pressures act on structures which move into the retained soil at failure. These limit analysis techniques can account for a battered and/or rough soil-wall interface and sloping retained soil. Limit equilibrium techniques④ may alternatively be used which consider the equilibrium of a wedge of failing soil behind the retaining structure, and can additionally be used with flow-net/FDM techniques from Chapter 2 to analyse problems where seepage is occurring in the retained soil.

2 In-situ horizontal stresses within the ground are directly related to the vertical effective stresses (calculated using methods from Chapter 3) via the coefficient of lateral earth pressure at rest (K_0). This can be determined analytically for any soil based on its drained friction angle ϕ' and overconsolidation ratio (OCR). K_0 conditions apply when there is no lateral strain in the soil mass. Under extension, earth pressures will reduce to active values ($K_a < K_0$); under compression, earth pressures will increase to passive values ($K_p > K_0$).

③ 极限分析下限法。

④ 极限平衡法。

3 Surcharge loading[①] on the surface of retained soil may be accounted for by modifying the vertical total or effective stress used in lateral earth pressure calculations in undrained and drained materials, respectively. If the wall is constructed before the soil it retains is placed (a backfilled wall), compaction induced lateral stresses may be induced along the wall.

4 Gravity retaining structures[②] rely on their mass to resist lateral earth pressures. The key design criteria for these structures is maintaining overall stability[③] (ULS). Earth pressure forces may be determined from the limiting earth pressures (point 1) and used to check the stability of the wall in bearing, sliding and overturning using the yield surface approach. Embedded walls are flexible, and resist lateral earth pressures from a balance of lateral earth pressure forces from the soil behind the wall acting to overturn or translate the wall (active) and those resisting failure from the soil in front of the wall (passive). ULS design may be accomplished using the free-earth support method. Ground anchors or propping may be used to provide additional support, and this must also be designed to resist structural or pull-out failure. If a flexible wall is braced more heavily it is described as a braced excavation, the operative earth pressures on such a construction being much lower. These earth pressures are used to structurally design the bracing system avoiding progressive collapse. Basal heave[④] may also occur in braced excavations, and must also be designed against. In flexible walls movements are also significant, imposing additional SLS criteria in design.

① 地表超载。
② 重力式挡墙。
③ 整体稳定。
④ 基底隆起。

Problems

9.1 The backfill behind a smooth retaining wall, located above the water table, consists of a sand of unit weight 17kN/m^3. The height of the wall is 6m and the surface of the backfill is horizontal. Determine the total active thrust on the wall if $c' = 0$ and $\phi' = 37°$. If the wall is prevented from yielding, what is the approximate value of the thrust on the wall?

9.2 Plot the distribution of active pressure on the wall surface shown in Figure 9.42. Calculate the total thrust on the wall (active + hydrostatic) and determine its point of application. Assume $\delta' = 0$ and $\tau_w = 0$.

9.3 A line of sheet piling is driven 4m into a firm clay and retains, on one side, a 3-m depth of fill on top of the clay. The water table is at the surface of the clay. The unit weight of the fill is 18kN/m^3, and the saturated unit weight of the clay is 20kN/m^3. Calculate the active and passive pressures at the lower end of the sheet piling (a) if $c_u = 50$kPa, $\tau_w = 25$kPa and $\phi_u = \delta' = 0$, and (b) if $c' = 0$, $\phi' = 26°$ and $\delta' = 13°$, for the clay.

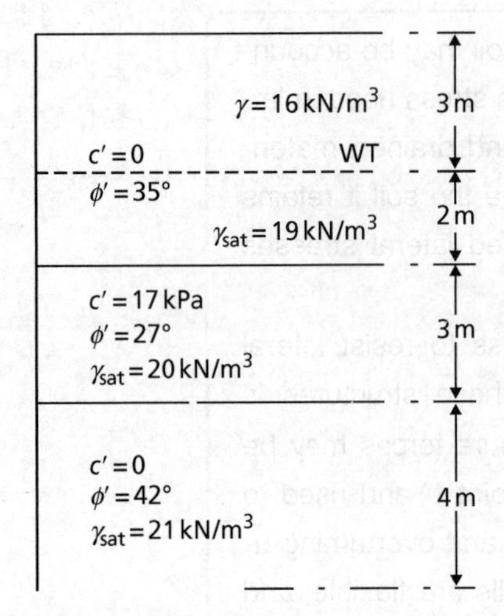

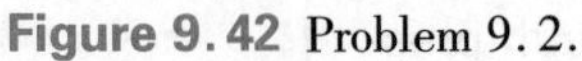
Figure 9.42 Problem 9.2.

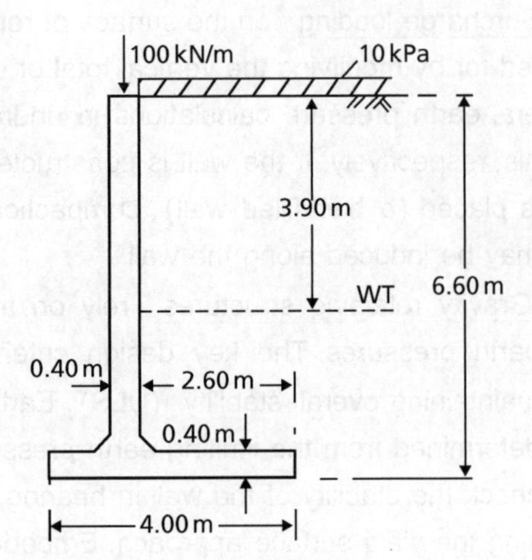

Figure 9.43 Problem 9.4.

9.4 Details of a reinforced concrete cantilever retaining wall are shown in Figure 9.43, the unit weight of concrete being 23.5kN/m^3. Due to inadequate drainage, the water table has risen to the level indicated. Above the water table the unit weight of the retained soil is 17kN/m^3, and below the water table the saturated unit weight is 20kN/m^3. Characteristic values of the shear strength parameters are $c'=0$ and $\phi'=38°$. The angle of friction between the base of the wall and the foundation soil is 25°. Check whether or not the overturning and sliding limit states have been satisfied to EC7 DA1b.

9.5 The section through a gravity retaining wall is shown in Figure 9.44, the unit weight of the wall material being 23.5kN/m^3. The unit weight of the backfill is 19kN/m^3, and design values of the shear strength parameters are $c'=0$ and $\phi'=36°$. The value of δ' between wall and backfill and between base and foundation soil is 25°. The ultimate bearing capacity of the foundation soil is 250kPa. Determine if the design of the wall is satisfactory with respect to the overturning, bearing resistance and sliding limit states, to EC7 DA1a.

Figure 9.44 Problem 9.5.

9.6 The sides of an excavation 3.0m deep in sand are to be supported by a cantilever sheet pile wall. The water table is 1.5m below the bottom of the excavation. The sand has a saturated unit weight of 20kN/m^3 and a unit weight above the water table of 17kN/m^3, and the characteristic value of ϕ' is 36°. Determine the required depth of embedment of the piling below the bottom of the excavation if the excavation is to be designed to EC7 DA1b.

9.7 An anchored sheet pile wall is constructed by driving a line of piling into a soil for which the saturated unit weight is 21kN/m^3 and the characteristic shear strength parameters are $c'=10\text{kPa}$ and $\phi'=27°$. Backfill is placed to a depth of 8.00m behind the piling, the backfill having a saturated unit weight of 20kN/m^3, a unit weight above the water table of 17kN/m^3, and characteristic shear strength parameters of $c' = 0$ and $\phi' = 35°$. Tie rods, spaced at 2.5-m centres, are located 1.5m below the surface of the backfill. The water level in front of the wall and the water table behind the wall are both 5.00m below the surface of the backfill. Determine the required depth of embedment to EC7 DA1b and the design force in each tie rod.

9.8 The soil on both sides of the anchored sheet pile wall detailed in Figure 11.45 has a saturated unit weight of $21 kN/m^3$, and a unit weight above the water table of $18 kN/m^3$. Characteristic parameters for the soil are $c' = 0$, $\phi' = 36°$ and $\delta' = 0°$. There is a lag of 1.5m between the water table behind the wall and the tidal level in front. Determine the required depth of embedment to EC7 DA1a and the design force in the ties.

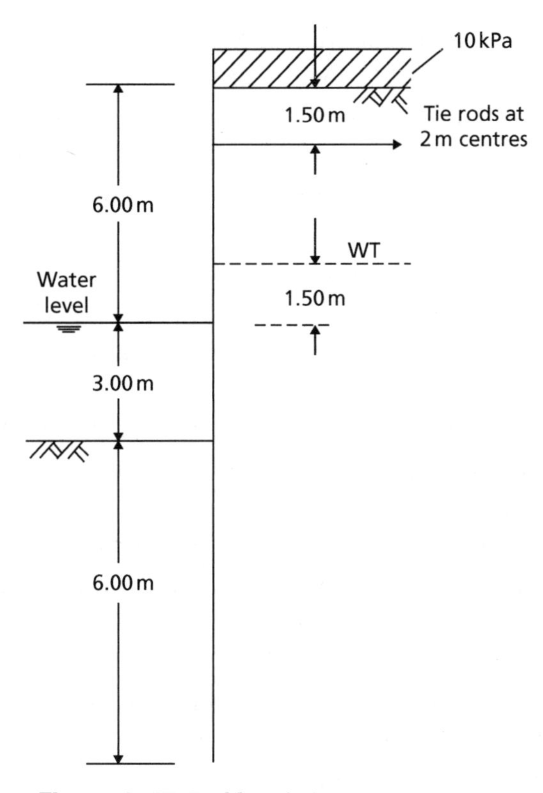

Figure 9.45 Problem 9.8.

9.9 A ground anchor in a stiff clay, formed by the gravel injection technique, has a fixed anchor length of 5m and an effective fixed anchor diameter of 200mm; the diameter of the borehole is 100mm. The relevant shear strength parameters for the clay are $c_u = 110 kPa$ and $\phi_u = 0$. What would be the expected characteristic ultimate load capacity of the anchor, assuming a skin friction coefficient of 0.6?

9.10 The struts in a braced excavation 9m deep in a dense sand are placed at 1.5-m centres vertically and 3.0-m centres horizontally, the bottom of the excavation being above the water table. The unit weight of the sand is $19 kN/m^3$. Based on design shear strength parameters $c' = 0$ and $\phi' = 40°$, what load should each strut be designed to carry? (Use EC7 DA1a.)

9.11 A long braced excavation in soft clay is 4m wide and 8m deep. The saturated unit weight of the clay is $20 kN/m^3$, and the undrained shear strength adjacent to the bottom of the excavation is given by $c_u = 40 kPa$, ($\phi_u = 0$). Determine the factor of safety against base failure of the excavation.

9.12 A reinforced soil wall is 5.2m high. The reinforcing elements, which are spaced at 0.65m vertically and 1.20m horizontally, measure 65 × 3mm in section and are 5.0m in length. The ultimate tensile strength of the reinforcing material is 340MPa. Design values to be used are as follows: unit weight of the selected fill = 18kN/m^3; angle of shearing resistance of selected fill = 36°; angle of friction between fill and elements = 30°. Using (a) the tie-back wedge method and (b) the coherent gravity method, check that an element 3.6m below the top of the wall will not suffer tensile failure, and that slipping between the element and the fill will not occur. The value of K_a for the material retained by the reinforced fill is 0.30 and the unit weight of this material is 18kN/m^3.

References

Berezantzev, V. G., Khristoforov, V. S. and Golubkov, V. N. (1961) Load bearing capacity and deformation of piled foundations, in *Proceedings of the 5th International Conference on Soil Mechanics and Foundation Engineering, Paris, France*, pp. 11 – 15.

Bjerrum, L. and Eide, O. (1956) Stability of strutted excavations in clay, *Géotechnique*, **6** (1), 32 – 47.

British Standard 8006 (1995) *Code of Practice for Strengthened Reinforced Soils and Other Fills*, British Standards Institution, London.

Coulomb, C. A. (1776). Essai sur une application des régeles des maximus et minimus a quelque problémes de statique rélatif à l'architecture, *Memoirs Divers Savants*, 7, Académie Sciences, Paris (in French).

EC7 – 1 (2004) *Eurocode 7: Geotechnical design – Part 1: General rules, BS EN 1997 – 1: 2004*, British Standards Institution, London.

Ingold, T. S. (1979) The effects of compaction on retaining walls, *Géotechnique*, **29**, 265 – 283.

Jaky, J. (1944). The coefficient of earth pressure at rest, *Journal of the Society of Hungarian Architects and Engineers*, Appendix 1, 78 (22) (transl.).

Juran, I. and Schlosser, F. (1978) Theoretical analysis of failure in reinforced earth structures, in *Proceedings of the Symposium on Earth Reinforcement, ASCE Convention, Pittsburgh*, pp. 528 – 555.

Mayne, P. W. (2007) *Cone Penetration Testing: A Synthesis of Highway Practice*, NCHRP Synthesis Report 368, Transportation Research Board, Washington DC.

Mayne, P. W. and Kulhawy, F. H. (1982) Ko-OCR (At rest pressure – Overconsolidation Ratio) relationships in soil, *Journal of the Geotechnical Engineering Division, ASCE*, **108** (GT6), 851 – 872.

Pipatpongsa, T., Takeyama, T., Ohta, H. and Iizuka, A. (2007) Coefficient of earth pressure at-rest derived from the Sekiguchi-Ohta Model, in *Proceedings of the 16th Southeast Asian Geotechnical Conference, Subang Jaya, Malaysia, 8 – 11 May*, pp. 325 – 331.

Potts, D. M. and Fourie, A. B. (1984) The behaviour of a propped retaining wall: results of a numerical experiment, *Géotechnique*, **34** (3), 383 – 404.

Potts, D. M. and Fourie, A. B. (1985) The effect of wall stiffness on the behaviour of a propped retaining wall, *Géotechnique*, **35** (3), 347 – 352.

Terzaghi, K. (1943) *Theoretical Soil Mechanics*, John Wiley & Sons, New York, NY.

Twine, D. and Roscoe, H. (1999) Temporary propping of deep excavations: guidance on design, *CIRIA Report C517*, CIRIA, London.

Further reading

Frank, R., Bauduin, C., Driscoll, R., Kavvadas, M., Krebs Ovesen, N., Orr, T. and Schuppener, B. (2004)

Designers' Guide to EN 1997 – 1 Eurocode 7: Geotechnical Design – General Rules, Thomas Telford, London.

This book provides a guide to limit state design of a range of constructions (including retaining walls) using Eurocode 7 from a designer's perspective and provides a useful companion to the Eurocodes when conducting design. It is easy to read and has plenty of worked examples.

Gaba, A. R., Simpson, B., Powrie, W. and Beadman, D. R. (2003) Embedded retaining walls – guidance for economic design, *CIRIA Report C580*, CIRIA, London.

This report provides valuable practical guidance on the selection of design and construction methodologies for flexible retaining structures, including procedural flowcharts. It also incorporates a large collection of case history data to inform future design.

Chapter 10

Stability of self-supporting soil masses[1]

① 自承式土体的稳定性。

Learning outcomes

After working through the material in this chapter, you should be able to:

1 Determine the stability of unsupported trenches[2], including those supported by slurry, and design these works within a limit state design framework (Eurocode 7);

2 Determine the stability of slopes, vertical cuttings and embankments[3], and design these works within a limit state design framework;

3 Determine the stability of tunnels and the ground settlements caused by tunnelling works, and use this information to conduct a preliminary design of tunnelling works within a limit state design[4] framework.

② 无支护沟槽

③ 路堤。

④ 极限状态设计。

10.1 Introduction

This chapter is concerned with the design of potentially unstable soil masses which have been formed through either human activity (excavation or construction) or natural processes (erosion and deposition). This class of problem includes slopes, embankments and unsupported excavations. Unlike the material in Chapter 9, however, the soil masses here are not supported by an external structural element such as a retaining wall[5]; rather, they derive their stability from the resistance of the soil within the mass in shear.

Gravitational and seepage forces[6] tend to cause instability in natural slopes, in slopes formed by excavation and in the slopes of embankments. A vertical cutting[7] (or trench, formed of two vertical cuttings) is a special case of sloping ground where the slope angle is 90° to the horizontal. Design of self-supporting soil systems is based on the requirement to maintain stability (ULS) rather than on the need to minimise deformation (SLS).[8] If deformation were such that the strain in an element of soil exceeded the value corresponding to peak strength, then the strength would fall towards the ultimate value. Thus, it is appropriate to use the

⑤ 挡土墙。
⑥ 渗流力。
⑦ 直立切削面。
⑧ 自承式土体的设计基于维持稳定性的需求（即承载力极限状态），而并非为了保证最小变形的要求（即正常使用极限状态）。

① 临界状态强度。
② 残余强度。

critical state strength[1] in analysing stability. However, if a pre-existing slip surface were to be present within the soil, use of the residual strength[2] would be appropriate.

Section 10. 2 will apply both limit analysis and limit equilibrium techniques to the stability of vertical cuts/trenches. These methods will then be extended to consider how fluid support may be used to improve the stability of such constructions (e. g. drilling of bored piles or excavation of diaphragm wall piles under slurry) . In Sections 10. 3 and 10. 4, the analytical methods will be further extended to the consideration of slope and embankment design, respectively. Finally, in Section 10. 5, an introduction to the design of tunnelling works will be considered, where the stability of a vertical cut face deep below the ground surface governs the design. This final section will also consider how the stability of tunnel headings[3] may be improved by pressurising the cut face (analogous to the use of drilling fluids in trench support) .

③ 隧道开挖面稳定性。

10. 2 Vertical cuttings and trenches

Vertical cuts in soil can only be supported when soil behaves in an undrained way (with an undrained strength c_u) or in a drained soil where there is some cohesion (c') . As, in the absence of chemical or other bounding[4] between soil particles, $c' = 0$, vertical cuts and trenches cannot normally be supported under drained conditions. This is because, from the Mohr-Coulomb strength definition[5] (Equation 5. 11), a drained cohesionless soil will always fail when the slope angle reaches ϕ'. Under undrained conditions, however, vertical cuts may be kept stable up to a certain limiting depth/height which depends on the undrained strength of the soil. This is very useful during temporary works in fine-grained soils (typically clays) which are fast enough that undrained conditions can be maintained. The excavation of trial pits/trenches and bored pile construction techniques are two examples of where this is used in engineering practice.

④ 注：此处原书有误，应为 bonding。
⑤ 莫尔-库仑强度准则。

Limiting height/depth using limit analysis[6]

⑥ 由极限分析确定的极限高度/深度。

Figure 10. 1 shows a simple upper bound failure mechanism UB-1 for a vertical cut in undrained soil having a unit weight γ. As the mechanism develops and the soil fails into the excavation/cut, work is input to the system from the potential energy recovered as the weight of the sliding block moves downwards with gravity. The vertical force due to the weight of the block (W) is given by

$$W = \frac{\gamma h^2}{2\tan\theta} \tag{10.1}$$

per metre length of the cut. If the block slides along the slip plane at velocity v, then the component in the vertical direction (in the direction of W) is $v\sin\theta$. From Equation 8. 2, the work input is then

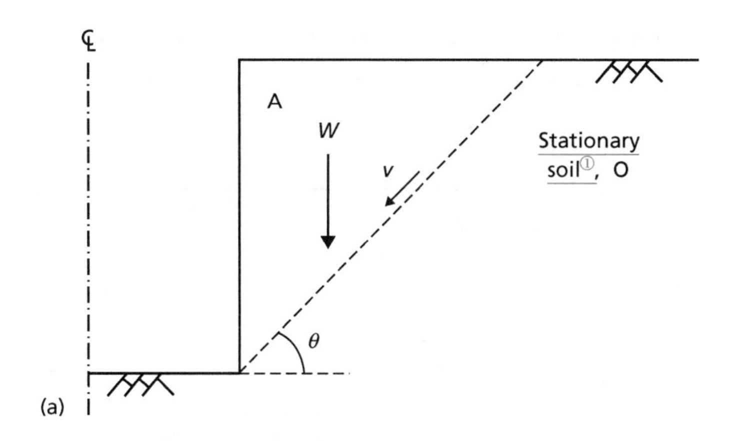

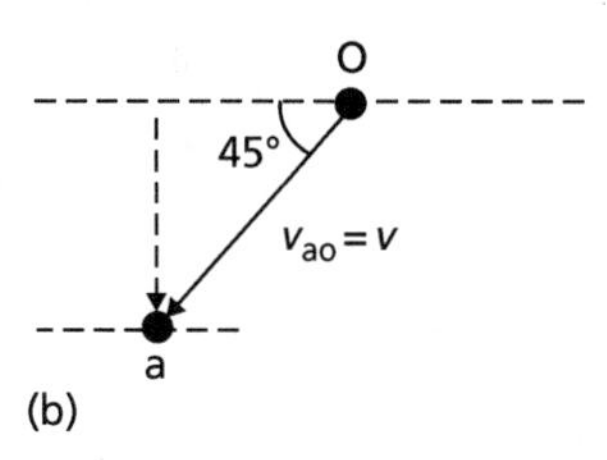

Figure 10.1 (a) Mechanism UB-1②, (b) hodograph.③

① 固定土体。
② UB-1 机理。
③ 速度矢量图。

$$\sum W_i = \frac{\gamma h^2 v\sin\theta}{2\tan\theta} = \frac{1}{2}\gamma h^2 v\cos\theta \tag{10.2}$$

As in Chapter 8, energy is dissipated in shear along the slip plane; the length of the slip plane $L_{OA} = h/\sin\theta$ per metre length of cut, the shear stress at plastic failure is c_u, and the slip velocity is v. Then, from Equation 8.1, the energy dissipated④ is

④ 消散的能量。

$$\sum E_i = \frac{c_u hv}{\sin\theta} \tag{10.3}$$

By the upper bound theorem⑤, if the system is at plastic collapse, the work done by the external loads/pressures must be equal to the energy dissipated within the soil, so from Equation 8.3:

⑤ 上限理论。

$$\sum W_i = \sum E_i$$

$$\frac{1}{2}\gamma h^2 v\cos\theta = \frac{c_u hv}{\sin\theta}$$

$$h = \frac{2c_u}{\gamma\sin\theta\cos\theta} \tag{10.4}$$

Equation 10.4 is a function of the angle θ. The cutting will fail when h is a minimum. The value of θ at which the minimum h occurs may be found from solving $dh/d\theta = 0$, giving $\theta = \pi/4$ (45°); substituting this value into Equation 10.4 gives $h \leqslant 4c_u/\gamma$ for stability.

A simple lower bound stress field⑥ for the vertical cutting is shown in Figure 10.2 (a). At plastic failure, the soil will move inwards towards the cutting, so the major principal stresses in the retained soil are vertical (active condition). Figure 10.2 (b) shows the Mohr circle for the soil; for undrained conditions, $\sigma_3 = \sigma_1 - 2c_u$ and $\sigma_1 = \sigma_v = \gamma z$. The horizontal stresses on the vertical boundary must then sum to zero for equilibrium,⑦ i.e.

⑥ 下限分析的应力场。

⑦ 垂直边界上的水平应力必须合力为零，从而满足平衡条件。

$$\int_0^h (\gamma z - 2c_u)\,dz = 0$$

$$\frac{1}{2}\gamma h^2 - 2c_u h = 0 \tag{10.5}$$

Solution of Equation 10.5 gives $h \leqslant 4c_u/\gamma$ for stability. This is the same as the upper bound, and therefore represents the true solution.

① 非摩阻型应力不连续面。

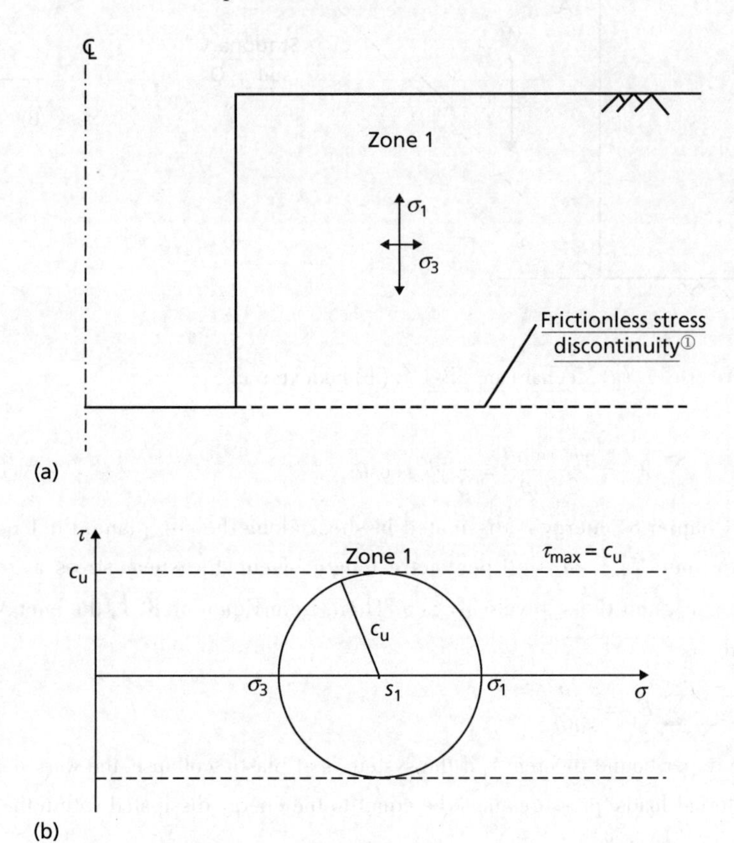

② 应力场 LB-1。

③ 应力莫尔圆。

Figure 10.2 (a) Stress field LB-1②; (b) Mohr circle.③

④ 由极限平衡法确定的极限高度/深度。

Limiting height/depth using limit equilibrium④ (LE)

The limit equilibrium (Coulomb) method considering a wedge of soil presented in Section 9.5 may also be used for assessing the stability of a vertical cut. Figure 10.3 shows a wedge at an angle θ in undrained soil. An additional force S is also included in this analysis to model the support provided by drilling fluid within a trench. The unit weight of the slurry is γ_s and that of the soil is γ, while the depth of the slurry is nh. The resultant resistance force along the slip plane (R_s from Section 9.5) is here split into normal and tangential components, denoted N and T, respectively. Considering force equilibrium

$$S + T\cos\theta - N\sin\theta = 0 \tag{10.6}$$

$$W - T\sin\theta - N\cos\theta = 0 \tag{10.7}$$

⑤ 泥浆护壁的作用力。

The resultant thrust from the slurry⑤ arises from the hydrostatic pressure distribution within the trench, i.e.

$$S = \int_0^{nh} (\gamma_s z)\,dz = \frac{1}{2}\gamma_s (nh)^2 \tag{10.8}$$

The weight of the wedge[1] is given by Equation 10.1 as before. The tangential force at failure is the shear strength of the soil multiplied by the slip plane area ($h/\sin\theta$ per metre length), i.e.

① 土楔。

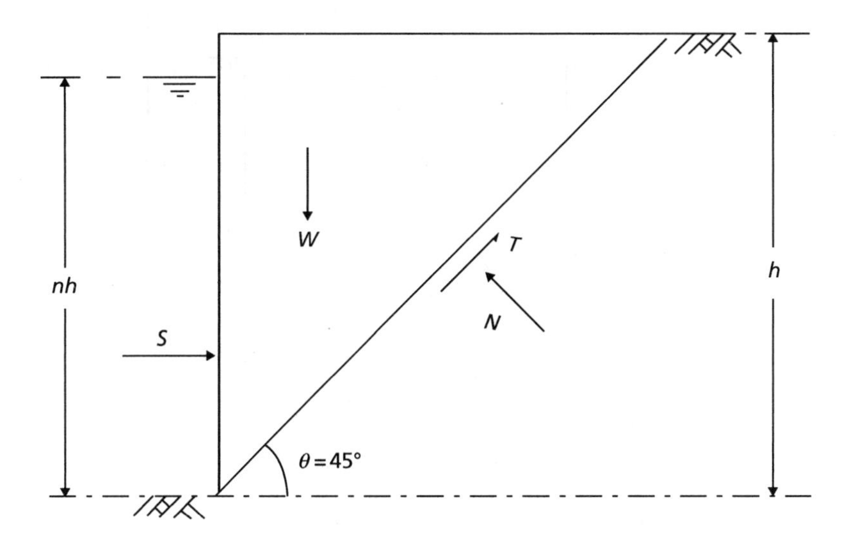

Figure 10.3 Stability of a slurry-supported trench in undrained soil.[2]

② 不排水土中泥浆护壁沟槽的稳定性。

$$T = c_u \cdot \frac{h}{\sin\theta} \tag{10.9}$$

Substituting for S and T in Equation 10.6 and rearranging gives

$$N = \frac{S}{\sin\theta} + \frac{c_u h}{\tan\theta\sin\theta} \tag{10.10}$$

Then, substituting for W, T and N in Equation 10.7 and rearranging gives

$$h = \frac{2c_u}{\gamma\sin\theta\cos\theta} \cdot \left[\frac{1}{1 - \left(\frac{\gamma_s}{\gamma}\right)n^2\tan\theta}\right] \tag{10.11}$$

If $\gamma_s = 0$ (i.e. there is no slurry in the trench), Equation 10.11 reduces to Equation 10.4.

As mentioned in Section 9.10, bentonite slurry[3] forms a filter cake[4] on the surface of the excavation allowing full hydrostatic pressures to be maintained, even against drained materials with no cohesion. Under drained conditions, the limit equilibrium analysis presented above may be modified to perform an effective stress analysis, accounting for the presence of the water table[5] (at a height mh above the bottom of the trench) as shown in Figure 10.4.

③ 膨润土泥浆。

④ 滤饼。

⑤ 地下水位。

Equations 10.6-10.8 may be used unchanged; however, the total shear resistance force T is now based on the effective stress along the slip plane, i.e.

$$T = c'\left(\frac{h}{\sin\theta}\right) + (N - U)\tan\phi' \tag{10.12}$$

where U, the boundary water force on the failure plane, is given by

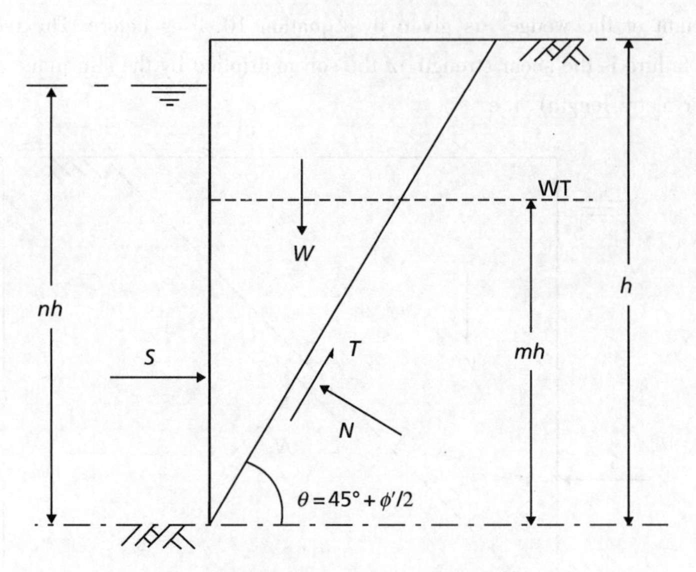

Figure 10.4 Stability of a slurry-supported trench in drained soil.[①]

① 排水土中泥浆护壁沟槽的稳定性。

$$U = \frac{1}{2}\gamma_w (mh)^2 \operatorname{cosec}\theta \tag{10.13}$$

② 当土楔体即将滑入沟槽时，即楔体内的土处于运动极限平衡条件，角度 θ 可假设为 45° + φ′/2。

③ 归一化后的安全开挖深度。

④ 冰碛土。

⑤ 裂隙黏土。

When the wedge is on the point of sliding into the trench, i. e. the soil within the wedge is in a condition of active limiting equilibrium, the angle θ can be assumed to be $45° + \phi'/2$[②]. The equations are solved using the same procedure as before; however the closed-form solution for the drained case is more complex than Equation 10.11. Instead, the equations may be straightforwardly programmed into a spreadsheet.

Figure 10.5 (a) plots the normalised safe depth of excavation[③] h as a function of slurry unit weight for excavation in an undrained soil. For the case of no slurry (unsupported excavation, $\gamma_s = 0$), $h = 4c_u/\gamma$ as before. In order to maintain workability for excavation, fresh slurry will typically have a density of 1150kg/m^3 (γ_s = 11.3kN/m^3). The data points in Figure 10.5 (a) represent the maximum depths of excavation for some of the clays described in Chapters 5 and 7, namely the NC organic clay at Bothkennar, the glacial till[④] at Cowden and the fissured clay[⑤] at Madingley. The value of γ_s/γ for each of these clays is based on typical unit weights of 15.5, 21.5 and 19.5kN/m^3, respectively, and that of fresh slurry given above. It will be seen that excavation under slurry is particularly beneficial in NC soils, where $h = 14c_u/\gamma$ may be achieved (i. e. three and a half times the depth of an unsupported excavation). Even in the heavier clays, the excavation depth can be at least doubled by using slurry support.

Figure 10.5 (b) plots the minimum slurry density required to avoid collapse as a function of the normalised water table height m in drained soil. It can be seen that excavation in such soils will only be problematic for situations where the water table is close to the ground surface.[①] As a result, when installing bored piles in drained materials (e. g. sands) it is common to use steel casing[②] towards the top of the excavation to prevent collapse.

① 可以认为，在这种土中开挖，问题只会出现在地下水位接近地表时。

② 钢护筒。

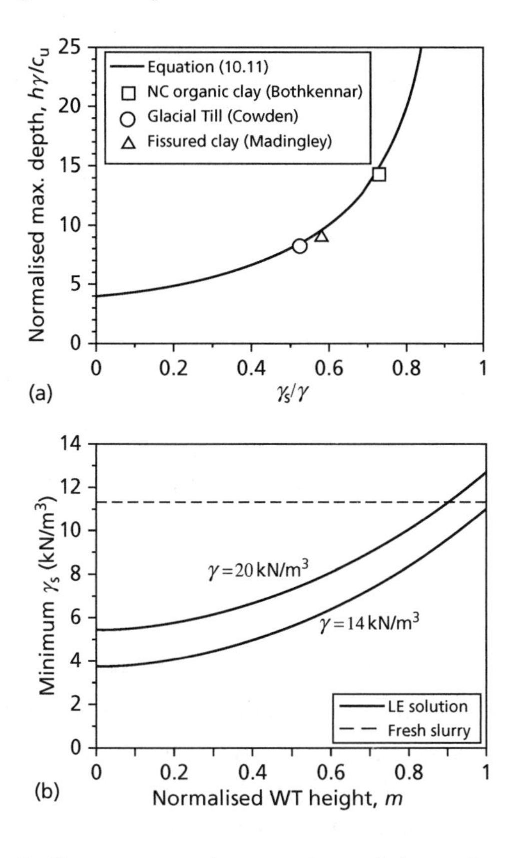

Figure 10. 5 Slurry-supported excavations: (a) maximum excavation depth in undrained soil[③], (b) minimum slurry density to avoid collapse in drained soil[④] ($\phi' = 35°$, $n = 1$).

③ 不排水土中的最大开挖深度。

④ 在排水土中为防止崩塌的最小泥浆密度。

10.3 Slopes

The most important types of slope failure are illustrated in Figure 10. 6. In **rotational slips**[⑤], the shape of the failure surface in section may be a circular arc or a non-circular curve. In general, circular slips are associated with homogeneous, isotropic[⑥] soil conditions, and non-circular slips with non-homogeneous conditions. **Translational** and **compound** slips[⑦] occur where the form of the failure surface is influenced by the presence of an adjacent stratum of significantly different strength, most of the failure surface being likely to pass through the stratum of lower shear strength. The form of the surface would also be influenced by the presence of discontinuities such as fissures and pre-existing slips.[⑧] Translational slips tend to occur where the adjacent stratum is at a relatively shallow depth below the surface of the slope, the failure surface tending to be plane and roughly parallel to the slope

⑤ 旋转滑动。

⑥ 均匀的，各向同性的。

⑦ 平动和组合滑动。

⑧ 破坏界面的形式受到不连续性的影响，如裂隙和之前已发生的滑动面。

surface. Compound slips usually occur where the adjacent stratum is at greater depth, the failure surface consisting of curved and plane sections. In most cases, slope stability can be considered as a two-dimensional problem, conditions of plane strain being assumed.

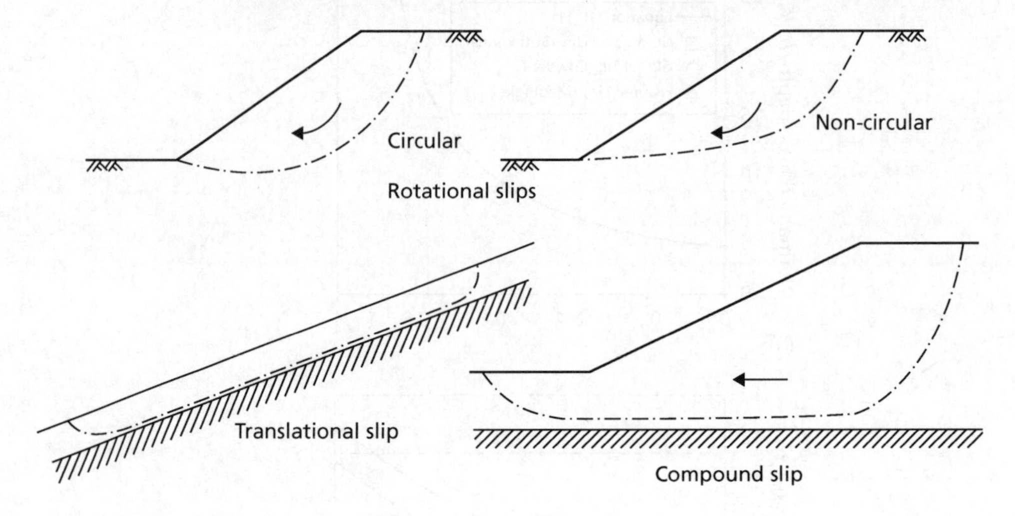

Figure 10.6 Types of slope failure.

An example of a rotational slip occurred 3-5 June 1993 at Holbeck, Yorkshire. Pore water pressure build-up as a result of heavy rain, coupled with drainage problems, was thought to be the cause of the failure, which involved approximately 1 million tonnes of glacial till as shown in Figure 10.7. The landslide caused catastrophic damage to the Holbeck Hall Hotel situated at the crest of the slope, as shown in Figure 10.7.

Figure 10.7 Rotational slope failure at Holbeck, Yorkshire.

Limiting equilibrium techniques[①] are normally used in the analysis of slope stability in which it is considered that failure is on the point of occurring along an assumed or a known failure surface. To check stability at the ultimate limiting state, gravitational forces driving slip (e. g. the component of weight acting along a slip plane) are considered as actions and factored accordingly; the forces developed due to shearing along slip planes are treated as resistances, along with any gravitational forces resisting slip, and factored down using a partial factor γ_{Rr}. In Eurocode 7, the normative value of $\gamma_{Rr} = 1.00$ for sets R1 and R3 and 1. 10 for set R2.

① 极限平衡手法。

Having analysed a given failure surface, the calculations should be repeated for a range of different positions of the slip surface. The failure surface that is closest to ULS is then the critical slip surface[②] along which failure will occur. This process is normally automated using a computer.

② 最危险滑动面。

Rotational slips in undrained soil

This analysis, in terms of total stress[③], covers the case of a fully saturated clay under undrained conditions, i. e. for the condition immediately after construction.
Only moment equilibrium[④] is considered in the analysis. If the soil is homogeneous, the failure surface can be assumed to be a circular arc in section. A trial failure surface (centre O, radius r and length L_a) is shown in Figure 10. 8. Potential instability is due to the total weight of the soil mass (W perunit length) above the failure surface. The driving (clockwise) moment about O is therefore $M_A = Wd$ (an action). The soil resistance is described by an anticlockwise moment $M_R = c_u L_a r$ aboutO. If the slip surface is a circular arc, then $L_a = r\theta$ from Figure 10. 8. The criterion of stability at ULS is then described by

③ 基于总应力。

④ 力矩平衡。

$$M_A \leqslant M_R \tag{10.14}$$

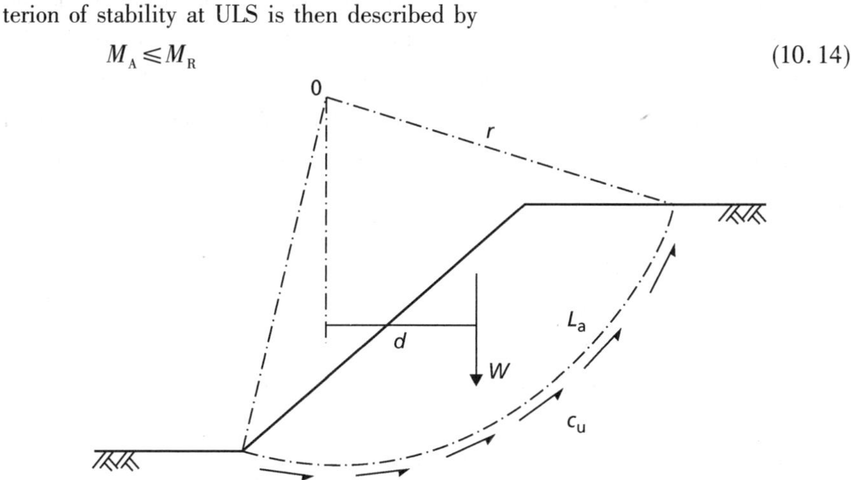

Figure 10. 8 Limit equilibrium analysis in undrained soil.[⑤]

⑤ 不排水土中的极限平衡分析法。

The moments of any additional forces (e. g. surcharge[⑥]) must be taken into account in determining M_A. In the event of a tension crack developing, the arc length

⑥ 超载。

① 静水压力。
② 试算破坏面。
③ 安全系数。

L_a is shortened and a hydrostatic force[①] will act normal to the crack if it fills with water. It is necessary to analyse the slope for a number of trial failure surfaces[②] in order that the most critical failure surface can be determined. In the analysis of an existing slope, M_A will be less than M_R (as the slope is standing) and the safety of the slope is usually expressed as a factor of safety[③], F where

$$F = \frac{M_R}{M_A} \tag{10.15}$$

From Equation 10.15 it can be seen that a stable slope will have $F > 1$, while an unstable slope will have $F < 1$. If $F = 1$, the slope is at the point of failure and the ULS stability criterion is regained.

④ 在新建边坡的设计中，目标一般是找到倾角为β的边坡所对应的最大高度。
⑤ 设计图表。
⑥ 稳定系数。

In the design of a new slope, the aim is normally to find the maximum height h to which a slope of a given angle β can be constructed[④]. To achieve this, the parameters W and d can be expressed in terms of the slope properties h and β and the properties describing the slip surface geometry (r and θ), though the derivation is not trivial. Fortunately, design charts[⑤] have been published by Taylor (1937) for the case of c_u uniform with depth and an underlying rigid boundary, and by Gibson and Morgenstern (1962) for the case of c_u increasing linearly with depth ($c_u = Cz$). Both of these solutions express the critical conditions leading to slope failure as a non-dimensional **stability number**[⑥] N_s, where

$$N_s = \frac{c_u}{F\gamma h} \tag{10.16}$$

⑦ 坡角β。
⑧ 分项系数。

In the case of increasing undrained shear strength with depth, $N_s = C/F\gamma h$. Values of N_s as a function of slope angle β[⑦] are shown in Figure 10.9. Rearranging Equation 10.16 to give $F = c_u/N_s\gamma h$ and comparing this to Equation 10.15, it can be seen that the numerator of Equation 10.16 represents the resistance of the soil to failure while the denominator represents the sum of the driving actions for applying partial factors[⑧]. These should be factored accordingly, along with the material properties (c_u and γ). In limit state design, the value of N_s is therefore determined from Figure 10.9 for a given slope angle; F is set to 1.0 such that the resulting equation governing ULS is

$$h \leqslant \frac{1}{N_s}\left[\frac{\left(\frac{c_u}{\gamma_{cu}}\right)}{\gamma_{Rr}\gamma_A\left(\frac{\gamma}{\gamma_\gamma}\right)}\right] \tag{10.17}$$

Equation 10.17 is then solved using the known soil properties and N_s to determine the maximum height of the slope.

⑨ 力矩平衡法。

Equation 10.16 may also be used to analyse existing slopes in place of the moment equilibrium method[⑨] (Equation 10.15), in which case the slope height h is known and the equation is rearranged to find the unknown factor of safety. A three-dimensional analysis for slopes in clay under undrained conditions has been presented by Gens *et al.* (1988).

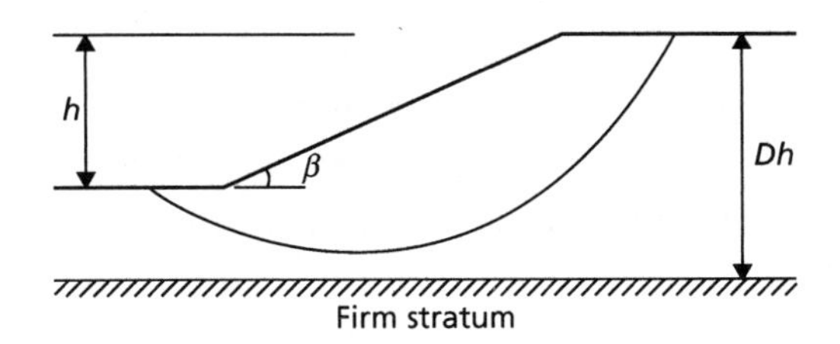

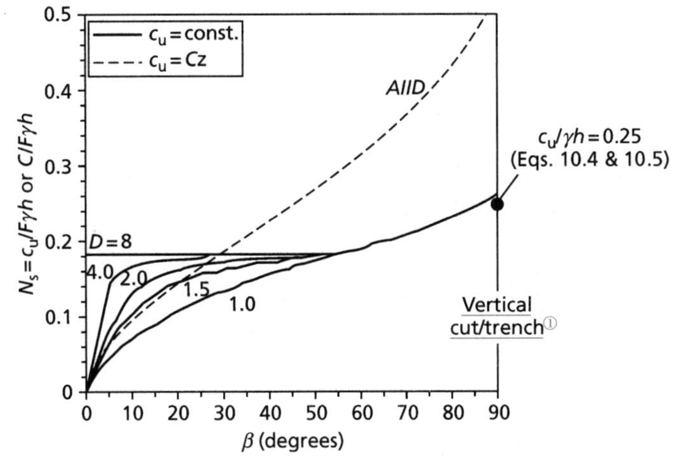

Figure 10.9 Stability numbers for slopes in undrained soil.[②]

① 垂直切削面/沟槽。

② 不排水条件下边坡的稳定系数。

Example 10.1

A 45° cutting slope is excavated to a depth of 8m in a deep layer of saturated clay of unit weight 19kN/m^3: the relevant shear strength parameters are c_u = 65kPa.

a Determine the factor of safety for the trial failure surface specified in Figure 10.10.

b Check that no loss of overall stability will occur according to the limit state approach (using EC7 DA1b).

c Determine the maximum depth to which the slope could be excavated if the slope angle is maintained at 45°.

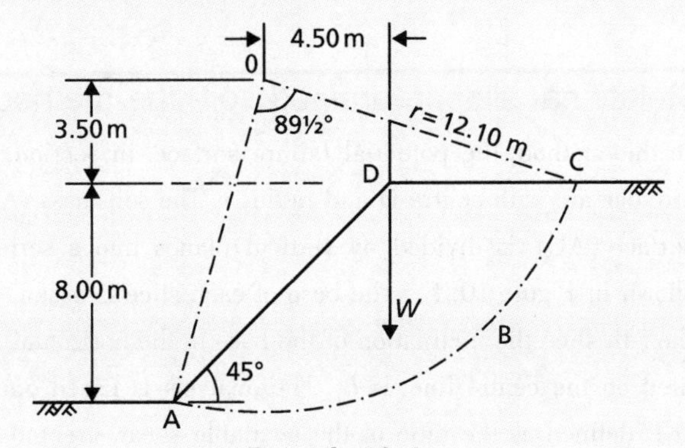

Figure 10.10 Example 10.1.

Solution

a In Figure 10.10, the cross-sectional area ABCD) is $70m^2$. Therefore, the weight of the soil mass $W = 70 \times 19 = 1330 kN/m$. The centroid of ABCD is 4.5m from O. The angle AOC is 89.5°and radius OC is 12.1m. The arc length ABC is calculated (or may be measured) as 18.9m. From Equation 10.15:

$$F = \frac{c_u L_a r}{Wd}$$

$$= \frac{65 \cdot 18.9 \cdot 12.1}{1330 \cdot 4.5}$$

$$= 2.48$$

This is the factor of safety for the trial failure surface selected, and is not necessarily the minimum factor of safely. From Figure 10.9, $\beta = 45°$ and, assuming that D is large, the value of N_s, is 0.18. Then, from Equation 10.16, $F = 2.37$, This is lower than the value found previously, so the trial failure plane shown in Figure 10.10 is not the actual failure surface.

b For DA lb, $\gamma_\gamma = 1.00$, $\gamma_{cu} = 1.40$, $\gamma_A = 1.00$ (slope self-weight is a permanent unfavourable action) and $\gamma_{Rr} = 1.00$, hence

$$M_A = \gamma_A \left(\frac{W}{\gamma_\gamma}\right) d = 1.00 \cdot \left(\frac{1330}{1.00}\right) \cdot 4.5 = 5985 kNm/m$$

$$M_R = \frac{\left(\frac{c_u}{\gamma_{cu}}\right) L_a r}{\gamma_{Rr}} = \frac{\left(\frac{65}{1.40}\right) \cdot 18.9 \cdot 12.1}{1.00} = 10520 kNm/m$$

$M_A < M_R$, so the overall stability limit state (ULS) is satisfied to EC7 DAlb.

c The maximum depth of the cutting is given by Equation 10.17, the partial factors being as given above (same design approach as in part (b)) and $N_s = 0.18$ for $\beta = 45°$:

$$h \leqslant \frac{1}{0.18}\left[\frac{\left(\frac{65}{1.40}\right)}{1.00 \cdot 1.00 \cdot \left(\frac{19}{1.00}\right)}\right]$$

$$\leqslant 13.58m$$

Rotational slips in drained soil- the method of slices[①]

① 排水条件下的旋转滑动——条分法。

In this method, the potential failure surface, in section, is again assumed to be a circular arc with centre O and radius r. The soil mass (ABCD) above a trial failure surface (AC) is divided by vertical planes into a series of slices of width b, as shown in Figure 10.11. The base of each slice is assumed to be a straight line. For the i-th slice the inclination of the base to the horizontal is α_i and the height, measured on the centre-line, is h_i. The analysis is based on the use of factor of safety (F), defined as the ratio of the available shear strength (τ_f) to the shear strength (τ_{mob}) which must be mobilised to maintain a condition of limiting equilibrium along the slip surface, i. e.

$$F = \frac{\tau_f}{\tau_{mob}}$$

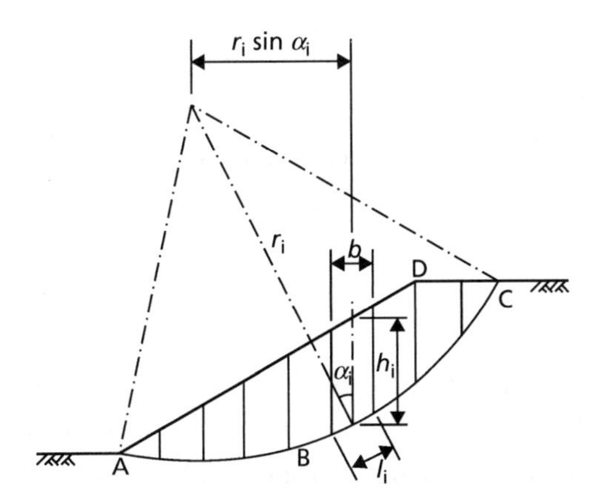

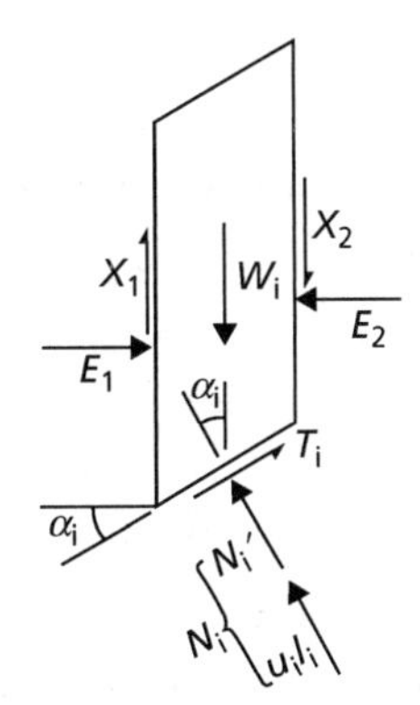

Figure 10.11 The method of slices.[①]

① 条分法。

The factor of safety is taken to be the same for each slice, implying that there must be mutual support between slices, i. e. inter-slice forces[②] must act between the slices (E_1, X_1, E_2 and X_2 in Figure 10.11). The forces (per unit dimension normal to the section) acting on a slice are:

② 条间力。

1 the total weight of the slice, $W_i = \gamma b h_i$ (γ_{sat} where appropriate);
2 the total normal force on the base, N_i (equal to $\sigma_i l_i$) – in general this force has two components, the effective normal force N_i' (equal to $\sigma_i l_i$) and the boundary pore water force U_i (equal to $u_i l_i$), where u_i is the pore water pressure at the centre of the base and l_i the length of the base;
3 the shear force on the base, $T_i = T_{mob} l_i$;
4 the total normal forces on the sides, E_1, and E_2;
5 the shear forces on the sides, X_1 and X_2.

Any external forces (e. g. surcharge, pinning forces from inclusions) must also be included in the analysis.

The problem is statically indeterminate[③], and in order to obtain a solution assumptions must be made regarding the inter-slice forces E and X; in general, therefore, the resulting solution for factor of safety is not exact.

③ 超静定的。

Considering moments about O, the sum of the moments of the shear forces T_i on the failure arc AC must equal the moment of the weight of the soil mass ABCD. For any slice the lever arm of W_i is $r_i \sin \alpha_i$, therefore at limiting equilibrium

$$\sum M_R = \sum M_A$$

$$\sum_i T_i r_i = \sum_i W_i r_i \sin \alpha_i$$

Now

$$T_i = \tau_{mob,i} l_i = \left(\frac{\tau_{f,i}}{F}\right) l_i$$

$$\therefore \sum_i \left(\frac{\tau_{f,i}}{F}\right) l_i = \sum_i W_i \sin\alpha_i$$

$$\therefore F = \frac{\sum_i \tau_{f,i} l_i}{\sum_i W_i \sin \alpha_i}$$

① 有效应力分析。

For an effective stress analysis[①] (in terms of parameters c' and ϕ') $\tau_{f,i}$ is given by Equation 5.11, so

$$F = \frac{\sum_i (c_i' + \sigma_i' \tan\phi_i') l_i}{\sum_i W_i \sin\alpha_i} \quad (10.18a)$$

② 均质土。

Equation 10.18a can be used in the general case of c' and/or ϕ' varying with depth and position in the slope, the appropriate average values being used for each slice. For the case of homogeneous soil[②] conditions, Equation 10.18a simplifies to

$$F = \frac{c' L_a + \tan\phi' \sum_i N_i'}{\sum_i W_i \sin\alpha_i} \quad (10.18b)$$

③ 临界状态强度。

where L_a is the arc length AC (i.e. length of the whole slip plane). Equation 10.18b is exact, but approximations are introduced in determining the forces N_i'. For a given failure arc, the value of F will depend on the way in which the forces N_i' are estimated. In many cases, the critical state strength[③] is normally appropriate in the analysis of slope stability, i.e. $\phi' = \phi'_{cv}$ and $c' = 0$, therefore the expression simplifies further to

$$F = \frac{\tan\phi'_{cv} \sum_i N_i'}{\sum_i W_i \sin\alpha_i} \quad (10.18c)$$

④ 费伦纽斯法（瑞典法）。

⑤ 这一方法假设每一条块的条间力的合力为零。

The Fellenius (or Swedish) solution[④]

In this solution, it is assumed that for each slice the resultant of the inter-slice forces is zero[⑤]. The solution involves resolving the forces on each slice normal to the base, i.e.

$$N_i' = W_i \cos\alpha_i - u_i l_i$$

Hence the factor of safety in terms of effective stress (Equation 10.18b) is given by

$$F = \frac{c' L_a + \tan\phi' \sum_i (W_i \cos\alpha_i - u_i l_i)}{\sum_i W_i \sin\alpha_i} \quad (10.19)$$

⑥ 分子。

⑦ 分母。

The components $W_i \cos \alpha_i$ and $W_i \sin \alpha_i$ can be determined graphically for each slice. Alternatively, the values of W_i and α_i can be calculated. Again, a series of trial failure surfaces must be chosen in order to obtain the minimum factor of safety. It can be seen from the derivation of Equation 10.19 that the numerator[⑥] represents the overall resistance, while the denominator[⑦] represents the overall action driving failure. Therefore, Equation 10.19 may be used to verify the ULS by setting

$F = 1$, factoring the numerator (resistance), denominator (action) and material properties appropriately as before, and ensuring that the numerator is larger than the denominator.

This solution is known to underestimate the true factor of safety[①] due to the assumptions which are inherent in it; the error, compared with more accurate methods of analysis, is usually within the range 5-20%. Use of the Fellenius method is not now recommended in practice.

① 这一解法通常会低估真实的安全系数。

The Bishop routine solution[②]

② 毕肖普常用方法。

In this solution it is assumed that the resultant forces on the sides of the slices are horizontal,[③] i. e.

③ 通常假设条块侧面的合力是水平的。

$$X_1 - X_2 = 0$$

For equilibrium, the shear force on the base of any slice is

$$T_i = \frac{1}{F}(c_i' l_i + N_i' \tan\phi_i')$$

Resolving forces in the vertical direction:

$$W_i = N_i'\cos\alpha_i + u_i l_i \cos\alpha_i + \left(\frac{c_i' l_i}{F}\right)\sin\alpha_i + \left(\frac{N_i'}{F}\right)\tan\phi_i'\sin\alpha_i$$

$$\therefore N_i' = \frac{W_i - \left(\frac{c_i' l_i}{F}\right)\sin\alpha_i - u_i l_i \cos\alpha_i}{\cos\alpha_i + \left(\frac{\tan\phi_i'\sin\alpha_i}{F}\right)} \tag{10.20}$$

It is convenient to substitute

$$l_i = b\sec\alpha_i \tag{10.21}$$

Substituting Equation 10.20 into Equation 10.18a, it can be shown after some rearrangement that

$$F = \frac{1}{\sum_i W_i \sin\alpha_i} \cdot \sum_i \left\{ [c_i' b + (W_i - u_i b)\tan\phi_i'] \frac{\sec\alpha_i}{1 + \left(\frac{\tan\phi_i'\tan\alpha_1}{F}\right)} \right\} \tag{10.22}$$

Bishop (1955) also showed how non-zero values of the resultant forces (X_1-X_2) could be introduced into the analysis, but this refinement has only a marginal effect on the factor of safety.

The pore water pressure can be expressed as a proportion of the total 'fill pressure' at any point by means of the dimensionless **pore water pressure ratio (r_u)**[④], defined as

④ 孔隙水压力比例系数。

$$r_u = \frac{u}{\gamma h} \tag{10.23}$$

Therefore, for the i-th slice

$$r_u = \frac{u_i b}{W_i}$$

Hence Equation 10.22 can be written as

$$F=\frac{1}{\sum_{i} W_{i}\sin\alpha_{i}}\cdot\sum_{i}\left[\left[c_{i}'b+W_{i}(1-r_{u,i})\tan\phi_{i}'\right]\frac{\sec\alpha_{i}}{1+\left(\frac{\tan\phi_{i}'\tan\alpha_{i}}{F}\right)}\right] \tag{10.24}$$

As the factor of safety occurs on both sides of Equations 10.22 and 10.24, a process of successive approximation must be used to obtain a solution, but convergence is rapid. Due to the repetitive nature of the calculations and the need to select an adequate number of trial failure surfaces, the method of slices is particularly suitable for solution by computer. More complex slope geometry and different soil strata can also then be straightforwardly introduced.

Again, the factor of safety determined by this method is an underestimate, but the error is unlikely to exceed 7% and in most cases is less than 2%. Spencer (1967) proposed a method of analysis in which the resultant inter-slice forces are parallel and in which both force and moment equilibrium are satisfied. Spencer showed that the accuracy of the Bishop routine method, in which only moment equilibrium is satisfied, is due to the insensitivity of the moment equation to the slope of the inter-slice forces.

① 无量纲的均质土坡稳定系数。

Dimensionless stability coefficients for homogeneous slopes[①], based on Equation 10.24, have been published by Bishop and Morgenstern (1960) and Michalowski (2002). It can be shown that for a given slope angle and given soil properties the factor of safety varies linearly with r_u, and can thus be expressed as

$$F=m-nr_{u} \tag{10.25}$$

where m and n are the stability coefficients. The coefficients m and n are functions of β, ϕ', depth factor D and the dimensionless factor $c'/\gamma h$ (which is zero if the critical state strength is used).

② 土坡三维极限分析方法。

A three-dimensional limit analysis for slopes[②] in drained soil has been presented by Michalowski (2010).

Example 10.2

Using the Fellenius method of slices, determine the factor of safety, in terms of effective stress, of the slope shown in Figure 10.12 for the given failure surface: (a) using peak strength parameters c' = 10kPa and ϕ' = 29°; and (b) using critical state parameter ϕ'_{cv} = 31°. The unit weight of the soil both above and below the water table is 20kN/m³.

Solution

a The factor of safety is given by Equation 10.19. The soil mass is divided into slices 1.5m wide. The weight (W_i) of each slice is given by

$$W_{i}=\gamma bh_{i}=20\cdot1.5\cdot h_{i}=30h_{i}\,\text{kN/m}$$

The height h_i, and angle α_i for each slice are measured from Figure 10.12 (which are drawn to scale), from which values of W_i are calculated using the expression given above, and values of l_i

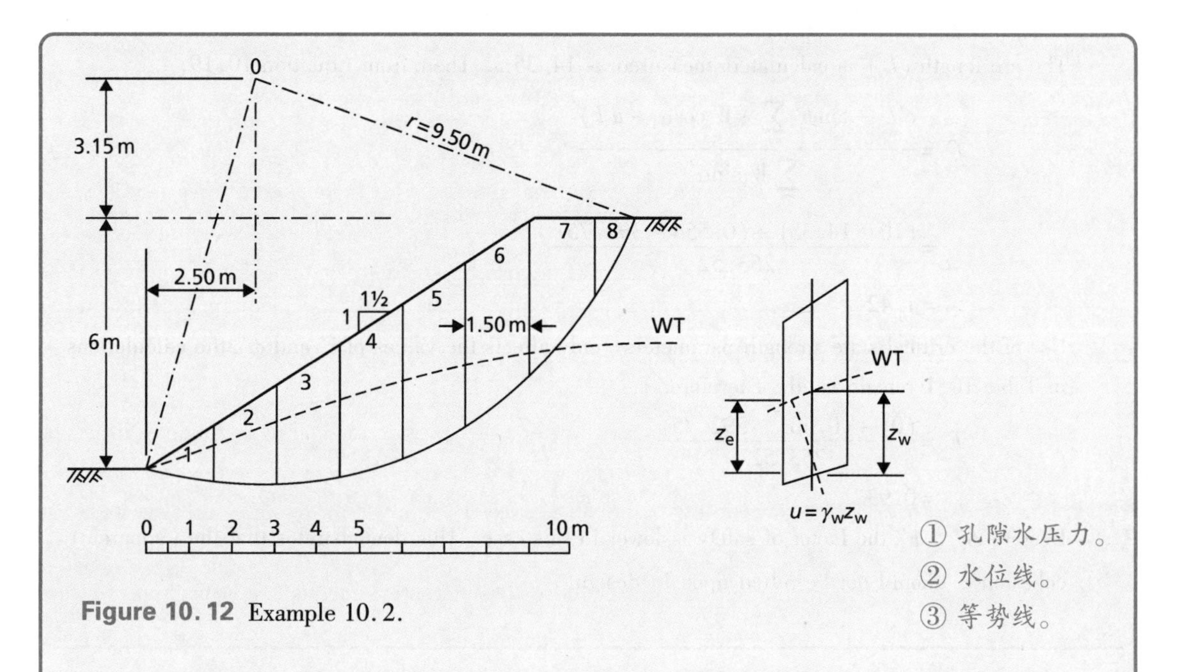

Figure 10.12 Example 10.2.

are calculated from Equation 10.21. The pore water pressure[①] at the centre of the base of each slice is taken to be $\gamma_w z_w$, where z_w is the vertical distance of the centre point below the water table (as shown in the figure). This procedure slightly overestimates the pore water pressure, which strictly should be $\gamma_w z_e$, where z_e is the vertical distance below the point of intersection of the water table[②] and the equipotential[③] through the centre of the slice base. The error involved is on the safe side. The derived values are given in Table 10.1.

Table 10.1 Example 10.2

Slice	h_i (m)	α_i (°)	W_i (kN/m)	l_i (m)	u_i (kPa)	$W_i \cos\alpha_i - u_i l_i$ (kN/m)	$W_i \sin\alpha_i$ (kN/m)
1	0.76	−11.2	22.8	1.55	5.9	13.22	−4.43
2	1.80	−3.2	54.0	1.50	11.8	36.22	−3.01
3	2.73	8.4	81.9	1.55	16.2	55.91	11.96
4	3.40	17.1	102.0	1.60	18.1	68.53	29.99
5	3.87	26.9	116.1	1.70	17.1	74.47	52.53
6	3.89	37.2	116.7	1.95	11.3	70.92	70.56
7	2.94	49.8	88.2	2.35	0	56.93	67.37
8	1.10	59.9	33.0	2.15	0	16.55	28.55
						392.75	253.52

The are length (L_a) is calculated/measured as 14.35m. Then, from Equation 10.19,

$$F = \frac{c'L_a + \tan\phi' \sum_i (W_i \cos\alpha_i - u_i l_i)}{\sum_i W_i \sin\alpha_i}$$

$$= \frac{(10 \cdot 14.35) + (0.554 \cdot 392.75)}{253.52}$$

$$= 1.42$$

b Use of the critical state strength parameters[①] only affects the values of c' and ϕ'; the calculations in Table 10.1 remain valid. Therefore,

$$F = \frac{(0) + (0.601 \cdot 392.75)}{253.52}$$

$$= 0.93$$

Despite $\phi'_{cv} > \phi'$, the factor of safety is lower in this case. This demonstrates that the (apparent) cohesion c' should not be relied upon in design.

Translational slips[②]

It is assumed that the potential failure surface is parallel to the surface of the slope and is at a depth that is small compared with the length of the slope[③]. The slope can then be considered as being of infinite length[④], with end effects being ignored. The slope is inclined at angle β to the horizontal and the depth of the failure plane is z, as shown in section in Figure 10.13. The water table is taken to be parallel to the slope at a height of mz $(0 < m < 1)$ above the failure plane. Steady seepage is assumed to be taking place in a direction parallel to the slope.[⑤] The forces on the sides of any vertical slice are equal and opposite, and the stress conditions are the same at every point on the failure plane.

① 临界状态强度参数。
② 平移滑动破坏。
③ 通常假设潜在的破坏面平行于边坡表面，且滑移土体深度与边坡长度相比较小。
④ 无限长。
⑤ 假设稳态渗流沿着平行于边坡的方向发生。
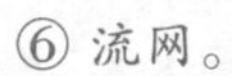
⑥ 流网。

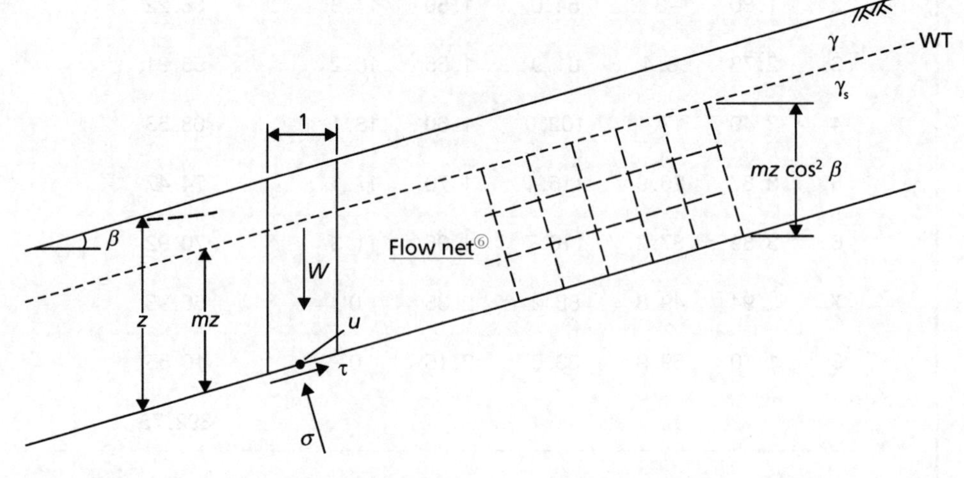

Figure 10.13 Plane translational slip.

In terms of effective stress, the shear strength of the soil along the failure plane (using the critical state strength) is

$$\tau_f = (\sigma - u)\tan\phi'_{cv}$$

and the factor of safety is

$$F = \frac{\tau_f}{\tau_{mob}} \tag{10.26a}$$

where τ_{mob} is the mobilised shear stress along the failure plane (see Chapter 9). The expressions for σ, τ_{mob} and u are

$$\sigma = [(1-m)\gamma + m\gamma_s] z\cos^2\beta$$

$$\tau_{mob} = [(1-m)\gamma + m\gamma_s] z\sin\beta\cos\beta$$

$$u = m\gamma_w z\cos^2\beta$$

giving

$$F = \frac{[(1-m)\gamma + m(\gamma_s - \gamma_w)]\tan\phi'_{cv}}{[(1-m)\gamma + m\gamma_s]\tan\beta} \tag{10.26b}$$

For a total stress analysis the $\tau_f = c_u$ is used, giving

$$F = \frac{c_u}{[(1-m)\gamma + m\gamma_s] z\sin\beta\cos\beta} \tag{10.26c}$$

As for rotational slips[①], the term in the numerator of Equation 10.26 represents the resistance of the soil to slip, while the denominator represents the driving action. For verification of the ULS, therefore, $F = 1.00$, and the numerator, denominator and material properties are factored appropriately.

① 注：此处原书有误，应为“translational slips”。

Example 10.3

A long natural slope in an overconsolidated fissured clay[②] of saturated unit weight 20kN/m³ is inclined at 12° to the horizontal. The water table is at the surface, and seepage is roughly parallel to the slope. A slip has developed on a plane parallel to the surface at a depth of 5m. Determine whether the ULS is satisfied to EC7 DA1b using (a) the critical state parameter $\phi'_{cv} = 28°$, and (b) the residual strength parameter $\phi'_r = 20°$.

② 超固结裂隙黏土。

Solution

a The water table is at the ground surface, so $m = 1$. For DA1b, $\gamma_\gamma = 1.00$, $\gamma_{\tan\phi} = 1.25$, $\gamma_A = 1.00$ (slope self-weight is a permanent unfavourable action) and $\gamma_{Rr} = 1.00$. The resistance τ_f is

$$\tau_f = \frac{\left[(1-m)\left(\frac{\gamma}{\gamma_\gamma}\right) + m\left(\frac{\gamma_s - \gamma_w}{\gamma_\gamma}\right)\right] z\cos^2\beta\left(\frac{\tan\phi'_{cv}}{\gamma_{\tan\phi}}\right)}{\gamma_{Rr}}$$

$$= \frac{\left[0 + \left(\frac{20 - 9.81}{1.00}\right)\right]\cdot 5 \cdot \cos^2 12 \cdot \left(\frac{\tan 28}{1.25}\right)}{1.00}$$

$$= 20.7\text{kPa}$$

while the mobilised shear stress τ_{mob}(action) is

$$\tau_{mob} = \gamma_A\left[(1-m)\left(\frac{\gamma}{\gamma_\gamma}\right) + m\left(\frac{\gamma_s}{\gamma_\gamma}\right)\right] z\sin\beta\cos\beta$$

$$= \left[0 + \left(\frac{20}{1.00}\right)\right] \cdot 5 \cdot \sin 12 \cdot \cos 12$$

$$= 20.3\text{kPa}$$

As $\tau_f > \tau_{mob}$, the ULS is satisfied and the slope is stable.

b Using ϕ'_r in place of ϕ'_{cv} changes the resistance to $\tau_f = 14.2$kPa, while τ_{mob} remains unchanged. In this case $\tau_r < \tau_{mob}$, so the ULS is not satisfied (the slope will slip if residual strength① conditions are achieved).

① 残余强度。

② 通用分析方法。

General methods of analysis②

Morgenstern and Price (1965, 1967) developed a general analysis based on limit equilibrium in which all boundary and equilibrium conditions are satisfied and in which the failure surface may be any shape, circular, non-circular or compound. Computer software for undertaking such analyses is readily available. Bell (1968) proposed an alternative method of analysis in which all the conditions of equilibrium are satisfied and the assumed failure surface may be of any shape. The soil mass is divided into a number of vertical slices and statical determinacy is obtained by means of an assumed distribution of normal stress along the failure surface. The use of a computer is also essential for this method. In both general methods mentioned here, the solutions must be checked to ensure that they are physically acceptable③. Modern computer-based tools are now available for analysing the ULS for slopes using limit analysis combined with optimisation routines④.

③ 物理上可接受的。

④ 优化程序。

⑤ 长期稳定性。

End of construction and long-term stability⑤

When a slope is formed by excavation, the decreases in total stress result in changes in pore water pressure in the vicinity of the slope and, in particular, along a potential failure surface. For the case illustrated in Figure 10.14 (a), the initial pore water pressure (u_0) depends on the depth of the point in question below the initial (static) water table (i.e. $u_0 = u_s$) Ch 8. For a typical point P on a potential failure surface (Figure 10.14 (a)), the pore water pressure change Δu is negative. After excavation, pore water will flow towards the slope and drawdown of the water table will occur. As dissipation proceeds the pore water pressure increases to the steady seepage value, as shown in Figure 10.14 (a), which may be determined from a flow net⑥ or by using the numerical methods described in Section 2.7. The final pore water pressure (u_f), after dissipation of excess pore water pressure⑦ is complete, will be the steady seepage value determined from the flow net.

⑥ 流网。

⑦ 超孔隙水压力消散。

If the permeability of the soil is low, a considerable time will elapse before any significant dissipation of excess pore water pressure will have taken place. At the end of construction the soil will be virtually in the undrained condition, and a total stress analysis will be relevant to verify stability (ULS). In principle, an effective stress analysis is also possible for the end-of-construction condition using the appropriate value of pore water pressure ($u_0 + \Delta u$) for this condition. However, because of its greater simplicity, a total stress analysis is generally used. It should be realised that the same factor of safety will not generally be obtained from a total stress and an effective stress analysis of the end-of-construction condition. In a total stress analysis it is implied that the pore water pressures are those for a failure condition (being the equivalent of the pore water pressure at failure in an undrained triaxial test); in an effective stress analysis the pore water pressures used are those predicted for a non-failure condition①. In the long term, the fully drained condition will be reached and only an effective stress analysis will be appropriate.

① 总应力分析时，孔隙水压力对应的是破坏时的孔隙水压力（即三轴不排水试验破坏时的孔隙水压力），而有效应力分析中孔隙水压力对应未破坏时的孔隙水压力。

On the other hand, if the permeability of the soil is high, dissipation of excess pore water pressure will be largely complete by the end of construction. An effective stress analysis is relevant for all conditions with values of pore water pressure being obtained from the static water table level or the steady seepage flow net.

Irrespective of the permeability of the soil, the increase in pore water pressures following excavation will result in a reduction in effective stress (and hence strength) with time such that the factor of safety will be lower in the long term, when dissipation is complete, than at the end of construction.

The creation of sloping ground through construction of an embankment results in increases in total stress, both within the embankment itself as successive layers of fill are placed, and in the foundation soil. The initial pore water pressure (u_0)② depends primarily on the placement water content of the fill. The construction period of a typical embankment is relatively short, and, if the permeability of the compacted fill is low, no significant dissipation is likely during construction. Dissipation proceeds after the end of construction, with the pore water pressure decreasing to the final value in the long term, as shown in Figure 10.14 (b). The factor of safety of an embankment at the end of construction is therefore lower than in the long term③. Shear strength parameters④ for the fill material should be determined from tests on specimens compacted to the values of dry density and water content to be specified for the embankment (see Chapter 1).

② 初始孔隙水压力。

③ 路堤在施工结束时的安全系数总要低于长期的安全系数。

④ 抗剪强度参数。

The stability of an embankment may also depend on the shear strength of the foundation soil. The possibility of failure along a surface such as that illustrated in Figure 10.15 should be considered in appropriate cases.

Slopes in overconsolidated fissured clays⑤ require special consideration. A number of cases are on record in which failures in this type of clay have occurred long after dissipation of excess pore water pressure had been completed. Analysis of these failures showed that the average shear strength at failure was well below the peak value. It is probable that large strains occur locally due to the presence of fissures, resulting in the peak strength being reached, followed by a gradual decrease

⑤ 超固结裂隙黏土。

towards the critical state value. The development of large local strains can lead eventually to a progressive slope failure[①].

① 边坡渐近破坏。

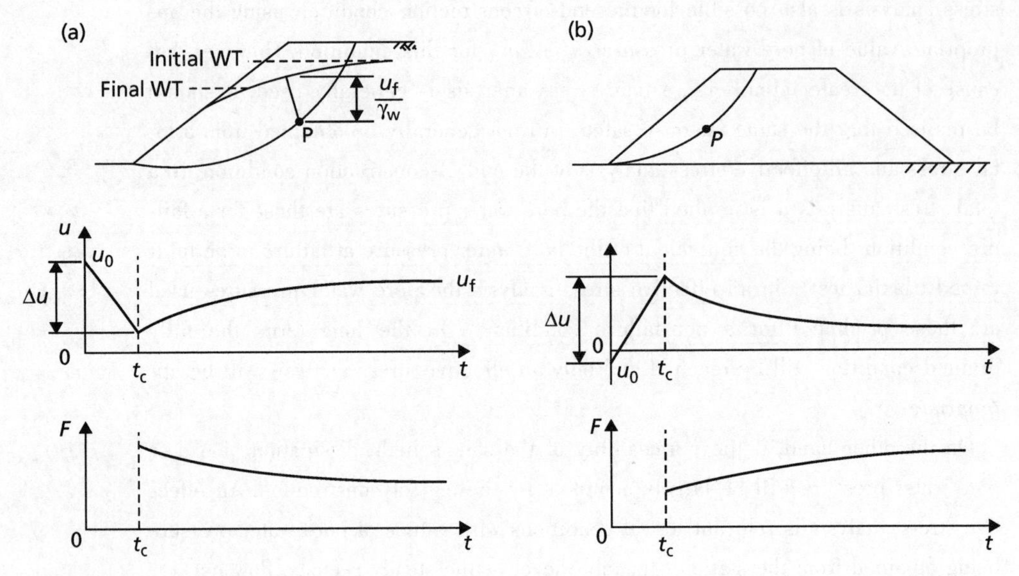

Figure 10.14 Pore pressure dissipation and factor of safety[②] (a) following excavation[③] (i. e. a cutting), (b) following construction[④] (i. e. an embankment).

② 孔隙水消散与安全系数。
③ 开挖。
④ 填筑。

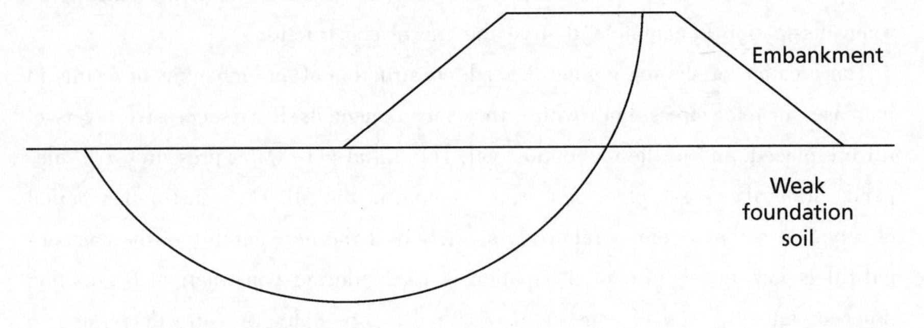

Figure 10.15 Failure beneath an embankment.

However, fissures may not be the only cause of progressive failure; there is considerable non-uniformity[⑤] of shear stress along a potential failure surface, and local overstressing[⑥] may initiate progressive failure. It is also possible that there could be a pre-existing slip surface in this type of clay and that it could be reactivated by excavation. In such cases a considerable slip movement could have taken place previously, sufficiently large for the shear strength to fall below the critical state value and towards the residual value.

⑤ 不均匀性。
⑥ 局部应力集中。

Thus for an initial failure (a 'first time' slip) in overconsolidated fissured clay, the relevant strength for the analysis of long-term stability is the critical state value[⑦]. However, for failure along a pre-existing slip surface the relevant strength is the residual value. Clearly it is vital that the presence of a pre-existing slip surface in the vicinity of a projected excavation should be detected during the ground in-

⑦ 临界状态值。

vestigation.

The strength of an overconsolidated clay at the critical state, for use in the analysis of a potential first time slip, is difficult to determine accurately. Skempton (1970) has suggested that the maximum strength of the remoulded clay① in the normally consolidated condition can be taken as a practical approximation to the strength of the overconsolidated clay at the critical state, i. e. when it has fully softened adjacent to the slip plane as the result of expansion during shear.

① 重塑土。

10.4 Embankment dams②

② 土石坝。

An embankment dam would normally be used where the foundation and abutment conditions are unsuitable for a concrete dam and where suitable materials for the embankment are present at or close to the site. An extensive ground investigation is essential, general at first but becoming more detailed as design studies proceed, to determine foundation and abutment conditions and to identify suitable borrow areas. It is important to determine both the quantity and quality of available material. The natural water content of fine soils should be determined for comparison with the optimum water content for compaction③.

③ 压实的最优含水量。

Most embankment dams are not homogeneous but are of zoned construction, the detailed section depending on the availability of soil types to provide fill for the embankment. Typically a dam will consist of a core of low-permeability soil with shoulders of other suitable material on each side. The upstream slope is usually covered by a thin layer of rockfill (known as **rip-rap**④) to protect it from erosion by wave or other fluid actions. The downstream slope is usually grassed (again, to resist erosion). An internal drainage system, to alleviate the detrimental effects of any seeping water, would normally be incorporated. Depending on the materials used, horizontal drainage layers may also be incorporated to accelerate the dissipation of excess pore water pressure. Slope angles should be such that stability is ensured, but overconservative design must be avoided: a decrease in slope angle of as little as 2-3° (to the horizontal) would mean a significant increase in the volume of fill for a large dam.

④ 抛石。

Failure of an embankment dam could result from the following causes: (1) instability of either the upstream or downstream slope; (2) internal erosion; and (3) erosion of the crest and downstream slope by overtopping⑤. (The third cause arises from errors in the hydrological predictions.)

⑤ (1) 上游或下游的边坡失稳; (2) 内部侵蚀; (3) 漫顶引起的坝顶及下游边坡的侵蚀。

The factor of safety for both slopes must be determined as accurately as possible for the most critical stages in the life of the dam, using the methods outlined in Section 10. 3. The potential failure surface may lie entirely within the embankment, or may pass through the embankment and the foundation soil (as in Figure 10. 15). In the case of the upstream slope, the most critical stages are at the end of construction and during rapid drawdown of the reservoir level. The critical stages for the downstream slope are at the end of construction and during steady seepage when the reservoir is full. The pore water pressure distribution at any stage has a dominant influence on the factor of safety of the slopes, and it is common practice to install a piezometer system (see Chapter 6) so that the actual pore water pressures can be measured and compared with the predicted values used in design (providedan effective stress analysis has been used). Remedial action could then be taken if, based on the measured values, the slope began to approach the ULS.

If a potential failure surface were to pass through foundation material containing fissures, joints or pre-existing slip surfaces, then progressive failure (as described in the previous section) would be a possibility. The different stress-strain characteristics of various zone materials through which a potential failure surface passes, together with non-uniformity of shear stress, could also lead to progressive failure.

Another problem is the danger of cracking due to differential movements between soil zones, and between the dam and the abutments. The possibility of **hydraulic fracturing**[①], particularly within the clay core[②], should also be considered. Hydraulic fracturing occurs on a plane where the total normal stress is less than the local value of pore water pressure. Following the completion of construction the clay core tends to settle relative to the rest of the embankment due to long-term consolidation; consequently, the core will be partially supported by the rest of the embankment. Thus vertical stress in the core will be reduced and the chances of hydraulic fracture increased. The transfer of stress from the core to the shoulders of the embankment[③] is another example of the arching phenomenon (Section 9.7). Following fracture or cracking, the resulting leakage could lead to serious internal erosion and impair stability.

① 水力劈裂。
② 黏土心墙。
③ 坝肩。

End of construction and long-term stability

Most slope failures in embankment dams occur either during construction or at the end of construction. Pore water pressures depend on the placement water content of the fill and on the rate of construction[④]. A commitment to achieve rapid completion will result in the maximisation of pore water pressure at the end of construction. However, the construction period of an embankment dam is likely to be long enough to allow partial dissipation of excess pore water pressure, especially for a dam with internal drainage. A total stress analysis, therefore, would result in an overconservative design. An effective stress analysis is preferable, using predicted values of r_u'.

④ 施工速度。

If high values of r_u are anticipated, dissipation of excess pore water pressure can be accelerated by means of horizontal drainage layers incorporated in the dam, drainage taking place vertically towards the layers: a typical dam section is shown in Figure 10.16. The efficiency of drainage layers has been examined theoretically by Gibson and Shefford (1968), and it was shown that in a typical case the layers, in order to be fully effective, should have a permeability at least 10^6 times that of the embankment soil: an acceptable efficiency would be obtained with a permeability ratio of about 10^5.

After the reservoir has been full for some time, conditions of steady seepage become established through the dam, with the soil below the top flow line in the fully saturated state. This condition must be analysed in terms of effective stress, with values of pore pressure being determined from a flow net (or using the numerical methods described in Section 2.7). Values of r_u up to 0.45 are possible in homogeneous dams, but much lower values can be achieved in dams having internal drainage[①]. Internal erosion is a particular danger when the reservoir is full because it can arise and develop within a relatively short time, seriously impairing the safety of the dam.

① 内部排水。

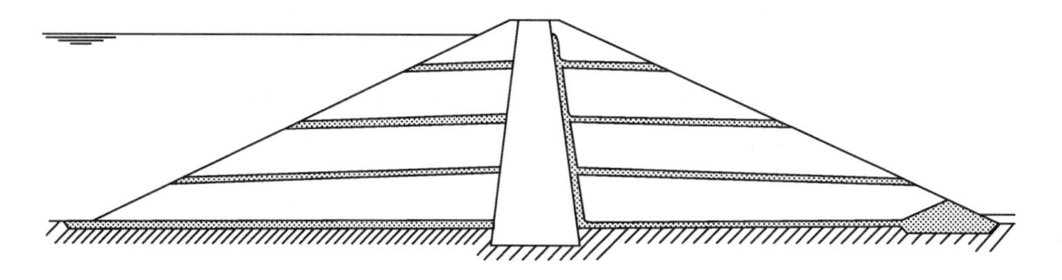

Figure 10.16 Horizontal drainage layers.[②]

② 水平排水层。

Rapid drawdown[③]

③ 水位骤降。

After a condition of steady seepage has become established, a drawdown of the reservoir level will result in a change in the pore water pressure distribution. If the permeability of the soil is low, a drawdown period measured in weeks may be 'rapid' in relation to dissipation time and the change in pore water pressure can be assumed to take place under undrained conditions. Referring to Figure 10.17, the pore water pressure before drawdown at a typical point P on a potential failure surface is given by

$$u_0 = \gamma_w (h + h_w - h') \qquad (10.27)$$

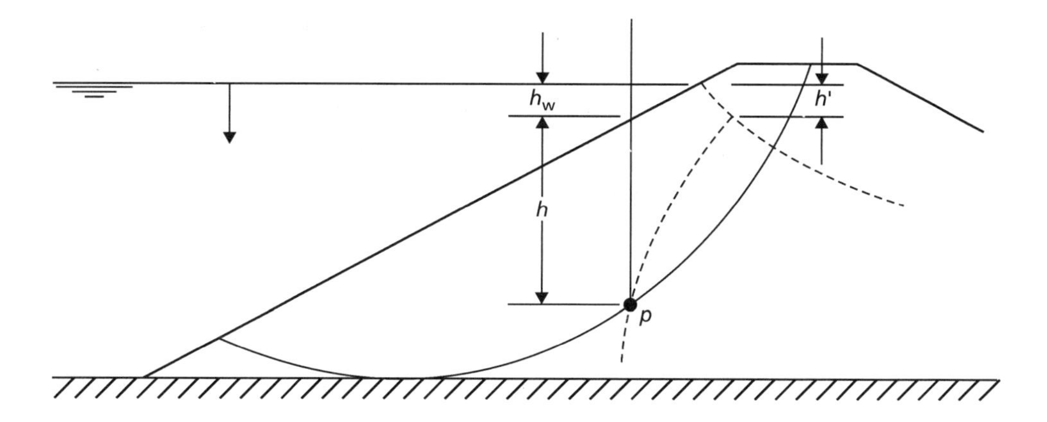

Figure 10.17 Rapid drawdown conditions.[④]

④ 水位骤降情况。

where h' is the loss in total head due to seepage between the upstream slope surface and the point P. It is again assumed that the total major principal stress at P is equal to the fill pressure. The change in total major principal stress[①] is due to the total or partial removal of water above the slope on the vertical through P. For a drawdown depth exceeding h_w,

① 最大主应力。

$$\Delta\sigma_1 = -\gamma_w h_w$$

the change in pore water pressure Δu can then be expressed in terms of $\Delta\sigma_1$ by

$$\begin{aligned}\frac{\Delta u}{\Delta\sigma_1} &= \frac{B[\Delta\sigma_3 + A(\Delta\sigma_1 - \Delta\sigma_3)]}{\Delta\sigma_1} \\ &= B\left[1 - (1 - A)\left(1 - \frac{\Delta\sigma_3}{\Delta\sigma_1}\right)\right] \\ &= \overline{B}\end{aligned} \tag{10.28}$$

Therefore the pore water pressure at P immediately after drawdown is

$$\begin{aligned}u &= u_0 + \Delta u \\ &= \gamma_w\{h + h_w(1 - \overline{B}) - h'\}\end{aligned}$$

Hence

$$\begin{aligned}r_u &= \frac{u}{\gamma h} \\ &= \frac{\gamma_w}{\gamma}\left[1 + \frac{h_w}{h}(1 - \overline{B}) - \frac{h'}{h}\right]\end{aligned} \tag{10.29}$$

The soil will be undrained immediately after rapid drawdown. An upper bound value of r_u for these conditions can be obtained by assuming $\overline{B} = 1$ and neglecting h'. Typical values of r_u immediately after drawdown are within the range 0.3 – 0.4.

Morgenstern (1963) published stability coefficients for the approximate analysis of homogeneous slopes after rapid drawdown, based on limit equilibrium techniques.

The pore water pressure distribution after drawdown in soils of high permeability decreases as pore water drains out of the soil above the drawdown level. The saturation line[②] moves downwards at a rate depending on the permeability of the soil. A series of flow nets can be drawn for different positions of the saturation line and values of pore water pressure obtained. The factor of safety can thus be determined, using an effective stress analysis, for any position of the saturation line. Viratjandr and Michalowski (2006) published stability coefficients for the approximate analysis of homogeneous slopes in such conditions, based on limit analysis techniques.

② 饱和水位线。

10.5 An introduction to tunnels[③]

③ 隧道。

Tunnels are the final class of problem that will be considered in this chapter, for which self-support of the soil mass controls the design. Shallow tunnels onshore[④] may be constructed using the **cut-and-cover** technique[⑤]; this is where a deep excavation is made, within which the tunnel is constructed, which is then backfilled to bury the tunnel structure. The design of such works may be completed using the

④ 陆上浅埋隧道。
⑤ 明挖法。

techniques described in Chapter 9, and this class of tunnel will not be considered further here. In marine and offshore applications, sections of tunnel structure are floated out to site, flooded to lower them into a shallow trench excavated on the riverbed/seabed and connected underwater, followed by pumping out of the internal water. These are known as **immersed tube** tunnels[1]. Some of the terminology related to tunnels is shown in Figure 10.18.

① 沉管隧道。

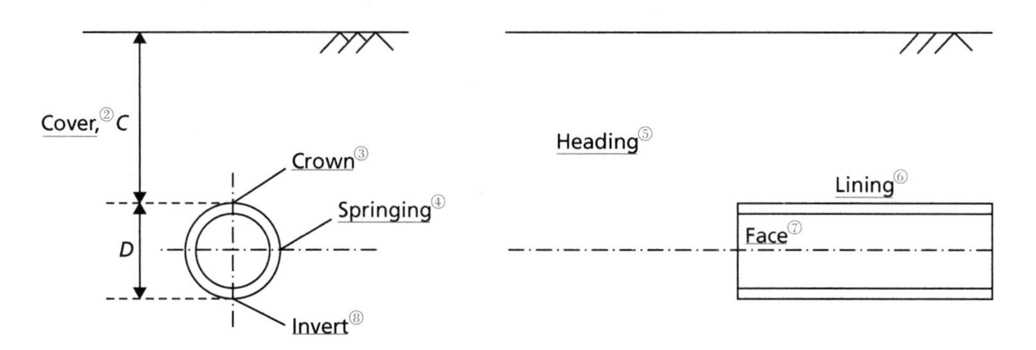

Figure 10.18 Terminology related to tunnels[9].

② 覆土。
③ 拱顶。
④ 起拱线。
⑤ 掌子面。
⑥ 衬砌。
⑦ 开挖面。
⑧ 仰拱
⑨ 与隧道相关的术语。
⑩ 运行深度。
⑪ 土压平衡。

In this chapter, the design of deep tunnels which are formed by boring deep within the ground are considered. In certain conditions (namely undrained soil response and a shallow **running depth**[10]) the tunnel may be self-supporting. For deeper excavations in undrained soil and for excavation in drained materials, the tunnel will need to be supported by an internal pressure to prevent collapse of the soil above the tunnel into the excavation (the ULS); this is known as **earth pressure balance**[11] construction. Once the tunnel is complete this internal support pressure describes the structural loading which the tunnel lining must be able to resist and is used in the structural design of the tunnel lining. In addition to maintaining the stability of the tunnel (ULS condition), the design of tunnelling works also requires consideration of settlements at the ground surface which are induced by the tunnelling procedure to ensure that these movements do not damage buildings and other infrastructure (SLS).

Stability of tunnels in undrained soil[12]

⑫ 不排水土中的隧道稳定性。

Figure 10.19 shows an element of soil above the crown of the tunnel (i.e. within the cover depth, C). This element of soil is loaded similarly to that around a pressuremeter test[13] (PMT, Figure 7.10 (b)), though tunnel collapse involves the collapse of the cylindrical cavity (tunnel) rather than expansion in the PMT. As the stresses and strains are now in a vertical plane (rather than the horizontal plane for the PMT), the weight of the soil must also be considered.[14] The volume of the soil element is given by

⑬ 旁压试验

⑭ 当应力-应变发生在垂直平面（而不是旁压试验中的水平面）时，必须考虑土重。

$$\left[\frac{(r+\mathrm{d}r)\,\mathrm{d}\theta + r\mathrm{d}\theta}{2}\right]\mathrm{d}r = r\mathrm{d}r\mathrm{d}\theta + \frac{d^2 r\mathrm{d}\theta}{2} \tag{10.30}$$

The second term in Equation 10.30 is very small compared to the first, and can be neglected. Resolving forces vertically then gives

$$(\sigma_r + d\sigma_r)(r + dr)\,d\theta + \gamma r dr d\theta = \sigma_r r d\theta + \sigma_\theta dr d\theta$$

$$\therefore r\frac{d\sigma_r}{dr} - (\sigma_r - \sigma_\theta) + \gamma r = 0 \qquad (10.31)$$

Equation 10.31 is similar to Equation 7.15; the sign of the $(\sigma_r - \sigma_\theta)$ term has changed (cavity collapse instead of expansion) and there is an additional unit weight term. As in Chapter 7, the associated maximum shear stress is $\tau = (\sigma_r - \sigma_\theta)/2$, and in undrained soil at the point of failure, $\tau = c_u$. Substituting these expressions into Equation 10.31 and rearranging gives

$$d\sigma_r = \left[\left(\frac{2c_u}{r}\right) - \gamma\right]dr \qquad (10.32)$$

① 隧道拱顶上方土中的应力情况。

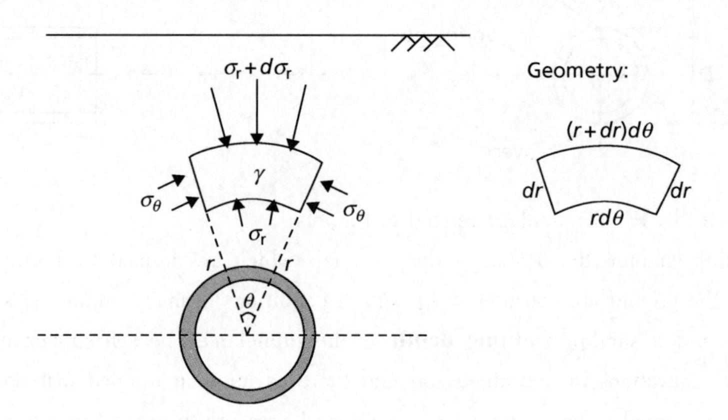

Figure 10.19 Stress conditions in the soil above the tunnel crown.①

② 隧道的内在压力。

At the tunnel (cavity) wall, $r = D/2$ and $\sigma_r = p$ (where p is any internal pressure within the tunnel②); referring to Figure 10.18, at the ground surface $r = C + D/2$ and $\sigma_r = \sigma_q$ where σ_q is the surcharge pressure. Equation 10.32 may then be integrated using these limits to give

$$\int_p^{\sigma_q} d\sigma_r = \int_{D/2}^{C+D/2}\left[\left(\frac{2c_u}{r}\right) - \gamma\right]dr$$

$$\sigma_q - p = 2c_u \ln\left(\frac{2C}{D} + 1\right) - \gamma C \qquad (10.33)$$

③ 所需支护压力。

④ 稳定系数。

Equation 10.33 may be used to determine the required support pressure③ based on the soil properties (c_u, γ), any external loading (σ_q) and geometric properties (C, D). Equation 10.33 is often expressed as a stability number④ N_t, where

$$N_t = \frac{\sigma_q - p + \gamma(C + D/2)}{c_u} \qquad (10.34)$$

For deep tunnels $C \gg D$, so that comparing Equations 10.33 and 10.34 gives an approximate expression for N_t, suitable for preliminary design purposes:

$$N_t = 2\ln\left(\frac{2C}{D} + 1\right) \qquad (10.35)$$

The foregoing analysis has considered collapse of the crown of a long tunnel (a plane-strain analysis[①] was conducted). While this is appropriate for the finished tunnel[②], during construction there may additionally be collapse of soil ahead of the tunnel (the heading) into the face. This involves a more complex three-dimensional failure mechanism/stress field. In the case of undrained materials, Davis *et al.* (1980) presented stability numbers for use in Equation 10.34 for the case of a circular tunnel heading where

① 平面应变分析

② 施工结束后的隧道。

$$N_t = \min\left\{2 + 2\ln\left(\frac{2C}{D} + 1\right), 4\ln\left(\frac{2C}{D} + 1\right)\right\} \tag{10.36}$$

Equations 10.35 and 10.36 are compared in Figure 10.20, which shows that the stability of the tunnel behind the excavation is usually critical (a lower N_t requires a higher support pressure to be supplied by the tunnel, from Equation 10.34)[③]

③ 隧道开挖面的稳定起控制作用（因为式（10.34）表明，低稳定系数需要更高的支护压力）。

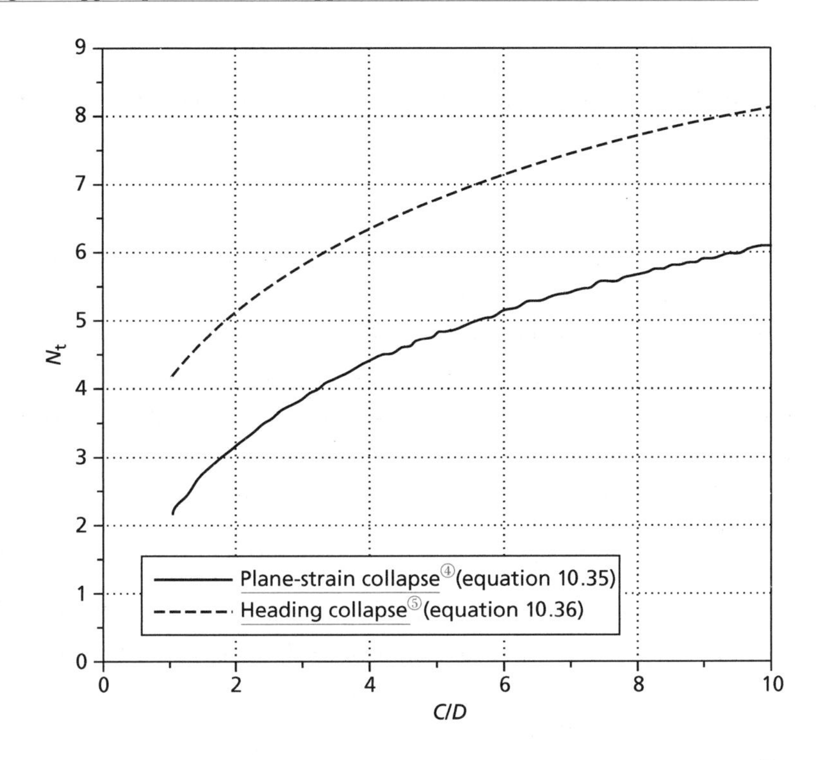

④ 平面应变破坏。

⑤ 掌子面破坏。

Figure 10.20 Stability numbers for circular tunnels in undrained soil.[⑥]

⑥ 不排水土中圆形隧道的稳定系数。

Stability of tunnels in drained soil[⑦]

⑦ 排水土中隧道的稳定性。

Under drained conditions, the relationship between the radial and horizontal effective stresses is σ'_q[⑧] $= K_p\sigma'_r$. Rewriting Equation 10.31 in terms of effective stresses and substituting then gives

⑧ 注：此处原书有误，应该为“σ'_θ”。

$$\frac{d\sigma'_r}{dr} - \frac{\sigma'_r}{r}(K_p - 1) + \gamma' = 0 \tag{10.37}$$

Equation 10.37 can be simplified for integration by substitution of $x = \sigma'_r/r$, so that $d\sigma'_r/dr = r(dx/dr) + x$, giving

$$r\frac{dx}{dr}+x=x(K_p-1)-\gamma$$

$$\frac{dx}{(K_p-2)x-\gamma}=\frac{dr}{r} \tag{10.38}$$

Equation 10.38 is integrated between the same limits as before, but with $\sigma'_r=\sigma'_q$ where σ'_q is the effective surcharge pressure. This gives $x=2p'/D$ at $r=D/2$ and $x=\sigma'_q/(C+D/2)$ at $r=C+D/2$, so that

$$\int_{2p'/D}^{\sigma'_q/(C+D/2)}\frac{dx}{(K_p-2)x-\gamma}=\int_{D/2}^{C+D/2}\frac{dr}{r}$$

$$\frac{(K_p-2)\left(\frac{2\sigma'_q}{\gamma(2C+D)}\right)-1}{(K_p-2)\left(\frac{2p'}{\gamma D}\right)-1}=\left(\frac{2C+D}{D}\right)^{(K_p-2)} \tag{10.39}$$

① 有效径向压力。

② 上覆土。

③ 浮重度。

④ 最大支护压力。

Equation 10.39 may be rearranged to find the effective radial pressure[①] applied by the overlying soil[②] which the tunnel must support (p'). If the soil is dry, the total support pressure $p=p'$, and γ in Equation 10.39 is that for dry soil. If the soil is submerged, then $p=p'+u$, where u is the pore water pressure at the level of the springing and $\gamma=\gamma'$ (buoyant unit weight[③]) in Equation 10.39 for a tunnel with an impermeable lining.

Equation 10.39 is plotted for various different values of ϕ' in Figure 10.21(a) for the case of $\sigma'_q=0$. As expected, as the shear strength of the soil increases (represented by ϕ'), the required support pressure reduces. It can also be seen that for most common values of ϕ' (>30°) a maximum support pressure[④] is reached even for shallow tunnels (low C/D). From Equation 10.39, this value is

$$\frac{p'_{max}}{\gamma D}=\frac{1}{2(K_p-2)} \tag{10.40}$$

⑤ 排水土中浅隧和深隧的支护压力。

⑥ 在极限承载条件下最大的支护压力。

Equation 10.40 is plotted in Figure 10.21 (b), which demonstrates that long-term stresses (after consolidation is complete in the case of fine-grained soils) in tunnel linings are generally very small and independent of tunnel depth.

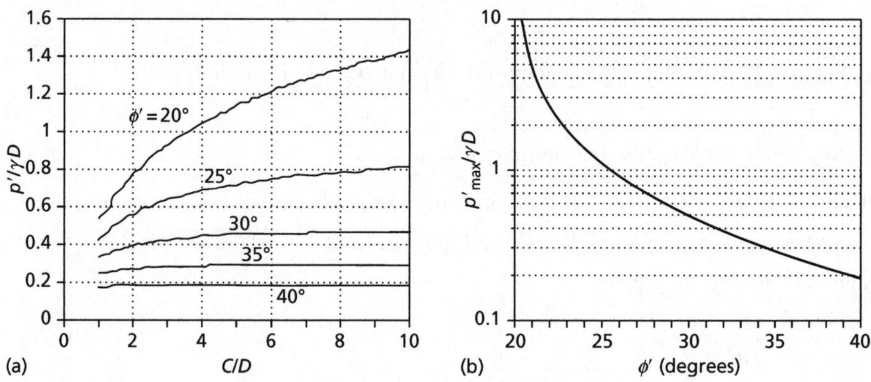

Figure 10.21 (a) Support pressure in drained soil for shallow and deep tunnels[⑤] ($\sigma'_q=0$), (b) maximum support pressure for use in ULS design[⑥] ($\sigma'_q=0$).

Further information regarding heading collapse in drained materials may be found in Atkinson and Potts (1977) and Leca and Dormieux (1990).

Serviceability criteria for tunnelling works[1]

① 隧道施工引起的变形情况。

As material is excavated from the face of a tunnel, the soil ahead of the tunnel will slump towards the tunnel face under the action of its own self-weight. This will tend to lead to over-excavation of material, which will generate a settlement trough[2] at the ground surface due to the loss of soil volume[3] over and above that of the tunnel (Figure 10.22). This trough will have a maximum settlement immediately above the crown of the tunnel, reducing with radial distance from the tunnel. Any buildings or other infrastructure will therefore be subject to differential settlement as the ground beneath them subsides. The minimisation of damage to existing infrastructure is the main serviceability consideration in tunnelling in an urban environment.

② 沉降槽。

③ 土层损失。

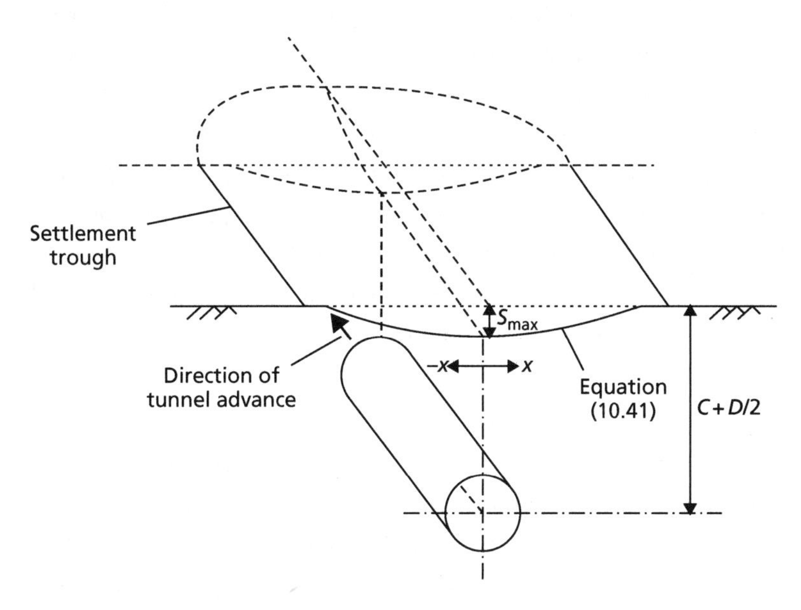

Figure 10.22 Settlement trough above an advancing tunnel.[4]

④ 隧道推进上方的沉降槽。

Observation of tunnelling works has shown that the settlement trough can be described by

$$s_g = s_{max} e^{-x^2/2i^2} \tag{10.41}$$

where s_g is the settlement of the ground surface at a point defined by position x, s_{max} is the maximum settlement (above the crown of the tunnel), and i is the trough width parameter governing the shape of the curve, which is a function of soil type. In clays, $i = 0.5\ (C + D/2)$; in sands and gravels, $i = 0.35\ (C + D/2)$. The parameter i can also be expressed as a function of depth, so that the settlement profile can be found at any point between the ground surface and the tunnel; this can then be used to check differential movements of pipelines[5] or other existing buried services.

⑤ 管线的差异位移。

① 高斯曲线。
② 超挖土体。
③ 地层损失比。
④ 土压平衡盾构机。

The class of equation described by Equation 10.41 is also known as a **Gaussian curve**[①]. The volume of over-excavated material[②] per metre length of the tunnel (V_{soil}) can be found by integrating Equation 10.41 from $x = -\infty$ to $x = \infty$, giving

$$V_{soil} = \sqrt{2\pi} \cdot is_{max} = 2.507 is_{max} \tag{10.42}$$

(this is a standard result for a Gaussian curve). The volume of over-excavation in Equation 10.42 is usually normalised so that it can be expressed as a percentage of the tunnel volume, $V_t = \pi D^2/4$ per metre length. The resulting percentage is known as the **volume loss**[③] (V_L):

$$V_L = \frac{2.507 is_{max}}{0.25\pi D^2} = 3.192 \frac{is_{max}}{D^2} \tag{10.43}$$

The volume loss depends on the tunnelling technique used and on the quality control that can be achieved during construction. A perfectly constructed tunnel would only excavate just enough soil for the tunnel so that $V_L = 0$ and $s_{max} = 0$ from Equation 10.43. In practice this is not possible, and volume loss is typically between 1% and 3% in soft ground. Volume loss may be minimised using modern **earth pressure balance (EPB) tunnel boring machines**[④]. In these computer-controlled machines the cutting face is pressurised with the aim of matching the in-situ horizontal stresses within the ground; however, even these are not perfect. In design to SLS, a conservative (higher) value of V_L is selected based on previous experience in similar soils. Equation 10.43 is then used to determine s_{max}, from which the ground settlement profile is found using Equation 10.41. These settlements are then applied to infrastructure within the area affected by the settlement trough and checked for damage from differential settlement.

Summary

1 Trenches and open shafts may be excavated to a limited depth in fine-grained soils under undrained conditions (i. e. for temporary works only) and in bonded/cemented coarse-grained soils (having $c' > 0$). These excavations may be extended in depth using fluid support in the trench (e. g. bentonite). Fluid support also allows for such excavations to be made in cohesionless soil. The overriding design criterion is preventing collapse of the excavation (ULS).

2 Limit analysis techniques may be applied to the stability of slopes and vertical cuttings in homogeneous soil. Limit equilibrium techniques may also be applied using the method of slices, which can also account for variable pore water pressure distribution and hence cases where seepage is occurring, and consider both rotational and translational slips. For both techniques, optimised failure surfaces must be found which give the most critical conditions. As for trenches, the overriding design criterion is preventing collapse of the slope and the occurrence of catastrophically large slip displacements (ULS).

3 In undrained materials, unsupported tunnels may be made to a limited depth in the short term. Under drained conditions and at deeper depths in cohesive material, internal support pressure must be applied along the axis of the tunnel (from the tunnel lining) and at the face while excavation is proceeding. This information may be used to determine the earth pressures acting on a tunnel lining when the tunnel is completed, for structural design of the lining system at both ULS and SLS. In addition to preventing collapse of the tunnel (ULS), the design must also consider the ground settlement profile above the tunnel due to volume loss and the potential damage to surface or buried infrastructure due to gross or differential settlement in this region (SLS).

Problems

10.1 A diaphragm wall is to be constructed in a soil having a unit weight of 18kN/m^3 and design shear strength parameters $c' = 0$ and $\phi'' = 34°$. The depth of the trench is 3.50m and the water table is 1.85m above the bottom of the trench. Determine whether the trench is stable to EC7 DA1b if the unit weight of the slurry is 10.6kN/m^3 and the depth of slurry in the trench is 3.35m. Determine also the maximum depth to which the trench could be excavated if the slurry is maintained at the same level below the ground surface.

10.2 For the given failure surface, determine whether the slope detailed in Figure 10.23 is stable in terms of total stress to EC7 DA1a. The unit weight for both soils is 19kN/m^3. The characteristic undrained strength (c_u) is 20kPa for soil 1 and 45kPa for soil 2. How would the answer change if allowance is made for the development of a tension crack?

10.3 A cutting 9m deep is to be excavated in a saturated clay of unit weight 19kN/m^3. The characteristic shear strength parameters are $c_u = 30\text{kPa}$ and $\phi_u = 0$. A hard stratum underlies the clay at a depth of 11m below ground level. Determine the slope angle at which failure would occur. What is the allowable slope angle if the slope is to satisfy EC7 DA1b, and what is the overall factor of safety corresponding to such a design?

10.4 For the given failure surface, determine whether the slope detailed in Figure 10.24, is stable to EC7 DA1b using the Fellenius method of slices. The unit weight of the soil is 21kN/m^3, and the characteristic shear strength parameters are $c' = 8\text{kPa}$ and $\phi' = 32°$.

10.5 Repeat the analysis of the slope detailed in Problem 10.4 using the Bishop routine method of slices.

10.6 Using the Bishop routine method of slices, determine whether the slope detailed in Figure 10.25 is stable to EC7 DA1-a in terms of effective stresses for the specified failure surface. The value of r_u is 0.20 and the unit weight of the soil is 20kN/m^3. Characteristic values of the shear strength parameters are $c' = 0$ and $\phi'' = 33°$.

10.7 A long slope is to be formed in a soil of unit weight 19kN/m^3 for which the characteristic shear

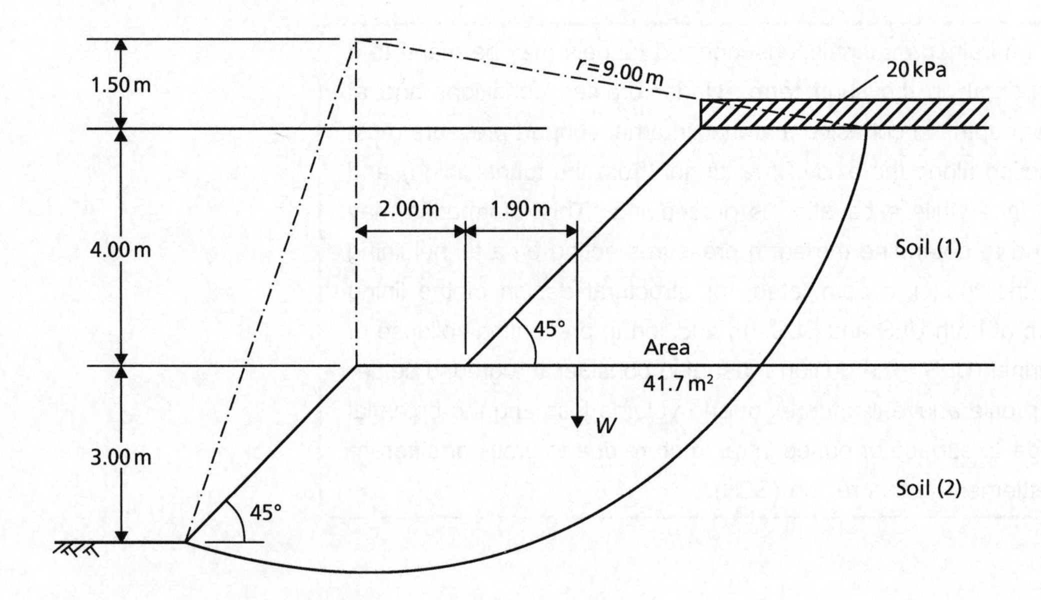

Figure 10.23 Problem 10.2.

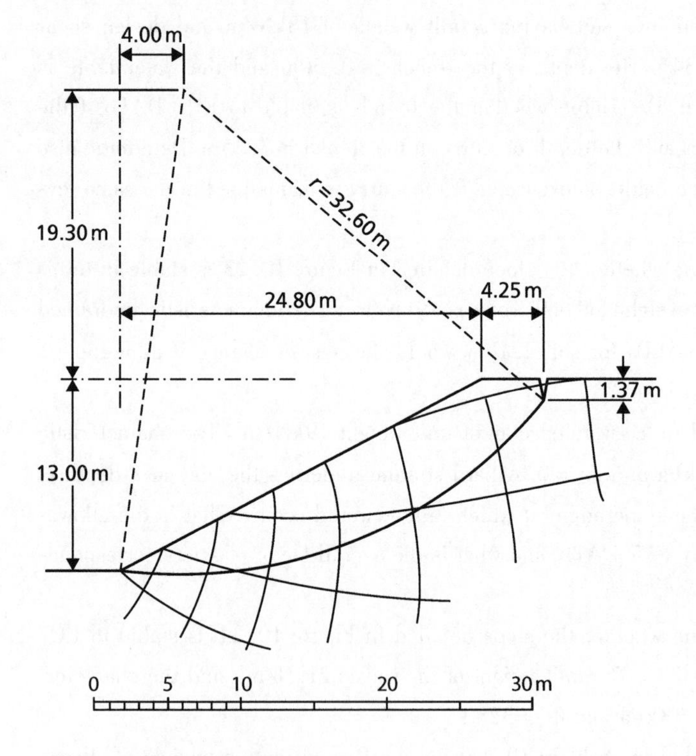

Figure 10.24 Problem 10.4.

strength parameters are $c' = 0$ and $\phi' = 36°$. A firm stratum lies below the slope. It is to be assumed that the water table may occasionally rise to the surface, with seepage taking place parallel to the slope. Determine the maximum safe slope angle to satisfy EC7 DA1-b, assuming a potential failure surface parallel to the slope. Determine also the overall factor of safety for the slope angle determined above if the water table were well below the surface.

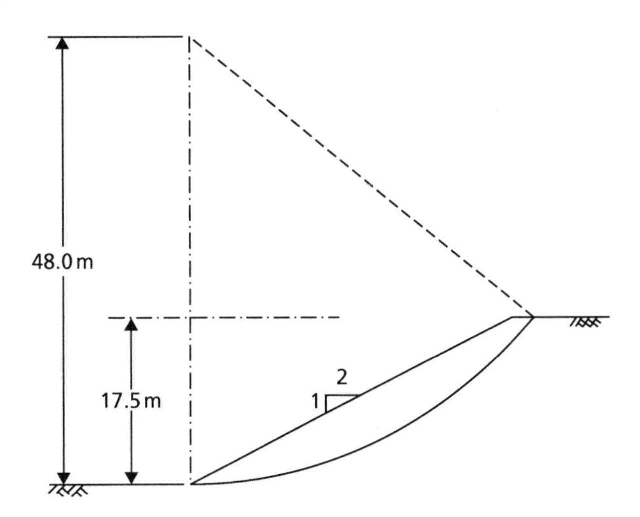

Figure 10.25 Problem 10.6.

10.8 A circular tunnel of diameter 12m is to be bored in stiff clay with $c_u = 200\text{kPa}$ and $\gamma = 19\text{kN/m}^3$ (both constant with depth). A structure with a footprint of $16 \times 16\text{m}^2$ is situated directly above the tunnel, which is 16m high and has masonry load bearing walls. The tunnelling method can be assumed to induce a volume loss of 2.5%. Determine a suitable running depth of the tunnel (at the tunnel centerline).

References

Atkinson, J. H. and Potts, D. M. (1977) Stability of a shallow circular tunnel in cohesionless soil, *Géotechnique*, **27** (2), 203 – 215.

Bell, J. M. (1968) General slope stability analysis, *Journal ASCE*, **94** (SM6), 1253 – 1270.

Bishop, A. W. (1955) The use of the slip circle in the stability analysis of slopes, *Géotechnique*, **5** (1), 7 – 17.

Bishop, A. W. and Bjerrum, L. (1960) The relevance of the triaxial test to the solution of stability problems, in *Proceedings of the ASCE Research Conference on Shear Strength of Cohesive Soils, Boulder, Colorado*, pp. 437 – 501.

Bishop, A. W. and Morgenstern, N. R. (1960) Stability coefficients for earth slopes, *Géotechnique*, **10** (4), 129 – 147.

Davis, E. H., Gunn, M. J., Mair, R. J. and Seneviratne, H. N. (1980). The stability of shallow tunnels and underground openings in cohesive material, *Géotechnique*, **30** (4), 397 – 416.

EC7 – 1 (2004) *Eurocode 7: Geotechnical design – Part 1: General rules, BS EN 1997 – 1: 2004*, British Standards Institution, London.

Gens, A., Hutchinson, J. N. and Cavounidis, S. (1988) Three-dimensional analysis of slides in cohesive soils, *Géotechnique*, **38** (1), 1 – 23.

Gibson, R. E. and Morgenstern, N. R. (1962) A note on the stability of cuttings in normally consolidated clays, *Géotechnique*, **12** (3), 212 – 216.

Gibson, R. E. and Shefford, G. C. (1968) The efficiency of horizontal drainage layers for accelerating consolidation of clay embankments, *Géotechnique*, **18** (3), 327 – 335.

Leca, E. and Dormieux, L. (1990) Upper and lower bound solutions for the face stability of shallow circular tunnels in frictional material, *Géotechnique*, **40** (4), 581 – 606.

Michalowski, R. L. (2002). Stability charts for uniform slopes, *Journal of Geotechnical and Geoenvironmental Engineering*, **128** (4) 351 – 355.

Michalowski, R. L. (2010) Limit analysis and stability charts for 3D slope failures, *Journal of Geotechnical and Geoenvironmental Engineering*, **136** (4), 583 – 593.

Morgenstern, N. R. (1963) Stability charts for earth slopes during rapid drawdown, *Géotechnique*, **13** (2), 121 – 131.

Morgenstern, N. R. and Price, V. E. (1965) The analysis of the stability of general slip surfaces, *Géotechnique*, **15** (1), 79 – 93.

Morgenstern, N. R. and Price, V. E. (1967) A numerical method for solving the equations of stability of general slip surfaces, *Computer Journal*, **9**, 388 – 393.

Skempton, A. W. (1970) First-time slides in overconsolidated clays (Technical Note), *Géotechnique*, **20** (3), 320 – 324.

Spencer, E. (1967) A method of analysis of the stability of embankments assuming parallel inter-slice forces, *Géotechnique*, **17** (1), 11 – 26.

Taylor, D. W. (1937) Stability of earth slopes, *Journal of the Boston Society of Civil Engineers*, **24** (3), 337 – 386.

Viratjandr, C. and Michalowski, R. L. (2006). Limit analysis of slope instability caused by partial submergence and rapid drawdown, *Canadian Geotechnical Journal*, **43** (8), 802 – 814.

Further reading

Frank, R., Bauduin, C., Driscoll, R., Kavvadas, M., Krebs Ovesen, N., Orr, T. and Schuppener, B. (2004) *Designers' Guide to EN 1997 – 1 Eurocode 7: Geotechnical Design – General Rules*, Thomas Telford, London.

This book provides a guide to limit state design of a range of constructions (including slopes) using Eurocode 7 from a designer's perspective and provides a useful companion to the Eurocodes when conducting design. It is easy to read and has plenty of worked examples.

Mair, R. J. (2008) Tunnelling and geotechnics: new horizons, *Géotechnique*, **58** (9), 695 – 736.

Includes some interesting case histories of tunnel construction, building on many of the basic concepts outlined in Section 12.5, and highlights current and future issues.

Michalowski, R. L. (2010). Limit analysis and stability charts for 3D slope failures. *Journal of Geotechnical & Geoenvironmental Engineering*, **136** (4), 583 – 593.

Provides stability charts similar to those in Section 12.3 for analysis of slopes under a range of soil conditions for the more complicated (but more realistic) cases of three-dimensional, rather than plane strain, failure (e.g. Figure 12.7).